The Correspondence of John Tyndall

VOLUME 8

THE CORRESPONDENCE OF JOHN TYNDALL

VOLUME 8

The Correspondence, June 1862–January 1865

Edited by
Michael D. Barton, Piers J. Hale,
Nathan Kapoor, and Elizabeth Neswald

General Editors
James Elwick, Roland Jackson, Bernard Lightman,
and Michael S. Reidy

UNIVERSITY OF PITTSBURGH PRESS

Published by the University of Pittsburgh Press, Pittsburgh, Pa., 15260

Manufactured in the United States of America
Printed on acid-free paper
10 9 8 7 6 5 4 3 2 1

Cataloging-in-Publication data is available from the Library of Congress

ISBN 13: 978-0-8229-4577-2

CONTENTS

Acknowledgments vii
List of Abbreviations ix
Introduction to Volume 8 xiii
Editorial Principles xxix
Note on Money xxxiii
Timeline of John Tyndall's Life xxxv
Timeline of Events in John Tyndall's Life Specific to Volume 8 xxxvii

The Correspondence, June 1862–January 1865 1

Biographical Register 457
Index 485

ACKNOWLEDGMENTS

The editors of this volume wish to thank the many people who have offered their advice and expertise as we worked on this volume. Roland Jackson was of particular help, not only in offering sage advice from his deep familiarity with the Tyndall archive but also in the form of his biography of Tyndall, *The Ascent of John Tyndall*, which was published as we were in the process of editing this volume. We owe Roland additional thanks for his diligent proof-reading of our volume in his capacity of general editor.

The project owes its success to the efforts of an expansive team of named and unnamed graduate students and scholars. The collection and collation of the letters for this and other volumes was the result of a considerable amount of work by the general editors, and in particular by James Elwick and the team at the York office, while the early stages of transcription are the result of hours of labor by dozens of graduate students from the universities associated with this project across the globe. The editors would also like to thank Erin Grosjean, who took over as manuscripts coordinator in 2018, and whose aid and organizational skills have been invaluable. We would like to thank, too, Caley McCarthy, Eleanor Coulter, and Emily Herring, each of whom provided us with translations of the letters from Tyndall's French correspondents, and Nicolas Sanchez-Guerrero and Sven Pinczewski, who, along with Elizabeth Neswald and Tatyana Shestakov, participated in the translation of letters from Tyndall's German correspondents. We owe a debt of gratitude also to Jane Harrison, who obtained and scanned particular letters from the Royal Institution archive; to Jennifer Allan, who obtained scanned images for us from the British Library and other London archives; and to Jan Mende, Stadtmuseum Berlin, for biographical information on Hermann Knoblauch.

LIST OF ABBREVIATIONS

ADB	*Allgemeine Deutsche Biographie*
ANB	*American National Biography*
Ascent of John Tyndall	Roland Jackson, *The Ascent of John Tyndall* (Oxford: Oxford University Press, 2018)
BAAS	British Association for the Advancement of Science
BL	British Library
Brit. Assoc. Rep.	*Reports of the British Association for the Advancement of Science*
Brock and MacLeod, *Hirst Journals*	William H. Brock and Roy M. MacLeod, eds., *Natural Knowledge in Social Context: The Journals of Thomas Archer Hirst FRS* (London: Mansell microfiche, 1980)
Cosmos	*Cosmos. Revue encyclopédique hebdomadaire des progrès des sciences et de leurs applications aux arts et a l'industrie*
CDSB	*Complete Dictionary of Scientific Biography* (New York: Charles Scribner's Sons, 2007)
CUL SC	Cambridge University Library, Stokes Correspondence
Family of Lubbock	Robert Birkbeck, *Notes on the History and Genealogy of the Family of Lubbock* (London: Mitchell and Hughes, 1891)
Faraday Correspondence	*The Correspondence of Michael Faraday*, ed. Frank A. J. L. James, 6 vols. (London: Institution of Electrical Engineers/Institution of Engineering and Technology, 1991–2012)
FRS	Fellow of the Royal Society of London
Glaciers of the Alps	John Tyndall, *The Glaciers of the Alps: Being a Narrative of Excursions and Ascents, an Account of the Origin and Phenomena of Glaciers and an Exposition of the Physical Principles to Which They Are Related* (London: John Murray, 1860)
Heat as a Mode of Motion	John Tyndall, *Heat as a Mode of Motion* (London: Longmans and Green, 1863)
HLS	*Historisches Lexikon der Schweiz*
IC HP	Imperial College of Science, Technology and Medicine, College Archives, Huxley Papers
Journal	John Tyndall's Journal (RI MS JT/2/1–22)

Les Mondes	*Les Mondes: Revue hebdomadaire des sciences et leurs applications aux arts et a l'industrie*
LT	Louisa Tyndall, John Tyndall's widow
LT, "Biography"	Louisa Tyndall's typescript biography of her husband (RI MS JT/5/9–11)
Modern English Biography	Frederic Boase, *Modern English Biography [Supplement]*, 3 vols. (Truro: Netherton and Worth, 1908–21)
Mountaineering in 1861	John Tyndall, *Mountaineering in 1861: A Vacation Tour* (London: Longman, Green, Longman, and Roberts, 1862)
ODNB	*Oxford Dictionary of National Biography* (Oxford: Oxford University Press, 2004)
OED	*Oxford English Dictionary*
Phil. Mag.	*Philosophical Magazine*
Phil. Trans.	*Philosophical Transactions of the Royal Society of London*
Poggend. Annal.	*Annalen der Physik und Chemie*, edited by J. C. Poggendorff
Remembrances	William Frederick Pollock, *Personal Remembrances of Sir Frederick Pollock, Second Baronet, Sometime Queen's Remembrancer*, 2 vols (London: Macmillan, 1887)
RI	Royal Institution of Great Britain
RI MS JT	Royal Institution, London, John Tyndall Papers
RS HS	Royal Society, Herschel Papers
RS RR	Royal Society of London, Referee Reports
Tyndall Correspondence, vol. 6	Michael D. Barton, Janet Browne, Ken Corbett, and Norman McMillan, *The Correspondence of John Tyndall, Volume 6: The Correspondence, November 1856–February 1859* (Pittsburgh, University of Pittsburgh Press, 2019)
Tyndall Correspondence, vol. 7	Roland Jackson, Nanna Katrine Lüders Kaalund, and Diarmid A. Finnegan, eds., *The Correspondence of John Tyndall, Volume 7, The Correspondence, March 1859–May 1862* (Pittsburgh: University of Pittsburgh Press, 2019)
Wellcome	Wellcome Library

John Tyndall in the early 1860s, by the London photographer Maull & Polyblank. This photograph of John Tyndall was taken in the early 1860s by the London-based commercial photographers Maull & Polyblank (later Maull & Co.), who had studios on Gracechurch Street and on Piccadilly. It shows Tyndall at the time of the correspondence in this volume. From the late 1850s Maull & Polyblank produced a series titled *Photographic Portraits of Living Celebrities*, which were widely collected. That Tyndall was among their subjects reflects his growing status in both science and society. The National Portrait Gallery, London, holds thirty-three portraits of Tyndall. Illustration used with permission of the National Portrait Gallery, London (NPG x45495; copyright, National Portrait Gallery, London)

INTRODUCTION TO VOLUME 8

This eighth volume of *The Correspondence of John Tyndall* consists of 318 letters covering the two-and-a-half-year period between 1 June 1862 and 31 January 1865. The letters reveal much about Tyndall's personality, the development of his career, and his role in attempting to better establish science as a respectable and professional enterprise. However, Tyndall was not above controversy, and, as became typical of his career and attitude, this volume includes more than one occasion in which Tyndall entered public disputes either in defense of his own or a colleague's priority claims. Tyndall's dispute with James Prescott Joule, William Thomson, and Peter Guthrie Tait regarding what he believed to be the German physicist Julius Robert Mayer's rightful claim to priority for the discovery of the mechanical equivalent of heat was ongoing for much of the period covered by this volume. In 1864 he became embroiled in another priority dispute, this time with the Hungarian physicist C. K. Akin. Further, and although more quietly, he fell into disagreement with the geologist Roderick Murchison over the action of glaciers.

Since his appointment to the Royal Institution (RI) in 1853, Tyndall had been increasingly busy. His work schedule was draining, and like many of his nineteenth-century contemporaries both in science and in literature, Tyndall frequently worked himself to the point of exhaustion. He lectured at the RI and, upon occasion, at the Royal Society (RS); he worked hard on revising his lectures for publication as well as pursuing his own research in the basement laboratory at the RI. He worked as editor for the *Phil. Mag.*, not only reviewing for content but helping revise the English of a number of his foreign colleagues whose work he wanted to see put before the English scientific community. By the 1860s Tyndall had become recognized as one of the most significant men of science in Europe, with translations and notices of his own work appearing in both French and German publications. His fame ultimately led to him being invited to lecture to the royal household, which only enhanced his reputation as a social lion, and thus his obligations. As Roland Jackson notes in his recent biography of Tyndall, "Tyndall was continually torn between enjoying social events and feeling that he ought to be working."[1] Further, though, Ian Hesketh has made the case in an important essay that what is seen here is the formation of the mid-nineteenth-century scientific persona in process. Building on the work of Paul White, who suggested in his 2003 biography of Thomas Henry Huxley that the mid-nineteenth century was an important period in the "making of the man of science," and Lorraine Daston and Peter Galison's 2007 book *Objectivity*, Hesketh shows that Tyndall was instrumental in cultivating an ethical persona for the man of science through his development of scientific practices and the assertion of standards of behavior, both for himself and as standards for others.[2] "Patience, industry, sacrifice, self-denial, and self restraint" were key elements in this fusion of epistemology and ethics,[3] which often saw Tyndall push himself to his limits, but also put himself forward to assert and defend the priority claims of those he believed to be the rightful claimant.

Tyndall had long been occupied with his own health and diet, and turned to Turkish baths, long walks in the countryside, and immersions in the sea off the coast of the Isle of Wight, as well as dramatic and daring excursions into the Swiss Alps for restoration. He suffered from severe headaches, although he did not experience the debilitating mental exhaustion that dogged Huxley, Herbert Spencer, Charles Kingsley, and other contemporaries.

Perhaps the most dramatic letters of this volume—if not those detailing the accounts of his cousin Hector Tyndale's courageous exploits in the American Civil War—are those relating Tyndall's mountaineering exploits. Tyndall had first visited the Alps in 1849, during his time as a student at Marburg, but first climbed there with Huxley in 1856, and he had returned every summer thereafter.[4] Tyndall climbed in pursuit of science, but it was also among the peaks and glaciers that he felt both most at peace with himself and most at one with the universe. He was without a doubt one of the most accomplished and daring mountaineers in the golden age of Alpinism; he often climbed with only a guide, made an attempt on the Matterhorn just days after Edward Whymper had failed in the effort, and wrote to the *Times* detailing accidents that he had witnessed and experienced.[5] He eulogized the landscape in almost spiritual terms to his most intimate correspondent, Juliet Pollock.

Toward the end of the period covered by this volume, Tyndall was involved in two endeavors that point to his interest in shaping the public face of British science and solidifying his own role within it. Although the letters in this volume give little indication of the significance of a dinner meeting that took place between Tyndall, Huxley, Joseph Dalton Hooker, Thomas Hirst, Edward Frankland, John Lubbock, George Busk, William Spottiswoode, and Spencer on 3 November 1864, this date marks the foundation of an informal network that came to be known as the X Club. Begun as a dining club and networking opportunity, the X Club became one of the most influential circles within the British scientific community for the next several decades.[6] X Club members were united by their commitment to scientific naturalism and the perception of science as a profession, and they differed from many of their colleagues not only in their more skeptical approach to organized religion but also in their more modest economic background and social roots.[7]

Shortly afterward, in November of the same year, Tyndall, along with a number of friends and other members of the X Club, purchased the *Reader*, a journal of science and literature that they hoped to turn to good use in promoting and publishing the works, society meetings, and correspondence of scientific men. Although the project was short-lived,[8] this episode shows the extent to which Tyndall was invested in the popularization of science and the need to better facilitate communication within the international scientific community.

Research Focus and Significant Publications

Tyndall's work in this period reflects his ongoing interests as well as some new ones. By 1862 he had begun to establish a name for himself in several areas of research—in particular, the study of radiant heat and a theory of glaciers. He published on both topics in 1860 and 1861, and the reception of his book, *Glaciers of the Alps*, and his early work on radiant heat continued into the following years.

Tyndall's first memoir on the absorption and radiation of heat by gaseous matter appeared in the *Phil. Trans.* in 1861 and received a generally positive reception.[9] Over the next few years, he expanded on this topic, publishing another four memoirs on various aspects of this research by 1865 and additional memoirs later. George Gabriel Stokes, who reviewed Tyndall's fifth memoir on the subject, provided detailed commentary for revisions, criticism that Tyndall clearly appreciated.[10]

Related to this was Tyndall's continuing work on the absorption of radiant heat by aqueous vapor. Tyndall had been working on this topic for several years and had been in a dispute with the German physicist Gustav Magnus for nearly as long.[11] Tyndall's third memoir, "On the Relation of Radiant Heat to Aqueous Vapour," was published in 1863.[12] Although Magnus was unable to replicate Tyndall's results, and the two continued to disagree, their relationship remained friendly. This was not the case, however, in some of Tyndall's other scientific disputes, as the next section shows.

Work on water vapor led Tyndall in another direction as well, to considerations of the blue color of the sky. In an exchange with John Herschel, he mused, "[W]e have in the vapour a body which absorbs the same rays as the water which produces it—in other words it shares the colour of water. Through the operation of this cause our atmosphere is certainly <u>a blue medium</u>. The quantity of vapour might not be sufficient to produce a sensible blue, if the rays went straight through it; but the reflections within the body of the atmosphere must be innumerable, and thus its <u>virtual depth</u> increased.

I wonder is this likely to throw any light on the blue of the sky!"[13] Herschel did not agree, and Tyndall eventually came to a different conclusion himself.[14]

Tyndall also frequently discussed the phenomenon of heat more generally in his correspondence, especially as he was revising his lecture series on heat as a mode of motion for publication. His correspondence with the German physicist Rudolf Clausius reveals the deep influence Clausius had on Tyndall's work. Clausius was, along with the Scottish physicist William Thomson, a leading expert on the theory of heat, but, in contrast to Thomson, there was no rivalry between him and Tyndall. Clausius reviewed Tyndall's manuscript chapters for the book on heat and offered many considerations on precision and terminology, as well as clarifying numerous points on the specific heat of gases. The two physicists had very different styles of presentation, and Clausius's insistence on strict precision of expression provided a necessary corrective to Tyndall's more enthusiastic popular style.[15]

The topic of glaciers and alpine geology, while not as prominent as that of heat in this period, continued to occupy Tyndall's time and form a regular topic in his correspondence. Several letters included here discuss the reception of his theory of glaciers. Controversially received when it was first published, it remained an object of contention in some circles.[16] His paper on the "Conformation of the Alps" was published in 1862, and it, along with considerable correspondence on the Alps, ice, and mountain geology, reflects his continuing interest in the topic and his recognized expertise. Tyndall's international reputation comes through clearly here, in his exchanges with scholars as far away as Switzerland and New Zealand.[17]

Newer was Tyndall's growing interest in the phenomenon of "luminous and invisible light." This research was a continuation of his work on thermal radiation, but now he studied "obscure" or "black" heat; that is, infrared radiation. It particularly fascinated him that the same invisible light could heat a platinum wire to a red-hot glow or set a piece of paper on fire. For Tyndall, this demonstrated that it could be transformed from one form to another. The topic formed the subject matter of several letters, as Tyndall discussed this research at length with Herschel, James Croll, and Stokes.[18] It was this research that was later to become the basis for Tyndall's dispute with Akin, which is discussed later in this introduction.

While these areas constitute Tyndall's main research activities, he had also begun to develop some new interests. Over several weeks from April through June 1863, he gave a number of lectures on sound, and also published a short notice on Laplace and sound.[19] Although this was not the first time he lectured on this topic, having already given the extensive series of lectures on sound in 1857, he was by now slowly beginning to think about publication.[20]

Tyndall's growing reputation and the wide range of his expertise are reflected in the international scope of his correspondents. Stokes and Herschel stand out among his British correspondents and Clausius among the German ones for the depth of their professional exchange. Correspondence with researchers in Germany, the United States, France, the Netherlands, Belgium, and New Zealand points to the extent of Tyndall's international networks even at this relatively early stage in his career.

Lectures on Heat and the Defense of Julius Mayer

In the years 1862–63 Tyndall became embroiled in one of the most contentious scientific disputes of his career when he became a public champion of Julius Robert Mayer. The dispute shaped his relationships with several leading British physicists for decades. In 1842 Mayer had published a short paper discussing a method to compare the transformation of one force into another through a universal equivalent, the mechanical equivalent of heat.[21] The paper, based on philosophical premises and written by a scientific outsider, received little attention. In 1843 the British brewer James Prescott Joule presented some experiments on transforming motion into heat and came to similar conclusions.[22] Joule had closer ties to the British scientific community, and his work attracted the attention of the then young physicist William Thomson. As a result, Joule was acknowledged as the discoverer of the mechanical equivalent of heat.

The letters in this volume begin in the middle of a discussion between Tyndall and Clausius about the merits of Mayer's work. As Tyndall's letter to Clausius of late 1859 indicates, he had long been interested in reading Mayer's papers,[23] and he was particularly eager to do so in advance of his series of lectures at the RI on "heat, regarded as a mode of motion." In these lectures, Tyndall engaged with the still relatively young science of thermodynamics, presenting numerous examples of the transformation of energy in machines, terrestrial nature, and the solar system. While preparing the lectures, it struck him that Mayer had made a greater contribution than had hitherto been acknowledged. In May of 1862 Tyndall wrote to both Clausius and the German physicist Herman Helmholtz, asking them for their opinion, as well as for copies of Mayer's publications.[24] Helmholtz replied a few days later, acknowledging Mayer's priority, while questioning his methods, but admitted that his results were approximately correct. Clausius, by contrast, was skeptical at first. He quickly changed his mind, however, once he had properly digested Mayer's papers. As he wrote to Tyndall a week later, "I must retract what I wrote in my previous letter that you would find nothing of significance in Mayer's writings . . . I am amazed at the number of fine and correct ideas that they contain."[25] This was enough for Tyndall, and on 6 June he delivered a Friday Evening Discourse at the RI, "On Force," in which he introduced Mayer's work to the London lecture-going public and sent one of Mayer's papers on to the *Phil. Mag.* recommending translation.[26]

Tyndall's active promotion of Mayer reverberated within both the German and British scientific communities and brought smoldering conflicts with some of his North British contemporaries into the open.[27] Joule reacted quickly and publicly to what he perceived as a questioning of his own priority in the formulation of the mechanical equivalent in a note published in the *Phil. Mag.*[28] He also expressed his concerns to Tyndall privately, remarking that "I think in a case like that of the Equiv[alen]ᵗ of Heat the experimental worker rather than the mere logical reasoner (however valuable the latter) must be held as the establisher of a theory. I have determined the mech[anica]ˡ equivalent in nearly a dozen ways, and the figure I arrived at in 1849 has not yet been altered or corrected."[29] Tyndall responded with a public letter of his own, in which he reiterated his respect for Joule, but strongly defended Mayer, whom he called "a man of genius, to whom the fates have been singularly unkind."[30] He disputed the claim that Mayer's derivation of the mechanical equivalent was speculative and presented the previous research on which it had been based. He concluded in a conciliatory manner by again praising Joule, writing, "I do not think that the public estimate of your labours can be in the least affected by any recognition which may be accorded to Mayer. There is room for both of you on this grand platform."[31]

By early 1863 Tyndall was preparing his lectures for publication and sending drafts to Clausius, a friend of his since his Berlin sojourn of 1851, for criticism and comments.[32] Since Clausius was a leading expert on the theory of heat and kinetic theory of gases, his input was particularly valuable. Importantly, Clausius also approved of the way Tyndall presented Mayer's and Joule's work.[33]

Clausius's position was not shared by either William Thomson, Britain's leading physicist for the theory of heat, or his colleague Peter Guthrie Tait. Tyndall had a very different background from these middle-class, Cambridge-educated, Scottish Presbyterian physicists, and they viewed him with suspicion. The feeling was mutual. Thomson had supported Joule since the 1850s,[34] and with Tyndall now publicly standing up for Mayer, Joule again turned to Thomson for support. The disagreement over the rival claims of Mayer and Joule escalated across a series of increasingly acrimonious publications.[35]

The dispute with his North British colleagues occupied Tyndall's mind, as a letter to Clausius makes clear: "Thomson has been unsparing in his censures, but has never had the manliness or frankness to attack me openly. . . . His putting Tait forward & keeping in the background himself is quite characteristic, but he has for once mistaken his man."[36] He continued, "Believe me I am most anxious to live in peace with Thomson and every body else. A man's life is too short to have any part of it devoted to such controversy, but with some people forbearance is interpreted as timidity."[37] Tyndall spent much of his life in science thus ensuring that no man could find reason to cast aspersions of timidity or a want of manliness against him.

The volume of lectures titled *Heat as a Mode of Motion* was published in February 1863. Tyndall distributed copies to his friends and colleagues, who were overwhelmingly positive in their responses.

Tyndall's Alpine Adventures

In the mid-1850s Tyndall was in pursuit of answers regarding the cleavage of slate, a topic that had developed from questions relating to how pressure acted in his experiments on diamagnetism. Having read James David Forbes's *Travels through the Alps* (1843), he was determined to see what Forbes had seen, and thus, in 1856, his lifelong love affair with the Alps began.[38] Although many of his scientific colleagues shared his passion—Huxley, Hooker, and Lubbock either traveled with him or met up with him in the Alps on a number of occasions—none came close to his skill, daring, or accomplishment.

By the summer of 1862 Tyndall was ready to escape to the peaks once more. The dispute with Joule had left him drained, and he needed, he told Faraday, to give his head a rest and restore his health, which "had been sadly broken in upon."[39] He met Huxley and Lubbock in Switzerland; he and Lubbock scaled the Galenstock; and then all three trekked across the Aletsch Glacier intent on making an attempt on the Jungfrau. They sent their porters ahead of them to set up a base camp, but caught up with them a mile short of their destination. Beilander, one of the porters, had fallen some forty feet into a crevasse, and his colleagues had presumed him dead, swallowed by snow and ice. In truth, had it not been for Tyndall and Lubbock's attempts to ascertain this fact, he might well have been given up and lost, but Lubbock heard the man's groans from the abyss and Tyndall directed a rescue operation that saved the man's life. Tyndall wrote home that it was a lack of caution rather than the innate danger of the glacier that had caused the accident, a sentiment that was characteristic of Tyndall's attitude to even the most challenging ascents.[40]

That same summer Tyndall made the attempt upon the Matterhorn. The young British mountaineer Edward Whymper had just failed to attain the summit, but Tyndall did not let this deter him. Much to his disappointment, the challenging terrain stopped him some distance short of the summit, and he wrote home to Juliet Pollock, bitter at his failure.[41] The peak was successfully summited in 1865, by a party of seven, led by Whymper and the guide Michel Croz, although several men, including Croz, fell to their deaths as they made their descent.

Tyndall was successful in his attempt on the Jungfrau in 1863, despite atrocious weather, which drove him home to England earlier than he had intended. After a trip to Ireland, he joined with Huxley for some lighter climbing in the English Lake District before heading on to the British Association for Advancement of Science (BAAS) meeting in Newcastle. In 1864 Tyndall suffered an accident of his own on the Piz Morteratsch.[42] A slight misstep by one of his climbing companions had brought on an avalanche, which swept the whole party off their feet and a thousand feet down the mountainside. More by luck than a result of their own desperate efforts, they came to a halt at the edge of a precipice, the overreach of which surely would have been their end. No one was seriously injured, and Tyndall climbed again the next day. On his way back to England, he stopped in Zurich to attend the meeting of the Naturforschende Gesellschaft, a Swiss scientific society of which Tyndall had been elected an honorary member in 1858. Upon his return, Tyndall happily dined at the Lubbocks' with Hirst and the Hookers on 27 August, where he regaled them with tales of his adventures.[43]

Dispute with Akin

Toward the end of 1864 Tyndall again became embroiled in controversy regarding the priority of discoveries in relation to heat. Shortly after Tyndall turned to the study of what was then called "black" or "obscure" heat—radiation in the infrared part of the spectrum, which is beyond the range of human vision—Akin accused him of seeking to forestall his own work in the study of ray transmission.

Prior to this point, the relations between Akin and Tyndall had been cordial; Akin first wrote to Tyndall in early September 1862 in the latter's capacity as an editor of the *Phil. Mag.*, to ask whether he would be willing to provide "some linguistic correction" for a manuscript he hoped to see published.[44] It was not until late 1864 that Tyndall received further communication from Akin, in which Akin enclosed proofs of an essay that he had just written up on the work he had presented at the BAAS meeting in Newcastle.[45] He wrote that he "would gladly avail himself of this opportunity to call on Prof. Tyndall, and have the pleasure of making his acquaintance."[46]

When they met, Tyndall had taken Akin into his laboratory at the RI, and shortly thereafter Tyndall wrote to the secretary of the RS, George Gabriel Stokes, that he had "succeeded in raising platinum to vivid redness by black heat," and that "the same black heat sets paper instantly ablaze," noting that "D^r^. Akin has seen some of the experiments."[47]

There is more than one important letter in this communication that is missing, but it is clear that Akin had claimed priority regarding an experiment showing "the glowing of platinum wire in a hydrogen flame," and Tyndall subsequently ceded Akin's priority, publishing a short note to this effect in the *Reader* in April 1864.[48] However, Akin was clearly not content with this accommodation, or perhaps with Tyndall's phrasing of it, and Akin asserted exclusive claim to this kind of ray transmission in terms that were clearly less than polite. He demanded time to complete and publish the results of his experiments, and Tyndall, although clearly offended by the tone that Akin had taken, responded that he could have a full year to make good on his work, promising that "from this 3rd of November 1864 to the 3rd of November 1865 I shall not make known, publicly or privately, any experiments on 'ray-transmutation.'"[49] He wrote directly to Stokes, telling him that "circumstances which have since occurred cause me to ask you to consider [his earlier note, in which he described his own experimental success on exactly this point] unwritten."[50]

Although Tyndall did go on to discuss obscure heat in his paper "On Luminous and Obscure Radiation," which appeared shortly thereafter in the *Phil. Mag.*, he was careful to note that "the rendering of metals incandescent by obscure rays has not yet been accomplished," adding, "This is a question on which Dr. Akin has been working for some years, and it is not my intention to publish anything relating to it until the very promising arrangements which he has devised have had sufficient trial."[51]

This was clearly too much for Akin, who wrote a deeply critical response in the *Phil. Mag.*, in which he publicly accused Tyndall not only of breaking his word regarding steering clear of another man's research but of quite evidently appropriating the results of his own BAAS paper. Tyndall was outraged by such a suggestion, noting in his journal, "I have read Akin's article—it is vile."[52] His friends shared his opinion, William Frederick Pollock writing to declare the man to be 'an unmitigated s—nd—l."[53]

In his reply, "On the History of Negative Florescence," Tyndall charted the history of his own work parallel to that of Akin's, and reminded his readers that he had already publicly ceded priority to Akin on the one point mentioned above, but then went on to note that Akin's claims, unlike his own, had not been adequately experimentally demonstrated. As Jackson notes in his account of this episode, Tyndall now saw no reason to hold back. He sent off a short paper to the RS, "On the Invisible Radiation of the Electric Light," in which, after consulting with Stokes, he used James Challis's term *calorescence* to describe the process of matter absorbing infrared energy and emitting radiant energy in its place, doing so again in a letter to the *Phil. Mag.*, which was published in February.[54] This effectively undermined Akin, who had ventured the term *calcescence* to describe this process in his own work.

Tyndall wrote a number of letters to colleagues at this time to ascertain in his own mind the timeline of his ideas, his discussions with Stokes about Akin's work, and his experiments, an account of which he sought to publish in order to vindicate himself. He appealed to friends as well to provide him with written testimony of their recollections of his discussions with them about his work. In letter 2183 he wrote to Debus suggesting edits he would like to make to Debus's statement in this regard, and although there is no evidence that Tyndall sought to manipulate the substance of Debus's testimony, this episode shows how Tyndall marshaled his colleagues and connections to successfully assert his priority. Hirst noted in his journal entry for 13 November 1864 that he had been to the RI to see Tyndall perform experiments in which he rendered platinum incandescent "by means of obscure rays," noting too that "I may possibly have to bear witness some day since an important difference has arisen between Tyndall and Akin who has long been trying to obtain the same effect."[55] In line with Hesketh's claims about Tyndall's concerns about propriety and the maintenance of a proper scientific demeanor, in letters 2198, 2201, and 2204 to William Francis, Tyndall sought to walk back the severity of his criticisms of Akin and his work before overstatement, written in anger and frustration, reached the printed page.[56] Tyndall and Francis met in person to discuss how best to phrase Tyndall's criticisms, and made the appropriate changes. Stokes, who was well aware of Tyndall's dispute with Akin, played an active role in taking Tyndall's side by providing him with the opportunity to cement his claim by suggesting that it would be "extremely interesting" for Tyndall to demonstrate his experiments at the RS at his convenience.[57] Tyndall proved to the assembly that platinum could indeed be raised to incandescence by the invisible rays, demonstrating the transmutation of rays from a longer to a shorter, and thus visible, wavelength.[58] Ever the showman, as a finale to the evening Tyndall used invisible heat to light the cigar of the Count de Paris, who had chaired the meeting.[59] Akin could not compete with such a performance. Indeed, nor was he invited to do so.

The Reader: A Journal for Literature, Science, and Art

In November of 1864, just as he was in the midst of his dispute with Akin, Tyndall, along with his friends Huxley, William Frederick Pollock, William Spottiswoode, Norman Lockyer, and others, purchased the *Reader: A Journal for Literature, Science and Art*, with the aim of using it to advance modern science. Tyndall and Huxley were the driving force behind the hurriedly put-together scheme. Pollock recalled: "Their object was to provide a better record than then existed of the progress of science, and of the labours of scientific men, with reports of the proceedings of scientific institutions, and space for correspondence,"[60] and Tyndall canvassed his friends to encourage them to buy into the enterprise. Thirty-five shares were offered at £100 each, and as Tyndall explained to Henry Bence Jones, "the proprietary will therefore be small and we shall have the matter in our own hands."[61] Despite his enthusiasm, however, from the start Tyndall struggled to find time for their new project, and Pollock recalled that he could not remember ever seeing Tyndall at the *Reader*'s Tavistock Street offices.[62]

Huxley, in light of his earlier experiences with both the *Scientific Review* and the *Natural History Review*, was well aware of what they were taking on, but it seems that this was less the case for the majority of the others involved.[63] Pollock was not alone in expecting the venture to turn a profit. Tyndall bubbled enthusiasm for the project, writing to Herschel, "There never was a better opportunity for establishing a healthy scientific organ," further noting that the venture was "exceedingly well supported."[64] And indeed it was, for in addition to the likes of Huxley, Herschel, Spottiswoode, and Francis Galton, a great many prominent figures lent their name to the cause. Names alone, however, did not pay the bills, and their venture came to a close after only a few issues. This was not Tyndall's last venture into publishing, and he was supportive of Lockyer and Huxley's efforts with the journal *Nature* in 1869.

Tyndall's Family and Friends, His Social and Scientific Circle

What comes to the fore throughout these letters and the drama and dispute that characterize this period of Tyndall's life is the depth and breadth of Tyndall's social circle. He formed deep and abiding friendships not only among his scientific colleagues but among artists and writers as well. Exchanges of photographic portraits, dinner invitations, and visits to notables such as Thomas Carlyle and Alfred Tennyson enlivened his working life and reveal the extent to which he was integrated into upper-middle-class Victorian society, despite his own comparatively humble background. Tyndall was particularly good friends with William Frederick Pollock, his wife, Juliet, and their children, and visited them frequently both in their London home and on the Isle of Wight. Tyndall's relationship with Juliet Pollock is particularly intriguing. There was clearly great admiration and deep emotional affection between them, with Tyndall even writing poetry inspired by Juliet. Tyndall, as noted, was also close to her husband, with whom he not only collaborated on the *Reader* but often went to the theatre.

Family was never far from Tyndall's mind. He supported his mother and sister in Ireland with a regular allowance, arranging for a friend, Annette Roche, to ensure that they were adequately clothed and accommodated, as well as to have a gravestone placed on his father's grave.[65] Tyndall corresponded regularly with his cousin Hector Tyndale in the United States, with whom he shared a keen interest in the genealogy of the Tyndale/Tyndall family.[66] Several letters contain extensive lists of numerous historical Tyndalls and Tyndales and describe visits to living representatives of the extended family.

Tyndall maintained close contact with his German colleagues, as professional relationships developed into private friendships. German colleagues were not only an important connection to the world of European physics but an important connection to German instrument makers, overseeing orders, shipping, and payments for specialized apparatus on his behalf. Meetings such as those of the BAAS, the Swiss scientific society Naturforschende Gesellschaft, and the Versammlung Deutscher Naturforscher und Ärtze (or the German Society for Natural Researchers and Physicians) offered opportunities to catch up on the latest gossip and send greetings to colleagues not in attendance. The professional and private melded, for example, when Tyndall offered his London lodgings to the editor of *Annalen der Physik und Chemie*, Johann Christian Poggendorff, and his daughter, or agreed to accept the Rumford Medal of the RS on behalf of his German colleague Gustav Kirchhoff, who was unable to come to the award ceremony himself.[67]

Tyndall's extensive correspondence with Rudolf Clausius in this period is particularly noteworthy. The two had met in Berlin in 1851 and held each other in great esteem. Their professional friendship became even stronger when Tyndall turned to Clausius for advice and criticism of his lectures on heat and was cemented by their shared position during the Mayer-Joule priority dispute. In October 1864 their friendship became even deeper, when Clausius not only invited Tyndall to be the godfather to his youngest child but also switched to using the informal German form of address *Du*, reserved for only the closest and most trusting of relationships. "The lad will be named John, after you," Clausius wrote.[68] Tyndall's delight at this expression of trust and friendship is evident in his response, when he begins his English-language letter with an expression in German addressing Clausius with *Du* himself.[69] Although, as Tyndall wrote, he had often been asked to be a godfather to colleagues' children and had always declined, he felt honored to take on the role for the youngest member of Clausius's family. Whether or not the infant John would be allowed to join Tyndall on his mountaineering adventures, however, was a subject that Adelheid Clausius reserved for later discussion.

The years from 1862 through 1865 thus show Tyndall becoming a force to be reckoned with in British and, increasingly, international science. With a principled moral code of conduct, and a work ethic that was both physically and mentally demanding, Tyndall was combative when he felt unfairly treated or maligned, but also generous and supportive to those whose work he admired. His deep private friendships reflect that capacity for generosity. Although Tyndall's research interests were still quite broad, he had developed clear areas of focus and had demonstrated his commitment to the public communication of science not only through his RI lectures—a requirement of his employment—but also through his popular publications and attempt to establish a popular science journal. This volume thus reflects an important period in the development of Tyndall as a man of science but also what it meant to be a man of science during the early years of the second half of the nineteenth century.

Notes

1. *Ascent of John Tyndall*, 162–63, on 163.
2. Paul White, *Huxley: Making the Man of Science* (Cambridge: Cambridge University Press, 2003); Lorraine Daston and Ian Galison, *Objectivity* (Cambridge, MA: MIT Press, 2007); Ian Hesketh, "Technologies of the Scientific Self.: John Tyndall and his Journal," *Isis* 110, no. 3 (2019): 4.
3. Daston and Galison, *Objectivity*, cited in Hesketh, "Technologies of the Scientific Self," 3.
4. Michael S. Reidy, "Evolutionary Naturalism on High: The Victorians Sequester the Alps," in *Victorian Scientific Naturalism: Community Identity, Continuity*, ed. Gowan Dawson and Bernard Lightman (Chicago: University of Chicago Press, 2014), 59.
5. Letter 2145.
6. Ruth Barton, *The X Club: Power and Authority in Victorian Science* (Chicago: University of Chicago Press, 2018); Roy M. McLeod, "The X-Club: A Social Network of Science in Late-Victorian Britain," *Notes and Records of the Royal Society of London* 24 (1970): 305–22.
7. It is important to note here, however, that Barton makes the case in *X Club* that the cohesion of this group was less tightly unified in support of scientific naturalism than many scholars have previously thought.
8. Ruth Barton, "Scientific Authority and Scientific Controversy in Nature: North Britain against the X-Club," in *Culture and Science in the 19th Century Media*, ed. Louise Henson et al. (London: Routledge, 2004), 223–35.
9. John Tyndall, "On the Absorption and Radiation of Heat by Gases and Vapours, and on the Physical Connexion of Radiation, Absorption, and Conduction," *Phil. Trans.* 151 (1861): 1–36. See, e.g., letter 1995.
10. John Tyndall, "Contributions to Molecular Physics. Being the 5th Memoir of Researches on Radiant Heat," *Phil. Trans.* 154 (1864): 327–68. See letters 2097, 2099, 2112, 2114, 2115, 2117, 2119, and 2136.
11. *Ascent of John Tyndall*, 166–68.
12. John Tyndall, "On the Relation of Radiant Heat to Aqueous Vapour. Third Memoir," *Phil. Trans.* 53 (1863): 1–12.
13. Letter 1958.
14. Letter 1960. For Tyndall's later work on the blue color of the sky, see *Ascent of John Tyndall*, 245–49.
15. Much of the correspondence between Tyndall and Clausius in this period discusses questions about the phenomenon of heat. See letters 1962, 1963, 1965, 1966, 1967, 1969, 1972, 2048, 2049, and 2093.
16. This topic was extensively discussed in *Tyndall Correspondence*, vol. 7. See also in this volume letters 2021, 2028, 2100, and 2152.
17. For New Zealand, see letters 1928, 2087, and 2120.
18. Letters 2119, 2176, 2177, 2179, 2195, 2214, 2215, and 2216.
19. Letters 2048 and 2057. John Tyndall, "Note on Laplace's Correction for the Theoretic Velocity of Sound," *Phil. Mag.* 26 (1863): 384–87.
20. Letter 2001. John Tyndall, *Sound: A Course of Eight Lectures Delivered at the Royal Institution of Great Britain* (London: Longmans, 1867).
21. The physical concept of energy, which eventually replaced the concept of force, was not yet established. Julius Robert Mayer, "Bemerkungen über die Kräfte der unbelebten Natur"

(1842), in Julius Robert Mayer, *Mechanik der Wärme, in gesammelten Schriften*, ed. Jacob J. Weyrauch, 3rd ed. (Stuttgart: J. C. Cotta, 1893), 23–30.
22. James Prescott Joule, "On the Calorific Effects of Magnetic Co-electricity, and on the Mechanical Value of Heat," *Phil. Mag.* 23 (1843): 263–76, 347–55, 435–43.
23. *Tyndall Correspondence*, vol. 7, letters 1664 and 1674 (Clausius's reply).
24. A number of Mayer's works were self-published and thus were, and remain, difficult to obtain in their original format. While references in this volume provide the publication details of the original texts, they were subsequently republished and made more widely accessible in a collected edition of Mayer's work. See Julius Robert Mayer, *Mechanik der Wärme, in gesammelten Schriften*, ed. Jacob J. Weyrauch, 3rd ed. (Stuttgart: J. C. Cotta, 1893).
25. For Tyndall's correspondence with Clausius and Helmholtz, see *Tyndall Correspondence*, vol. 7, letters 1890, 1891, 1892, 1893, and 1896.
26. Letter 1905.
27. Crosbie Smith, *The Science of Energy: A Cultural History of Energy Physics in Victorian Britain* (Chicago: University of Chicago Press, 1998), chapter 9.
28. James Prescott Joule, "Note on the History of the Dynamical Theory of Heat," *Phil. Mag.* 24 (1862), 121–23.
29. Letter 1914.
30. John Tyndall, "On Mayer, and the Mechanical Theory of Heat," *Phil. Mag.* 24 (1862), 173–75. See also letter 1920.
31. Letter 1920.
32. Letters 1962, 1963, and 1969; and replies 1965, 1966, and 1967.
33. Letter 1966.
34. William Thomson, "On the Dynamical Theory of Heat, with Numerical Results Deduced from Mr. Joule's Equivalent of a Thermal Unit, and M. Regnault's Observations on Steam (1851)," in *Mathematical and Physical Papers* (Cambridge: Cambridge University Press, 1882), 1:174–233.
35. The opening salvos can be found in: William Thomson and Peter Guthrie Tait, "Energy," *Good Words* 3 (1862): 601–7; John Tyndall, "Remarks on an Article Entitled 'Energy 'in *Good Words*," *Phil. Mag.* 25 (1863): 220–24; and Peter Guthrie Tait, "Reply to Prof. Tyndall's Remarks on a Paper 'Energy' in *Good Words*," *Phil. Mag.* 25 (1863): 263–66. The dispute continued into 1864, continuously generating further publications. See Smith, *Science of Energy*, 186–91, and 338n68.
36. Letter 2006.
37. Letter 2006.
38. *Ascent of John Tyndall*, 224.
39. Letter 1915.
40. Letter 1915.
41. Letter 1919.
42. Letter 2145.
43. *Ascent of John Tyndall*, 234.
44. Letter 1925.
45. C. K. Akin, "Account of Preliminary Experiments on Calcesence," *Brit. Assoc. Rep. 1863*, 11.
46. Letter 2075.
47. Letter 2162.
48. Letter 2077.
49. Letter 2167.
50. Letter 2168.

51. John Tyndall, "On Luminous and Obscure Radiation," *Phil. Mag.* 28 (1864): 329–41. For a more detailed account of Tyndall's paper, see *Ascent of John Tyndall*, 184.
52. See Tyndall's journal entry of 1 December 1864 (RI MS JT/2/13c/1270).
53. Letter 2190.
54. *Ascent of John Tyndall*, 185.
55. Brock and Macleod, *Hirst Journals*, 1709.
56. Hesketh, "Technologies of the Scientific Self," 3–4.
57. Letter 2215.
58. *Ascent of John Tyndall*, 185–86.
59. *Ascent of John Tyndall*, 186.
60. *Remembrances*, 2:128–29.
61. Letter 2181.
62. *Remembrances*, 2:129.
63. "Science in Grub Street," *Nature* 224 (1 November 1969): 423–27.
64. Letter 2184.
65. Letter 2113.
66. Letters 1912, 1946, 1953, 1956, 1988, 2010, 2019, 2026, and 2095.
67. For Poggendorff, letters 1905 (n. 8), 1910, and 1931. For Kirchhoff, letters 1943 and 1949.
68. Letter 2147. In German, Clausius's words were: "Der Knabe soll dann nach Dir 'John' heissen." See also letters 2191 and 2192.
69. Letter 2153.

EDITORIAL PRINCIPLES

Our aim is to include all letters to and from John Tyndall that are currently extant. Some of the letters are from published sources, such as letters written by Tyndall to newspapers or journals, but the majority are either the originals or typescripts produced by Louisa Tyndall from originals. One central editorial principle has been to transcribe, wherever possible, from original letters rather than Louisa's typescripts. This was because of Louisa's conscious and unconscious editorial interventions, a point discussed in more detail below. We have also aimed to reproduce, as accurately as possible, the text of the letters as they were written. Spelling mistakes that have appeared in the handwritten letters have therefore been preserved; mistakes appearing where only the typescript letter has survived are silently corrected, as in our judgement they are almost always typographical errors. But, in general, nonstandard spellings (e.g., the use of dialect) has been retained even when in a typescript-only letter (though there is the possibility that the spelling is Louisa's, not her husband's). Where the transcription is based on a typescript transcription by Louisa (or one of her assistants), this will be indicated by "LT Transcript Only." For those cases in which our transcription comes from typescript letters not done by Louisa, or it is unclear if she was involved, this will be indicated by "Typed Transcript Only." For the rare case when only a handwritten transcript of a letter exists, this will be indicated by "Transcript Only."

We have included illustrations that are part of the letter. In cases where it was possible to reprint the original handwritten image, we have done so. In other cases, volume editors have reconstructed drawings because it was not possible to reprint an image from the handwritten letter (due, for example, to deterioration of the original document).

While quotation marks within some handwritten and typescript letters use double quotation marks, for consistency we have used single marks in the transcriptions. However, if our source for a letter was a publication such as a journal, newspaper, or book, we have retained the quote marks as they appear in that publication.

The letters are presented in chronological order of writing. Where more than one letter has the same date, letters from Tyndall (in alphabetical order of addressee) come first, followed by letters to Tyndall (in alphabetical order of writer), with the exception that if the order of writing can be determined, that order takes precedence. Where letters cannot be dated precisely they are placed at the latest likely date of writing. (In some cases if it is possible, but unlikely, that they were written later, notes indicate such possibilities.) To indicate the beginning of a new year, the date of that year has been inserted in a large font before the first letter from that year. Readers should note that although volumes are arranged in chronological order, letters that are discovered from years already covered in published volumes will appear in one of the final volumes of the correspondence.

For non-English-language letters, both the transcription of the original and the English translation are presented. The transcription of the non-English letter appears first, followed by the English translation in a smaller font. Editorial notes are inserted in the English translation. Some foreign letters are in an original hand, while others exist only as transcriptions by Louisa or others who often did not have a full command of the language. In the latter case these letters contain frequent spelling, grammatical, and interpretation errors that do not seem to have been committed by the original author. In these typescript-only cases, and only in these cases, spelling and grammatical errors have been corrected, as have poor interpretations where a better word choice was obvious from the context.

There is a standard format for each letter. The first line lists the author, if it is to Tyndall, or the recipient, if it is from Tyndall; the date; and the letter number. The second line lists any information supplied by the writer in their salutation, such as the place, date, or time of day when the letter was written. (In some cases, though, where the writer put the date at the foot of the letter, we have left it there in order to reproduce the text accurately.) The opening salutation, the text of the letter, and closing then follow. Then come the source and editorial notes, the latter indicated in the text by Arabic numerals.

Editorial notes are intended to provide the contextualization necessary to understand the contents of each letter. There are basically six types of editorial notes. Many of them are informative notes on persons mentioned in the letters. If a person referred to in a letter is not described in a note, they have been mentioned more than twice in the volume and they will have an entry in the biographical register at the back of the volume. For the rest—those mentioned once or twice—biographical information will appear in the editorial note appended to the letter in which he or she is first mentioned. When there is a second reference to that person, a note will refer back to the letter in which he or she was first mentioned. Biographical notes on well-known figures such as Faraday or Huxley stress what that person was doing at the time when the letter was written, rather than providing a full overview of the person's life.

A second type of editorial note identifies allusions and quotations. In the case of quotations, where possible, the first edition of the work is cited unless a different edition is mentioned in the letter. If a quote is not in English a translation is given. Glosses on obscure places, abstruse words, and words used in an unfamiliar sense make up the third type of editorial note. The fourth type provides information about the context of the letter, drawing on such sources as contemporary publications, scholarly sources, Tyndall's personal journals and field notes, and other letters in the *Tyndall Correspondence*. The fifth type of editorial note is cross-references to other letters in the correspondence, referring to the letter number. The last type is information on significant textual changes that Tyndall made to the original that is thought to be relevant to the meaning of the letter.

There are a series of conventions for indicating when editors have given additional information about the text of a letter:

When Louisa Tyndall has made an annotation or marginal insertion that the editors have deemed important enough to retain, we have included her annotation in an editorial footnote.

When editors other than she have made insertions, they are indicated by unitalicized square brackets; for example, [word].

When certain words are ambiguous and we have had to conjecture their meaning, they are indicated by italicized square brackets; for example, *[*words*]*; when we have considered such words illegible the number of illegible words is indicated in full italicization and square brackets; for example, *[3 words illeg]*.

If a word or series of words are not present or have been destroyed—by inkblots, holes, or mold, for example—they are indicated as missing, italicized and in angular brackets; for example, *<3 words missing>*.

In extremely rare cases where it is clear that Louisa herself has intentionally destroyed a few words, they are indicated as excised, italicized and in angular brackets; for example, *<3 words excised>*.

Finally, a brief word about Louisa. She was responsible for collecting together many of Tyndall's letters in order to write a biography of her late husband. This biography was not completed. It is important to recognize that Louisa was not always reliable as a transcriber. She often threw out the manuscript letter when she had made her transcription, but some original manuscripts remain, which indicate her editing processes. Like any transcriber, she was sometimes inaccurate, but corrections show that she proofread her transcriptions. However, she also edited Tyndall's writing to make his letters seem more consistent or grammatically correct. For example, she made punctuation more formal, corrected grammar, and, occasionally, corrected quotations. She consistently corrected Tyndall's usual (deliberate) lowercase for "god" to "God." Her typewriter was unable to insert the superscripts that

Tyndall used in abbreviations, or the "&c" symbols that stood for an address at the foot of the letter. (Modern typefaces are unable to deal with all the variant scrawls that the "etc." symbol took, and they are all formalized here to "&c.") Louisa also left out details that she considered too private or too salacious for public consumption. She equated postmark and date of writing and overlooked the time taken to write some letters. When she inferred dates from internal evidence, her reasoning was sometimes faulty. Similarly, her identifications of people are sometimes in error. Independent evidence has always been sought for her dates and identifications. Lastly, the patterns of punctuation differ between LT sources and manuscript sources. Tyndall himself used dashes more often than commas, semicolons, and stops, and he used dashes of many different lengths, which are all standardized here to em dashes. It is often difficult to distinguish between very short dashes and commas or very short dashes and full stops. Similarly, capitalization is often ambiguous; distinguishing S/s, for "science," for example, is a particular problem. Therefore, scholars for whom punctuation and capitalization are significant will need to consult the manuscripts. Users should also be aware that throughout this project the transcribers and editors have generally worked from scans and photographs rather than the original manuscripts.

NOTE ON MONEY

In the nineteenth century the currency used in England was the pound sterling, usually abbreviated to £ (or, occasionally, to L). A pound consisted of 20 shillings (abbreviated *s*) and each shilling contained 12 pennies (*d*). A guinea was 21 shillings; that is, £1 1s. A crown was 5 shillings (5s), and half a crown equaled two shillings and sixpence (2s 6d). The smallest coin was a farthing, one-quarter of a penny (¼d).

Several forms of representation were in contemporary use. Sometimes the numbers for pounds, shillings, and pence were separated by dots, forward slashes, or spaces; thus, for example, 3 pounds, 4 shillings, and 5 pennies could be written as £3.4.5, £3/4/5, or £3 4 5. A zero could be represented by a dash; for example, 7 shillings was often written as 7/- or just 7/. A half crown was often written as 2/6.

Currencies mentioned in this volume are French francs, United States dollars, and German Thaler. The franc was the national currency of France from 1795 until 1999, when France adopted the Euro. In 1864 the exchange rate between British and French currencies was approximately 25 francs per £.[1] A centime was a coin worth one-hundredth of a franc. In the mid-nineteenth century, one pound sterling was worth approximately 10 US dollars.[2] The Reichsthaler, a silver coin issued by Prussia from 1750, became a common currency throughout much of Germany in the mid-nineteenth century before being replaced by the Vereinsthaler in 1857. In midcentury the exchange rate between the German and British currencies was approximately 7 Thaler per pound sterling.[3]

---

1. Rodney Edvinsson, Historical Currency Converter (test version 1.0), accessed 29 September 2018, http://www.historicalstatistics.org/Currencyconverter.html.
2. Lawrence H. Officer, "Exchange Rates between the United States Dollar and Forty-One Currencies," MeasuringWorth, accessed 29 September, 2018, http://www.measuringworth.com/exchangeglobal.
3. Henry Vizetelly, *Berlin under the New Empire: Its Institutions, Industry, Monuments, Social Life, Manners* (London: Tinsley Brothers, 1879), 1:166.

TIMELINE OF JOHN TYNDALL'S LIFE

Year	Event
c. 1822	John Tyndall born at Leighlin Bridge, County Carlow, Ireland
c. 1836	Attends John Conwill's National School in Ballinabranna
1839	Begins employment as civil assistant with Ordnance Survey of Ireland in Carlow
1840	Becomes civil assistant in the Ordnance Survey Office at Youghal, Cork County
1841	First appearance in print (poem in *Carlow Sentinel* under pseudonym "W[alter] S[nooks]")
1842	Joins English Ordnance Survey in Preston
1843	Leads written criticism of Ordnance Survey in *Liverpool Mercury*; dismissed November
1844	Home in Ireland, mostly unemployed, until obtaining a position with the firm of Nevins and Lawton, Surveyors, of Manchester
1845–47	Works for Richard Carter, Surveyor, of Halifax
1845	Meets Thomas Archer Hirst in Halifax
1847	Begins teaching mathematics and surveying at Queenwood College, Hampshire
1848–50	At University of Marburg, working for PhD with Robert Bunsen and Friedrich Stegmann
1850	Returns to England; meets Michael Faraday and William Francis and attends his first British Association meeting, before returning to Marburg
1851	Moves from Marburg to Berlin (April–June); then returns to Queenwood
1852	Elected fellow of the Royal Society
1852	Friendship developed with Thomas Henry Huxley
1853	First lecture at Royal Institution (RI)
1853	Meets Herbert Spencer
1853	Appointed professor of natural philosophy at RI
1853	Awarded Royal Medal for magnetic work; declines when controversy arises over award
1856	Meets Thomas Carlyle
1857	Climbs Mont Blanc for first time with Hirst
1859	Demonstrates existence of greenhouse gases and climatic implications
1859	Joins Government School of Mines as professor of natural philosophy
1860	*The Glaciers of the Alps* published
1860	Meets John Lubbock and his wife, Ellen Lubbock
1861	First ascent of the Weisshorn

Year	Event
1861	Delivers his first Christmas Lectures at RI
1862	*Mountaineering in 1861* published
1862	Begins to champion Mayer's priority in discovery of conservation of energy
1863	*Heat Considered as a Mode of Motion* published
1864	Awarded Royal Society's Rumford Medal
1864	First meeting of X Club
1865	*On Radiation* published
1865	Engages in public controversy over the efficacy of prayer
1866	Succeeds Michael Faraday as scientific adviser to Trinity House
1867	Faraday dies; Tyndall becomes superintendent of the RI
1867	*Sound: A Course of Eight Lectures* published
1868	Successfully climbs Matterhorn
1868	Discovers the cause of light scattering, to be known as "Tyndall Effect"
1868	*Faraday as a Discoverer* published
1868	"Scientific Materialism" lecture at the British Association
1868	Resigns from Royal School of Mines
1869	Joins Metaphysical Society
1870	"On the Scientific Use of the Imagination" lecture at British Association
1870	*Three Scientific Addresses* published
1870	*Researches on Diamagnetism and Magne-Crystallic Action* published
1871	Meets Louis Pasteur for first time while in Paris
1871	*Hours of Exercise in the Alps* published
1871	*Light and Electricity* published
1871	*Fragments of Science* published
1872	Debates over how to measure the efficacy of prayer, aka the "Prayer-Gauge Debate"
1872	*Contributions to Molecular Physics in the Domain of Radiant Heat* published
1872	*The Forms of Water in Clouds and Rivers, Ice and Glaciers* published
1872–73	USA Lecture Tour
1873	*Six Lectures on Light* published
1874	As president of the BAAS, delivers the "Belfast Address"
1876	Marries Louisa Charlotte Hamilton
1877	Develops "Tyndallization" (discontinuous heating and cooling process to destroy heat-resistant spores)
1877	Tyndall and Louisa build summer cottage, Alp Lusgen, at Belalp, northern side of Valais, above Brig
1877	*Fermentation and Its Bearings on Phenomena of Disease* published
1879	*Fragments of Science*, which had gradually expanded, first published in two volumes
1881	*Essays on the Floating Matter of the Air* published
1885	Tyndall and Louisa build Hindhead retreat, Surrey Downs
1887	Resigns from RI
1890	Fight with Gladstone in the *Times* over Irish Home Rule
1892	Hirst dies
1892	*New Fragments* published
1893	Dies at Haslemere, Surrey, accidentally poisoned

TIMELINE OF EVENTS IN JOHN TYNDALL'S LIFE SPECIFIC TO VOLUME 8

Month and year	Biographical details	Notable social, political, and cultural events
April 1862		Victor Hugo publishes *Les Misérables* Louis Pasteur and Claude Bernard begin pasteurization experiments
May 1862		International Exhibition of Industry and Science opens in South Kensington New Westminster Bridge, designed by Thomas Payne, opens in London John Ruskin's *Unto This Last* published Creation of US Department of Agriculture
June 1862	Delivers Friday Evening Discourse at the RI, "On Force" (6th) Introduces Mayer to British public Sends Mayer's "Remarks on the Forces of Organic Nature" to *Phil. Mag.* for translation Begins revising lectures "On Heat, Regarded as a Mode of Motion" for publication	John William Draper publishes *History of the Intellectual Development of Europe* "Revised Code" of "payment by results" introduced in English schools
July 1862	Visits Switzerland with Huxley for mountaineering in the Alps Helps rescue porter after climbing accident Almost makes it to Matterhorn summit Offered honorary membership of Berlin Society of Friends of Natural Science Mayer-Joule priority dispute begins to heat up	Marriage of Princess Alice, second daughter of Queen Victoria, to Prince Ludwig of Hesse and by Rhine George Eliot's novel *Romola* serialized in the *Cornhill Magazine*

Month and year	Biographical details	Notable social, political, and cultural events
August 1862	In good health after month of mountaineering Publishes short paper on "Conformation of the Alps" in *Phil. Mag.* Publishes open letter to Joule titled "Mayer and the Mechanical Theory of Heat" in *Phil. Mag.*	Confederate "ironclad" *Arkansas* badly damaged in Union attack Berlioz's opera *Beatrice et Benedict* premieres in Baden-Baden
September 1862		Namamugi incident takes place, prelude to the Anglo-Satsuma War Battle of Antietam (Sharpsburg), American Civil War Otto von Bismarck becomes prime minister of Prussia
October 1862	Attends Cambridge meeting of BAAS Visits Isle of Wight	BAAS meets in Cambridge
November 1862	Learns that cousin Hector Tyndale wounded in Battle of Antietam (Sharpsburg) Works on "On the Relation of Radiant Heat to Aqueous Vapour. Third Memoir" Kirchhoff asks Tyndall to receive for him the Rumford Medal from the RS	Civil engineer Joseph Bazalgette begins construction of the Thames embankment Notts County Football Club founded Gatling gun patented
December 1862	Works on genealogy of Tyndall family Visits the Isle of Wight	
January 1863	Asks Clausius to review chapters of manuscript *Heat as a Mode of Motion* Delivers Friday Evening Discourse at the RI, "Radiation through the Earth's Atmosphere" (23rd) Publishes "Radiation through the Earth's Atmosphere" in *Phil. Mag.*	Abraham Lincoln delivers the Emancipation Proclamation Yorkshire County Cricket Club founded First section of London Underground, from Paddington to Farringdon Street, opens to the public Beginning of January Uprising in Poland, Lithuania, Belarus Jules Verne publishes *Five Weeks in a Balloon*
February 1863	Publishes *Heat as a Mode of Motion*	International Red Cross Founded Sergi-i Umumî-i Osmanî (Ottoman General Exposition)

Month and year	Biographical details	Notable social, political, and cultural events
February 1863		Charles Lyell's *Antiquity of Man* published Thomas Huxley's *Man's Place in Nature* published First land grant college founded in United States
March 1863	Dispute with Thomson and Tait about Mayer priority intensifies Publishes response to Thomson in *Phil. Mag.* Organizes RI lecture for Alexander Herschel Publishes *Heat as a Mode of Motion*	Marriage of Elbert Edward, Prince of Wales (later Edward VII), to Princess Alexandra of Denmark at Saint George's Chapel, Windsor Castle US National Academy of Sciences founded
April 1863	Visits the Isle of Wight Discusses American and French editions of *Heat as a Mode of Motion* with publishers and translators Delivers a course of lectures at the RI on "Sound"	
May 1863	Dispute with Thomson and Tait becomes more rancorous	
June 1863		Samuel Butler's "Darwin among the Machines" published in the *Press* (New Zealand)
July 1863	Visits Switzerland for mountaineering in the Alps	Slavery abolished in Surinam Resumption of New Zealand Wars
August 1863	Hikes in Lake District with Huxley Attends BAAS meeting in Newcastle upon Tyne	Bayer Chemical Manufacturing founded by Friedrich Bayer
October 1863		Foundation of the International Committee for Relief of the Wounded, precursor to the International Red Cross
November 1863		Abraham Lincoln delivers the Gettysburg Address
December 1863	Experiments on radiant heat Gives RI Christmas Lectures on "Electricity at Rest and Electricity in Motion" Gives instruction to prince and princesses	First steam railway opens in New Zealand British novelist William Thackeray dies

Month and year	Biographical details	Notable social, political, and cultural events
January 1864	Lectures on "Experimental Optics" at the RI (January–February)	Tauranga Campaign begins in New Zealand Charing Cross Station opens
March 1864	Visits the Isle of Wight	Great Sheffield flood Rebecca Lee Crumpler becomes first African American woman to receive medical degree
April 1864	Acknowledges Akin's priority in explaining the glowing of a platinum wire in a hydrogen flame Mother falls ill Delivers course of lectures at the School of Mines (April–May)	Italian general Giuseppe Garibaldi visits England
May 1864	Travels to Ireland Visits mother and sister Asks Annette Roche to administer the allowance he sends his mother Continues genealogical research Bennen, his trusted alpine guide, dies in an avalanche Health is poor in late May Writing "Notes on Scientific History"	
June 1864	Sends portraits to Roche Delivers Friday Evening Discourse at the RI, "On a Magnetic Experiment" (10th) Travels to Ledbury Works on "On the Absorption and Radiation of Heat by Gaseous and Liquid Matter"	End of the Russo-Circassian War
July 1864	Visits Switzerland for mountaineering in the Alps and Italy for the Dolomites Health improving Makes observations on glaciers and mountain formation	
July 1864	Revising *Heat* for 2nd edition Narrowly escapes an avalanche	
August 1864	Huber requests permission to translate *Glaciers of the Alps* into French Returns to site of avalanche to look for his watch	British chemist John Newlands produces the first periodic table First Geneva Convention signed Taiping Rebellion ends

Month and year	Biographical details	Notable social, political, and cultural events
August 1864		Italian astronomer Giovanni Donati first observes spectrum of a comet Prussian-German philosopher and activist Ferdinand Lassalle dies
September 1864	Purchases tombstone for father's grave in Ireland	Charlottetown Conference Napoleon III and Victor Emmanuel II sign the September Convention BAAS meets in Bath International Workingmen's Association founded in London
October 1864	Accepts request from Clausius to be godfather to new son Clausius begins addressing Tyndall with German *Du* Working on "On the History of Negative Fluorescence" Invited to lecture at Lowell Institute in United States but declines Publishes "On the Formation of the Alps" in *Phil. Mag.* Informs Stokes he has succeeded in making platinum incandescent through radiation	Beginning of Quebec Conference for establishment of Dominion of Canada Cyclone destroys Calcutta Treaty of Vienna ends Second Schleswig War French engineer Louis Lumière born
November 1864	First meeting of X Club Reaches agreement with Akin to allow him research priority on radiation and platinum Receives RS's Rumford Medal "for his researches on the Absorption and Radiation of Heat by Gases and Vapours" (30th) Continues genealogical research Discusses radiation and incandescence with Herschel	HMS *Victoria* enters active service Abraham Lincoln reelected
November 1864	Purchases *Reader* with friends Publishes "On Luminous and Obscure Radiation" in *Phil. Mag.*	

Month and year	Biographical details	Notable social, political, and cultural events
December 1864	Dispute with Akin becomes rancorous Working to establish *Reader*	Great Fire of Brisbane Clifton suspension bridge across the River Avon, designed by Isambard K. Brunel, opens at Bristol English mathematician and philosopher George Boole dies
January 1865	Sends portrait to Knoblauch and Clausius Speaks to RS on "History of Calorescence"	New York Stock Exchange opens on Wall Street United States ratifies the Thirteenth Amendment, abolishing slavery

The Correspondence of John Tyndall

1862

To Richard Dawes [3 June 1862][1] 1902

London | Tuesday.

My dear Mr Dean,

I have had much pleasure in sending the card to your friend.[2]

With regard to my visit, I leave entirely in Mrs Dawes's[3] hands. I will either go to you after my lecture,[4] or I will join you immediately after my vacation tour is ended: that will be about the middle of August.[5] I certainly should have a weight upon my mind which no protestations could take away, if I thought I had hastened your visit to Ledbury[6] by a single day.

The Malvern Hills[7] I suppose will not be less beautiful towards the end of August than they are now. They will be beginning to show the tinting of their trees, and I like the colouring of autumn. But as I said before I will do exactly as Mrs Dawes wishes. She shall have me palefaced as I am, or she shall have me sunbrowned as I hope to be a couple of months hence.

Believe me, dear Mr Dean, | Yours very truly, | John Tyndall.

To my shame be it spoken, I have not yet sent Mrs Dawes my little book on mountaineering.[8] I will write to Longmans[9] forthwith about it.

RI MS JT/1/TYP/1/323
LT Typescript Only

---

1. *[3 June 1862]*: June 1862 was suggested by LT. Tyndall delivered the lecture referred to on 6 June, the previous Tuesday being 3 June. Tyndall did not visit the Dawes until August (see letter 1923).
2. *your friend*: not identified.
3. *Mrs Dawes*: Mary Helen Dawes.
4. *after my lecture*: Tyndall delivered a Friday Evening Discourse, 'On Force', at the RI on 6 June 1862 (*Roy. Inst. Proc.*, 3 (1858–62), pp. 527–36).
5. *middle of August*: Tyndall returned from his vacation through the Alps on 16 August.
6. *Ledbury*: a town in Herefordshire in the west-midlands, to the west of the Malvern Hills.
7. *Malvern Hills*: a range of hills that run through Worcestershire, Herefordshire, and northern Gloucestershire.

8. *my little book on mountaineering*: *Mountaineering in 1861.*
9. *Longmans*: letter missing; to either Thomas Longman (1804–79) or William Longman (1813–77), brothers who ran the publishing firm Longman, Green, Longman, and Roberts.

From Elizabeth Dawson Steuart 6 June 1862 1903

June 6th 1862.

My dear John,

I thank you very much for the photograph,[1] which is an excellent likeness. I am so very glad to have it in my Album, amongst those of many other highly valued friends and relatives. It gave me great pleasure to hear from you, and I truly hope your excursion to the Continent[2] will renovate your health, and give you increased strength for the pursuits you are so much engaged in. I always feel anxious about you, lest you may be overworked, knowing your constitution is not a very robust one, and fearing you do not spare yourself as much as you might. Thanks for your kind wishes about my health: it has been very much broken for the last 18 months, but I have many undeserved blessings to be thankful for, and must not murmur at the loss of one, even though it be the greatest and richest of all.[3]

With respect to your last publication,[4] I could not fancy you writing any thing contrary to the precepts and commands of Him who has so distinctly desired us to "Ask and we shall have, seek and we shall find, knock and it shall be opened to us",[5] and it would grieve me more than words could express, to think for one moment that you could neglect the one thing needful: We should not live as if this earth were our resting place, for "here we have no abiding city",[6] and we know not the moment we may be called on to give an account of our thoughts, words, and actions. Dear John, let us seek the Lord while He may be found, and He will not forsake us in the hour of trial and sorrow. The deep and anxious interest I feel in your eternal welfare, must plead my apology for these few words, and hoping you will think of them sometimes,

I remain as ever | your most sincere friend | E. D Steuart.

RI MS JT/1/TYP/10/3345
LT Typescript Only

---

1. *photograph*: photograph not identified.
2. *excursion to the Continent*: Tyndall planned to tour the Swiss Alps beginning in July 1862; and would return in August.

3. *must not murmur at the loss . . . the greatest and richest of all*: i.e., her health.
4. *your last publication*: [J. Tyndall], 'Physics and Metaphysics', *Saturday Review*, 4 (1860), pp. 140–1. This was Tyndall's first substantial statement of materialism in print, although it appeared anonymously, as was standard practice with the *Saturday Review*. As Roland Jackson points out, Tyndall had intended to publish it as a part of his *Glaciers of the Alps*, but his friends discouraged him from doing so (*Ascent of John Tyndall*, pp. 158–9).
5. *"Ask and we shall have . . . and it shall be opened to us"*: Matthew 7:7–8.
6. *"here we have no abiding city"*: Hebrews 13:14.

From Rudolf Clausius 7 June 1862 1904

Zürich, 7 Juni 62

Lieber Tyndall,

Unsere letzten Briefe haben sich gekreuzt. Ich glaube, Sie werden meine letzten Bemerkungen über Mayer gerechtfertigt finden; seine beiden Bücher über "die Organische Bewegung" und die "Dynamik des Himmels" sind wohl nur wenig bekannt geworden, sonst würde er mehr Anerkennung gefunden haben.

Für Ihr Porträt danke ich Ihnen sehr, es hat mich sehr erfreut. Wenn es Ihnen angenehm ist, dass ich die Correcturbogen von Ihrem Buche über die Wärme durchlese, so bin ich dazu sehr gern bereit, und es wird mir sehr interessant sein, dadurch Ihre Vorträge kennen zu lernen, die ich später doch jedenfalls lesen würde. Wenn ich etwas darin finde, was mit meinen Ansichten nicht ganz übereinstimmt, so werde ich es Ihnen offen mittheilen. Die kleine Mühe, welche mir daraus erwächst, darf unter Freunden gar nicht in Betracht kommen.

Vor einigen Tagen ist mir die Mittheilung gemacht, dass der Schweizerische Bundesrath wünscht dass einige Professoren des Polytechnikums, unter denen auch ich mich befinde, nach London zur Industrieausstellung gehen, um die neuesten Fortschritte der Technik kennen zu lernen, und dass er zu dem Zwecke einen Geldbetrag für die Reise bewilligt hat. In Folge dessen gedenke ich in der zweiten Hälfte des August nach London zu gehen, und werde mich sehr freuen, wenn ich Sie dort sehe. Freilich ist das dieselbe Zeit, wo auch die Londoner Professoren auf Reisen zu gehen pflegen, um sich von ihren Anstrengungen zu erholen, und wenn Sie eine solche Erholungsreise vorhaben so dürften Sie sich davon nicht abhalten lassen. Ich gedenke etwa 4 Wochen in England zu bleiben, und hoffe dabei auch die Versammlung der British Association in Cambridge besuchen zu können. Da ich nicht weiss, wenn diese Versammlung beginnt so würde es mir lieb sein, wenn Sie mir dieses gelegentlich einmal mittheilen wollten, damit ich meinen Reiseplan danach einrichten kann,

Mit herzlichen Grüssen, | Ihr | Clausius

Zurich, 7 June 62

Dear Tyndall,

Our last letters[1] crossed. I believe you will find my last remarks about Mayer[2] justified; both of his books on "die Organische Bewegung"[3] and the "Dynamik des Himmels"[4] are indeed little known, otherwise he would have received more recognition.

I thank you very much for your portrait,[5] I was quite delighted. If it suits you, I would be happy to read through the proofs of your book on heat,[6] and I would be very interested in acquainting myself with your lectures in this way, although I will certainly be reading them later as well. When I find something that does not completely agree with my views, I will communicate this to you directly. The small effort it will cost me is of no significance between friends.

A few days ago I received the information that the Swiss Federal Council wants some professors from the Polytechnic Institute,[7] including myself, to go to the industrial exhibition in London[8] in order to acquaint ourselves with the latest technological innovations, and that he has approved a sum of money to travel for this purpose. As a result, I plan to go to London in the second half of August and would greatly look forward to seeing you there. Admittedly, it is at the same time that the London professors tend to travel to recover from their exertions, and if you have planned such a leisure trip, please do not let me keep you from it. I am considering remaining in England for about 4 weeks and hope while there, to attend the meeting of the British Association in Cambridge.[9] Since I do not know when this meeting begins, I would be grateful if you could communicate this to me at your convenience so that I can make my travel arrangements accordingly.

With fond regards, | Your | Clausius

RI MS JT/1/TYP/7/2244–5
LT Typescript Only

---

1. *Our last letters*: letter 1896, *Tyndall Correspondence*, vol. 7; Tyndall's last letter to Clausius has not been found.
2. *my last remarks about Mayer*: although Clausius had stated that he doubted that there would be much of significance in Mayer's works (see letter 1892, *Tyndall Correspondence*, vol. 7), he soon after confessed a change of heart; having read Mayer's papers he now found much of his work of great interest and original (see letter 1896, *Tyndall Correspondence*, vol. 7).

3. *"die Organische Bewegung"*: J. R. Mayer, *Die organische Bewegung in ihrem Zusammenhang mit dem Stoffwechsel. Ein Beitrag zur Naturkunde* (Heilbronn: C. Drechsler, 1845). This work was later reissued in a collected edition of Mayer's work. See J. R. Mayer, *Die Mechanik der Wärme. in gesammelten Schriften*, ed. Jacob J. Weyrauch, 3rd edn (Stuttgart: J. G. Cotta, 1893), pp. 45–128.
4. *"Dynamik des Himmels"*: J. R. Mayer, *Beiträge zur Dynamik des Himmels, in populärer Darstellung* (Heilbronn: John Ulrich Landherr, 1848). This self-published paper is also more commonly cited from the Weyrauch edition, on pp. 151–223. This was also reprinted as J. R. Mayer, 'On Celestial Dynamics', *Phil. Mag.*, 25 (1863), pp. 241–8, 387–409, and 417–28.
5. *For your portrait*: Clausius had sent Tyndall his own portrait and asked for one of Tyndall in return (see letter 1892, *Tyndall Correspondence*, vol. 7). Tyndall thanked Clausius for sending his portrait and enclosed one of himself in return (see letter 1898, *Tyndall Correspondence*, vol. 7).
6. *your book on heat*: *Heat as a Mode of Motion.*
7. *Polytechnic Institute*: in Zurich, where Clausius worked.
8. *industrial exhibition in London*: the International Exhibition of London or the Great London Exhibition took place in the gardens beside the Royal Horticultural Society Gardens in London from 1 May to 1 November 1862.
9. *meeting of the British Association in Cambridge*: the BAAS held its thirty-second annual meeting in Cambridge from 1–7 October 1862 (*Brit. Assoc. Rep. 1862;* 'Analysis of Attendances at the Annual Meetings, 1831–1906', *Brit. Assoc. Rep. 1907*, p. lxxxviii). From letter 1972, we know that while Clausius did travel to London, he did not attend this meeting in Cambridge.

To Rudolf Clausius 12 June [1862][1] 1905

Royal Institution of Great Britain | Thursday 12th. June

My dear Clausius.

I return you my best thanks for Mayer's papers which I have read with astonishment[2] On Friday evening last I gave a discourse in our theatre[3] which consisted entirely in a resumé of what Mayer had done, and gave to him all the credit which he merits. Dove was there and many Noblemen and gentlemen—some of them eminent in Science. Faraday, Wheatstone, and the President of the Royal Society[4] were among the audience and they all seemed greatly interested.

I have sent his last pamphlet to the Philosophical Magazine[5] for translation and shall probably translate the others in due time.

Your papers are also in hand, and partly in the press.

The British Association meets very late this year, the 1st. of October at

Cambridge[6]—I go to Switzerland early in July and hope to return before the end of August.[7] We shall then have many pleasant meetings—I would offer you a bed had I not already offered mine to Poggendorff.[8] But we must arrange to spend a good deal of time together when you come to London.

If you go to Cambridge we will go together then.

Many thanks for your kind promise to look at my proof sheets. I have been compelled to defer the writing out of the lectures my head is so bad.

Yours ever | John Tyndall

RI MS JT/1/T/236

---

1. *[1862]*: the year is given based on relation to letter 1904 and reference to the 1862 annual meeting of the BAAS in Cambridge .
2. *Mayer's papers which I have read with astonishment*: Tyndall had asked Clausius to send him copies of the German physicist Julius Robert Mayer's works (see letter 1896, *Tyndall Correspondence*, vol. 7), which he was amazed to find were of high quality, original, and importantly, had anticipated the results of James Prescott Joule, William Thomson, and Peter Guthrie Tait regarding the mechanical equivalent of heat. As a result, Tyndall took up Mayer's cause as having the rightful claim to priority. Tyndall's efforts and the acrimony it fostered were ongoing for much of the period covered by this volume. See the Introduction for further commentary. Tyndall had also asked for and received papers of Mayer's from Hermann Helmholtz (see letter 1893, *Tyndall Correspondence*, vol. 7).
3. *I gave a discourse in our theatre*: in Tyndall's Friday Evening Discourse delivered at the RI on 6 June 1862, 'On Force' (*Roy. Inst. Proc.*, 3 [1858–62], pp. 527–36), he publicly asserted Mayer's priority in the discovery of the mechanical equivalent of heat.
4. *President of the Royal Society*: Edward Sabine.
5. *his last pamphlet to the Philosophical Magazine*: J. R. Mayer, 'Remarks on the Forces of Inorganic Nature', *Phil. Mag.*, 24 (1862), pp. 371–7.
6. *The British Association . . . at Cambridge*: the BAAS held its thirty-second annual meeting in Cambridge from 1–7 October 1862 (see letter 1904, n. 9). Clausius had asked Tyndall for the dates (see letter 1904).
7. *I go to Switzerland early in July . . . the end of August*: Tyndall travelled to Switzerland on 4 July and returned on 16 August.
8. *Poggendorff*: Poggendorff accepted Tyndall's invitation to London and stated his intention to bring his daughter with him (see letter 1900, *Tyndall Correspondence*, vol. 7). In letter 1910 of this volume, he confirmed his intentions to come to London with his daughter to attend the International Exhibition of London (see letter 1904, n. 8). In letter 1931, Poggendorff thanked Tyndall for hosting them.

From Rudolf Clausius 15 June 1862 1906

Zürich 15. Juni 62

Lieber Tyndall!

Ich habe mich sehr gefreut in Ihrem letzten lieben Briefe zu lesen, dass Sie nach der Schweiz kommen, und ich beeile mich Ihnen bevor Sie einen anderen Reiseplan machen, zu schreiben, dass Sie dann jedenfalls zu uns nach Zürich kommen müssen. Wir wohnen hier in einer Vorstadt die noch ganz ländlich ist, wo man schöne frische Luft, und die Aussicht auf den See und in die Berge hat. Da wollen wir Ihnen ein Zimmer welches wir sehr gut entbehren können, einräumen, und dann können Sie ganz nach Ihrer Bequemlichkeit leben. Sie sind bei uns als Mitglied der Familie, und wenn Sie einmal allein sein und Ruhe haben wollen, so können Sie sich in Ihr Zummer zurückziehen. Ferner können Sie von hier aus kleinere und grössere Excursionen machen, indem die nächste Umgebung sehr schön ist, und viele besuchungswerthe Puncte hat, und man von Zürich auch alle übrigen Theile der Schweiz leicht erreichen kann. Meine Frau, der ich schon viel von Ihnen erzählt habe, freut sich auch sehr darauf, Sie kennen zu lernen, und lässt Sie ebenfalls frendlich einladen, einige Zeit bei uns zuzubringen, um sich hier von Ihren Arbeiten zu erholen. Ich werde dann während der Zeit meine Vorlesungen im Polytechnicum wie gewöhlich halten, und meine laufenden Geschäfte abmachen, und in der Zeit die mir dann übrig bleibt, können wir zusammen sein, und können vielleicht manche Excursionen machen. Auch Mousson, Escher, und die anderen Mitglieder der naturforschenden Gesellschaft werden sich sehr freuen, wenn Sie herkommen.

Sie müssen von der Eisenbahn aus gleich zu uns kommen. Unsere Wohnung ist: neun Plattenstrasse, das erste Haus in Hottingen. (Hottingen ist nämlich eine Vorstadt von Zürich, von welcher die Plattenstrasse nach einer anderen Vorstadt "Platte" führt.)

Die Nachricht dass ich Poggendorff in London treffen werde, ist mir sehr angenehm gewesen. Ich werde mir dann eine Wohnung möglichst nahe bei der Ihrigen nehmen, damit wir oft zusammen sein können.

Dass Sie schon einen Vortrag über Mayer's Arbeiten gehalten haben hat mich sehr überrascht, und ich habe auch darin Ihre Arbeitskraft bewundert. Ich habe mir vorgenommen einmal an Mayer zu schreiben, und werde ihm das dann auch mittheilen. Er wird sich gewiss sehr darüber freuen und verdient solche Anerkennung auch.

Mit herzlichen Grüssen von meiner Frau und mir, und in der festen Hoffnung dass Sie unsere Bitte erfüllen,

Ihr | Clausius

Zurich 15 June 62

Dear Tyndall!

I was very pleased to read in your last dear letter[1] that you are coming to Switzerland, and I hasten to write to you before you make other travel plans, that you must by all means come to Zurich to us. We live here on the outskirts, which are still very rural, where one has fresh air and a view of the lake and the mountains. We will furnish you a room that we can easily do without, and where you can enjoy comfort and privacy. At our home you will be a member of the family, and if you want to be alone and have your peace, you can withdraw to your room. In addition, you can go on long and short excursions from here, as the area is very beautiful and has many sights worth visiting, and from Zurich you can easily reach all other parts of Switzerland. My wife,[2] whom I have already told much about you, is very much looking forward to meeting you and extends her friendly invitation as well to spend some time here with us to recover from your exertions. During this time, I will be lecturing as usual at the Polytechnic Institute[3] and taking care of my current business, and in my remaining time, we can be together and perhaps go on some excursions. Mousson,[4] Escher[5] and the other members of the Society of Natural Science[6] will also be very pleased, if you come.

You must come directly from the train to us. Our house is: nine Plattenstrasse, the first house in Hottingen. (Hottingen is namely a neighbourhood on the outskirts of Zurich, the Plattenstrasse leads to another settlement, "Platte".)

I am glad to learn that I will meet Poggendorff in London. I will take an apartment as close as possible to yours, so that we can meet often.

I was quite surprised that you have already given a lecture on Mayer's work,[7] and I admire your capacity for work. I intend to write to Mayer sometime and will let him know that. I am certain that he will be very pleased and he deserves such recognition too.

With fond regards from my wife and I, and in the sincere hope that you will fulfill our request,

Your | Clausius

RI MS JT/1/TYP/7/2246–7
LT Typescript Only

---

1. *your last dear letter*: letter 1905.
2. *My wife*: Adelheid Clausius.
3. *Polytechnic Institute*: see letter 1904, n. 7.
4. *Mousson*: Johann Rudolf Albert Mousson (1805–90), a Swiss physicist and malacologist (*HLS*).
5. *Escher*: Arnold Escher von der Linth.

6. *Society of Natural Science*: the Naturforschende Gesellschaft in Zürich was a Swiss scientific society founded in 1746 to promote the study of the natural sciences, of which Tyndall had been recently elected an Honorary Member in 1858. See 'Verzeichniss der Mitglieder der naturforschenden Gesellschaft in Zürich', *Vierteljahrsschrift der Naturforschenden Gesellschaft in Zürich*, 4 (1859), p. 4.
7. *a lecture on Mayer's work*: see letter 1905, n. 3.

To Rudolf Clausius [*c.* 25 June 1862][1] 1907

My dear Clausius.

I have been prevented from replying to your last kind letter[2] by a very bad head. Since last September my brain has been continuously grinding, and now I am feeling the effect. For the last six weeks I have not been able to visit any of my friends—not even able to go up and have a little dinner with Faraday being quite unfit for the conversation. I am now very low indeed, and my friend Huxley[3] is equally so. Five weeks ago we agreed to bear each other company to Switzerland, and we start next week. Were my health not so broken I should undoubtedly accept your kind invitation,[4] but I really shall be unfit for all intercourse with my fellow men until the mountain air restores my energies. We might perhaps be able to arrange matters so that we might come to England together. I shall return about the 20th of August.

My intention is to go with Huxley straight to Grindelwald,[5] and to remain there until both of us get a little strength. Afterward I shall make my way to the Aggishorn[6] & thence to Zermatt

In London you will find it a great convenience to be able to visit the Athenaeum Club;[7] I shall have your name put down upon the list of the foreign celebrities that are to be invited. The club house is a fine building & furnished with everything calculated to promote the Physical & Intellectual well-being of a man.

Will you present my kindest regards to your wife.[8] I do not know yet how my vacation is to end, but if I can manage to take in Zürich on my return I shall be delighted to do so.

I beg of you also to present my kind remembrances to Professors Mousson,[9] Escher,[10] and any other of the Naturforschende Gesellschaft.[11]

Believe me dear Clausius | Yours ever truly | John Tyndall

RI MS JT/1/T/238

---

1. *[c. 25 June 1862]*: dated with reference to Tyndall's departure for Switzerland on 4 July. '25 Juni 1862' is written in pencil at the top of the letter, probably not in Tyndall's hand.

2. *your last kind letter*: letter 1906.
3. *Huxley*: Tyndall's friend Thomas Huxley wrestled with depression throughout his life; the death of his first-born son Noel in 1860 had exacerbated his troubles.
4. *your kind invitation*: in letter 1906, Clausius had invited Tyndall to visit him in Zurich.
5. *Grindelwald*: a village and municipality in the Interlaken-Oberhasli administrative district in the canton of Berne in Switzerland. The landscape surrounding the village is dominated by the Mittellegi ridge of the Eiger.
6. *the Aggishorn*: the Eggischhorn, a mountain of the Bernese Alps in the Swiss canton of Valais.
7. *Athenaeum Club*: a private members' club located at 107 Pall Mall in London, particularly for those with intellectual interests, and accomplishments in art, science, or literature. Tyndall was elected a member on 31 January 1860. The club was founded in 1824 by John Wilson Croker (1780–1857) and others. On the club, see M. Wheeler, *The Athenaeum: More Than Just Another London Club* (New Haven, CT: Yale University Press, 2020).
8. *your wife*: Adelheid Clausius.
9. *Mousson*: see letter 1906, n. 4.
10. *Escher*: Arnold Escher von der Linth.
11. *Naturforschende Gesellschaft*: see letter 1906, n. 6.

To George Gabriel Stokes 27 June 1862 1908

27th June 1862

My dear Sir.

I have read the abstract of Mr. Hopkins's paper[1]—unhappily I am at present unable to read[2] the whole paper. The abstract is able and clearly written, and I do not therefore doubt that the paper itself ought to be published in the Philosophical Transactions.

The older a scientific subject grows the more accuracy is needed in its treatment, and to the future student of glacial phenomena Mr Hopkins's paper will be extremely valuable, giving as it does, a definition to the subject which has not even been aimed at by any other writer.

A further claim to publication is founded on the fact that the Philosophical Transactions already contain an elaborate paper on this subject[3] many portions of which require the revision which Mr. Hopkins's has bestowed upon them.

In expressing my opinion that the paper ought to be printed I do not say that my views coincide in every particular with those of Mr Hopkins's. I look at the paper broadly and in this aspect regard it as an extremely valuable contribution to glacier science.

Yours very sincerely | John Tyndall.

Prof. Stokes | &c &c &c

1862 | Hopkins | by | Tyndall[4]

RS RR/5/118 Tyndall 27 June 1862

1. *Hopkins's paper*: W. Hopkins, 'On the Theory of the Motion of Glaciers', *Phil. Trans.*, 152 (1862), pp. 677–745.
2. The sense of the sentence suggests that Tyndall was in fact unable to read the paper and based his thoughts on the abstract.
3. *the Philosophical Transactions already contain an elaborate paper on this subject*: in reference to J. Tyndall and T. Huxley, 'The Structure and Motion of Glaciers', *Phil. Trans.*, 147 (1857), pp. 327–46.
4. *1862 | Hopkins | by | Tyndall*: written either on an envelope or reverse side of letter.

To John Lubbock [30][1] June 1862 1909

Monday night | June 1862

My dear Lubbock

I was aiming to see you & to arrange with you the character of your baggage and the mode of carrying it. I had arranged with Huxley to go to Grindelwald.[2] He starts tomorrow, I start on Friday & we hoped that you would join us there three days after my arrival:[3]—Then we had arranged to cross the Scheideck together march on to the Grimsel, scramble over the Rhone glacier up the Galenstock and afterwards to cross the Oberaarjoch to the Aeggishorn—whence our way would lead towards the Monte Rosa chain. But this general arrangement blots out all—We thought that *[1 word illeg]* you would go straight to Bale[4] thence to Thun and instead of being at Geneva, that you would be at Grindelwald on the 10th. There are many chances against our meeting at all, or being able to do any thing together. No plan save the vaguest general general[5] one can be made.

In the first half of July we intend to accomplish what I have stated & about the 20th I shall be at Breuil on the Italian side of the Alps. —you love the lakes & I love the mountains & thus I fear our places are not going to harmonize.

Kindest regards to Mrs Lubbock[6]—I wish she had put an extinguisher on Mr Morlot![7]

Yours ever | John Tyndall

BL Add MS 49639, ff. 103–4

1. *[30]*: the day is given as Tyndall left London on Friday, 4 July 1862, with the preceding Monday being 30 June.
2. *Grindelwald*: see letter 1907, n. 5.
3. *three days after my arrival*: date uncertain, but around 8 July.
4. *Bale*: Basle, or Basel, a town in Switzerland, close to the borders with France and Germany (Bale is the French spelling).

5. *general*: word repeated by Tyndall.
6. *Mrs Lubbock*: Ellen Lubbock.
7. *Mr Morlot*: possibly Charles Adolph Morlot (1820–67), a Swiss archeologist (*HLS*).

From Johann Christian Poggendorf 30 June 1862 1910

Berlin 30 Juni 62

Liebster Freund!

Ich habe lange überlegt, ob ich so dreist seyn dürfe, Ihr gütiges Anerbieten anzunehmen, doch da Sie so freundlich und inständig bitten, sind am Ende meine Bedenken gewichen. Ich werde also kommen mit meiner Tochter u. Ihre Wohnung in Beschlag nehmen für die Dauer Ihrer Abwesenheit. Wir werden, denke ich, Ihrer Haushälterin nicht zu sehr zur Last fallen!

Aber wann werden Sie reisen? Fast fürchte ich, daß dieser Brief Sie nicht mehr treffen wird. —Ich habe daher schon den Dr. Becker, der vor etwa 8 Tagen nach London gereißt ist, gebeten, Ihnen unsere Ankunft zu melden. Haben Sie ihn gesprochen? In Ungewißheit darüber, will ich hier wiederholen, daß ich wahrscheinlich am 3', 4' oder 5' August von Berlin abreisen u. den geraden Weg über Ostende nach London nehmen werde.

Es sollte mir übrigens leid thun, wenn ich Sie gar nicht sehen sollte. Ich hoffe indeß, daß Sie Ende August's zurückkehren werden u. sich nicht durch uns abhalten lassen zu dieser Zeit wirklich zurückzukehren. Sind wir dann noch in London, so räumen wir Ihre Wohnung u. ziehen in einen Gasthof!

Ihnen nochmals für Ihre Freundlichkeit dankend, will ich wünschen daß das schlechte Wetter Sie noch bis heute in London zurückgehalten habe, damit diese Zeilen noch in Ihre Händen gelangen. Dann mögen Sie aber zu Ihrer Reise die doch wahrscheinlich wieder nach der Schweiz gerichtet ist, stets den schönsten blauen Himmel haben.

Mit diesem herzlichen Wunsche wie immer | der Ihre | Poggendorff

N. Der 2' Theil Ihrer Abhandlung, den Dr. Becker noch vor seiner Abreise übersetzt hat, ist gedruckt!

Berlin, 30 June 1862

Dearest friend!

I pondered for a long time, whether I might be so brazen as to accept your kind offer,[1] but since you were so kind and persistent, my reservations vanished in the end. Thus, I will come with my daughter and occupy your residence for the duration of your absence. I do not think that we will be a great burden for your housekeeper!

But when do you travel? I fear that this letter will not reach you in time. —I have therefore asked Dr Becker,[2] who travelled to London nearly 8 days ago, to inform you of our arrival. Have you spoken with him? Not knowing this, I repeat here that I will most likely depart from Berlin on the 3rd, 4th or 5th of August and take the direct route via Ostende to London.

Incidentally, I would be sorry if I did not get to see you at all. I have hope that you will return at the end of August and that our presence will not keep you from coming back at that time. If we are still in London at that time, we will vacate your residence and move to a guesthouse!

Thanking you again for your friendliness, I will hope that the bad weather has forced you to stay in London until today, so that these lines will reach you. But may you then have the most beautiful blue skies during your travels, which will probably take you again to Switzerland.

With warm wishes, as always | your | Poggendorff

N. The 2nd part of your article,[3] which Dr Becker was able to translate before his departure, has been printed!

RI MS JT/1/P/132

---

1. *your kind offer*: Tyndall had offered to host Poggendorf and his daughter Marie (b. 1838) when they came to London. See letter 1900, *Tyndall Correspondence*, vol. 7, in which Poggendorf first acknowledged Tyndall's offer. Tyndall's letter in which he made the offer is missing. See also letters 1921 and 1931.
2. *Dr Becker*: Ernst Becker.
3. *your article*: J. Tyndall, 'On Dynamic Radiation and Absorption of Heat by Gaseous Matter. Second Memoir', *Phil. Trans.*, 152 (1862), pp. 59–98. Poggendorf discusses the article in more detail in a letter of 24 May 1862 (see letter 1900, *Tyndall Correspondence*, vol. 7). As editor of *Poggend. Annal.*, Poggendorf was preparing to translate and publish the article.

From Heinrich Gustav Magnus 1 July 1862 1911

Berlin 1 July 62.

Mein theurer Tyndall

Wie ich hoffe hat Dr. Becker Ihnen gesagt dass ich ihn gebeten hatte ein Diplom für Sie mitzunehmen. Die Gesellschaft Naturforschender Freunde hierselbst, eine Gesellschaft zu der Klaproth, A. v. Humboldt, kurz die ausgezeichnetsten Naturforscher Berlin's, nicht Chemiker und Physiker allein, sondern Botaniker, Zoologen u.d.m gehört haben, und auch jetzt gehören hat Sie und Th. Graham zu Ehren Mitgliedern ernannt. Das Diplom ist mir aber so eben erst zur Beförderung an Sie zugesendet worden. Ich werde nun eine Gelegenheit suchen um es Ihnen zuzufertigen, wenn nicht früher so doch durch Poggendorff.

Sie werden sich wundern daß ich Ihre Einwände gegen meine Untersuchungs Methode noch nicht beantwortet habe. Ich bin indeß in diesem Sommer sowenig zum Arbeiten gekommen daß ich einige andere Versuche die ich vorhatte, nicht habe vollenden können.

Es thut mir Leid daß, wenn es mir noch vergönnt sein sollte zur Ausstellung nach London zu kommen, ich Sie nicht finden werde. Es ist so lange her daß wir uns nicht gesprochen haben und ich fürchte es ist wenig Hoffnung daß Sie uns in Berlin besuchen, um so mehr bedaure ich daß auch diese Gelegenheit vorüber geht ohne daß ich hoffen darf Sie zu sehn. Faraday wird doch wohl wahrscheinlich in England bleiben. Ich würde ihn aufsuchen wo er auch wäre, nur um ihm die Hand zu drücken. Bitte grüßen Sie ihn und alle die sich meiner erinnern herzlichst von

Ihrem | G Magnus.

Berlin, 1 July ,62

My dear Tyndall

I hope Dr. Becker[1] has told you that I asked him to bring a certificate for you. The Society of Friends of Natural Science[2] of this city, a society to which Klaproth,[3] A. v. Humboldt, in short the most outstanding natural researchers in Berlin, not only chemists and physicists, but also botanists, zoologists and many more have belonged and still belong, has chosen you and Th. Graham[4] to become honorary members. The certificate has only now been sent to me for dispatch to you. I will look for an opportunity to send it to you, if not earlier, then with Poggendorff.

You may be wondering why I still have not replied to your objections to my research method. I have had so little opportunity to work this summer that I was unable to finish several other experiments that I had planned.

I am sorry that even if I am fortunate enough to be able to come to London for the exhibition,[5] I will not find you there. It has been a long time since we last spoke and I fear that there is little hope that you will visit us in Berlin, all the more I regret that this occasion will pass as well without any hope for me to see you. Faraday will presumably stay in England. I would look for him wherever he may be just to shake his hand. Please greet him and all who remember me fondly from

Your | G Magnus.

RI MS JT/1/M/28

1. *Dr. Becker*: Ernst Becker.
2. *Society of Friends of Natural Science*: the Gesellschaft Naturforschender Freunde zu Berlin, or Berlin Society of Friends of Natural Science, is a German scientific society founded in 1773.
3. *Klaproth*: Martin Heinrich Klaproth (1743–1817), a German chemist known for his discovery of Uranium, Cerium, Titanium, and Tellurium (*CDSB*).
4. *Th. Graham*: Thomas Graham.
5. *the exhibition*: see letter 1904, n. 8.

To George Hector Tyndale 2 July 1862 1912

London

My dear Hector,

For many a day I have been thinking of writing to you, but my head has been in so bad a state that I could hardly muster power to do so.[1] I go to Switzerland on Friday[2] in the firm hope of being able there to restore my strength, but I must write to you before I go. What a fearful business this is in which you are engaged[3]—History I suppose has no parallel to it. I heard with alarm that Harper's Ferry[4] was involved in recent actions, knowing that you were there. I never doubted that wherever you might be you would behave as a man & a soldier, but I wish that instead of being where you are you were here with me. Reflection would perhaps cause me to cancel this wish; for you no doubt are where your duty calls you, that is to say where each of us ought to be. Is there any hope of its being settled? any prospect of this awful shedding of blood being about to cease? Your first brush with the confederates was mentioned in the English newspapers—old Anderson,[5] our assistant at the R.I., asked me soon afterwards whether I had any relatives in America, as he had seen your name. It is idle now to ask whether it would not have been better at first to let the South go her way—the fire has been lighted and must burn itself out.

Well my dear Hector however my general sympathies may tend in this quarrel be assured of it that there is at least one man in England who would glory in your success, and sympathise with you in reverse. My whole interest in the grounds of the quarrel is I confess merged in its actual consummation. Excuse this rambling note—I am unfit to write a better one. Hirst & Debus often speak of you, and with the strongest desire to learn something of your fate:—I know Hirst has been purposing to write to you for many weeks; but like myself his head has been confused & his work heavy. For a good while after your leaving[6] England Col. Yorke[7] enquired after you. Do write to me like a good fellow—Your last letter[8] was a great delight to me. Were I a praying man I would say may God guard you. Give me credit for a wish equivalent to the prayer.

Affection[atel]y yours | John Tyndall

RI MS JT/1/T/1445
RI MS JT/1/TYP/1635–6

---

1. *my head has been in so bad a state*: Tyndall suffered from headaches, dyspepsia, and insomnia from the late 1850s until his death (U. DeYoung, *A Vision of Modern Science: John Tyndall and the Role of the Scientist in Victorian Culture* (New York: Palgrave Macmillan, 2011), p. 16). They are a recurrent theme in his letters throughout this and other volumes.
2. *I go to Switzerland on Friday*: Tyndall took a tour of the Swiss Alps from 4 July until mid-August.
3. *What a fearful business this is in which you are engaged*: a reference to Tyndale's military service as a Union General in the American Civil War.
4. *Harper's Ferry*: Harpers Ferry is a small town in West Virginia. Prior to and during the American Civil War, the town was the site of a famous raid on an arsenal by abolitionist John Brown (1800–59) on 16 October 1859. During the war, the town changed hands several times because of strategic location, particularly for railroad access.
5. *Anderson, our assistant at the R.I.*: Charles Anderson (1790–1866) served in the Royal Artillery, achieving the rank of sergeant. In 1827 he was first employed at the RI to assist Faraday in the project (on behalf of the Board of Longitude) of improving optical glass. With the termination of that project in 1830 Anderson became Faraday's laboratory assistant and also assisted him in his RI lectures. He remained in this position until his death in 1866.
6. *after your leaving*: Tyndale visited London in 1854, staying at 142 Strand, the boarding house and publishing business of John Chapman, proprietor of the *Westminster Review* (*Ascent of John Tyndall*, p. 90).
7. *Col. Yorke*: Philip James Yorke (1799–1874), a British army officer and keen amateur chemist and mineralogist. He was one of the founding members of the Chemical Society of London (*ODNB*).
8. *last letter*: see letter 1860 (23 January 1862), *Tyndall Correspondence*, vol. 7.

From Henry Barnett[1] 19 July [1862][2] 1913

Vevay[3] | July 19th | Saturday night.

My dear Tyndall,

Here I be with my party, George,[4] Col. Greathed[5] & his nephew,[6] & I received joyfully today y[ou]r letter[7] (without date or time) while eating a hasty dinner at Geneva to catch the 2 o'clock steamer up the lake—We left London with a crowd, chiefly French, on Thursday, made a prosperous but not speedy journey to Paris, which we left again last night at 8.40, & arrived at this charming place about six this evening, & the lake never looked more lovely than during our steam over it today—In fact we left cold unsettled weather in England, for warm temperature & bright sun on this side the Channel, & I trust by this time you have experienced a similar change[8]—I am sorry to hear you have not recovered y[ou]r ancient vigour yet. It shows the elasticity of bodily powers does not increase as months & years roll by, & the machine must be treated more carefully in consequence—I don't care, if you do not again attempt the Matterhorn,[9] but I do hope we may have a nice jolly walk of moderate ascent together, as we are likely to be somewhere near each other ere long. I am disappointed at not finding Couttet[10] here, but only a letter from his daughter to say her father has been from home with some medicines, whose course has been prolonged beyond his expectation, but he is sure to turn up in a day or two, if I will only amuse myself about here in the meantime, as the said Couttet w[oul]d be greatly grieved to lose the "precieux avantage"[11] of guiding me! I suppose we shall do so, but I hope by Sunday the 27th we may be at or about Gressonay,[12] whence we may make several excursions according to weather & time. I propose to go hence to Aosta by the Gt. St Bernard,[13] thence to Cogne[14] & if possible direct to St. Martin[15] & the Val de Lys.[16] It w[oul]d be very pleasant, if you could meet us some where in those parts with John Lubbock—My plan is to return over the Moro[17] to Visp[18] & home either by Coire[19] & Zurich or else Grimsel[20] & Meyringen[21] &c. It is all new to two of my fellow travellers. George has just gone to bed with a sore throat, & a wet handkerchief & oil silk over it, which I hope will cure him, & I must go to my downy to make up for the deficiencies of last night on the rail. My kind regards to your companion

Ever y[ou]r sincere friend | Harry Barnett

RI MS JT/1/B/79

---

1. *Henry Barnett*: probably Henry Barnett (1815–96), a banker, landowner, and conservative politician (*Modern English Biography*).

2. *[1862]*: the year is given by reference to Tyndall's second attempt on the Matterhorn (see n. 9); 19 July was a Saturday in 1862.
3. *Vevay*: a small town in Switzerland on the north shore of Lake Geneva, near Lausanne.
4. *George*: possibly George Barnett, his son.
5. *Col. Greathed*: probably William Wilberforce Harris Greathed (1826–78), who at this point, as a Lieutenant-Colonel, was assistant military secretary at the Horse Guards (*ODNB*).
6. *his nephew*: not identified.
7. *Y[ou]^r letter*: letter missing.
8. *I trust by this time . . . a similar change*: Tyndall had left for Switzerland on 4 July.
9. *if you do not again attempt the Matterhorn*: Tyndall first attempted the Matterhorn, one of the high peaks of the Alps located on the border between Switzerland and Italy, in August 1860 (*Hours of Exercise in the Alps*, pp. 27–52), and again, unsuccessfully, in August 1862, (*Hours of Exercise in the Alps*, pp. 153–65).
10. *Couttet*: Francois Couttet was the proprietor of Couttet's Hotel.
11. *precieux avantage*: precious advantage (French).
12. *Gressonay*: a town in the Aosta Valley in northern Italy.
13. *St Bernard*: Great St. Bernard Pass, an Alpine pass on the border between Italy and Switzerland.
14. *Cogne*: a town in the Aosta Valley in northern Italy.
15. *S^t. Martin*: a town in the Aosta Valley in northern Italy.
16. *Val de Lys*: magnificent views of glaciers could be seen at the head of the Val de Lys. Many of these locations were later highlighted in the several guidebooks and handbooks for travellers to the Alps that were published in England in the following decades.
17. *the Moro*: the Monte Moro pass.
18. *Visp*: a town in the Rhône valley; the capital of the district of Visp in the Swiss canton of Valais
19. *Coire*: a town in the Swiss Alps, later noted in guide books as a location for tolerable accommodation and good wine.
20. *Grimsel*: a mountain pass in southern Switzerland that crosses the Bernese Alps.
21. *Meyringen*: a town at the foot of the Grimsel pass.

From James P. Joule 21 July 1862 1914

Thorncliff Old Trafford | Manch[este]^r 21 July 1862

Dear Sir,

I thank you for your kindness in sending me the second part of your admirable researches on radiation & absorption of heat by gases & vapours.[1] It is almost impossible to exaggerate their interest as setting forth original facts of most extensive application to the phenomena of nature. I thank you also for your R[oyal]. Inst[itutio]^n lecture[2] on the subject of which you will see I have made one or two remarks in the Phil[osophical] Mag[azi]ne.[3] I think in a case like that of the Equiv[alen]t of Heat the experimental worker rather than the mere logical reasoner (however valuable the latter) must be held as the establisher of a theory.[4] I have determined the mech[anica]l equivalent in nearly a dozen ways, and the figure I arrived at in 1849 has not yet been altered or corrected.[5] Mr Dyer[6] of this town was I believe the first man who noticed the rise of temp[eratur]e on shaking water, but he applied it to nothing—in fact explained away what he saw.

Believe me Dear Tyndall | Yours always truly | J. P. Joule

RI MS JT/1/J/138
RI MS JT/1/TYP/2/662

---

1. *second part of your admirable researches*: J. Tyndall, 'On the Absorption and Radiation of Heat by Gaseous Matter. Second Memoir', *Phil. Trans.*, 152 (1862), pp. 59–98.
2. *your R[oyal]. Inst[itutio]n lecture*: see letter 1902, n. 4.
3. *Phil[osophical] Mag[azi]ne.*: James P. Joule, 'Note on the History of the Dynamical Theory of Heat', *Phil. Mag.*, 24 (1862), pp. 121–3.
4. *the experimental worker rather than the mere logical reasoner . . . theory*: Joule here reasserts his priority over that of Mayer, who Tyndall had advanced as in fact the first to arrive at the mechanical equivalent of heat. This was to become a long-running dispute between Tyndall and Joule that occupied Tyndall throughout the period covered by this volume. See the Introduction for further commentary.
5. *figure I arrived at in 1849 has not yet been altered or corrected*: a reference to Joule's paper, 'On the Mechanical Equivalent of Heat', in *Phil. Trans.* delivered in 1849 and published in 1850. In this paper he specified the numerical value for the amount of mechanical work required to produce a unit of heat. J. P. Joule, 'On the Mechanical Equivalent of Heat', *Phil. Trans.*, 140 (1850), pp. 61–82.
6. *Mr Dyer*: probably Joseph Chessborough Dyer (1780–1871), an American-born inventor who moved to England in 1811, introducing several American inventions to England. He was a member of the Institution of Civil Engineers and the Literary and Philosophical Society of Manchester, and was involved in the foundation of the Royal Manchester Institution in 1823. In 1842, following the death of his wife, Chessborough sold his machine works in Manchester to Parr, Curtis, & Madeley. Later, he published pamphlets for the abolitionist cause (*ODNB*).

To Michael Faraday 22 July 1862 1915

22nd July 1862

My dear Mr. Faraday

I quitted London with the desire above all things to give my head a rest, and to strengthen my general health which has been sadly broken in upon. My friend Huxley, tired as myself, started three days before me. I joined him at Grindelwald,[1] remained there a week then walked along the ordinary route to the Rhone glacier,[2] at the foot of which we stayed two or three days in a capital hotel recently erected there. Here we were joined by Mr. John Lubbock, according to an arrangement made previous to leaving England. Huxley, though suffering from neuralgia,[3] and wishing to be quiet, accompanied us in our little excursions, and the only thing we did without him was to ascend the Galenstock,[4] a mountain somewhat over eleven thousand feet high. We reached the summit from the hotel in five hours, so that the labour was by no means excessive. It was Lubbock's first visit to Switzerland; he had never previously trodden on a snow mountain, and both he and myself were delighted with the glimpses which we obtained of the splendid névé[5] of the Rhone. But the air thickened, and intermittent gushes of dense fog swept over us. We did not hope to see any thing from the summit, nevertheless we attained it and were rewarded:—not by the clearness of the prospect, but by the changes of the atmosphere, which were quite marvelous—sometimes shrouding all sometimes melting as if by magic and revealing the mountains. The atmosphere is a wondrous factory; the grand origin of all its power being overhead, lifting the snows and driving the clouds by his individual might.

From the Rhone glacier we passed on to that of the Aletsch;[6] and here the presence of the Jungfrau[7] suggested the thought of ascending it. Lubbock had showed himself so good a mountaineer on the Galenstock that our guide considered him quite fit to attack the Jungfrau. We made our preparations, proposing to spend the night previous in a grotto on the flank of an adjacent mountain. To this grot[to] we sent some food, firewood and covering. Our porters started at 12½ P.M. and we, accompanied by our guide Bennen, started at about 2 P.M. The weather was very promising, and we looked forward with confidence to a successful expedition. The stretch of the Aletsch Glacier between the AEggischhorn,[8] which was our starting point, and the place where we purposed to spend the night, presents no difficulty; at least any mountaineer would say so; certainly an English lady of average walking powers, and properly guided would be able to accomplish it. Yet an accident upon this portion of the glacier, which threatened at one time to be of the most shocking character, defeated all our hopes and arrangements regarding the Jungfrau. Bennen, Lubbock, and myself marched securely along the ice, calculating on finding our two porters at the grotto. When about a mile distant from it, we saw a man standing on the lateral moraine, whom we at first imagined to belong to a party which had started the day before for the Jungfrau. On approaching him however we saw that he was one of our own porters. He did not move to meet us, but stood stolidly upon the moraine until we came up to him. Bennen spoke to him and he answered in the Swiss patois.[9] A bewildered expression crossed Bennen's countenance: "My God!" he exclaimed turning to me "Walters[10] is dead!" Walters was a guide belonging to the party which had preceeded us. "No," said the porter, "it is not Walters." "Who then?" cried Bennen "Bielander"[11] he replied, "my comrade who started with me from the hotel is killed." Bielander was our second porter—a strong and daring fellow who had been brought up from Laax[12] by Bennen that morning. I was utterly stunned, and for five minutes seemed to undergo the tortures of a lifetime. Bennen broke out into a frenzy of lamentation. Bielander had quitted his mother to accompany Bennen, and the thought of returning to the old woman without her son totally overpowered him. Ten minutes I suppose passed, before we could gather up the precise conditions of the case.

"How does he know that the man is dead?" demanded Lubbock. "Where is he?" I asked. "There," said he, pointing to the glacier, "in a crevasse; but he is certainly dead." We were speedily beside the fissure, and on looking into it, saw it at one place packed with ice and snow while adjacent to this the darkness of the cleft cut the vision short. We stood still for a moment; a low moan issued from the depths and was repeated. Here Bennen's frenzy revived. "He lives, he lives, quick! quick! Let me down to him." He was about to dart into the fissure at the imminent risk of ruin to himself. Our rope was in the chasm with the lost man. I tried to quench somewhat sternly Bennen's excitement and to make clear to him that all now depended upon presence of mind.

All the coats of the party were instantly taken off and knotted together. Bennen I saw in the intensity of his emotion was neglecting to make the knots sure, and I flatly refused to trust his connexions until I had tested them. One half of them gave way, partly through bad material, and partly through bad tying. All being secured we let him down, and I was let down after him, mainly by my friend Lubbock, for the porter was too stupefied to be trusted. The moaning continued, and we at first rooted about here and there amid the frozen débris, to make sure of the place whence the sounds issued. A layer of about four feet of ice and snow was cleared away, and here a hand was seen. It looked like the hand of a corpse, perfectly bloodless, but the fingers moved at intervals. I shall never forget the look of it. Bennen's excitement again rose, and I was again forced to act the tyrant. The jammed fragments of ice and snow seemed terribly intractable, and Bennen's strength was hampered by the narrowness of the space in which he had to wield his axe. The knapsack of the man was next cleared, and soon afterwards his head emerged from the debris. My brandy flask was immediately at his lips, and the stimulant seemed to revive him, his mutterings became intelligible. His head and shoulders being freed, we passed the rope under his arms and tried to draw him out bodily; but he was too firmly wedged. We repeated this attempt four or five times, but were compelled to clear away the debris to the very feet before the man could be extricated. It was a great relief to find him perfectly free, for we feared his dying though inanition.[13] With Lubbock, the porter and myself straining at the rope, and Bennen's strength exerted below, we raised him from platform to platform of the chasm, and finally placed him on the glacier. We lost no time in getting him to the side. Bennen's idea was to strike for home instantly; he thinking that he could carry the man a good part of the way upon his back. I dissented from this and had him carried to the grotto. Bennen had miscalculated his strength. We had been in the chasm nearly an hour; during which time the dripping of the melting ice had drenched us. Bennen's efforts moreover had been prodigious and on quitting the crevasse he was seized by a painful trembling fit.

After carrying the man for ten minutes his force failed, and I then got him on my back: we took him thus by turns: Bennen's strength returned, and we finally had the satisfaction of seeing our charge on a bed of grass in the grotto, not knowing however how the matter was likely to end. His wet clothes were changed; his wrists were rubbed, a hot water bottle was placed against his feet, and he had a copious draught of warm wine. For a long time he shook violently, and continued to moan: by degrees this subsided, and once Bennen said to me with some alarm, "We cannot hear him." I placed my ear to his face and found him breathing quietly. I felt his pulse, it was striking quietly but regularly. We piled all our clothes upon him, and tended him through the night. Next morning he was able to walk home with us, and he is at this moment in good health and spirits and without a broken bone. He fell through a space of about 40 feet, and we found him quite upright packed closely in the ice.

Depend upon it there is less to be feared from the presence of danger than from the absence of caution. There was certainly no more real danger of falling into a crevasse on the Aletsch Glacier, than there is of being run over by a cab in crossing from Albemarle Street to James's Street.[14] Recklessness however makes both positions dangerous, and recklessness amid the Alps and glaciers is far more to be deprecated than the confronting of their dangers where true caution is exercised. Give my kindest regards to Mrs Faraday and Miss Barnard.[15] I hope in a day or two to write to Dr Bence Jones. Excuse this rambling narrative and believe me

Ever yours most sincerely, | John Tyndall

Would you kindly allow Dr Bence Jones, Mr Barlow[16] and Mr Pollock[17] to see this letter. It is badly written but may interest them.

Faraday Correspondence, 6:4202
RI MS JT/1/TYP/12/4114

1. *Grindelwald*: see letter 1907, n. 5.
2. *Rhône glacier*: the largest glacier in the Urner Alps and the source of the Rhône River and one of the primary contributors to Lake Geneva.

3. *neuralgia*: pain, typically stabbing or burning, in the area served by a nerve (*OED*).
4. *Galenstock*: the fourth highest mountain in the Urner Alps in Switzerland. Its summit ridge lies on the border between the cantons of Uri and the Valais.
5. *névé*: compacted or hardened snow, as found on the upper part of a glacier (*OED*).
6. *Aletsch*: the largest glacier in the Swiss Alps and the source of the Massa River.
7. *Jungfrau*: one of the main summits of the Bernese Alps, located between the northern canton of Bern and the southern canton of Valais, between Interlaken and Fiesch.
8. *the AEggischhorn*: see letter 1907, n. 6.
9. *patois*: a dialect spoken by the people of a particular region (esp. of France or French-speaking Switzerland), and differing substantially from the standard written language of the country (*OED*).
10. *Walters*: Anton Walters, one of the guides who ascended the Matterhorn with Tyndall and Bennen. He climbed with Tyndall a number of times, notably on his ascent of the Weisshorn.
11. *Bielander*: one of the porters traveling with Tyndall's party.
12. *Laax*: a village in the Valais, near Fiesch; where Bennen lived.
13. *inanition*: the action or process of emptying; the condition of being empty; spec. the exhausted condition resulting from want or insufficiency of nourishment (*OED*).
14. *from Albemarle Street to James's Street*: two major streets in London, Albemarle being the location of the RI, i.e. across Piccadilly and its busy traffic, which separates them.
15. *Miss Barnard*: Jane Barnard.
16. *Mr Barlow*: John Barlow.
17. *Mr Pollock*: William Frederick Pollock.

To Thomas Archer Hirst 24 July 1862 1916

24th. July 1862.

My dear Tom

Huxley will soon see you and explain to you all that I have been doing of late: it is not much. This day I reached Zermatt and tomorrow I cross to Breuil.[1] I know not what the future may afterwards bring forth. Huxley will tell you of the accident on the Aletsch glacier.[2] I never was so shocked in my life, but the lost man was rescued, and is now in perfect health. The accident is to be ascribed mainly to carelessness.

My health is growing very strong. In fact I am nearly as strong at present as ever I have been in my life. Huxley also improved greatly. I wish sincerely that when you commence your holiday your recovery may be as quick and as complete.

Yours affectionately | John

RI MS JT/1/HTYP/520
LT Typescript Only

1. *Breuil*: a village on the Italian side of the Matterhorn which serves as a base for climbing the mountain from that side, and is reached from Zermatt via the Théodule pass.
2. *Aletsch glacier*: the largest glacier in the Swiss Alps; Tyndall's climbing party suffered an accident when, Bielander, one of their number, fell into a crevasse (see letter 1915).

To Juliet Pollock 24 July [1862][1] 1917

24th. July

My dear Mrs. Pollock.

I send you just one line to say that after dawdling about for the last fortnight I find my strength almost wholly restored. I reached Zermatt today & tomorrow I intend to cross the frozen boundary of Italy & Switzerland—What I shall do afterwards I know not. The Matterhorn[2] is at hand; but I shrink from augmenting the excitement that already exists regarding it.

I wrote a letter to Mr Faraday & asked him to send it to three of my friends in London.[3]—one of whom is your husband—It will inform you of an accident[4] which threatened to be very serious—Huxley takes the letter home with him & he will reach London on the 27th.[5]

BL Add MS 63092, f. 101
RI MS JT/1/TYP/6/2093

1. *[1862]*: the year is given by relation to letter 1915.
2. *Matterhorn*: see letter 1913, n. 9.
3. *three of my friends*: Henry Bence Jones, John Barlow and William Frederick Pollock (see letter 1916).
4. *an accident*: see letter 1915.
5. *he will reach London on the 27th*: fragment only; letter incomplete.

To Thomas Henry Huxley 29 July 1862 1918

29th. July 1862

My dear Huxley,

I left you at Visp[1] trying to kill time, not dreaming that the enemy[2] would be so soon upon myself. Well here he is: and I write not through love of you but through hate of him. I joyfully reached Breuil[3] and found on my arrival a note from Mr Whymper[4] informing me that he had started on another expedition up the Matterhorn.[5] Next evening he arrived, having satisfied himself that without ladders the rocks could not be scaled. On Sunday morning I started with my two guides and two porters from this place. That night I lay under my little tent upon the rocks, and heard the roaring of the stones as they thundered down an adjacent gully from the heights of the Matterhorn. We started at 4. The heavens all clear and the air fresh. Crept up chimneys, squeezed ourselves through slits, wound round ledges and cut our way across the scanty snow patches which are competent to retain their hold upon the rock. All heavy work. We came to the base of a succession of precipices which were topped by the serrated ridge of the mountain. Here with Bennen in front and Walters[6] behind him we worked up a bit by bit; holding on almost by our eyelids. It was the best and hardest bit of rock climbing that I have ever known man to accomplish. It was these precipices that stopped Whymper. Well we scaled them and got upon the arête. Up it, and in a very short time both summits of the mountain were in sight. We scorned the lower, just paused upon it to look down at Breuil and advanced towards the final cap. We were as sure of it as you are of your dinner. "There shall our flag be planted"—Bennen laughed uproariously: "What will the people in Zermatt say when they see our flag floating yonder?" Now along a notched and narrow and ugly arête almost horizontal to the base of the rocks which constitute the true summit. Elation gradually subsided. The summit was within a stone's throw of us; but how to get there! To the doubt of our being able to do so succeeded the dark certainty that we must give it up. Had the crags been planted on a plain we should no doubt have attempted them, though the chance of slipping approached almost to a certainty, but we looked around and saw the fate which awaited us should the footing of any one give way. A fall of at least 5000 feet, tumbling from precipice to precipice would be the consequence. The thing looked horrid. And though by a little management I caused the party to move up almost to contact with the peril I could not conscientiously push them further. Bennen and myself were certainly the two least daunted men of the party: but even Bennen at length exclaimed "Here we must be content." We stuck a pole on the lower summit[7] and unless splintered by lightening it will be visible for many a year. The defeat is a great one, for the efforts which we made for victory were great. It was impossible to descend the cliffs which we had ascended, so we attached a rope to a piece of our ladder which we fixed in the rocks, and by means of it let ourselves down the precipices. Any climber attacking the mountain this year will find one grand difficulty set aside, for he can ascend the rocks by the rope which we were compelled to leave behind. This defeat has fallen upon me like the chill of age, and I must "mix myself with action"[8] to shake it off. On quitting the mountain it poured down hail like grapeshot upon us. Indeed soon after quitting the summit we were enveloped in fog, and the fog ended by this cannonade. The mountain is one of the most savage grandeur: no-where have I seen the effects of time and weather so grandly exhibited. He seems a suffering old Titan,[9] unconquered still, but still wearing away. His rocks all red with the oxide of iron, and all spotted with lichens which give them an appearance of indescribable hoariness. Certainly on the whole—both near and far—the grimmest and grandest object in the Alps. Well goodbye to him and goodbye to you and goodbye to my climbing. for there is nothing else in the Alps that I should care to do. I cannot frank this.

IC HP 8.43
Typed Transcript Only

---

1. *Visp*: see letter 1913, n. 18.
2. *the enemy*: a reference to Edward Whymper, a rival climber, and ultimately the first to summit Matterhorn in July 1865 (see n. 4).
3. *Breuil*: see letter 1916, n. 1.
4. *Mr. Whymper*: Edward Whymper (1840–1911), a British mountaineer and wood-engraver. In 1860 the publisher William Longman commissioned Whymper to make illustrations of the Alps, for which he visited Saas Fee, Zermatt, Chamonix, and the Dauphiné. Whymper centered his climbing ambitions on the Matterhorn, and from 1861 to 1865 he made seven attempts from the Italian side, sometimes in partnership with his rivals. Tyndall viewed him as his main competitor in his alpine pursuits (*ODNB*).
5. *Matterhorn*: see letter 1913, n. 9.
6. *Walters*: Anton Walters (see letter 1915, n. 10).
7. *stuck a pole on the lower summit*: this was the highest point yet reached and now called the Pic Tyndall. However, it is more than a 'stone's throw' from the summit as it is some 800ft lower in elevation.
8. *mix myself with action*: here Tyndall is quoting from Alfred Tennyson's poem 'Locksley Hall' (1830): 'I myself must mix with action, lest I wither with despair' (published in A. Tennyson, *Poems* (London: Edward Moxon, 1842).
9. *suffering old Titan*: possibly referring to the Titans, the pre-Olympian Gods of Greek mythology.

To Juliet Pollock [29 July 1862][1] 1919

Gressoney Val du Lys[2] | 1st. Aug. 1862.

My dear Mrs. Pollock.

I have been here for the last two days[3] and I have many times thought of you during this period. I know you would like to be here in this lovely Italian valley. —one of the fairest that I have ever seen. There are green meadows, lichened rocks, a rushing river, a cascade a clear blue sky & noble peaks aspiring to it; while at a distance the slopes of the snow mountains appear & temper the warmth by their mere presence. It is a lovely spot; and one which

I know you would enjoy. There is moreover one particular peak called the Grayhead[4]—a little over 10000 feet high, to the summit of which I know you could climb, I being your guide: and thence the whole panorama of the Alps unfolds itself. I have thought that you might come here on some festive occasion, bringing Maurice[5] with you & live in pension here for three weeks. I tried the Matterhorn[6] on the 28th, and after some desperate climbing reached a great height. —The summit was in view & I felt as sure of winning it as if I already stood upon it. Cheerily we urged our way towards it, but when close to it came to a stand. The peril was so deadly that we sat down, gazed, debated, and finally gave it up.[7] We planted flags far higher than human foot had previously attained and there they now flutter, emblems of failure instead of success.[8] My companions were all caved & I dare not urge them—all except Bennen—but we could not attempt the thing alone. As far as I am concerned therefore the Matterhorn is inaccessible and may raise its head defiant as it has hitherto done. —the only unconquered & unconquerable peak in the whole Alps. I crossed the mountains to the place two days ago, & tomorrow I walk to Alagna.[9] Afterwards I purpose steeping myself in the softer beauty of the Italian lakes. I look towards them with a kind of longing—my spirit is prepared for their repose by having dwelt so long amid the sterner glories of the Alps. Thence I return by Zurich & Bale.[10] I have to spend a day or two in Paris & by the 20th of August at the latest I shall be in London.

My health is good at present—In fact I do not think I was ever in a better condition for mere physical labour. My lungs are perfect: Bennen has stronger limbs than mine but I have stronger lungs than he has. He pants sooner than I do. As a pair we are well matched. I wish we had you here, one at each side raising you up the Grauhaupt.[11] Goodbye. Kindest regards to your husband[12] & love to all the boys.[13]

Yours as usual | John Tyndall

BL Add MS 63092, ff. 208–9

---

1. *[29 July 1862]*: the postmark on the envelope indicates that this letter was sent on 29 July 1862, which suggests Tyndall was a few days off in his indication that he wrote this on 1 August.
2. *Gressoney Val du Lys*: in the Aosta valley of northwestern Italy.
3. *I have been here for the last two days*: in *Hours of Exercise in the Alps*, Tyndall recalled that 'Towards the end of 1862 Bennen and myself made "The Tour of the Monte Rosa", halting for a day or two at the excellent hostelry of Delapierre, in the magnificent Val du Lys. We scrambled up the Grauhaupt, a point exceedingly favourable to the study of the formation of the Alps' (p. 151).
4. *Grayhead*: Grauhaupt (also known as Testa Grigia), a rock prominence in the Swiss Alps on the border between Switzerland and Italy. .

5. *Maurice*: Maurice Pollock.
6. *Matterhorn*: see letter 1913, n. 9.
7. *finally gave it up*: see letter 1918.
8. *We planted flags far higher. . . . instead of success*: see letter 1918, n. 7.
9. *Alagna*: a village (also known as Alagna Valsesia) in Piedmont, situated on the Italian side of Monte Rosa; a long day's walk from Gressoney.
10. *Bale*: see letter 1909, n. 4.
11. *Grauhaupt*: see n. 4.
12. *your husband*: William Frederick Pollock.
13. *the boys*: Frederick, Walter, and Maurice Pollock.

To James Joule [between 12 and 19 August 1862][1] 1920

Royal Institution, London

My Dear Joule,

On my return from Switzerland two days ago,[2] I became acquainted with the note which you have published in the last Number of the Philosophical Magazine.[3] Would you allow me to make the following remarks in connexion with the subject of it?

During the spring of the present year I gave at the Royal Institution a course of lectures "On Heat, regarded as a kind of Motion."[4] During the early portion of the course, I had engaged a short-hand writer to report the lectures, with a view to their subsequent publication; and from this gentleman's notes of my second lecture I make the following extract, which refers to the mechanical theory of heat:—"It is to Mr. Joule, of Manchester, that we are almost wholly indebted for the experimental treatment of this subject. With his mind firmly fixed upon a principle, and undismayed by the coolness with which his first labours appear to have been received, he persisted for years in his attempts to prove the invariability of the relation between heat and ordinary mechanical force. He placed water in a suitable vessel, agitated it by paddles moved by measurable forces, and determined the elevation of temperature; he did the same with mercury and sperm-oil. He also caused disks of cast iron to rotate against each other, and measured the heat produced by their friction. He urged water through capillary tubes, and measured the heat thus generated. The results of his experiments leave no doubt upon the mind that under all circumstances the absolute amount of heat produced by the expenditure of a definite amount of mechanical force is fixed and invariable." Such has been my language regarding you; and to it I still adhere. I trust you find nothing in it which indicates a desire on my part to question your claim to the honour of being the experimental demonstrator of the equivalence of heat and work.

It was not my object in the lecture to which you refer to give a history of the mechanical theory of heat, but simply to place a man of genius, to whom the fates had been singularly unkind, in a position in some measure worthy of him. I was quite aware of all that you have stated regarding Locke,[5] Rumford,[6] Davy, and others: you might have added Bacon[7] to your list. Probably no great generalization was ever established without having first simmered in the minds of many thinkers. But the writings of Mayer form an epoch in the history of this subject; and I certainly should not feel disposed to retract a single sentence that I have written in his favour. I believe he deserves more praise than I have given him. It was he who first used the term "equivalent" in the precise sense in which you have applied it; he calculated the mechanical equivalent of heat from data which, as I have said, "a man of rare ingenuity alone could turn to account;" and his calculation is in striking accordance with your own experimental determinations*. You worked independently of Mayer, and in a totally different way. You brought the mechanical theory to the test of experiment, and in this way proved its truth.

Mayer calculated correctly the mechanical equivalent of heat; but you say that, at the time he wrote, there were no known facts to warrant the hypothesis which he adopted. If by this you mean to say that he made a haphazard guess which had no basis of physical probability, I cannot agree with you. The known constitution of an elastic fluid is, in my opinion, quite sufficient to justify Mayer's proceeding. His hypothesis was this:—Let the quantity of heat required to raise the temperature of gas, preserved *at a constant volume*, $t°$, *be* x, and let the heat required to raise the same gas, under *constant pressure*, $t°$, *be* $x + y$. The weight raised by the expanding gas in the latter case being P, and the height to which it is raised h, then, according to Mayer,

$$y = \mathrm{P} \times h\,;$$

that is to say, the excess of heat imparted in the latter case is precisely equivalent to the mechanical work performed.

It is undoubtedly implied in this equation that the quantity of heat y is expanded wholly in *external* work, and that none of it has been consumed in overcoming internal molecular attractions. This, I think, on the face of it is an extremely probable hypothesis—so probable, indeed, as to amount, in my estimation, almost to a certainty. Clausius makes the same assumption with no better authority than Mayer; and I believe (for I here trust my memory merely) that the assumption has been completely verified by the experiments of the very philosophers who once questioned it. "The law," says Mayer, "'Heat=Mechanical effect,' is independent of the nature of an elastic fluid, which only serves as the apparatus by means of which the one force is converted into the other."

The law of Mariotte[8] was an old principle when Mayer wrote; and the fact of its holding good for gases generally renders the conclusion exceedingly probable that, in yielding to compression, the attractions of the gaseous molecules were insensible; otherwise it is hardly conceivable that the same results could have been obtained with gases so differently constituted: the attractions of the hydrogen atoms, for example, would in all probability be different from those of oxygen. Mayer was further justified in his hypothesis, as to the absence of interior work in the case of a true gas, by the experiments of Ørsted[9] and Despretz,[10] which showed that the law of Mariotte was departed from by the liquefiable gases—the amount of departure depending on the proximity of the gas to its point of condensation. Where, therefore, no departure from the law had been observed (in the case of air for instance), Mayer, I submit was perfectly warranted in assuming that the molecular attractions were insensible, and that the quantity of heat (y) before referred to was entirely expended in raising the weight, and had its true mechanical equivalent in the weight so raised.

With reference to the application of the mechanical theory of heat to cosmical phenomena, if it were not a liberty, I would ask whether you have ever read the essay of Mayer entitled *Beiträge zur Dynamik des Himmels?*[11] If so, then I have good reason to suspect my competence to come to a correct conclusion as to what constitutes a scientific right.

Knowing that the original memoirs of Mayer would be the true court of appeal in connexion with this subject, I some months ago urged the responsible editor of the Philosophical Magazine to publish translations of them. This I hope he will do; for I quite agree with you in thinking that they would interest many of the readers of the Magazine. Let me add, in conclusion, that I do not think the public estimate of your labours can be in the least affected by any recognition which may be accorded to Mayer. There is room for both of you on this grand platform. Certainly, had Mayer never written a syllable on the mechanical theory of heat, I should not deem your work a whit nobler than I now hold it to be.

Believe me, | Yours, &c., | John Tyndall | Royal Institution, | August 1862

J. Tyndall, 'Mayer, and the Mechanical Theory of Heat', *Phil. Mag.*, 24 (1862), pp. 173–5

---

1. *[between 12 and 19 August 1862]*: the date for this open letter to the *Phil. Mag.* is given for the reasons detailed in n. 2.
2. *On my return from Switzerland two days ago*: according to Hirst's journal entry of 12 July, Tyndall had left London for Switzerland on 4 July (Brock and MacLeod, *Hirst Journals*, p. 1613); in letter 1928, dated 17 September, Tyndall indicated that he had been away for five or six weeks, placing his return approximately between 8 and 15 August. Tyndall thus read Joule's note between 10 and 17 August, most likely towards the latter date, and replied,

as this letter indicates, two days later, which dates the letter as written between 12 and 19 August. Tyndall wrote to Hirst on 20 August that he had the proof of a letter to Joule in his hand (see letter 1921).

3. *in the last Number of the Philosophical Magazine*: J. P. Joule, 'Note on the History of the Dynamical Theory of Heat', *Phil. Mag.*, 24 (1862), pp. 121–3.
4. *a course of lectures "On Heat, regarded as a kind of Motion"*: Tyndall delivered 'Twelve Lectures on Heat' at the RI in 1862, on Thursdays at 3pm from 23 January to 10 April (RI Guard Book, vol. 3, p. 11). He would publish these lectures as *Heat as a Mode of Motion* in 1863.
5. *Locke*: John Locke (1632–1704), an English philosopher and physician, widely regarded as the one of the most influential thinkers of the Enlightenment for his writings on liberalism, theories of the mind, and introspection (*ODNB*).
6. *Rumford*: Benjamin Thompson (1753–1814), an American-born British physicist and inventor known for experiments on heat and weaponry. One of his most noted contributions centered on the nature of heat. Using a cannon, he concluded that heat generated was not caloric but due to friction. He was made Count Rumford in 1791 (*ODNB*).
7. *Bacon*: Francis Bacon (1561–1626), a British statesman and philosopher, author of *Novum Organum* (1640). He argued that scientific knowledge could only be produced by inductive reasoning and direct observation (*ODNB*).
8. *Mariotte*: Edme Mariotte (1620–84), a French physicist best known for his recognition in 1679 of Boyle's law about the inverse relationship of volume and pressures in gases (*CDSB*).
9. *Ersted*: Hans Christian Ørsted (1777–1851), a Danish physicist and chemist credited with the discovery that electric currents create magnetic fields (*CDSB*).
10. *Despretz*: César-Mansuète Despretz (1791–1863), a French chemist and physicist who taught at the Faculté des Sciences in Paris. His research covered a wide range of topics, including the conduction of heat and the functioning of the voltaic pile. He also published textbooks in both physics and chemistry (*CDSB*).
11. *Beiträge zur Dynamik des Himmels*: J. R. Mayer, *Beiträge zur Dynamik des Himmels in populärer Darstellung* (Heilbronn: Vorlag von Johann Ulrich Landherr, 1848).

To Thomas Archer Hirst 20 August 1862 1921

1862 | 20th. Aug.

My dear Tom,

I was very glad to hear from you. I am happy to be able to report myself in excellent health. Up the Matterhorn[1] I had some desperate climbing, and for an hour was as sure of the summit as I am of finishing this letter, but the last 200 feet are deadly and from these we recoiled.[2] We of course moaned and groaned & I believe swore; but it was all to no purpose—the Matterhorn sent us home again.

I almost had a fearful accident on the Aletsch glacier.[3] We had to dig a man out of a crevasse in which he had been jammed for nearly two hours[4]

I have been writing a letter to Joule about Mayer—the proof of which is now in my hand.[5] I have also sent a short paper to the Phil. Mag. on the conformation of the Alps.[6]

Where do you intend to go. Take my advice Tom & do not choose a hot climate: I shrank from Italy this year—I dared not even visit the Lakes.

Goodbye | Yours affectionately | John

Poggendorff & his daughter[7] seem very happy. The house is very nice.

RI MS JT/1/T/920
RI MS JT/1/HTYP/520

1. *Matterhorn*: see letter 1913, n. 9.
2. *from these we recoiled*: see letter 1918.
3. *Aletsch glacier*: see letter 1915, n. 6.
4. *I almost had a fearful accident . . . for nearly two hours*: Tyndall recounted this tale in detail in his earlier letter to Faraday (1915). The man who had fallen into the crevasse was Bielander, one of the porters for their trip.
5. *I have been writing a letter to Joule . . . now in my hand*: letter 1920.
6. *I have also sent a short paper to the Phil. Mag. on the conformation of the Alps*: J. Tyndall, 'On the Conformation of the Alps', *Phil. Mag.*, 24 (1862), pp. 169–73.
7. *Poggendorff & his daughter*: Johann Christian Poggendorff and his daughter Marie (b. 1838) visited London to attend the International Exhibition of London. Tyndall had offered them the use of 14 Waverley Place while he was away climbing (see letter 1900, *Tyndall Correspondence*, vol. 7; and letters 1910 and 1931 of this volume).

To Richard Dawes [23 August 1862][1] 1922

Saturday

My dear Mr Dean,

I picked up your note[2] last night at the Athenaeum,[3] had it come to the Institution,[4] I should have replied to it earlier.

On Monday evening I am fixed. But on Tuesday I am free. On Thursday I am also fixed. So I propose starting from Paddington at 9½ on Tuesday morning and hope to reach Ledbury[5] by 2.30.

From Ledbury I see I must start on Thursday at 9.16. All the other trains reach London too late.

Don't ask anybody. Let me run wild myself over the hills. I shall be a bore to your friends in my present state of dullness.

Yours ever, | John Tyndall.

I hope Mrs Dawes[6] will not give me up as a heretic after she has read my chapter on Miracles.[7]

RI MS JT/1/TYP/1/328
LT Typescript Only

---

1. *[23 August 1862]*: the date is given since Tyndall returned to London from Switzerland on 16 August, a Saturday, making this Saturday letter fall on the next. Also by relation to letter 1902 in which Tyndall discussed his intentions to visit Dawes and his wife in mid-August after his return.
2. *your note*: letter missing; but see letter 1902 in which Tydnall discussed his intentions to visit Dawes and his wife in mid-August.
3. *Athenaeum*: see letter 1907, n. 7.
4. *Institution*: the RI.
5. *Ledbury*: a town in Herefordshire in the west-midlands; with a railway station on the line from Paddington.
6. *Mrs Dawes*: Mary Helen Dawes.
7. *my chapter on Miracles*: chapter five of Tyndall's *Mountaineering in 1861*, titled 'Reflections' (pp. 30–40).

To Juliet Pollock [23 August 1862][1] 1923

Saturday

My dear Mrs. Pollock

Hurrah for Walter![2] It does not surprise me, holding as I do the general steadfastness & transmissibility of the molecular forces of the human brain; I am sure I feel the blood of 300 years ago in my own veins.[3] Walter Pollock follows from his antecedents as naturally as any other effect follows its cause.

Yours ever | John Tyndall

I am off to Hampshire for a day to rest my weary head

RI MS JT/1/T/1213
RI MS JT/1/TYP/6/2025

---

1. *[23 August 1862]*: the date is given based on the content of the letter. Pollock's son Walter had been elected a King's Scholar at Eton, 'No. 11 on a chosen list of twenty-one, out of eighty-three candidates', on Monday 28 July 1862 (*Remembrances*, vol. 2, p. 107). This letter appears to be one Tyndall wrote following his return from Switzerland on Saturday, 16 August. Unless he started replying to letters on the same day he returned, this letter would likely be the Saturday following, 23 August.
2. *Hurrah for Walter!*: see n. 1.

3. *the blood of 300 years ago in my own veins*: presumably a reference to William Tyndale. Tyndall had a deep interest in the possible genealogy of his extended family.

To John Murray 2 September 1862 1924

Royal Institution of Great Britain | 2nd Sept. 1862

Dear Sir,

On the other leaf of this sheet I have written a receipt for £13–14–0 which you have been good enough to send me.

If the book has not fulfilled my hopes it has at all events given me some experience & this is always valuable.

Yours very sincerely | John Tyndall

Received from John Murray Esq[ui]re thirteen pounds 14/- on account of the "Glaciers of the Alps".[1]

£13–14–0 | John Tyndall | 2nd. Sept. 1862

[Verso:] Sept 2/62 | Receipt | Prof. Tyndall | £13–14–

National Library of Scotland, John Murray Archive, Ms. 41214

1. *"Glaciers of the Alps"*: *Glaciers of the Alps*.

From C. K. Akin 8 September 1862 1925

5 St. George's Terrace, Ryde I. W.[1] | September 8th. /62

Sir,

Will you kindly excuse the liberty I am taking in writing to you these lines?[2]

I have lately finished a few pages of M[anu]m. Scr[iptu]m.[3] with a view of submitting it for publication in the Phil[osophic]al. Mag[azi]ne.; but am afraid its contents will have to undergo some linguistic correction, at the hands of a person better competent than myself, for to be fit to appear before the public. I, in consequence, venture to address to you Sir, as an editor of the Magazine, beforehand the question whether it would be too bold to expect to be obliged in the matter in question on the part of its managers or contributors: Professor Stokes who had offered me his kind services being now, I believe, abroad in Ireland, whilst other's good offices are not now within my reach, and I am anxious to see my paper published, if possible, with the next (October) number of your valuable journal.

I have the honour to be | Sir | Your most obed[ie]^nt^. serv[an]^t^. | C. K. Akin | from Hungary

Professor Tyndall, Fell[o]^w^. Roy[a]l. Soc[iet]^y^. | &c &c &c

CUL SC MS Add.7656/A615

---

1. *Ryde I. W.*: Ryde, Isle of Wight.
2. *these lines?*: this is the first correspondence between Akin and Tyndall (other letters between them are 2075, 2076, 2077, 2165, 2167, and 2170). They quickly fell out over their respective researches, beginning another long-lasting and acrimonious dispute which reverberates throughout several other letters as well. See the Introduction for further commentary. The Akin dispute will continue in the forthcoming ninth volume of *The Correspondence of John Tyndall.*
3. *M[anu]^m^. Scr[iptu]^m^.*: written by hand (Latin).

To Heinrich Debus 11 September [1862][1] 1926

Royal Institution, 11th. Sept.

My dear Heins

We[2] have just got home—come and dine with us tomorrow (Friday) at 7.

We can then arrange about the South Kensington payments.[3]

I was rejoiced to find that you were here in advance of me

Yours ever | John Tyndall

RI MS JT/1/TYP/7/2401
LT Typescript Only

---

1. *[1862]*: the year is given by reference to 'tomorrow (Friday)'. 12 September fell on a Friday in 1862 but no other year in the 1860s. Also, Tyndall and Hirst shared accommodation at 14 Waverley Place from June 1861 until 25 January 1868.
2. *We*: Tyndall and Thomas Hirst lived together (see n. 1).
3. *South Kensington payments*: in 1859, Tyndall became an Examiner for Science for the Department of Science and Art (sometimes Science and Art Department) at the South Kensington Museum. He had arranged for Debus, and from 1862 Hirst, to grade examinations for him. This British government department, a subdivision of the Board of Trade, was responsible for promoting education, art, science, technology, and design in Britain and Ireland (from 1853–99). See B. Robertson, 'The South Kensington Museum in context: an alternative history', *Museum and Society*, 2 (2004), pp. 1–14.

From Jacques-Louis Soret[1] 11 September 1862 1927

Genève 11 Sept. 1862

Cher Monsieur

Je viens de corriger les épreuves d'un article pour les Archives des Sc. Phys. et Nat relatif à la discussion entre M. Magnus et vous sur l'absorption de la chaleur rayonnante par la vapeur d'eau. Toute cette question est éminemment intéressante et il n'y a pas de doute dans mon esprit que vous n'ayez tout à fait raison. —L'élaboration de cet article m'a amené à voir qu'il y avait encore une partie de vos recherches dont les Archives n'ont pas rendu compte. J'ai vu dans les Annales de Poggendorff un mémoire de vous qui fait suite, je pense à votre premier travail publié dans le Philosophical Magazine et les Transactions *[de raison]* et je suis indiscret en vous demandant si un second mémoire a paru en anglais et dans ce cas s'il vous serait possible de me le faire parvenir; et même si ce n'était pas abuser de votre bonté, je vous demanderai l'envoi de votre premier mémoire que je ne possède pas (celui que j'ai eu entre les mains appartient à Mr de la Rive[2]) et qui me serait fort utile pour la compréhension du second.

J'ai à vous remercier, Monsieur, de plusieurs envois que vous avez bien voulu me faire; vous avez vu que j'ai cherché à analyser dans les Archives votre beau travail sur les glaciers qui m'a vivement intéressé. Beaucoup de nos lecteurs m'ont remercié de les leur avoir fait connaître. —En lisant vos travaux sur ce sujet j'ai cherché sans rien trouver si vous aviez mentionné un détail sur la structure de la glace des glaciers que j'ai eu l'occasion d'observer pendant l'été de 1861. Permettez-moi de vous dire à quelle occasion.

J'ai eu l'honneur de vous envoyer un mémoire de Mr Thury sur les glacières naturelles qui contient une petite notice que j'avais écrite sur la glacière de Schafloch (Voyez <u>Archives</u> 1861 No de *[1 word illeg]*). Je suis retourné visiter cette glacière le 3 Sept. 1861. Je ne vous entretiendrai pas de quelques observations que j'ai faites sur les dimensions de la grotte, sur la température etc. La théorie de la formation de la glace est je crois suffisamment expliqué dans le mémoire de M. Thury. Mais il reste encore différentes choses à étudier sur la constitution de la glace et la structure. —Un premier fait remarquable c'est que souvent les pilastres de glace présentent des cavités de forme variée; *[ainsi] [l'année] [dernière]* il y avait un énorme pilastre pyramidal sous lequel se trouvait une cavité assez grande pour que j'ai pu y pénétrer; on remarquait à l'intérieur de cette petite caverne des stalactites de glace. L'intérieur de ces cavités n'est pas très facile à expliquer. —Un autre point qui m'a frappé c'est la ressemblance très grande de la glace de cette glacière avec cette des glaciers dans les parties inférieures. Ainsi les blocs de glace de la partie inférieure du glacier supérieur de Grindelwald ou du glacier de Rosenlaui ont une ressemblance frappante de structure avec les pilastres des Schafloch (au moins dans l'été de 1861) On remarquait une sorte de stratification tout à fait analogue et la même structure aréolaire. —De plus j'avais remarqué que dans la glace de Schafloch, la surface extérieure des aréoles est striée, c'est-à-dire couverte de petites rides, que je ne puis mieux comparer pour l'apparence et la dimension, qu'avec les petites rides que présente la peau des doigts à leur extrémité du côté opposé aux ongles. Dans chaque aréole, c.-à-d. dans chaque morceau compacte de glace ces stries présentent une sorte de parallélisme mais elles ne sont point parallèles dans deux aréoles contigües. —Or, et c'est à ce détail de structure que je faisais allusion au commençant, j'ai observé *[distinctement]* la même chose dans la glace du glacier de Grindelwald et aussi quoique moins nettement dans le glacier de Rosenlaui. Cette structure striée était parfaitement nette dans la caverne creusée artificiellement dans le glacier supérieur de Grindelwald où l'on conduit toujours les touristes depuis quelques années. A Rosenlaui elle était aussi visible dans des grottes naturelles. *[En un mot]* il m'a semblé que l'on observait ce phénomène dans les points où la glace est en fusion lente sans être exposée aux injures du temps (when not weathered, comme vous dites). —Je ne sais si je m'exprime assez clairement pour vous faire comprendre ce que j'entends par cette structure striée; pour mieux y parvenir, jetez un coup d'œil sur la figure 19 de votre mémoire <u>On the Veined Structure of Glaciers</u> (p. 300); vous avez la représentation d'un morceau de la structure de glace composée de *[cinq]* aréoles. Ce sont chacune de ces aréoles dont la surface *[est]* couverte de petites rides—J'ajouterai que Mr Thury m'a dit avoir observé le même fait, la même année à la glacière de St. Georges.

Cette grande analogie entre la glace des glaciers et celle d'une glacière m'a paru curieuse et c'est ce qui m'a engagé à vous parler puisque j'avais l'occasion de vous écrire. Comment ce fait-il que la glace formée dans des circonstances si différentes, et ayant subi des actions diverses, présente une possible ressemblance? Je n'ai pas eu le temps malheureusement de faire une étude plus complète de ce fait, que vous expliquez sans doute beaucoup plus facilement que moi.

Excusez, cher Monsieur, cette longue lettre, et croyez à mes sentiments les plus dévoués.

L. Soret

P.S. Je vais bientôt partir pour Heidelberg où je compte passer l'hiver; si je pouvais vous y être de quelque utilité disposez de moi je vous prie.

Monsieur le Professeur Tyndall.

Geneva, September 11 1862

Dear Sir,

I have just corrected the drafts of an article for the Archives des sciences physiques et naturelles[3] relating to the discussion between Mr Magnus and yourself on the absorption of radiant heat by water vapour.[4] This entire issue is outstandingly interesting and there is no doubt in my mind that you are completely right. —In preparing this article, I discovered that there is still a part of your research unreported by the Archives.[5] In the Annals of Poggendorff[6] I saw one of your articles that I think follows on your first work published in the Philosophical Magazine and the Transactions,[7] and I must ask you, if I may be so bold, if a second article was published in English, and if so, whether it would be possible to send it to me.[8] If I am not abusing your kindness, I would ask you to send me the first article, which I do not have (the one that I consulted belonged to Mr de la Rive) and which would be extremely helpful for me in order to understand the second one.

I must thank you, Sir, for the several communications that you were kind enough to send me.[9] You saw that I intended to analyze your wonderful work on glaciers,[10] which greatly interested me, in the Archives.[11] Many of our readers have thanked me for making it known. —While reading your work on this subject, I could not find if you mentioned in detail the structure of the ice of glaciers that I had the opportunity to observe during the summer of 1861. Allow me to tell you on what occasion.

I had the honour of sending you Mr Thury's[12] article about natural ice caves, which included a small note that I wrote on the ice cave of Schafloch[13] (See Archives 1861, *[*issue*] [1 word illeg]*). I went back to visit this ice cave[14] on September 3rd 1861. I will not bore you with the observations that I made on the dimensions of the cave, its temperature, etc. The theory of the ice formation is, I think, sufficiently explained in Mr Thury's article. But various aspects of the constitution and structure of the ice still remain to be studied. —The first remarkable fact is that often the pillars of ice develop cavities of a various forms. *[*Thus last year*]* there was an enormous pyramidal pillar under which there was a cavity large enough that I could enter it. One could see that inside that small cavern there were ice stalactites. The interior of these cavities is difficult to explain. —Another point that stuck me was the very great similarity of the ice of this ice cave with that of the glaciers in the lower parts. Thus, the blocks of ice of the lower part of the upper Grindelwald glacier,[15] or of the Rosenlaui glacier,[16] have a striking structural similarity to the pillars of Schafloch (at least in the summer of 1861). You could see an absolutely matching kind of stratification and the same areolar structure. —Also, I noticed that in the ice of Schafloch, the exterior surface of the areolas is striated. In other words, it is covered in small creases, which are best compared, based on their appearance and size, to the small wrinkles on the skin of the fingertips opposite the nails. In each areola, i.e. in each compressed piece of ice, these striations display a kind of parallelism, but they are not parallel in two adjacent areolas. —Yet, and it is this structural detail that I was referring to in the beginning, I have observed distinctly the same thing in the ice of the Grindelwald glacier, as well as less clearly in the Rosenlaui glacier. This striated structure was perfectly clear in the artificially dug cavern in the upper Grindelwald glacier, which has been open to tourists for the past several years. At Rosenlaui it was also visible in natural caves. *[In short]*, it seemed to me that we saw this phenomenon in places where the ice is slowly melting without being exposed to the weather (when not weathered, as you put it). —I do not know if I am expressing myself clearly enough for you to understand what I mean by this striated structure. To better understand it, take a glance at the figure 19 of your article On the Veined Structure of Glaciers[17] (p. 300). You have an image of a piece of the ice's structure, composed of *[five]* areolas. It is on each of these areolas that the surface is covered in small creases. —I would add that Mr Thury told me that he observed the same fact in the same year at the St. Georges glacier.

This striking similarity between the glacier ice and that of an ice cave seemed curious to me, and this is what urged me to bring it to your attention, since I had the chance to write to you. How is it that ice formed under such different circumstances, and having undergone different things, could have a *[possible]* resemblance? Unfortunately, I did not have the time to do a more complete study of this phenomenon, which can no doubt explain more easily than I could myself.

Forgive me, dear Sir, for this long letter, and accept my best and most devoted wishes.

L. Soret

P.S. I will soon leave for Heidelberg where I plan to stay the winter. If I could be of any use to you there, please do not hesitate to let me know.

Professor Tyndall.

RI MS JT/1/S/86

---

1. *Jacques-Louis Soret*: Jacques-Louis Soret (1827–90), a Swiss chemist who, in 1878, discovered the chemical element holmium. Soret also recognised that the chemical composition of ozone is three oxygen atoms bound together (*HLS*).
2. *Mr de la Rive*: Arthur-Auguste de la Rive (1801–73), a Swiss physicist and one of the founders of the electrochemical theory of batteries. He served as professor of natural philosophy at the Academy of Geneva and worked principally on electricity, although he also investigated the specific heats of gases and calculated the temperature of the Earth's crust. His principal book was *Traité d'électricité théoretique et appliquée*, 3 vols (Paris: J.-B. Baillière et H. Baillière, 1854–8).
3. *Archives des sciences physiques et naturelles*: a printed reproduction of scientific lectures edited in Geneva by Auguste de la Rive. Soret ensured that significant aspects of Tyndall's work were summarised in this publication throughout the 1860s.
4. *the discussion . . . heat by water vapour*: this appeared as 'Sur la Propagation de la Chaleur Rayonnante Dans L'air Humide par M. Magnus et M. Tyndall', *Archives des sciences physiques et naturelles*, 15 (1862), pp. 20–40.

5. *part of your research unreported by the Archives*: *Archives* had previously published summaries of Tyndall's lectures at the RI and a lengthy summary of his book *Glaciers of the Alps.* See J. Tyndall, 'De L'Action Des Gaz et Des Vapeuresur Le Calorique Rayonnant (Extract d'un discours prononcé à l'Insitution royale de la Grande-Bretagne, le 18 janvier 1861)', *Archives des sciences physiques et naturelles*, 10 (1861), pp. 373–7; L. Soret, 'Les Glaciers par M. le prof. John Tyndall', *Archives des sciences physiques et naturelles*, 13 (1862), pp. 199–240; and J. Tyndall, 'Sur L'Absorption et le Rayonnement de la Chaleur Par les Matieres Gazeuses (Extact d'une communication faite à l'Insitution royale de la Grand-Bretagne, le 17 janvier 1862)', *Archives des sciences physiques et naturelles*, 13 (1862), pp. 260–4.
6. *Annals of Poggendorff*: *Poggend. Annal.*, edited by Johann Christian Poggendorff.
7. *your first work published in the Philosophical Magazine and the Transactions*: Soret was here referring to: J. Tyndall, 'On the Absorption and Radiation of Heat by Gases and Vapours, and on the Physical Connexion of Radiation, Absorption, and Conduction', *Phil. Trans.*, 151 (1861), pp. 1–36 (which was also published in the September issue of the *Phil. Mag*)..
8. *if a second article was published in English, and if so, whether it would be possible to send it to me*: Tyndall clearly did send Soret his second memoir (see letter 1914, n. 1), because Soret gave a detailed summary of the paper in a further article, 'Sur L'Absorption et le Rayonnement de la Chaleur par les substances Gazeuses par M. J. Tyndall', *Archives des sciences physiques et naturelles*, 16 (1863), pp. 5–46.
9. *the several mails that you were kind enough to send me*: letters missing.
10. *your wonderful work on glaciers*: *Glaciers of the Alps.*
11. *in the Archives*: L. Soret, 'Les Glaciers par M. le prof. John Tyndall', *Archives des sciences physiques et naturelles*, 13 (1862), pp. 199–240.
12. *Mr Thury*: Louis-Étienne François Héricart-Ferrand vicomte de Thury (1776–1854), a French politician and scientist who specialized in mining. Thury, who authored more than 350 articles and was a member of a dozen learned societies, was the owner of a significant tract of land of great arboreal and horticultural richness and co-founded of the National Horticultural Society of France (*Dictionnaire de biographie française*).
13. *Mr. Thury's article . . . the ice cave of Schafloch*: M. Thury, 'Études sur Les Glacières Naturelles', *Archives des sciences physiques et naturelles*, 10 (1861), pp. 97–153. This included Soret's note, 'Note sur le Schafloch, ou galciere naturelle du Rothhorn par M. L. Soret', on p. 130.
14. *ice cave*: the ice cave of Schafloch was near Lake Thun in the Swiss Alps.
15. *Grindelwald glacier*: two glaciers, the Lower Grindelwald (Unterer Grindelwaldgletscher) and the Upper Grindelwald (Oberer Grindelwaldgletscher) south of the town of Grindelwald, on the north side of the Swiss Bernese Alps.
16. *Rosenlaui glacier*: a glacier in the Swiss Bernese Alps.
17. *your article On the Veined Structure of Glaciers*: J. Tyndall, 'On the Veined Structure of Glaciers; with observations upon White Ice-seams, Air-bubbles and Dirt-bands, and remarks upon Glacier Theories', *Phil. Trans.*, 149 (1859), pp. 279–307.

To [Julius Haast][1] 17 September 1862 1928

17th Sept[embe]r. 1862

My dear Sir.

Your letter[2] reached me a few minutes ago and I hasten to tell you that I read it with great gratification. I had already, through the kindness of Mr Holmes[3] made myself acquainted with your recent expeditions, which are described in an extremely interesting style.[4] Certainly you must have made good use of your time as a student of English, and but few Englishmen could put their adventures in a more captivating form than you do yours.

I shall be most happy to take charge of any papers or maps that you may send and to deliver them to the Geographical Society.[5] Sir Roderick Murchison, a former president, who takes great interest in the proceedings of the Society is my personal friend, & I am sure he will have due acknowledgement made of your contributions.

I have had an excursion of 5 or 6 weeks in the Alps this year my chief object of scientific interest being the extension of ancient glaciers. From an inspection of the Alps it must occur to any man with a power of mechanical reasoning that the present valleys are in almost all cases, valleys of erosion. But by what agent? The traces of the ancient glaciers, clearly stamped as if they had occurred last year, may be traced to the bottom of the present valleys and their presence proves how comparatively feeble ordinary atmospheric denudation is; since during the stretch of ages which have intervened since the glacier epoch the said denudation has been unable to obliterate their superficial marks. But what a mighty plough a glacier must have been. With a thousand feet of depth every square foot of the glacier's bed is pressed upon by a force of 25 tons! Imagine a tool thus pressed & at the same time urged downward by the vis à tergo and you have an agent competent to scoop out the valleys of the Alps.

Wishing you every success in your researchers

Believe me dear Sir | Yours most truly | John Tyndall

Your letter bears upon its envelope the words "TOO LATE."

Alexander Turnbull Library, National Library of New Zealand, MS-Papers Turnbull 0037–143

---

1. *[Julius Haast]*: probably Julius Haast, a German-born New Zealand explorer and geologist (H. F. Haast, *The Life and Times of Sir Julius von Haast: Explorer, Geologist, Museum Builder* (Wellington: H. F. von Haast, 1948), based on the content of the letter and New Zealand archive address.

2. *Your letter*: letter missing.
3. *Mr Holmes*: not identified.
4. *your recent expeditions . . . interesting style*: between 1860–61, Haast carried out several geological surveys in New Zealand, namely near Auckland and Nelson (J. Haast, *Topographical and Geological Exploration of the Western Districts of the Nelson Province, New Zealand* (Nelson, New Zealand: C. and J. Elliott, 1861).
5. *Geographical Society*: the Geographical Society of London.

From Charles Lyell 26 September 1862 1929

Sept. 26th./ 62

My dear Sir

I have never seen Mortillet's[1] paper[2]—or you sh[ou]ld have had it—He sent me his map alone which is not at hand. I had a small part of it made into a wood cut[3] in which operation it seems to have been mislaid but may turn up in a day or two—If the map of the glaciers w[ou]ld be of any use I will try & find it for you if you will write & say so by post.

Ramsay I think told me he had some paper by Mortillet explanatory of the map.

[al[wa]y[s] y[ou]r[s] | Cha Lyell

RI MS JT/1/L/48
RI MS JT/1/TYP/3/847

1. *Mortillet's*: Louis Laurent Gabriel de Mortillet (1821–1898), a French archeologist and anthropologist best known for his clarification and ordering of the archeology of the Paleolithic (*Dictionnaire de biographie française*, 21 vols (Paris: Letouzey et Ané, 1932–2016).
2. *paper*: G. de Mortillet, 'Carte Des Anciens Glaciers du Versant Italien Des Alpes. Memoria letta nell' adunanza del 23 decembre 1860', *Atti della Società Italiena di Scienze Naturali*, 3 (1861), p. 44–81.
3. *made into a wood cut*: Lyell included this as figure 43 in his *Antiquity of Man* (London: John Murray, 1863), with the heading, 'Map of the Moraines of Extinct Glaciers Extending from the Alps at the Plains of the Po near Turin' (p. 306).

From John Couch Adams 29 September 1862 1930

Cambridge Observatory | Sept. 29th 1862

Dear Dr. Tyndall,

I am sorry that you are not coming to us on Wednesday. I had accepted an invitation for you to dinner on that day. We would have been very moderate and would not have given you any reason to fear for the state of your nerves on Thursday.[1]

I think as you have your preparations to make on Thursday morning, it will be most convenient for you to drive at once to my rooms at Pembroke College[2] and leave your things there. Then when your preparations are done, you can return there, and get ready for dinner, as we are going to dine with a friend a fellow of that College. The time of going to and returning from the Observatory will thus be saved, which is an object considering how little you allow yourself between your preliminary arrangements and the giving of your lecture.[3]

If you will tell me by what train you will come on Thursday, I will be at my rooms to meet you.

Believe me, | Yours very truly | J. C. Adams.

RI MS JT/1/TYP/1/1
LT Typescript Only

1. *your nerves on Thursday*: Tyndall was to give a lecture (see letter 1932, n. 6).
2. *Pembroke College*: at the University of Cambridge.
3. *your lecture*: see letter 1932, n. 6.

From Johann Christian Poggendorff 29 September 1862 1931

Berlin 29 Sept. 62

Mein lieber Tyndall!

Wohlbehalten und munter von der Reise zurückgekehrt, sitze ich nun seit einer Woche etwa in meiner gewohnten Klause, und gedenke, so weit es die Arbeiten zulassen, der schönen Tage, die ich in London verlebt. Daß darunter die Erinnerung an Waverley Place und seine freundliche Umgebung die erste Stelle einnimmt, brauche ich Ihnen wohl nicht zu sagen. Der Aufenthalt bei Ihnen und die unverdient gastfreie Aufnahme welche Sie uns so lange geschenkt, haben wesentlich dazu beigetragen uns den Besuch von London zu einem höchst angenehmen zu machen. In dankbarer Erinnerung der Freundlichkeiten, die Sie uns erwiesen, sprechen wir, meine Tochter und ich, noch oft von Ihnen, und erzählen meiner Frau wie sehr Sie sich unserer angenommen haben. Uns Allen beschleicht dabei freilich das demüthigende Gefühl, daß wir Ihnen Aehnliches in Berlin nicht bieten können, aber wir wünschen doch sehr, daß Sie uns einmal in nicht gar langer Zeit das Vergnügen Ihres Besuches machen und somit Gelegenheit gäben, unsere Erkenntlichkeit wenigstens einigermaaßen thatsächlich zu bekunden. Berlin ist freilich kein London, aber es lebt sich schon ganz gut darin, und wenn es auch weiter keinen Vorzug hat als den, daß die Entfernungen kleinere sind, so spart man doch dadurch an Zeit und "Time is money". Mir kommt es jetzt kolossal vor, wenn ich bedenke, daß Sie jährlich mehre Wochen auf Omnibusfahrten in der Stadt verwenden,—das wäre hier nur möglich wenn man in Charlottenburg wohnte u täglich hereinkommen müsste. —Wie viele mutton chops gehören dazu, um solchen Kraftaufwand zu ersetzen? —Doch Scherz bei Seite! Kommen Sie nur bei Ihrem nächsten Ausfluge in die Alpen durch Berlin: Sie werden finden, daß Sie hier, trotz einiger Differenz in wissenschaftlichen Ansichten, noch viele wahre Freunde haben, unter welchen ich nicht der schlechteste zu seyn glaube

Meine Frau läßt sich Ihnen bestens empfehlen u. sagt Ihnen schönsten Dank, daß Sie mich und ihre Tochter so gut gepflegt haben

Nun leben Sie wohl, überarbeiten Sie sich nicht und seyen Sie versichert, dass ich Ihnen stets in Dankbarkeit gedenken werde

Ihr | Poggendorff

Auch einen schönsten Gruß an Ihren Hausgenossen, Dr. Hirst. —Sobald ich von Magnus, der übrigens auch wohl zurück gekehrt ist, seine letzte Notiz bekommen und gedruckt habe, werde ich sie Ihnen senden.

Berlin 29 Sept. 62

My dear Tyndall!

Having returned safe and sound from the journey, I have now been sitting for a week or so in my usual hermitage and reminisce, as far as work permits, of the pleasant days that I spent in London.[1] I probably do not need to tell you that the memories of Waverley Place and its friendly surroundings occupy the first place. The stay at your home and the undeservedly hospitable reception that you offered us for so long contributed significantly to make our visit to London a most pleasant one. In grateful memory of the friendliness that you showed us, my daughter and I speak about you often and tell my wife how well you took care of us. We all feel humbly that we cannot offer you anything similar in Berlin, but we still hope that we will have the pleasure of your visit not too long from now, and that you will give us the opportunity to at least truly show our gratitude. Berlin is certainly not London, but one lives quite well here and if it has no other advantage over London than that the distances are shorter, one nevertheless saves time and "time is money". I find it incredible when I think that you spend several weeks per year travelling around the city on omnibuses,—this would only be possible here if one lived in Charlottenburg[2] and had to come into the city every day. —How many mutton chops does one need to replace such effort? —But joking apart! Do travel through Berlin on your next excursion to the Alps: despite some differences in scientific views, you will find that you have many loyal friends here amongst whom I do not think I am the worst.

My wife[3] asks to give you fond greetings and thanks you much for having taken such good care of me and her daughter.

Farewell for now, do not overwork yourself and rest assured that I will always think of you with gratitude.

Yours | Poggendorff

Kind greetings also to your housemate, Dr Hirst. —As soon as I have received his last report from Magnus[4]—who by the way has also returned safely—and printed it, I will send it to you.

RI MS JT/1/P/133

1. *of the pleasant days that I spent in London*: Poggendorff and his daughter Marie (b. 1838) had just returned from London, where they stayed with Tyndall at 14 Waverley Place. Many continental physicists were in London for the International Exhibition of London, which took place in the gardens beside the Royal Horticultural Society Gardens in London from 1 May to 1 November 1862 (see letter 1910, and *Tyndall Correspondence*, vol. 7, letter 1900). Tyndall had taken Poggendorff to the RI and had reproduced his experiment to show the absorption of ordinary air compared to dry air in front of both Poggendorff and Faraday (*Ascent of John Tyndall*, p. 171).
2. *Charlottenburg*: a neighborhood in Berlin.
3. *My wife*: Charlotte Eleonore Poggendorff.
4. *last report from Magnus*: probably referring to G. Magnus, 'Ueber die Diathermansie trockner und feuchter Luft', *Poggend. Annal.*, 118 (1863), pp. 575–88, which was soon after published as G. Magnus, 'On the Diathermancy of Dry and Moist Air', *Phil. Mag.*, 26 (1863), pp. 21–30.

To Juliet Pollock [September 1862][1] 1932

Friday

My dear M[rs]. Pollock.

I have been leading such a life of hard work & dissipation—acting the host almost daily & leading my guests to drink wine by showing them the good example of drinking cognac myself—that I have found my whirling brain quite out of order for writing. Across this Getümmel,[2] your letter[3] reached me. I was quite at your side. London vanished & I was listening to the ripples of Grassmere.[4]

I have been calculating the time it would take me to get down & back & the result is not favourable. Two days would be spent on the journey—two or three ought to be spent there. Five days are absoluty[5] beyond my power to afford.

I shall be compelled to go to Cambridge to the Brit[ish] Association.[6] This makes a gap in my working time. & there is so much before me that I must work like a Titan[7] to get through it

We have lovely weather here. How lovely it must be among the lakes just now. But I must draw a screen over the picture lest my frail nature should be too strongly tempted.

Goodbye then. Kindest regards to your husband[8]—love to all the boys,[9] & all kindly wishes to yourself

Yours ever | John Tyndall

RI MS JT/1/T/1208
RI MS JT/1/TYP/6/2049

---

1. *[September 1862]*: the content of this letter suggests it was written some weeks after Tyndall returned from the Alps on 16 August, and prior to the BAAS meeting which began on 1 October (see n. 6).
2. *Getümmel*: turmoil (German).
3. *your letter*: letter missing.
4. *Grassmere*: Grasmere, a lake and village in the Lake District.
5. *absoluty*: misspelt by Tyndall.
6. *I shall be compelled to go to . . . the Brit Association*: the BAAS held its thirty-second annual meeting in Cambridge from 1–7 October 1862 (see also letter 1904, n. 9). At this meeting, on 2 October at 8pm Tyndall delivered a 'Discourse on the Forms and Action of Water' (*Brit. Assoc. Rep. 1862, 1863*), p. 1). The lecture impressed the American science editor Edward Youmans who was visiting England at the time (*Times*, 4 October 1862 p. 7). For his thoughts about the lecture, see J. Fiske, *Life and Letters of Edward Livingston Youmans: Comprising Correspondence with Spencer, Huxley, Tyndall, and Others* (London: Chapman & Hall, 1894), pp. 139–40.
7. *Titan*: possibly referring to the Titans, the pre-Olympian Gods of Greek mythology.
8. *your husband*: William Frederick Pollock.
9. *the boys*: Frederick, Walter, and Maurice Pollock.

From John Couch Adams 8 October 1862 1933

Cambridge Observatory, | Oct 8th 1862.

My dear Tyndall,

I meant before now to have written a line to ask how you are getting on, and to tell you how sorry we all are that you were obliged to leave so soon, especially for such a cause. I hope, however, that the unpleasant consequences of the accident will soon pass away.[1]

Stokes tells me that he was once in an exactly similar case, and was not long in getting right again, though he had slight occasional reminders of the strain to which he had been subjected, for some time afterwards.

Our meeting[2] is now almost over, and has, I think, passed off tolerably well. We had a great crush yesterday in Section A,[3] listening to Glaisher's[4] account of his Balloon ascent,[5] which he gave very well and clearly.

Please give us a line as soon as convenient, and tell me how you are.

Yours very truly, | J. C. Adams.

RI MS JT/1/TYP/1/2
LT Typescript Only

---

1. *accident will soon pass away*: Tyndall had injured his back while he was at Cambridge, while trying to lift a one hundred pound bag of ice when preparing the experiments for his lecture (see letter 1932, n. 6 and letter 1937). Afterward he went to the Isle of Wight to recuperate.
2. *Our meeting*: the BAAS held its thirty-second annual meeting in Cambridge from 1–7 October 1862 (see letter 1904, n. 9).
3. *Section A.*: Section A is the Mathematical and Physical Science section of the BAAS.
4. *Glaisher*: James Glaisher (1809–1903), a British meteorologist, aeronaut, and astronomer. He is most noted for his balloon ascents to measure temperature and humidity (*ODNB*).
5. *Balloon ascent*: this refers to Glaisher's 5 September ascent, in which he is estimated to have risen as much as 10,900 meters above sea level, and passed out around 8,800 meters. Tyndall was a member of the BAAS Balloon Committee, which had been established at the 1858 meeting and ran for several years ('Recommendations of the General Committee', *Brit. Assoc. Rep. 1858*, p. xxxix; and 'An Account of Meteorological and Physical Observations in Eight Balloon Ascents, made, under the Auspices of the Committee of the British Association for the Advancement of Science at Manchester, by James Glaisher, F.R.S., at the request of the Committee, consisting of Colonel Sykes, Mr. G B. Airy, Lord Wrottesley, Sir D. Brewster, Sir J. Herschel, Dr. Lloyd, Admiral FitzRoy, Dr. Lee, Dr. Robinson, Mr. Gassiot, Mr. Glaisher, Dr. Tyndall, Mr. Fairbairn, and Dr. W. A. Miller', *Brit. Assoc. Rep. 1862*, pp. 376–503).

From Joseph Tairraz[1] 8 October 1862 1934

Chamonix le 8 8bre 1862

Monsieur le professeur Tyndall

Je vous prie de me pardonner la liberté que je prends de vous adresser ces quelques lignes—pour vous informer que les meubles de l'ami Balmat Auguste, mort à Sixt, doivent se vendre par voie de justice dans une quinzaine de jours. Ses affaires financières n'étant pas dans le meilleur état, je vous écris ceci parce que ce matin je fus appelé par les huissiers procédant à l'inventaire de ses biens, pour [dénocer] ses objets photographiques, et étant là j'ai vu plusieurs objets vous appartenant. Tels que votre tente votre trouelle et accessoires. Je les ai désignés à l'huissier en le priant de ne pas les porter sur l'inventaire comme vous appartenant.

Je croirais que vous pourriez réclamer vos objets avant la vente en vous adressant soit à M. le maire, soit à Monsieur le notaire Dupuis.

Je vous prierai seulement, Monsieur le professeur, de ne pas faire connaitre que je vous en ai informé.

Votre très humble serviteur | Tairraz Joseph photographe

Chamonix, October 8th 1862

Professor Tyndall,

Please forgive me for taking the liberty of sending you the following few lines to inform you that the possessions of our friend Auguste Balmat,[2] who died at Sixt, will be sold by the court in a fortnight. As his financial affairs are not in good order, I am writing to you because this morning I was summoned by the bailiffs conducting the inventory of his property, in order to [report on] his photographic items. While I was there, I saw some items belonging to you, such as your tent, your trowel, and accessories. I pointed them out to the bailiff to ask him not to count them into the inventory since they belong to you.

I think that you could claim your items before the sale by writing either to the mayor or to the notary, Mr Dupuis.[3]

I only ask you, Professor, not to tell them that I have informed you of it.

Your very humble servant | Tairraz Joseph, photographer

RI MS JT/1/T/3

---

1. *Joseph Tairraz*: Joseph Tairraz (1827–1902), a French pioneer of alpine photography (M. Dessemond, *Tairraz père et fils: photographes de montagne* [Paris: Contrejour, 1992]).
2. *Auguste Balmat*: Auguste Balmat (1808–62), a Swiss mountaineer and guide born in Chamonix, Savoy, who at the age of twenty-two made his first ascent of Mont Blanc. He met James D. Forbes in 1842 and assisted him during his studies of the movement of glaciers and survey work leading to the important map of the Mer de Glace published in Forbes's *Travels through the Alps of Savoy* (Edinburgh: A. and C. Black, 1843). In 1854 he guided the much-celebrated ascent of the Wetterhorn by Alfred Wills (1828–1912). Wills later dedicated his *Wandering among the High Alps* (London: Bentley, 1856) to Balmat. In 1858, while guiding Tyndall and Willis's ascent of Mont Blanc, he nearly lost his hands in the process of placing thermometers at the summit (C. D. Cunningham and W. Abney, *The Pioneers of the Alps*, 2nd ed. [London: Low, Marston, Searle and Rivington, 1888]).
3. *Mr Dupuis*: not identified.

From T. Zeno[1] 8 October 1862 1935

LII. *On the Changes in the Apparent Size of the Moon.* | *By* T. Zeno. | *To Professor Tyndall.*

Guy's Hospital, October 8th, 1862.

SIR,

MY attention having been called to a letter in the May Number of the Philosophical Magazine, "On the Changes in the Apparent Size of the Moon,"[2] where reference is made to my explanation of the phenomenon (Recreative Science, No. 24), I find it necessary to ask your indulgence while I make some allusion to that reference, as well as to the novel and ingenious experiment of your correspondent Mr. Lewis.[3]

First, with regard to the reference to my explanation, it is obvious that, to satisfy the "popular mind" as to the subjective character of the phenomenon, it would have been better to have quoted my second experiment, that of looking at the rising sun till an ocular spectrum was obtained, and then directing the eye to the zenith, where this spectrum immediately diminished, to enlarge again on looking to the opposite horizon. Secondly, with regard to the experiment of Mr. Lewis, it must be noticed that, while it *necessarily* involves the condition of binocular vision, the phenomenon it proposes to illustrate is equally exhibited to monocular vision; the experiment nevertheless, so far as I can judge, very well illustrates that law by which, when two bodies subtend equal angles, both bodies being of the same size, but one assumed to be nearer than the other, the (assumed) nearer appears to be smaller. And though, consequently, the moon in the zenith would appear smaller if *the senses assumed it to be nearer*, the experiment of your correspondent does not hint at the latter condition. Now as the gravamen of the evidence lies here, you will perhaps allow me to state briefly why it is that the luminary (seen with one eye or with both eyes) in the zenith is felt to be nearer than it is when seen in the horizon. I say *felt*, not *reasoned*, inasmuch as this is a sensory perception which no intellection can alter, seeing the setting or rising sun is no less enlarged to the eye of an astronomer (whose reason rejects the illusion) than it is to the eye of the most uninformed rustic.

I propose that the luminary appears nearer at zenith than horizon, because, first (A), the visible sky appears nearer at zenith than horizon, and, secondly (B), the apparent distance of the luminary accords with that of the sky.

A. Everything visible is seen by rays proceeding from it*.[4] Whatever, then, constitutes the visible sky is thus seen. Now sky would appear black if atmospheric particles did not reflect rays; it would appear white if intervals of black space beyond did not leave points of the retina unexcited; and it appears blue by a law (experimentally deduced by Da Vinci) that retinal excitation (by white light) interspersed with retinal points of rest (derived from the blackness of space) *is the sensory condition blue.* But the (partial) blueness of sky at horizon shows that the blackness of space is not altogether shut out by intervening reflecting particles: whence it follows that the *furthest* reflecting particles of our atmosphere cannot have their light wholly intercepted by the *nearer;* and that the whole atmospheric area contributes, with the blackness of space beyond, to constitute the visible sky. The whole atmosphere, then, give the *real* locality of that we call visible sky—seen by rays proceeding from it (its varying shades, from blue to nearly black, in various latitudes, showing that colour is its visible accident and not its visible essence). Now, whether we draw a line from the spectator to the limit of the atmosphere or to any assumed point within it, the line in direction of horizon will be longer than in that of zenith; and in whatever form any part, or the whole, of this transpicuous extent of atmosphere is manifested to vision by the aforesaid rays, such manifestation in direction of horizon is more or less *obscured,* in consequence of the rays depicting it coming from the greater distance and under a smaller angle; in other words, every visible quality and attribute of the sky will appear more distinct at the nearer point (zenith) than at the further point (horizon)[†5].

But if this necessarily curtailed demonstration should not be fully conclusive, the *facts,* reasoned to, are obvious; the sky *is* more definite and more blue in appearance at the zenith: moreover we naturally express our *sensation* by naming it the "near sky," and that at the horizon the "far sky."

The synthetic argument, finally, seems irrefragable: engraver and painter succeed in presenting, the one the form, and the other the form and colour of the sky as we really perceive them by the sense, their success depending solely on an *observed law of perspective,* which mathematically demonstrates the sky they depict at the horizon to be further considerably than that they represent at the zenith. Now the senses, in recognizing the truth of this representation of the sense-perceived form of the sky, assert unequivocally what the sense-perception is.

It remains to show, B, that the apparent distance of the luminary coincides with that of the sky. It may be argued, since there is no sensory evidence of the *actual distance* of the luminary, 1st, that the sense can only refer it to the ground which surrounds it; 2nd, that, by the universal law determining visual distance with one or both eyes, no object is felt to be beyond or in advance of another surrounding it, unless, the size, parts, texture, or colour of both being known or assumed, the sight detects that one is more obscured than the other, in which case the less obscured is inferred to be less distant, and *vice versa;* 3rd, a strong light, placed at any distance behind a hole in a semi-illuminated screen, appears to fill the aperture and be there situate; or a sunbeam piercing an interstice appears to emanate from the object in which the interstice exists.

If, then, the positions A and B are established, it follows that the luminary in the zenith appears nearer than at horizon; and, subtending (practically) the same angle in both cases, it is hardly necessary to add (what my paper referred to sets forth) that when the luminary is supposed to be nearer it is necessarily perceived to be smaller.

Trusting you will excuse the unforeseen length of this explanation of my views, perhaps somewhat misunderstood by your correspondent,

Believe me to be, Sir | Yours very respectfully, | T. ZENO.

T. Zeno, 'On the Changes in the Apparent Size of the Moon', *Phil. Mag.*, 24 (1862), pp. 390–92

---

1. *T. Zeno*: the sender of this open letter in the *Phil. Mag.* is not identified, and the name is possibly a pseudonym.
2. *letter in the May Number of the Philosophical Magazine*: see letter 1871, *Tyndall Correspondence*, vol. 7; or R. T. Lewis, 'On the Changes in the Apparent Size of the Moon', *Phil. Mag.*, 23 (1862), pp. 380–2.
3. *May Number of the Philosophical Magazine . . . your correspondent Mr. Lewis*: Zeno is referencing Lewis's article, 'The Rising of the Sun Physiologically Considered', *Recreative Science*, 3 (1862), pp. 110–20. Richard Thomas Lewis (1839–1916) was an amateur microscopist, entomologist, and temperance advocate (E. J. Lewis, 'Obituary notices: Richard Thomas Lewis FRMS', *Journal of the Quekett Microscopical Club*, 13 (1916–18), pp. 201–03).
4. The following footnote appeared here in the original: '* For, though perfect blackness, with its retinal equivalent non-excitement, is often an instrument of vision, an object from which no rays proceed is, *per se*, invisible'.
5. The following footnote appeared here in the original: '† Another source of obscuration, owing to the greater density of the lower strata of atmosphere (a consideration that has misled some writers on this subject), is intentionally omitted, as also (further on) obscuration of the luminary itself, from the same cause. For, first, the senses (which are much more wakeful observers than the reason) are conversant with the fact of the bases of high buildings, trees, vertical cliffs, &c. being more obscured than their summits, *in which case* the sense knows that greater obscuration does not mean greater distance; secondly, my second experiment, already stated, proves that our phenomenon is, or may be, independent of this obscuration'.

From Juliet Pollock 9 October [1862][1] 1936

AUDACTER ET STRENUE[2] | Rigg's Hotel. Windermere | Oct. 9th

My dear Mr. Tyndall

The room continued to be called yours, till we left Lake View[3] and all its charms, and gave up our nomadic wild life for the civilization of Greta Bank,[4] where we were again in society, with regular hours, a dressing bell and a change of costume for dinner. Still, nature had not deserted us—the sun still shone upon our days, and we still wandered through green meadows, rambled over mountain paths, & enjoyed our luncheon by the side of loitering hills; only we climbed less and talked more than before. We spent one very delightful afternoon as spectators to a rifle match between Keswick[5] and Cockermouth[6]—Young John Spedding,[7] whom Moss[8] distinguishes as "Mr. Spedding, the rifleman", being Captain of the Skiddaw Greys;[9] a grand mountain background, a fine echo of the sound, doubling each shot among the hills, a pleasant foreground of green fields; the men in uniform, picturesque as they knelt to fire; forming capital unconscious groups; a lively interest in the match with a generous open hearted good humour among the men,—a bright sun making the hills transparent (you know that curious effect) the shades of evening gathering over them till they changed to a heavy opaque purple & the character of the scene altering under the waning light but only to show a new and deeper beauty; & we going home, all very happy because Keswick had won, and because the day was so charming;—You would have enjoyed such a day, and we should, all the more for your presence, but you live with the gases and Bromine absorbs all your being! I am not sure whether you know that from Grasmere[10] (from Wythburn[11]) I climbed the dark brow of the mighty Helvellyn,[12] and reached the summit, and came down again with no other help than that of my own feet and Walters[13] dear little guiding hand. We had perfect weather for it, a sun that made every little blade of grass, every precious tuft of moss sparkle under our feet, and that threw a full distinct light upon the wide map of beauty that unfolded itself to us from the highest point. This was my most important performance for Greta Bank was not disposed for climbing, and there we spent our days in less ambitious enjoyments. We had a pleasant diversity in our programme after Greta Bank was done in a visit to the J. Marshalls[14] at Coniston,[15] taking Sarah Spedding[16] with us. There we had a fair variety of pictures without doors, scenery like, yet different from what we had been delighting in before; new combinations of rock and stream and mountain pass, and within doors, literature grace and poetry. Two poets the other visitors: Coventry Patmore[17] and Aubrey de Vere.[18] Poor Mr. Patmore was suffering from the recent loss of his wife[19] (The Angel in the House[20]) and was disposed to be silent but Mr de Veres has a very full flow of graceful discourse so that the silence of his friend was masked. We knew him only slightly before, but before our three days were finished, the acquaintance had grown into an intimacy which is not likely to be suspended again. M[rs]. J. M. is clever with an interesting appearance and would be a delightful woman if she were not affected. But the studied attitudes and the conscious speech are blemishes. She had the opportunity in one afternoon of pelting rain, of showing her powers of entertainment and the flow of conversation made us all indifferent to the flow of rain; till two hours before dinner when we judged it best to take a constitutional in waterproof suits, and we set forth through some of the charming country that we had been introduced to in sunlight, now so darkly veiled in mist, that we could not recognize its face. All the life had gone from the scene; we might have been exploring vaults of death, and now and then a shrouded shape gleamed a sudden whiteness upon us,—a ghost of the tombs—well—a waterfall—and it made us think of the divine Undine's[21] most bitter journey. Our last day smiled upon us, and by a device of M[r] de Veres, pleasure accompanied our departing steps, for while Moss and the maids were sent with the luggage in the heavy carriage by the high road to Grasmere—we walked among the mountains through passes of exquisite beauty to join them—Sarah S. and myself having the occasional help of a lift in a poney carriage—Through Tilberthwaite[22] by Colwith[23] and Skelwith[24] past Loughrigg Tarns[25] we rambled on (our party consisting of Fred[26] Sarah S. M[r]. Marshall M[r]. de Vere and myself) till we come over High Close[27] upon Grasmere looking inviting as it lay at our feet with a smiling caressing aspect recalling our holiday with our boys which was I think the happiest holiday of their young lives.

Partings must come; our pleasant party broke up; we saw Sarah S. start away for Greta Bank, and we said "good bye" and came on hither. We are very comfortable here and the weather is still friendly and we are enjoying a new honeymoon in the quiet and beauty of the place—only interrupted by the vivacity of Moss who likes this place best of all because of the railway station.[28] We go to town on the 14th or 13th for two or three days, afterwards, pay a visit of four or five days & the Charles Merivales[29] in Essex, and then return for our final settling in London. We look forward to seeing you and we have read with great interest all accounts of your brilliant lecture at Cambridge[30]

Yours ever most truly, | Juliet Pollock

RI MS JT/1/P/201

---

1. *1862*: the date is given by reference to Coventry Patmore, whose wife died mid-1862, by reference to Tyndall's lecture in Cambridge (see n. 30), and by relation to letter 1937 (Tyndall's reply to this letter).
2. *AUDACTER ET STRENUE*: the motto of the Pollock clan, meaning 'Boldly and

Earnestly', shown on a family crest bearing the image of a boar pierced by an arrow (B. Burke, *The General Armory of England, Scotland, Ireland, and Wales* (London: Harrison & Sons, 1884), p. 812).

3. *Lake View*: probably the name of the rented cottage they were staying in overlooking the lake.
4. *Greta Bank*: an estate located in the town of Keswick in the county of Cumberland in northwestern England, a Spedding family home about 5 miles from the Spedding estate at Mirehouse. The Pollocks were good friends with the family of James Spedding (1808–81), an English literary editor and biographer known especially for editing the works of Francis Bacon (*ODNB*). James dined often with the Pollocks, and on at least one occasion with Tyndall as well, later in 1864 (*Remembrances*, vol. 2, p. 117).
5. *Keswick*: a market town in historic Cumberland, present day Cumbria, in northwestern England.
6. *Cockermouth*: a small town to the northwest of Keswick.
7. *John James Spedding*: John James Spedding (1834–1909), only son of John Spedding (1806–39), who was educated at Harrow and Trinity College, Cambridge (E. Walford, *The County Families of the United Kingdom or Royal Manual of the Titled and Untitled Aristocracy of England, Wales, Scotland, and Ireland* (London: Chatto & Windus, 1893), p. 943). His uncle was James Spedding (see n. 4).
8. *Moss*: Maurice Pollock.
9. *Skiddaw Greys*: the Skiddaw Grey Rifles, a volunteer rifle unit based at Keswick.
10. *Grasmere*: a village and lake in the Lake District of northwest England, southeast of Keswick.
11. *Wythburn*: a village in the Lake District of northwest England, southeast of Keswick.
12. *Helvellyn*: a mountain in northwest England, the third highest in the country.
13. *Walter*: Walter Pollock.
14. *J. Marshalls*: possibly John Marshall (1818–91), a British politician, Member of Parliament for Leeds.
15. *Coniston*: a town in the Lake District.
16. *Sarah Spedding*: possibly the mother of James Spedding (see n. 6); or, more likely, an unmarried sister of Thomas Story Spedding (1800–70), a Cumberland landowner and elder brother of James.
17. *Coventry Patmore*: Coventry Patmore (1823–96), an English poet and critic, best known for his poem 'Angel in the House', which subsequently defined the Victorian ideal of middle-class womanhood (*ODNB*).
18. *Aubrey de Vere*: Aubrey Thomas de Vere (1814–1902), an Irish poet and critic (*ODNB*).
19. *the recent loss of his wife*: Emily Augusta Andrews (1824–62) died on 5 June 1862.
20. *Angel in the Home*: Patmore's wife Emily had been the inspiration for his lengthy narrative poem 'The Angel in the House', the first volume of which was published in 1854; the complete four-part poem being published in 1863. Emily was the model for the English painter John Everett Millais' (1829–96) painting *Angel in the Home*.
21. *Undine*: a German fairy story published in 1811 by Friedrich de la Motte Fouqué (1777–

1843) that received popularity in England throughout the nineteenth century, in which a water spirit named Undine marries a knight in order to gain a soul.

22. *Tilberthwaite*: a small village to the south of Keswick.
23. *Colwith*: a small Lake District village southeast of Keswick.
24. *Skelworth*: Skelwith Bridge is a small village in the southern area of the Lake District in Cumberland, present day Cumbria.
25. *Loughrigg Tarns*: a tarn situated underneath Loughrigg fell, which overlooks Grasmere.
26. *Fred*: Frederick Pollock.
27. *High Close*: an estate near Grasmere, now an Arboretum under the National Trust.
28. *the railway station*: Windermere station is the terminus of the Kendal and Windermere railway. 'The Terrace', a row of cottages, built for railway executives in 1849, is reputed to have been designed by the architect Augustus Pugin.
29. *Charles Merivales*: Charles Merivale (1808–93), an English historian and clergyman (*ODNB*).
30. *your brilliant lecture at Cambridge*: see letter 1932, n. 6.

To [Juliet Pollock][1] 11 October 1862 1937

5 Cliff Terrace | Black Gang Chine[2] | Isle of Wight | 11th. Oct. 1862

My dear Friend,

Surely you deserve my blessing—I have been here a solitary hermit for some days at times obliged to fly to a novel to kill the hankering after work which I was unable to satisfy in any other way. I went to Cambridge on the morning of my lecture[3]—I had among other things a hundred-weight of ice in a sack. I set my assistants to take it out, looking on at the time, and imagining that I could do the work quicker myself, I suddenly seized the sack, lifted it and tried to shake out the ice. I strained my back[4]—probably tore some of the muscular fibres. The pain grew very bad and took half my heart away. Nevertheless I must get through the lecture and I did get through it; but it was more an escape than a victory.

I could not remain in Cambridge, so came away. Stewed myself in a Turkish Bath, which did me great good, and to finish matters came down here. I was getting on famously when, in an evil hour, I ventured in among the high rolling waves of Niton Bay.[5] As long as I kept out among the rollers the pendulous swing as they passed under me was extremely pleasant and exciting—but on returning to the shore one of them overtook me, and just as it reached me and as I looked up at its crest thinking of Merlin and Vivien,[6] the villainous mass broke and pitched itself with all its weight upon my half disabled loins. All the mischief done at Cambridge was renewed and it was augmented by a 10 mile walk which I unwisely took afterwards.

Yesterday I was very stiff and sore—in bed till noon. Today I am better—

able to be up and spurred by the inspiration of your letter,[7] able to write. I read your letter with delight—it is perfectly charming, giving me hill dale and streamlet tinged with the sun and beauty of your own soul.

I will remain here for two or three days longer. Once I was on the point of surprising you at Keswick.[8] I wanted a bag of perfectly fresh air and the thought occurred to me to bring it from the top of Skiddaw.[9] Had a railway passed Keswick[10] I should surely have done so, but on consulting Bradshaw[11] I found there were 15 miles of coach work: this was too much for my patience so I furnished myself in the Isle of Wight.

I am proud of you and Walter:[12] it is really a fact worth remembering—could we only cap the performance one summer in the Alps!

Ever | J. Tyndall

RI MS JT/1/T/1140
RI MS JT/1/TYP/6/1999–2000

---

1. *[Juliet Pollock]*: recipient indicated by content and relation to letter 1936.
2. *Black Gang Chine*: now Blackgang Chine, an amusement park and holiday resort on the south coast of the Isle of Wight.
3. *my lecture*: see letter 1932, n. 6.
4. *I strained my back*: see letter 1933.
5. *Niton Bay*: also known as Reeth's Bay, Niton Bay is a small curved bay on the southernmost tip of the Isle of Wight.
6. *Merlin and Vivien*: 'Merlin and Vivien' is one of the poems that make up Alfred Tennyson's narrative poem *Idylls of the King* (1859–85).
7. *your letter*: letter 1936.
8. *Keswick*: a small English town in Cumberland, present-day Cumbria.
9. *Skiddaw*: a mountain just north of the town of Keswick, the sixth highest peak in England.
10. *Had a railway passed Keswick*: a line was built from Penrith to Keswick, which opened in 1865.
11. *Bradshaw*: colloquial designation of 'Bradshaw's Railway Guide', a timetable of all railway trains running in Great Britain (*OED*); published by W. J. Adams of London from 1839–1961.
12. *I am proud of you and Walter*: Walter Pollock, the second son of William Frederick and Juliet Pollock, had been elected a King's Scholar at Eton (see letter 1923, n. 1).

From Charles Lyell 13 October 1862 1938

13 Oct. 1862 | 53 Harley S[t].

My dear Sir,

I have just found Mortillets[1] pamphlet[2] explaining his map of glaciers of N. Italy. He seems to have sent it to me long after his map.[3]

Would it now be of any use to you? Have you published anything about glacial erosion—somebody told me there was a letter of yours in the newspaper on the subject?[4]

ever truly y[ou]rs | Cha. Lyell. | J. Tyndal Esq.

RI MS JT/1/L/49
RI MS JT/1/TYP/3/848

---

1. *Mortillets*: see letter 1929, n. 1.
2. *pamphlet*: see letter 1929, n. 2.
3. *He seems to have sent it to me long after his map*: see letter 1929.
4. *a letter of yours in the newspaper on the subject?*: possibly J. Tyndall, 'An Alpine Observation', *Times*, 22 October 1861, p. 10, which discussed the sliding of ice.

From John Couch Adams 25 October 1862 1939

Cambridge Observation | Oct. 25th 1862.

My dear Tyndall,

The shirt[1] duly arrived, and very fortunately before it had been missed, so that I had taken no steps towards procuring a search warrant. I am delighted to hear that you have so completely got rid of the effects of your accident,[2] but let me counsel you to be cautious for some little time to come not to expose your muscles to any extra strain, as time is an important element in the reconsolidation of the tissues. Therefore defer your challenge to Deerfoot,[3] and when the contest comes off may I be there to see. Yours very truly, | J. C Adams.

RI MS JT/1/TYP/1/3
LT Typescript Only

---

1. *shirt*: Tyndall had borrowed a shirt from Adams while in Cambridge for a lecture (see letter 1932, n. 6; and for an indication of Adams' hospitality toward Tyndall, see letter 1930).
2. *your accident*: see letter 1933, n. 1.
3. *Deerfoot*: Deerfoot Red-Jacket (1828–97), or Hut-goh-so-do-neh in his native Seneca language, and also known as Lewis Bennet, was a talented long-distance runner from the United States. He visited England on a running tour in 1861 becoming something of a sensation (E. P. Tanner, 'Deerfoot', in A. Johnson and D. Malone (eds), *Dictionary of American Biography* (New York: Charles Scribner's Sons, 1930), vol. 5, p. 194).

From Johann Friederich Christian Hessel[1] 29 October 1862 1940

Marburg den 29 October 1862.

Euer Hochwohlgeboren

Beehre ich mich anliegend eine Abhandlung zu übersenden, von der ich glaube dass sie in mehrfacher Hinsicht Interesse für Sie hat. —Ich lege ein Paar weitere Exemplare bei, um Sie in den Stand zu setzen, dieselben an solche Mathematiker, Physiker, oder Krystallographen Englands abzugeben, von denen Sie erwarten, daß sie sich dafür interessiren. —Eine weitere Ausführung des dritten Abschnittes dieser Abhandlung erscheint in der Kürze in Journalen.

Auch eine ältere Schrift lege ich bei, die vielleicht dadurch einiges Interesse für Sie hat, daß ich vor langer Zeit schon die, später durch die Entdeckung des Planeten Neptun bestätigte, Behauptung aufstellte, in welcher ich sagte: „wenn das Gesetz, von dem in der Abhandlung die Rede ist, Gültigkeit hat, so muß jenseits der Uranusbahn noch mindestens ein Planet vorhanden seyn, weil das Gesetz, wenn es richtig verstanden wird, noch eine ungerade Anzahl von weiteren Planeten fordert."—Die Frage ob trotz der abermaligen Bestätigung das Gesetz wirklich gültig ist oder nicht gehort nicht in mein Gebiet.

Da mein Sohn, der Dr. Phil. C. Hessel der bereits seit einer Reihe von Jahren als Lehrer der mathematischen und naturwissenschaftlichen Disciplinen in England lebt, seit einigen Jahren in London wohnt, und ich Gelegenheit habe, ein Paquet an ihn zu befördern, so habe ich die Abhandlungen ihm zugesandt. Er wird daher es sich zur Ehre rechnen, sie Ihnen persönlich zu überweisen.

Möge diese Sendung dazu dienen, freundliche Erinnerungen an unser Marburg in Ihnen rege zu halten und Sie zu überzeugen, daß in Marburg freundlich an Sie gedacht wird.

Mit ausgezeichneter Hochachtung | der Ihrige | Hessel

Marburg,[2] 29 October 1862.

Dear Sir,

Allow me the honour of sending you the enclosed paper,[3] which I believe will be of interest to you in many regards. —I include a few additional copies to enable you to pass them on to such mathematicians, physicists, or crystallographers in England, whom you believe would be interested. —An elaboration of the third paragraph of this paper will appear shortly in journals.

I am also including an older piece,[4] which will probably be of some interest to you in that a long time ago I made the claim—later confirmed by the discovery of the planet Neptune—that: "if the law stated in the paper is valid, then there must be at least one planet beyond the orbit of Uranus, as the law, if understood correctly, requires an uneven number of additional planets."—The question whether despite the renewed confirmation the law is indeed valid or not does not belong to my field.

As my son, Dr. Phil. C. Hessel, who has lived in England for a number of years as teacher of mathematical and scientific disciplines, has lived in London for some years now, and I have occasion to send him a package, I sent the articles to him. He will therefore have the honour of passing them on to you in person.

May this dispatch serve to keep friendly memories of our Marburg alive and convince you that we think of you fondly in Marburg.

With the greatest respect | Your | Hessel

RI MS JT/1/H/122

---

1. *Johann Friederich Christian Hessel*: Johann Friedrich Christian Hessel (1796–1872), a German mineralogist, crystallographer, and professor at the University of Marburg (*CDSB*).
2. *Marburg*: a university town north of Frankfurt, where Tyndall completed his doctorate at the University of Marburg in 1850, studying under Karl Knoblauch and Robert Bunsen, among others.
3. *the enclosed paper*: enclosure missing. Possibly his paper on crystallography that had not received much attention in Germany, published in 1830 in Gehler's *Physikalischen Worterbuche*, and republished much later as J. F. C. Hessel, *Krystallometrie oder Krystallanomie und Krystallographie* (Leipzig: Wilhelm Engelmann, 1897) (J. G. Burke, *Origins of the Science of Crystals* (Berkeley: University of California Press, 1966), p. 4).
4. *an older piece*: possibly a reference to J. F. C. Hessel, *Die merkwürdigen arithmetischen Eigenschaften der wichtigsten Näherungsreich für die Sonnenabstände der Planeten* (Marburg: N.G. Elwert'sche Universitäts-Buchhandlung, 1859).

To Joseph Dalton Hooker [October 1862][1] 1941

Friday

My dear Hooker

Would you, like a dear good fellow, cast your eye over the paper[2] which I send you by this post. It will give you an idea of some of my notions regarding the action of aqueous vapours. Possibly you may be able to write me a letter from your own experience pointing to some illustration of the experimental results. I should be greatly pleased to publish such a letter with the paper.[3] That aqueous vapour plays a most important part in meteorology is undoubted. Herschel[4] and Airy[5] have both said to me that it is a grand step in meteorology! Magnus[6] denies the action; but he has not a foot to stand upon.

Yours ever | John Tyndall

RI MS JT/1/TYP/8/2561
LT Typescript Only

---

1. *1862*: the month and year are given based on relation to Tyndall's paper on radiant heat and aqueous vapor that was submitted on 20 November 1862. This third memoir (see n. 2) is Tyndall's first significant paper on aqueous vapour. In the paper he mentioned Hooker's Himalayan observations (p. 9) and Hooker on hail (p. 12).
2. *the paper*: J. Tyndall, 'On the Relation of Radiant Heat to Aqueous Vapour. Third Memoir', *Phil. Trans.*, 53 (1863), pp. 1–12.
3. *I should . . . with the paper*: Tyndall's published paper does not include a letter from Hooker.
4. *Herschel*: for Herschel's letter from 12 April 1862, see letter 1878, *Tyndall Correspondence*, vol. 7.
5. *Airy*: if Airy shared his view in a letter, it is missing.
6. *Magnus*: Gustav Magnus.

To Joseph Dalton Hooker [October 1862][1] 1942

Monday

My dear Hooker,

I have devoured your notes[2] in the intervals of my soup. You have certainly opened my eyes on many points. What you have said regarding the ancient glaciers quite determines me to give the speculation up—I shall merely shew how the radiation phenomena may come into play by the deflection upwards of moist winds by mountains masses.

I was amazingly struck by your speculation on hail, but the whole thing is full of 'wrinkles', and I am extremely obliged to you for the trouble you have so kindly taken.

Yours ever & always | John Tyndall

RI MS JT/1/T/1024

---

1. *[October 1862]*: on the dating of this letter, see letter 1941, n. 1.
2. *your notes*: notes missing; there is likely at least one missing letter in this correspondence between letter 1941 and this one.

From Gustav Robert Kirchhoff 10 November 1862 1943

Heidelberg, den 10ten Nov. 1862.

Verehrter Freund!

Zuversichtlich irre ich nicht, wenn ich vermuthe, dass ich Ihnen zum grossen Theile der Ehre verdancke, die die Königl Gesellschaft mir hat angedenken lassen, indem sie mir für meine Untersuchungen über das Sonnenspectrum und über die Umkehrung der hellen Linien der Flammenspectren die Rumford-Medaille zuerkannt hat. Es hat mich diese Auszeichnung in hohem Grade überrascht und erfreut. Aber nicht allein, um Ihnen das zu sagen, schreibe ich jetzt an Sie, ich will noch eine Bitte an Sie richten. Schon meiner Vorlesungen wegen ist es mir unmöglich bei der feierlichen Sitzung der Gesellschaft am 1ten December persönlich zu erscheinen, um die Medaille in Empfang zu nehmen; würden Sie die Güte haben, das an meiner Stelle zu thun, und dann gelegentlich die Medaille mir zukommen zu lassen? Für den Fall, dass Sie das können und wollen, habe ich noch zu fragen, ob es der Sitte gemäss ist, wenn Sie bei dieser Gelegenheit mündlich meinen Dank der Gesellschaft aussprechen, oder ob ich ein Dankschreiben an diese senden soll? Hr Professor W. H. Miller hat, als Secretair der Gesellschaft, mir von der schmeichelhaften Anerkennung, die ich erhalten soll, Anzeige gemacht; ich will für jetzt nur diesem persönlich meinen Dank aussprechen, ohne ihn zu bitten, auch der Gesellschaft meinen Dank zu sagen.

Bunsen und ich denken oft noch an die schönen und interessanten Stunden, die wir mit Ihnen in London verlebt haben.

Mit herzlichen Grüssen von Bunsen | Ihr | G Kirchhoff.

Heidelberg, 10 Nov. 1862.

Dear friend!

I am confident that I am not mistaken in supposing that I owe to you to a large extent the honour that the Royal Society bestowed upon me by awarding me the Rumford Medal[1] for my research on the solar spectrum and on the inversion of the bright lines in the spectra of flames. This recognition surprised and pleased me greatly. But I write to you not only to say this, I also have a request to make of you.[2] Because of my lectures, it is impossible for me to be present at the Society's ceremonial meeting on December 1st to accept the medal; would you be so kind as to do so in my place and send me the medal when you have occasion? If you can do this and want to, I further ask whether it would be in accordance with custom for you to express my gratitude verbally to the Society on that occasion, or whether I should send a letter of thanks? Prof W. H. Miller as secretary of the Society[3] notified me of the flattering recognition that I am about to receive; for now, I would like only to thank him personally without asking him to extend my thanks to the Society.

Bunsen and I frequently remember the pleasant and interesting hours that we spent with you in London.[4]

With warm greetings from Bunsen | Yours | G Kirchhoff.

RI MS JT/1/K/7

1. *awarding me the Rumford Medal*: Kirchhoff was awarded the RS's Rumford Medal in 1862 for his researches on the fixed lines of the solar spectrum and on the inversion of the bright lines in the spectra of artificial light ('Anniversary Meeting', *Roy. Soc. Proc.*, 11 (1862), pp. 291–4). The medal is awarded every alternating year for 'an outstandingly important recent discovery in the field of thermal or optical properties of matter made by a scientist working in Europe'. Kirchhoff was indeed correct in his supposition that Tyndall had been instrumental in his receiving the award (see *Ascent of John Tyndall*, pp. 171–2).
2. *request to make of you*: Tyndall duly wrote to the RS secretary William Sharpey about Kirchhoff's requests (see letter 1949).
3. *Prof W. H. Miller as secretary of the Society*: William Hallowes Miller (1801–80), a Welsh mineralogist who held the professorship of mineralogy at the University of Cambridge (1832–70) and foreign secretary of the RS (1856–73) (*ODNB*). He was known for his *Treatise on Crystallography* (Cambridge: Deighton, 1839).
4. *Bunsen and I . . . spent with you in London*: Bunsen and Kirchhoff traveled to England for the International Exhibition of London (see letter 1904, n. 8) (they are listed, among other foreign scientists who had visited the Kew Observatory when in the country for the exhibition, in 'Report of the Kew Committee', *Brit. Assoc. Rep. 1862*, pp. xxxii–xxxix, on p. xxxvii).

To Mary Anne Coxe[1] [23 November 1862][2] 1944

Sunday night

My dear Miss Coxe,

How are you this cold weather? I am here in a still room of the Athenaeum,[3] dimly lighted with no companion, & no sound save the ticking of the clock.

'How shall I pass the next 10 minutes?' thought I—Your idea rose before me, & then & there I exclaimed 'why in writing to M^rs^. Coxe to be sure!'

So here I am committed to my task but with nothing to say.

How did you get home from London?

I broke my back at Cambridge & went to the Isle of Wight to recover.[4] I was almost well when, like a fool, I went to bathe amid huge rolling waves: one of them broke upon me as I was returning, tossed me about like a feather

& broke my back a second time—I was quite doubled up for several days, but am now all right. Alick[5] called but I did not see him, & I do not know where the boy now is.

I am working hard at a big book which I hope to get out early next year.[6]

I am told the Duke[7] has been "pulling me up"—those are the words in the 'Edinburgh.'—Well I have read his article[8] which is extremely clever; but he touches a point which his son's son, with all the accumulated advantages of the next two generations will fail to solve. I have thought of it more or less for years, and sometimes for whole weeks in succession, but always to retire baffled;—to feel, in fact, that man lacks the organ which can grasp this great question.

Now don't cry 'botheration!' & I will pester you with this subject no further.

I thought of you when at Cambridge. You said when I last met you that I should never 'lose heart.' But as I sat ten minutes before my lecture with my aching back, and my thoughts all muddled by the accident I felt that you had been mistaken, & that whatever heart I once possessed had become mere gizzard! I was a bird with a very white feather.[9]

Now the ten minutes are up—Goodbye. K[ind]. R[egards]. to the Commissioner.[10]

Ever & always yours | J. Tyndall

RI MS JT/1/T/253
RI MS JT/1/TYP/1/281

---

1. *Mary Anne Coxe*: Mary Anne Coxe (née Cumming, 1811–*c.* 1875), a daughter of Lesley Baillie and Robert Cumming of Logie, Co Moray. Her mother was the subject of Robert Burns's 'Bonnie Lesley', as noted in the novelist Charles Gibbon's dedication of *In Honour Bound*, vol. 1 (London: Richard Bentley and Son, 1874) to Lady Coxe. In 1841 she married the Scottish physician James Coxe (1811–78), and became Lady Coxe after he was knighted in 1863. Tyndall claimed to have toasted her on the summit of Mont Blanc in 1857. Coxe sought advice from Tyndall about the future educational and employment opportunities for her nephew Alexander (Alick) Cumming.
2. *[30 November 1862]*: the date is written on the letter, but in a different hand than the author's. This letter clearly prompts letter 1947 from Mary Anne Coxe. It is likely she received this letter on or around 25 November, and replied immediately. It is possible that she showed the letter to the Duke of Argyll the following Sunday, who wrote on the top right of the first page the following: 'Prof: Tyndall—on my article | on the | "Supernatural" | Nov: 30/62'. Tyndall's mention of his trip to Cambridge where he injured his back also confirms 1862.
3. *the Athenaeum*: see letter 1907, n. 7.
4. *I broke my back at Cambridge & went to the Isle of Wight to recover*: see letter 1933, n. 1.

5. *Alick*: Alexander (Alick/Alix) Cumming (d. 1865), nephew of Mary Anne and James Coxe. Alick was one of Tyndall's students at Queenwood College, who later trained at the East India Company Military Seminary in Addiscombe, Surrey (1855–7). Tyndall took a great interest in Alick's education and was keen for him to continue his studies in Germany. Tyndall referred to him several times in his journal during the 1850s (RI MS JT/2/13c/725, 746, 751, 808).
6. *a big book which I hope to get out early next year*: *Heat as a Mode of Motion.*
7. *the Duke*: George Douglas Campbell (1823–1900), Duke of Argyll, a Scottish politician and scientist. In the peerage of Scotland he was eighth Duke of Argyll (and first in the peerage of the United Kingdom). Campbell served over ten years as Lord Privy Seal and from December 1868 to February 1874 was Secretary of State for India. He was a proponent of educational reform and led the Argyll Commission (1865–8) into the Scottish school system. As an amateur scientist he wrote on ornithology, geology, and evolution, and defended the special creation of humans apart from other species (*ODNB*).
8. *been "pulling me up"—those are the words in the 'Edinburgh.'—Well I have read his article*: [G. Campbell], 'The Supernatural', *Edinburgh Review*, 116 (1862), pp. 378–97. In this anonymous article, Campbell reviewed several books about miracles as well as Charles Darwin's 1862 book on orchids.
9. *a bird with a very white feather*: a phrase used in cockfighting indicating that a bird was of inferior breeding.
10. *the Commissioner*: Mary Anne's husband James Coxe (1811–78), a Scottish physician who studied at Göttingen, Heidelberg, Paris, and received his MD in 1835 from the University of Edinburgh, where he then practiced medicine. In 1841 he married Mary Anne Cumming. In 1855 he was appointed to the Royal Commission that examined the conditions of the lunatic asylums in Scotland and, following the 1857 act that established the General Board of Lunacy for Scotland, served as one of its two paid commissioners. He was knighted in 1863 for his services. ('Sir James Coxe', *Illustrated London News*, 18 May 1878, p. 471.)

To John Lubbock 24 November 1862 1945

The Athenaeum[1] | 24th Nov 1862

My dear Lubbock

I saw your name on our Friday Evening programme a day or two ago and I was glad to see it there;[2] for I know that once pledged to us you will manfully redeem the pledge. Now I hope you will remember that if I can do any thing to help you, or to furnish in any way a little spice for your wholesome fare in the way of experimental illustration you will do me a favour in letting me know. I can guarantee almost any thing in the way of magnified projection on a screen. But you will probably scorn all such factitious aids and throw yourself back on the pure merits of your subject: this is the noblest way; though it is not at all times that the heroic frame of mind necessary to the task is attainable.

Will you give my love to little Amy & Johnny & Conny[3]—I hardly know which of them I like best. I think however Johnny has made the deepest impression on me. I often think of the boy. All three were at my knees on the evening of my arrival at Lamas;[4] we had been romping together, and Amy & Conny had entangled themselves in my legs and arms. But Johnny wanted more play & I caught at intervals the expression of entreaty in his eye—He was silent at intervals, but he never lost that expression—never once allowed his eyes to sink from mine—I was very much struck with him; though I cannot describe the little man's influence. Children however had always a great & I think good influence upon me. Without them I should be a cold stern hardhearted fellow, but they fuse the hardness & banish the chill. I am a great old fool of course for allowing myself thus to gossip, but here I am alone, with one dim lamp making the darkness of this room visible, and no sound save the intermittent ticking of the clock—here I am I say wanting 20 minutes occupation and did I not employ them in harmless twaddle I should probably employ them in something worse—.

—Kind regards to Mrs. Lubbock[5]—

Busk is here glorying in his bachelorhood—goodbye. | always yours | J. Tyndall

RI MS JT/1/T/1038

---

1. *The Athenaeum*: see letter 1907, n. 7.
2. *our Friday Evening programme*: Lubbock delivered a Friday Evening Discourse on 27 February 1863 at the RI 'On the Ancient Swiss Lake-Habitations of Switzerland' (*Roy. Inst. Proc.*, 4 (1862–6), pp. 29–40).
3. *Amy & Johnny & Conny*: Amy Harriet Lubbock (1857–1929), John Birkbeck Lubbock (1858–1929), and Constance Mary Lubbock (1860–92); three of Lubbock's four children at this time (another son was Norman Lubbock (1861–1926). Two more children would come later, Gertrude Lubbock (1863–1934) and Rolfe Arthur Lubbock (1865–1909) (R. Birkbeck, *Notes on the History and Genealogy of the Family of Lubbock* [London: Mitchell and Hughes, 1891], pp. 36–7).
4. *Lamas*: the Lubbocks' house at Chislehurst in southeast London; in 1865 they would take up residence at High Elms (Lubbock's parents' home, very near Charles Darwin's Down House).
5. *Mrs. Lubbock*: Ellen Frances Lubbock. (R. Birkbeck, *Notes on the History and Genealogy of the Family of Lubbock*).

From George Hector Tyndale 23–5 November 1862[1] 1946

Philadelphia Nov[em]b[e]r 23[d] 1862

My dear John

Yours of the 2[d] July last,[2] (before your last Swiss journey)[3] came to me while marching "on to Richmond" under Gen[era]l. Pope,[4] and just before the battle of Cedar Mountain,[5] where I held the right undisturbed. Your last, of the 5[th]. Oct., to Mrs Tyndale[6] came while she was up with me day and night, together with my sisters, caring for a bad wound and sickness, with which I came home from the battle of Antietam[7]—last September. I am yet at home, the wound healed but still weakened by the long lingering fever from privation and exposure, which would have prostrated me soon, even without the wound. The latter was bad,—received while reforming my command after falling back to a new position. It was exactly in the centre of the back head—on the occipital bone—thankful that my head is so hard or for the slight turning of it which glanced the ball along the skull, passing it under the large muscles back and side of neck, passing it between the jugular vein and carotid artery—grazing both and lodging the ball in front of throat: from whence it was cut out, flattened to a broadened hemisphere, a concave-convex form! Curious freak.

I was brought home and wife[8] and sisters[9] became angels to me watching me through all of it to this point—recuperating and strengthening as I now am. My experiences, subjective & objective too have been very beautiful lately—at least since this last battle. I was carried off from the face of the near enemy, under a terrible fire, by a Lieutenant and Sergeant of my own regiment (in the brigade I commanded that day) who thought, brave men as they are, I was dead, as I had fallen lifelessly without motion or sound, as indeed the reception of the wound was entirely unknown to me and still so remains. But these men had the rare beauty of love and the gallant courage of high duty—perhaps higher than that, as a dead body may be left in such a fire of hell as swept that field around all our heads, 'till the hour of my wound, for nearly nine hours from 6½ AM to 3¼ PM. And then the kindness I received, in coming home, from strangers! And afterwards at home, day and night, always wife & sisters with me, caring me back into life, and I earnestly hope usefulness. And then so many friends have come—have sent me fruits and kind things—the best they could emblematize of Heaven.

While dwelling on self I might as well say that on the 17[th] Sept[em]b[e]r, at Antietam, I commanded the first brigade of second Division of Mansfield's Corps, formerly Banks'.[10] All my seniors having succumbed to the privations, exposure, hardships, and battle casualties of our march into and from

Virginia. To be without food often for two days, to march every night, to stand under arms or engaged all day was the work for nearly forty days—to say nothing of cares and responsibilities—against which however I should say nothing, for these or, rather, the sense of these, kept me up, when otherwise I, too, would have yielded.

Thank God for these experiences, thank Him devoutly, for from Him alone they came as does the courage which makes these fields of blood a glory and a grandeur of recollection.

You should have seen these men of my brigade in action. Men whom I had commanded but three (3) weeks—except those of my own regiment of which I am now Lieut[enant]: Colonel & commander, (I was promoted last June) These men lit up like grand fires and went forward into the sharp teeth of Death like—like men. They were what we call old Soldiers, veterans, like myself in the Service for nearly eighteen (18) months, and that all Service & action. We Americans live long in a short time you know. But I acknowledge more than ever the Magnificence of Gods—Humanity—Godlike of Homer—as shown by these men who went quietly, firmly and constantly forward against the human enemies, the rebellious demons of Slavery, the unchained low passions, urged by Hell itself against the Right.

These men, so determinedly brave, so justified and glorified in their cause, did not hate the cruel, almost savages who fought them so terribly and hotly, for they gave the same attention to their wounded as to ours. But they advanced upon them in a fire I can't describe, except by trite words, raining, hailing, storming, burning shot and bullets, and drove them from three (3) positions, until our advanced line had got so far forward as to need strong buttressing on the flanks to fill the long gaps between our right and left and the nearest of our friends. In truth John that first brigade did nobly, advanced beyond all the near by lines of ours, for ours was the right flank and was the turning point for the rebels, who contested stubbornly and fought most dashingly to turn us there, or to stop us in our attempt to turn them there. I can never forget how from the small, painful grumblings of privations and distress, how from the debris of suffering, Men, the real, great, strong men—sons of God—arose, consuming with ashes those petty things, how those men's faces shone to me like flaming swords, gleaming with a higher sunlight, as we went on for God and our Faith.

I know how much of courage, bravery is included in this, but there is something else and I am glad to say that now more than ever I feel the sustaining hand which prayer opens, and it's need.

You will be glad to know that the General commanding my Division sent and then came himself the next day after the battle when I had sufficiently recovered, to thank me for the brigade's skill and courage; and I told him truly, that such men would make any man brave. But I am not at all unfeeling of the compliment and it gratifies me as I think it will you, and Hirst too and perhaps my long sight absent friend Debus—whom you name. In fact there has been praise enough for me. I had but four regiments in the battle, one I had detailed in the rear as guard. These four from numbering originally the full complement of about 4500 men (my own regiment having had 1500 men or 15 Companies an unusual size here) went into the battle only about 1050 strong in all, so cut up had they been in actions, movements, by disease &c. In this last battle that brigade lost of killed and wounded, 375, my own regiment which went in about 650 strong losing 264 men of those, or about forty per cent. And they were quite carefully handled too—I mean, were not exposed to any unnecessary fire. But we were advancing all the time, until two raw regiments, sent as supports & not my own, gave way before the rebel brigade when my whole command fell back for three hundred (300) yards, and this will account for the immense difference between my losses (I use the word to distinguish only that brigade) and the enemies. Nearly as I can judge (very carefully) I think we killed and wounded one thousand (1000) of the rebels—I mean my own brigade did—this is the lowest estimate, it running from that to an impossible number. Whole regiments went down. 69 (Sixty Nine) men were counted (I am told by an officer who counted) in one rebel regiment as dead on the field in the places they had stood, that is 69 fell from one regiment without one escape and all dead in line. Almost incredible. I saw this last line myself while charging and had to jump over it. That regiment (a North or South Carolina one; I do not know the particulars yet) I have been told, gathered only eight (8) men out of some hundreds after the battle. It was horrible, horrible.

On that same charge we drove several lines of the rebels at once. That brigade took Seven (7) regimental flags, took one battery and silenced another in front which I think we could have taken but for the untoward and untimely break of the two regiments spoken of which guarded my right. (There were other Union (Union means Right and Liberty, Education and Self Respect) other Union troops on our right and rear but at least half a mile away and they did not (perhaps could not) get forward, for the fire was awful there.

Nov[em]b[e]r 25th 1862

I was worn out writing the foregoing and now resume to finish, which will be grateful to you for this is very long and, I fear, tiresome. Besides it is almost unreasonable to send such chirography—"a damned cramped piece of penmanship" as Tony Lumpkin said.[11] But then in my present health &c it is doubtful if tremulous nerves can do any better.

In a few words—I left Harper's Ferry,[12] where I commanded four months, last February and entered, in Bailey's[13] Corps, into Virginia and remained ever since, at least until Pope was driven out & have been marching, counter-marching and so on.

The life was hard—full of exposure privation and dangers. Had I been surrounded by altogether congenial friends the life would not have been wholly unpleasant, nor indeed was it so. Like Tennyson's dispirited man the life of action was well for me—else would I "wither in despair."[14] I have found that for a large circulation, the mountain life of active duty is best. The spiritual blood needs your Alpine heights as much as the physical—yes more. Looked at so, what a grand, lovely typification, what a magnificent parable, "Nature" becomes. Looked at thus, all <u>mere</u> skepticism, all superficial causation—doubts are childs-questions of <u>things</u>—but never mind speculation. How are you getting on? How is Hirst, and what do he and you <u>do</u> now? Is Debus well? By the way do you remember my telling Hirst and yourself that before six months from that time English opinion, English sympathies and England's yearnings would be on the rebel's side—actually on the unhuman, slave master's side? Well so it has proved. It is not nor will not be well for England. We expected that greatest of all aids, the moral force of natural friendships, but were disappointed.

England has chosen her side and the future is dark for, either America or England. Old family feuds were dying out and the two open-eyed Nations were drawing together nearer but that has changed and it may be that England's name, as a standing bitter enemy, may close some American Orator's speech, demanding for the world's sake that England be destroyed. This is not a pleasant but it is a general feeling here now. We too are a commercial people, but ours is a commerce based upon the <u>needs</u> of the world, upon <u>all</u> it's needs, and not <u>solely</u>, or nearly so, upon the acquisition of wealth for the sake of a social position. At any rate we feel that human rights have been ridiculed and weakened and that Wrong has been strengthened by a love of wealth, at least by the temporary need of food and raiment. The thing this Nation is now doing has called for all sacrifices of that kind, for the sake of the future World and it is content to have made these.

We expected more from England and have been disappointed. The piratical building of pirate ships and the sly shipment of necessaries to the rebels by Englishmen, done openly as the day, are not nearly the chief source of indignation and hatred, that lies in the unconcealed joy of the English in our straits and troubles, in their want of real understanding of the greatest and most <u>caused</u> war ever known; or, if they understand, as does seem almost certain, their base pleasure in the separation of a State for the reason that it was growing too powerful! In truth if we were not a manufacturing people and a commercial, If we took English wares freely had no protective tariffs or other mere national exigencies, England would side with us and lay great stress upon the holiness of the contest.

This rebellion will be crushed I believe; at any rate, the Northern States. Like the Grecian States too Asia will remember things for centuries, and await European embroilments to show their quiet and bitter hatreds awakened by this measured jealousy and by the mean wants of the present hour.

I very seriously ask excuse from you as a man, English or other, for this unpleasant allusion. But I speak from fullness—and the very air is full here.

The General Mitchell[15] you name is no relation of my brother in law[16] (he had a chair of astronomy in Ohio and was somewhat known as an Astronomer—a graduate of West Point. —that General is since dead.

Mrs Tyndale is the best woman and the worst correspondent I know. I owe my life and health and happiness to her, but I do not owe her many letters! She sends regards to you as to Mr Hirst and says referring to your allusion to our old discussions, that "both were earnest and honestly covered any unpleasantnesses, which, however, were not excessive."

This is a miserably long letter and I can't find any excuse for it, save in Dogberry's full-hearted burst, who could find it in his heart "to bestow all his tedium upon his worships."[17]

John please drink my health with Hirst and Debus and don't forget to include Mrs Tyndale's.

Mrs Mitchell[18] sends her respects to you. My two brothers Sharon & Harold[19] are in the West. Wishing you my dear John all happiness and the possession (firm in hand and soul) of the endless beginning of Nature's truth and God's, I am really affectionately your cousin

Hector Tyndale

RI MS JT/1/T/52
RI MS JT/1/TYP/5/1637–43

---

1. *23–5 November 1862*: Tyndale began writing this letter on 23 November and finished it on 25 November.
2. *Yours of the 2nd July last*: letter 1912.
3. *your last Swiss journey*: this is in reference to Tyndall's trip to Switzerland in the summer of 1862 (he left London on 4 July and returned on 16 August).
4. *Gen[era]l. Pope*: John Pope (1822–92), a Union General in the American Civil War. He is remembered for some success in the Western Theater of the war, but is most remembered for his defeat at the Second Battle of Bull Run (or Battle of Second Manassas) from August 28–30 1862 (J. C. Frederiksen, 'John Pope', in D. S. Heidler and J. T. Heidler (eds), *Encyclopedia of the American Civil War: A Political, Social, and Military History* (New York: W. W. Norton & Company, 2000), pp. 1541–3).

5. *Battle of Cedar Mountain*: an American Civil War battle fought just prior to the Second Battle of Bull Run (see n. 4) on 9 August 1862. General Pope was ordered to attack Gordonsville, Virginia, where Thomas J. 'Stonewall' Jackson (1824–63) was based. It was an eventual Union victory.
6. *Your last, of the 5th Oct., to Mrs Tyndale*: letter missing.
7. *Battle of Antietam*: also known as the Battle of Sharpsburg, an American Civil War battle on 17 September 1862 that was one of bloodiest single-day battles of the entire war, leaving a tallied total of 22,717 dead. The battle halted General Robert E. Lee's (1807–70) northern advance but allowed Confederate forces to retreat into Virginia.
8. *wife*: Julia Tyndale.
9. *sisters*: one of Tyndale's sisters often referred to in correspondence with Tyndall was Elizabeth Mitchell; other sisters were Laura Tyndale (1822–88) and Clara Tyndale (1827/29–1910, who married Isaac Mickle in 1847.
10. *Banks*: Nathaniel Banks (1816–94), an American politician and Civil War general. After the Second Battle of Bull Run, Banks was relieved of command in Virginia and given command of the Army of the Gulf (M. J. Connolly, 'Nathaniel Banks', in D. S. Heidler and J. T. Heidler (eds), *Encyclopedia of the American Civil War: A Political, Social, and Military History* (New York: W. W. Norton & Company, 2000), pp. 174–6).
11. *"a damned cramped piece of penmanship" as Tony Lumpkin said*: a line delivered by Anthony Lumpkin in Irish playwright Oliver Goldsmith's (1728–74) 'She Stoops to Conquer' (1773).
12. *Harper's Ferry*: see letter 1912, n. 4.
13. *Bailey*: Joseph Bailey (1825–67), a Union general in the American Civil War, known for his service in the Army of the Gulf and the Red River Campaign (D. S. Heidler and J. T. Heidler, 'Joseph Bailey', in D. S. Heidler and J. T. Heidler (eds), *Encyclopedia of the American Civil War: A Political, Social, and Military History* (New York: W. W. Norton & Company, 2000), pp. 158–9).
14. *Like Tennyson's dispirited man the life of action was well for me—else would I "wither in despair."*: quoted from Alfred Tennyson's poem 'Locksley Hall' (1835).
15. *General Mitchell*: John Grant Mitchell (1838–94), a lawyer and general in the American Civil War for the Union Army (J. H. Eicher and D. J. Eicher (eds), *Civil War High Commands* (Palo Alto, CA: Stanford University Press, 2001), p. 392).
16. *brother in law*: Ormsby Macknight Mitchel (1809–62), an astronomer and major general in the American Civil War (C. Dee, 'Ormsby Macknight Mitchel', in D. S. Heidler and J. T. Heidler (eds), *Encyclopedia of the American Civil War: A Political, Social, and Military History* (New York: W. W. Norton & Company, 2000), pp. 1341–2; also see P. S. Shoemaker, 'Stellar impact: Ormsby MacKnight Mitchel and astronomy in antebellum America' (PhD diss, University of Wisconsin at Madison, 1991)).
17. *in Dogberry's full-hearted burst, who could find it in his heart "to bestow all his tedium upon his worships"*: quoted from Shakespeare's *Much Ado About Nothing* III.5.22.
18. *Mrs Mitchell*: Elizabeth Mitchell.
19. *Sharon & Harold*: two of Hector's brothers. Sharon Tyndale (1816–71), a civil engineer and surveyor who was Secretary of State for Illinois (1865–69) ('Biographical Sketch of

Sharon Tyndale', in N. Bateman, P. Selby and B. C. Allensworth (eds), *Historical Encyclopedia of Illinois and History of Tazewell County* (Chicago: Munsell Publishing Company, 1905), vol. 2, p. 532). Harold Tyndale (1826–92), who married Ann Cartlidge in 1847, continued with his father's china and glass business in Philadelphia. The third brother was Troilus Tyndale (1817–38).

From Mary Anne Coxe[1] 25 November [1862][2] 1947

Who[3] will deny the existence of natural affinity when I tell you, I was in the very act of putting up for you a copy of the boys "Poems" when your welcome letter[4] was put into my hands.

Yet such was the fact—First *[about]* these said "Poems"—you would *[yea]*, die of laughing if you only saw the pride and pleasure the authors take in them! They are both under the hallucination that they are born—poets—and that of the very first water; nothing would serve them but to get them printed for private circulation among their friends (with perhaps the latent hope that there might be a call from the public so vociferous that they would be compelled to yield to the pressure from without.)

Meanwhile they have despatched a copy to—Now just guess—No other than the great Alfred's self!!![5]

They had 100 copies thrown off—which cost them £2–7- The printer told them that another 50 would cost very little—as the types are meanwhile suspended!

They are perpetually sending off "tales" to different Magazines which if rejected by one, are straightways despatched to another—and as they are sometimes accepted, (5/. a page has been given by the "Boys Own Magazine" for "Outdone by a Yankee") On the strength of which they are now busy with "The Man Eater" and the "Confessions of a Murderer".

They are really fine boys—and take being laughed at in the most perfect good humour.

I am grieved to hear of your broken back,[6] but assuredly it did not affect your head, for Mr Heathfield[7] told me that you came out with all your wonted vigour, and that your lecture[8] was as admirable in manner as in matter. Neither were the "Aqueous (cant spell it) vapours" forgotten—!

The Argyles[9] were with us the beginning of this month for a couple of days—and were as ever most charming, from their easy simple happy ways. How wonderfully youthful they are in appearance—and how unlike the parents of 11. children! I wonder when—if ever—you really mean to come to Kinellan.

There is no denying the fact that you are a gay Deceiver—and no mistake and do lead one on from year to year "with promise false and vain".

The "Struck Eagle" you <u>may</u> be—but the "white-feathers"[10] you will never show.

The Spanish proverb will never apply to you—"Who loses money loses much—Who loses friends loses more—Who loses heart loses everything".

Interrupted.

Very many thanks for your thinking of me—You can think of no one who appreciates the compliment more.

Yours always and | very much | M. A. C.

I would fain have your Carte,[11] my book[12] is incomplete without it.

RI MS JT/1/TYP/1/1/270
LT Typescript Only

---

1. *Mary Anne Coxe*: see letter 1944, n. 1.
2. *[1862]*: the year is given as this letter is a reply to 1944.
3. *Who*: the letter begins without a salutation and thus there may be one or more pages missing.
4. *your welcome letter*: letter 1944.
5. *great Alfred's self!!!*: Alfred Tennyson.
6. *your broken back*: see letters 1933, n. 1.
7. *Mr Heathfield*: not identified.
8. *your lecture*: see letter 1932, n. 6.
9. *The Argyles*: George Douglas Campbell, Duke of Argyll (see letter 1944, n. 7), his wife Elizabeth Georgiana Campbell (née Leveson-Gower, 1824–78), Duchess of Argyll and an abolitionist, and their five sons and five daughters (a sixth daughter would be born in 1864) .
10. *The "Struck Eagle" you <u>may</u> be—but the "white-feathers"*: see letter 1944, n. 9.
11. *your Carte*: Cartes de Visite, a small photographic portrait mounted on a card, 3½ by 2¼ inches (*OED*). Perhaps Tyndall's Carte de Visite referred to here is the frontispiece of this volume.
12. *my book*: it was fashionable in this period to collect Cartes de Visite of notable persons in a scrap book.

To [Bernard Henry] Becker[1] [26 November 1862][2] 1948

Royal Institution of Great Britain | <u>Wednesday</u>

My dear Becker,

It gives me pleasure to learn that you are once more on English soil: you ought never to have quitted it; you have here the best air, the best beef, the best mutton which the world can produce, and these are the substratum of all manly achievements & of all happiness attainable by man. Why then should you quit this land to bury yourself in that sandy waste on which frowns the shadow of the Kreuzberg[3] and according to Dove,[4] "Der Staub bisweilen zum Meteor wird."[5]

Well I am happy to find you here again. I send you the required information.[6] My first Friday evening[7] will be something about the atmosphere after Easter I purpose giving 7 Lectures on sound[8]

Yours ever | John Tyndall

Don't be angry at my discriminations of Germany—Old Marburg[9] is still dear to me & so are many other memories of 'der Vaterland'[10]

RI MS JT/1/T/1401

---

1. *[Bernard Henry] Becker*: Bernard Henry Becker (1833–1900), a journalist and science writer, most famous for his publication of *Scientific London* (1874), in which he detailed many of the famous meetings of Britain's scientific academies and their most noted members, such as John Tyndall (*Modern English Biography*).
2. *[26 November 1862]*: the date '26/11 62', indeed a Wednesday, is penciled in on the original letter, presumably by LT.
3. *that sandy waste on which frowns the shadow of the Kreuzberg*: presumably Berlin.
4. *Dove*: Heinrich Wilhelm Dove (1803–79), a Prussian meteorologist and experimental physicist. He was educated at the University of Berlin and became an ordinary professor at Berlin's Friedrich-Wilhelms-Universität in 1845. In his research, he focused on meteorology and was awarded the RS's Copley Medal in 1853 for his contributions to understanding the distribution of heat in the atmosphere (*CDSB*).
5. *Der Staub bisweilen zum Meteor wird* : a speck of dust sometimes becomes a meteor (German).
6. *required information*: presumably Tyndall sent an enclosure detailing his RI lectures.
7. *My first Friday evening*: Tyndall delivered a Friday Evening Discourse, 'On Radiation through the Earth's Atmosphere', at the RI on 23 January 1863 (*Roy. Inst. Proc.*, 4 (1862–6), pp. 4–8). This was also published as J. Tyndall, 'Radiation through the Earth's Atmosphere', *Phil. Mag.*, 25 (1863), pp. 200–6.
8. *Lectures on sound*: Tyndall delivered seven lectures on the topic of sound at the RI in 1863, on Tuesdays at 3pm from 28 April through 9 June (RI Guard Book, vol. 3, p. 26).
9. *Marburg*: see letter 1940, n. 2.
10. *'der Vaterland'*: the Fatherland (German).

To William Sharpey[1] [November 1862][2] 1949

Friday

My dear Sharpey,

Kirchhof[3] has written to me[4] expressing the great pleasure which the receipt of the medal[5] has caused him. He regrets that on account of his lectures he will be prevented from attending at the meeting on the 30th[6] and he requests me to receive the medal for him and to return his acknowledgments for the great honour which the Society[7] has conferred upon him.

He also has asked me whether it is usual to return a formal letter of thanks to the Council. In fact he wishes to do what is right & simply needs instruction.

I of course am willing to comply with his request though perhaps Stokes or Grove would be the fitter person. Will you tell me how I am act under the circumstances?

I, of course, should have been his proposer had I been present soon enough on the occasion when his name was brought forward—In fact it was at your own suggestion that I withdrew his name as a candidate for the Copley medal[8] last year, in order to rescue him for the Rumford.[9]

Yours very truly | J. Tyndall

Wellcome, MS. 7777/75

---

1. *William Sharpey*: William Sharpey (1802–80), a Scottish anatomist, physiologist, and secretary of the RS (1854–72). From 1836 until his retirement from the post in 1874, Sharpey was chair in anatomy and physiology at the University College, London, where he taught the first course in systematic physiology in England. Sharpey was an influential figure in British science, introducing the work of other scholars to the Society, and refereeing papers for the *Phil. Trans.*, but also through his teaching (notable students include John Marshall and Joseph Lister). He also established a lectureship in practical physiology at University College in 1855, a position that was held, in turn, by J. B. Hayes, George Harley, Michael Forster, Edward Schafer, and John Burden Sanderson. Through these students and colleagues Sharpey was responsible for the re-establishment of physiology across Britain, as they took up chairs and lectureships not only in London, but also in Oxford, Cambridge, and Edinburgh.
2. *[November 1862]*: the date is given since the RS announced its awards at the first Council meeting in November each year, with the awards being presented at the Anniversary Meeting on 30 November. Kirchhoff received the Rumford Medal in 1862 (see letter 1943, n. 1).

3. *Kirchhof*: Gustav Robert Kirchhoff.
4. *has written to me*: letter 1943.
5. *receipt of the medal*: see letter 1943, n. 1.
6. *the meeting on the 30th*: the RS awards were announced at the first Council meeting in November and medals presented at the Anniversary Meeting on 30 November.
7. *the Society*: the RS.
8. *the Copley medal*: the Copley Medal, considered the most prestigious award of the RS—for the most important scientific discovery or for the greatest contribution made by experiment—was awarded in 1861 to the naturalist Louis Agassiz for his researches in paleontology and other branches of science, and particularly for his great works the *Poissons Fossiles*, and his *Poissons du Vieux Gres Rouge d'Ecosse* ('Anniversary Meeting', *Roy. Soc. Proc.*, 11 (1862), pp. 454–72, on pp. 461–4).
9. *the Rumford*: see n. 2.

To John Couch Adams 4 December 1862 1950

4th Dec. 1864

My Dear Adams,

I have a little crib which is scarcely ever occupied and which though far too homely for a philosopher like you will I am persuaded hold you comfortably. I shall guarantee that you shall be fearfully warm, and fairly out of the range of cabs & omnibus growlings. Remember the crib is very modest but sufficient to afford a human being all necessary rest. You shall be moreover as welcome as the flowers of May. So remembering the proverb "better a dinner of herbs where love is than a stalled ox with hatred therewith."[1] I hope you will make the crib your home whenever you come to London. Do not moreover be afraid that I shall feed you herbs. I long ago learned that they were awfully indigestible & that mutton chops are far to be preferred. On this knowledge I shall act in your case.

I live at 14 Waverly Place St. Johns Wood, within a minute's walk of the Eyre Acres.[2] The Atlas from Charing X and the City Atlas[3] from the city journey thitherward every four minutes. If you rush on at any hour of the day and demand food or drink from my housekeeper Mrs. Tunstall she will promptly place it before you.

Yours ever | John Tyndall

Cornwall DD.AM 580, Cornwall Record Office, Truro

---

1. *"better a dinner of herbs where love is than a stalled ox with hatred therewith"*: Proverbs 15:17.

2. *Eyre Acres*: the estate of Walpole Eyre (1773–1856), at this time leased as a Barracks for various detachments of Foot Guards.
3. *The Atlas from Charing X and the City Atlas*: Omnibus lines in London.

To Sarah Tyndall 16 December 1862 1951

My dear mother

I have not yet written to Dr. Tyndall[1] as I have been extremely busy of late, but I shall write to him when I have a little time to spare. I do not know that you are quite right in affirming that the Tyndall's have degenerated even supposing them to have sprung from the stock of the Martyr.[2] No martyr could endure more than some of them are now enduring, and no martyr could endure it more bravely. I had a letter[3] from Hector Tyndale[4] a few days ago such as old Martyr Tyndale might and would have written had he been a soldier. Hector has been through those late terrible battles.[5] He is Colonel of a Federal regiment, and all his seniors having fallen off from disease and death he commanded a brigade at the last great battle. His division suffered fearfully and he was carried apparently dead from the field by a Lieutenant and Sergeant of his own corps. He was hit by a bullet in the back of the head; it glanced along his neck, cutting up the muscles and finally lodged in front of his throat. Thence it was cut a flattened mass. His letter is like a battle itself, it is so grand and fiery. So you must not call all the Tyndall's degenerate. I believe that they belong to the Martyr's stock, for regardless of everything else I see the same stuff in the good specimens as that which made the martyr.

A merry Christmas to you and a happy new year.

Your affectionate son | John

RI MS JT/5/16a (unpaginated folder)
LT Typescript Only

---

1. *Dr Tyndall*: probably John Tyndall (of Gorey).
2. *stock of the Martyr*: a reference to being of the lineage of William Tyndale.
3. *letter*: letter 1946.
4. *Hector Tyndale*: George Hector Tyndale.
5. *late terrible battles*: battles of the American Civil War (1861–5).

From Heinrich Debus 16 December 1862 1952

This has just come as I was about to post my letter.[1]
Dec[em]b[e]r 16th. 1862

My dear John,

Many thanks for Hectors[2] letter,[3] which I have been reading with the greatest interest. I admire his great courage, perseverance under great sufferings and privations, as well as his great intentions, yet after all one cannot help wishing he had got out of this unfortunate war business. I have often thought about the causes of the displeasure harboured by Americans against England. Whatever these causes may be, certainly the two Hector mentions in his letter, cannot be the true one's. I have been all the time in England, whilst this unfortunate war has been going on, and I never heard any body express, or saw printed in any respectable newspaper, anything which could be taken as satisfaction at the reverses of the North. And as for powder, guns and ships, well, the North must expect this. All commercial communities contain plenty of members who will sell anything to anybody, even to the devil himself, if he can pay. Nay, the Americans themselves sold to the Russians during the Crimean War,[4] anything the Russians could find money for.

I wish your cousin lived in England. He is a fine specimen of a man and would do honor anywhere. One point is your cousins letter I cannot press quite without notice. His argument is: slavery is an inhumane institution, hence the N.S. in putting down, or attempting to put down slavery, by making war on the South, are doing a piece of holy work. —If it did not involve more writing then I can do this morning, it could easily be shown on religious as well as political grounds, that the best policy of the United States would be—to let the South quietly go.

Ever yours | H Debus

Read with care the Examination paper[5] I send you.

RI MS JT/1/D/30

1. *This has just come as I was about to post my letter*: this is Tyndall's hand, as he forwarded this letter from Debus as an enclosure with his own letter to Hector Tyndale (letter 1953).
2. *Hector's*: George Hector Tyndale.
3. *letter*: letter 1946.
4. *Crimean War*: the Russian Empire lost the conflict with an alliance of Britain, France, the Ottoman Empire, and Sardinia (1853–6).
5. *Examination paper*: see letter 1926, n. 3.

To George Hector Tyndale 17 December 1862 1953[1]

London Dec. 17th/1862

My dear Hector.

It would be a poor expression of my feelings to say that I was gratified and interested by the perusal of your letter.[2] I had long thought of you, and my thoughts regarding you were sometimes dark in the extreme, knowing as I did the fearful chances to which you were exposed. But your letter crossed this darkness like a sunbeam and made my spirit leap for joy. Even the recognition of your handwriting did this. Still after having seen it I did not expect the intelligence which your letter conveyed, and it is a long time since I read anything which affected me more powerfully. However we may differ on minor points the constitutional affinity between you and me is very strong, and you know I am not insensible to any act in which the greatness of the human heart comes into play: I walked with your brigade into that fearful fight, I heard the roar of cannon and the whistle of the shot, and I think had I been there in the body I should have struck my stroke as well as another. Your account is thrilling;—somewhat wild, but it is grand in its wildness, and I love to see the fiery blood of old Vilvordan[3] rushing in this fashion. If those two brave men who bore you from the field should ever come to England pray send them to me: the best a poor man can offer shall be at their service. I never doubted that you would fight if called upon to do so, but I often inarticulately prayed that you might never have the chance. The gods declined my prayer, but answered it in their own fashion. For surely the next best thing to answering it by granting it, is the answer actually vouchsafed. It is a great thing—a thing to be remembered for generations—to have come so near to death under such grand conditions and still not to have crossed the bourne. This day I repeated your account of the behaviour of your men to a member of the Athenaeum[4]—a son of old Dr. Prout,[5] the Bridgewater Treatises[6] man—His eye glistened, for he is a great Northerner: and he exclaimed "such men will lick the South to bits!"—<u>I do not agree with him.</u> Still I never entertained a doubt that there are thousands in the North who pursue this way on the same high grounds that animate yourself, but are they majority?

I cannot describe to you the joy of poor Hirst when I put your note into his hands and told him of your escape. He will soon be free for a few weeks from his present entanglements and will then write to you. We complied with your request and drank in full bumper the health of yourself and your wife. I can well believe that she tended you like an angel. Her depth of feeling—her love for you—were not hidden from me under her quiet demeanour in London. But hang it, you never gave <u>me</u> time to be quiet, or to show any sympathy with her gentleness. You were continually firing hot shot into me so that when near you I was compelled to encase my heart in sheet iron. Still old fellow much as I opposed you, much as I should oppose you tomorrow on many points, I cannot shut my eyes to that grandeur of soul which enables you to appreciate the noble behaviour of your men. I am going to pray that I might not have to meet such men when the Americans come to "destroy" England. I say let them come,—They will find a home-Stroke in the arm of England yet. And I trust when they 'come to grief', as they are sure to do if they come to England, that English hearts will be found to imitate their noble example & to treat them as vanquished heroes. For the present unhappy situation of England & America,[7] the American press is primarily to blame. This country has no sympathy with slavery, but you have goaded us to madness by your bullying and vituperation. You think we wanted to quarrel about the Trent[8]—To us the case was one of black necessity—99 out of every 100 Englishmen believed at the time that America wanted to fasten a quarrel upon us. You know Faraday,—he is as gentle as a child, as tender as a woman. On the occasion of the Trent affair he came down to me in the laboratory and said "Well Tyndall, they are determined to quarrel with us. Let them have their wish, let us go at it." And your cousin said the same; for with every one of us it was the profound and sad conviction that you wished to fasten a quarrel upon us. No doubt you hate us as you have so often & so indignantly proclaimed and that blood will be the end of this feeling is only too probable. This perhaps is the only way in which England can be taught her weakness, or America be made to feel that this land of "old and just renown"[9] is not to pass like a scroll at the demand of a Yankee orator. My dear Hector it is the earnest wish of our hearts to live peaceably with you, but we are not in the least afraid of you. Remember that & let us stand in each other's presence as men who mutually respect each other, but who feel that fear is felt on neither side. You know my friend Huxley—he has the reputation of a fighting man, and he appears to have worsted Owen[10] in many a conflict—Huxley & I have always been in harmony & one element of this harmony is that each knows that the other is not afraid of him. Our relations therefore are those of free men, and they are likely to be permanent. So it is with nations. Let America feel that England is not afraid of her,—Let England feel that America is not afraid of her and all will be right. I would be shot nine times over sooner than submit to your dictation, and I think you would be shot <u>ten</u> times over before you would submit to mine. This is as it ought to be, and let <u>both</u> of us recognize the fact.

But hang such differences—my main object is to express the relief from anxiety which your letter has brought me. I shall cease to look upon your portrait with the gloomy enquiry "Is he alive or dead." Debus has your letter at present & I doubt not he and others will read it with an interest almost equal to my own. But I do not think they can feel the pride that I feel in finding a man of my blood acquitting himself so honourably. Good bye dear Hector—give my kindest regards to your wife[11] and sister,[12] and believe me ever

Your affectionate cousin | John Tyndall

This has just come as I was about to post my letter.[13]

RI MS JT/1/T/1446
RI MS JT/1/TYP/1644–8

---

1. Tyndall enclosed letter 1952 with this letter, which he had just received from Heinrich Debus.
2. *your letter*: letter 1946.
3. *old Vilvordan*: probably William Tyndale.
4. *Athenaeum*: see letter 1907, n. 7.
5. *D^r^. Prout*: William Prout (1785–1850), an English chemist, physician, and natural theologian. Prout was author of one of the *Bridgewater Treatises*, worked on atomic weights, and developed the classification of foodstuffs into protein, carbohydrates, and fats (*ODNB*). He had six children; the identity of the son mentioned here remains unclear.
6. *Bridgewater Treatises*: in order to support the scientific ideas of the time, which explored the natural world within Paley's framework of a divine designer, Francis Henry Egerton (1756–1829), 8th Earl of Bridgewater, a gentleman naturalist, commissioned eight Bridgewater Treatises upon his deathbed to explore 'the Power, Wisdom, and Goodness of God, as manifested in the Creation' (*ODNB*). They were first published between 1833–40. Prout's was the eighth in the series, *Chemistry, Meteorology, and the Function of Digestion, considered with reference to Natural Theology* (1834).
7. *the present unhappy situation of England & America*: Britain remained officially neutral during the American Civil War (1861–5). However, public opinions varied with elites favoring and sympathizing with the Confederacy. In 1862, Britain ceased trade with the Confederacy which officially put a great strain on the British economy.
8. *about the Trent*: often referred to as the 'Trent Affair' in which the Union Navy captured, against treaty rules, two Confederate diplomats from British ships. The affair caused a great public outcry between the two sides and Britain.
9. *"old and just renown"*: a misquote of line 10 from Alfred Tennyson's poem 'You Ask Me, Tho Ill at Ease' (1833). The correct line is 'A land of just and old renown' (A. Tennyson, *Poems*, 2 vols (London: Edward Moxon, 1842), vol. 1, p. 219–20).
10. *Owen*: Richard Owen.
11. *wife*: Julia Tyndale.
12. *sister*: Elizabeth Mitchell.
13. *This has just come as I was about to post my letter*: see n. 1.

To Juliet Pollock [29 December 1862][1] 1954

Monday morning

My dear Mrs Pollock,

I quitted London on Friday week, having picked up my friend Huxley on my way to the railway station. As I drove northward it was part of my arrangement to pass through Montagu Square[2] and call upon you—but the time was too short. We reached Sandown[3] that night—walking from Ryde,[4] and next day we visited Whitecliff Bay[5]—an extremely instructive part of the Island. We returned to Sandown, and walked thence to Ventnor;[6] from which place we took a carriage to Black Gang Chine.[7] Thence on Sunday morning we started for Brook,[8] and were fortunate enough to find the fossil forest which there cambers the beach. It is extremely remarkable. Trees 2 and 3 feet in diameter, some perfectly detached, others embedded in the rock, and all silicified—turned to complete stone—are strewn in numbers over the sand. Thence we went on to Plumley's Hotel Fresh-water-Gate,[9] and the whole of the next day we spent about the Needles,[10] Alum Bay,[11] and the wave-forsaken cliffs (for the tide was out) about Freshwater. I thought many times of calling upon Tennyson,[12] and proposed two or three times to the party to pay our respects en masse; but we did not like to bother him. On the Tuesday we walked to Newport[13] and that evening reached London.

I took it for granted that you were at St. Julians,[14] otherwise I should assuredly have called upon you on Wednesday. On the afternoon of that day I went down to my friend Lubbock[15]—who has a house near Chiselhurst,[16] and with him I spent a most genial and happy xmas. I wish very much his wife[17] could know you—she is an extremely gifted person, I am quite persuaded that you would love each other.

I am now hard at work at my book:[18] it is tremendous uphill work, but having put the hand to the plough, it is not the want of either you or me to 'look back'. It must be done, and when done let us trust that it will be worth the labour expended on it.

The Juveniles[19] commenced yesterday: Frankland lectured very nicely, but he is not so Flexible, nor so much of a boy as Mr Faraday.

You know that I wish you a happy New Year—why then should utter what you know so well. Kiss Moss[20] for me, and give my love to Fred,[21] Walter[22] and to your husband[23]

Good bye | ever yours | John Tyndall

RI MS JT/1/TYP/6/2007
LT Typescript Only

1. *[28 or 29 December 1862]*: the date is given based on Tyndall spending Christmas 1862 at the Lubbocks' and the reference to the first of Edward Frankland's juvenile Christmas Lectures at the RI, which occurred on Saturday, 27 December 1862. Tyndall wrote in this letter that Frankland's lectures 'commenced yesterday', which implies that Tyndall wrote this on Sunday, 28 December, although he wrote 'Monday morning' (29 December) at the top. Tyndall also had left for the Isle of Wight on Friday, 19 December 1862 with Huxley, Busk, and Lubbock, as Hirst noted in his journal entry for 21 December 1862 (Brock and Macleod, *Hirst Journals*, p. 1627).
2. *Montagu Square*: a square in Marylebone, London, where the Pollocks had their primary residence at no. 59.
3. *Sandown*: a small seaside town on the southeast coast of the Isle of Wight.
4. *Ryde*: a small seaside town on the northeast coast of the Isle of Wight, six miles from Sandown.
5. *Whitecliff Bay*: a sandy bay on the easternmost point of the Isle of Wight.
6. *Ventnor*: a small seaside town on the south coast of the Isle of Wight, six miles from Sandown.
7. *Black Gang Chine*: now Blackgang Chine, an amusement park and holiday resort on the south coast of the Isle of Wight.
8. *Brook*: a small seaside town on the southwest coast of the Isle of Wight, 30 miles from Blackgang Chine.
9. *Plumley's Hotel Fresh-Water-Gate*: Freshwater is a large village on the western coast of the Isle of Wight, surrounded by Freshwater Bay, which used to be called Freshwater Gate. Plumey's Hotel was one of the many hotels at the resort.
10. *Needles*: the Needles is a row of distinctive chalk stacks that rise 30 meters out of the sea off the west coast of the Isle of Wight, near Alum Bay.
11. *Alum Bay*: westernmost point of the Isle of Wight.
12. *Tennyson*: Alfred Tennyson lived at Farringford, near Freshwater, from 1853.
13. *Newport*: a small town in the centre of the Isle of Wight.
14. *St. Julians*: the Herries family home in the town of Sevenoaks in Kent, England, southeast of London; the manor was built by John Herries between 1836–7.
15. *Lubbock*: John Lubbock.
16. *Chiselhurst*: their house Lamas.
17. *wife*: Ellen Frances Lubbock.
18. *my book*: *Heat as a Mode of Motion.*
19. *The Juveniles*: Edward Frankland gave the RI's 1862 juvenile Christmas Lectures on 'Air and Water' on 27 and 30 December 1862 and 1, 3, 6, and 8 January 1863 ('Royal Institution of Great Britain', *The Examiner: A Weekly Paper on Politics, Literature, Music and the Fine Arts,* 20 December 1862, p. 814). This lecture series for children began in 1827 and still continues. Tyndall himself delivered the Christmas Lectures a dozen times over his time with the RI, his first in 1861 (F. A. J. L. James (ed.), *Christmas at the Royal Institution: An Anthology of Lectures* (Hackensack, NJ: World Scientific, 2007)).

20. *Moss*: Maurice Pollock.
21. *Fred*: Frederick Pollock.
22. *Walter*: Walter Pollock.
23. *your husband*: William Frederick Pollock.

To [Josiah Parsons Cooke][1] 30 December 1862 1955

Royal Institution of Great Britain | 30th. Dec. 1862

My dear Mr. Cooke

I prize your gift[2] & the warmheartedness with which you offer it to me more than I can well express. Thank you a thousand times; Long may you live to delight your friends by the works of your hands, and the perennial freshness of feeling which encompasses your heart.

always yours most truly | John Tyndall

Wellcome, MS. 7777/04

---

1. *[Josiah Parsons Cooke]*: probably Josiah Parsons Cooke (1827–94), an American scientist who worked at Harvard University and was instrumental in the measurement of atomic weights (*CDSB*). His 1854 paper on atomic weights foreshadowed the periodic law developed later by Mendeleev and others. America's first Nobel laureate in chemistry Theodore Richards (1868–1928) was inspired by Cooke to pursue similar research.
2. *I prize your gift*: not identified, but possibly Cooke's recently published book, *Elements of Chemical Physics* (Boston: Little, Brown & Co., 1860).

To George Hector Tyndale 30 December 1862 1956

30th December 1862

Dear Hector,

I write in part to wish you a "happy new year" and in part to tell you that I have had some correspondence lately in which you will feel some interest. I was informed that a physician of our name and family resided at Gorey Co. Wexford, and I wrote a short note to him,[1] asking him for whatever information he could give me regarding the Irish Tyndalls, he wrote to me promptly and informed me that all he could learn regarding the emigration of the Tyndalls into Ireland was that Six of them accompanied Cromwell,[2] and after the war settled down in the Co. Wicklow, between Arklow and Wicklow,[3]—whence they branched off—some to Dublin, some to Co. Carlow, and some to the west. He had a brother practising as an M.D. in Montreal,[4] who moved up from that place to another called Mona Mills.[5] He tells me he has three sons,[6] one of whom has passed all his examinations as a surgeon, and the two others he intends for mercantile pursuits. He sent me a curious old printed document, quite brown with smoke, printed 40 years ago.

It was headed

> "Genealogy of William Tyndall[7]
> born 29 of September 1760"

But perhaps you would like to see the actual thing so here it is.

> "Gardiner Tindall, born 6th August 1789
> David Tindall, born 22nd September 1791
> Jane Tindall, born 14th December 1793
> She died in Cork in the year of the Rebellion[8] 1798
> William Tindall & Elizabeth Marks, married 16th August 1802.
> John Tindall born 24th June 1803
> Jane Tindall born 30th March 1805
> John Tindall, died 7th June 1805
> Jane Tindall, died 1st August 1805
> William Tindall born 9th June 1806
> Anne Tindall born 4th January 1808
> Anne Tindall died 4th February 1808
> Gavin Tindall born 16th September 1809
> Elizabeth Tindall, born 21st April 1811
> Gavin Tindall died 8th March 1813
> Hanna Tindall, born 10th April 1813
> M. Tindall born 22nd March 1815
> Died and was buried same day.
> Catherine Tindall born 9th December 1818
> Elizabeth Tindall died 10th March 1819
> I. Tindall born 14th May 1821
> died and buried same day.
> William Tindall grandson born 3rd Feb. 1817
> Thomas Tindall born 1st April 1819.
> Jane Tindall granddaughter, born 25th July 1819
> Amelia Tindall born 11th December 1820.

On the back of the document is written these words:—I copy the bad spelling.

> "This is the Father of Gardiner Tindall, and the Reason of him being caled Gardiner was after his grandmother, as her Maden name was Gardiner and First Cosin to Lord Mountjoy[9] that was killed in the Batell of Ross[10] in the year 1798. which laves the present Errell Mountjoy[11] and Gardiner Tindall second cosins that *[1 word excised]*s family John Tindall"

The name of my correspondent is also John Tyndall, I have written to him for further information regarding this William Tindall.[12] He also mentions an Alderman Tyndall of Dublin as belonging to our family.

And now dear Hector goodbye—I hope you are getting strong again, and that you are trying to get strong and not careless of yourself. It made my flesh creep to read this last news from America—that brave men should be thus swept down is fearful. Kindest regards to your wife[13] & sister.[14]

Your affectionate Cousin | John.

RI MS JT/1/TYP/1649–50
LT Typescript Only

---

1. *I wrote a short note to him*: John Tyndall (of Gorey); letter missing.
2. *Cromwell*: Oliver Cromwell (1599–1658), an English military and political leader who led the army of the Parliament against King Charles I in the English Civil War (1642–51). He was the key signatory of the order for the execution of the King and was subsequently installed as 'Lord Protector' (1653–58). In 1649 he led an invasion of Ireland, overseeing the massacre that followed the Siege of Drogheda and the brutal suppression of the residents of the town that followed the Siege of Wexford (*ODNB*).
3. *war settled down in the Co. Wicklow, between Arklow and Wicklow*: the English Civil War (1642–51); County Wicklow is in eastern Ireland, just south of County Dublin.
4. *brother practising as an M.D. in Montreal*: not identified.
5. *Mona Mills*: a village in Canada West (now the province of Ontario).
6. *three sons*: not identified.
7. *William Tyndall*: not identified.
8. *Rebellion*: the Irish Rebellion of 1798, when the United Irishmen rose up against British rule, from 24 May to 12 October.
9. *Lord Mountjoy*: Luke Gardiner (1745–98), 1st Viscount Mountjoy, an English politician and landowner (*ODNB*).
10. *Batell of Ross*: the Battle of New Ross occurred in County Wexford in southeastern Ireland during the Irish Rebellion of 1798 and was fought between the United Irishmen and the British. The attack on the town of New Ross was an attempt by the Irishmen to spread their rebellion into County Kilkenny and the province of Munster.
11. *Errell Mountjoy*: not identified.
12. *William Tindall*: not identified.
13. *wife*: Julia Tyndale.
14. *sister*: Elizabeth Mitchell.

1863

To Mary Helen Dawes 1 January 1863 1957

1st January, 1863.

My dear Mrs Dawes,

Had I been less beset by work which fills every cranny of my brain, I should have written to you long ago. Indeed, I had commenced a letter to you some time ago but was deflected from it by a sudden call and I have been under pressure ever since. I cannot however allow this day to pass without wishing you and the Dean[1]

A Happy New Year, | and many of them. | Yours ever, | John Tyndall.

RI MS JT/1/TYP/1/324
LT Typescript Only

1. *the Dean*: Richard Dawes.

To John Frederick William Herschel January 4 [1863][1] 1958

The Athenaeum | Royal Institution | 4th. Jan.

Dear Sir John,

At the time that you rendered me your debtor by sending me a mass of rock-salt your gift was accompanied by a note[2] containing a short history of the piece of salt, and directing my attention to the heat spots of the spectrum which you discovered in 1841.[3]

I have not yet had time to touch the spectrum, and therefore have nothing to communicate regarding these extremely remarkable spots. I shall attack the subject in due time. Meanwhile, as I am writing a book on Heat,[4] and have had occasion to refer to your piece of salt; it occurs to me that the printing of the few lines in your note would be a proper way of enabling you to take date should anything important arise out of the examination of the extra-red end of the spectrum.

Have you any objection to my printing the necessary extract?

Some time ago D^r. Bence Jones, our honorary Secretary, made me acquainted with a note from your son[5] which encouraged the hope that we might hear him on some Friday evening at the R. I.:[6]—Would you kindly say to him that any assistance I could render him in the way of illustration should be rendered with the utmost pleasure.

Assuming that the action of aqueous vapour upon radiant heat is established beyond doubt—(which it is)[7] we have in the vapour a body which absorbs the same rays as the water which produces it—in other words it shares the colour of water. Through the operation of this cause our atmosphere is certainly a blue medium. The quantity of vapour might not be sufficient to produce a sensible blue, if the rays went straight through it; but the reflections within the body of the atmosphere must be innumerable, and thus its virtual depth increased. I wonder is this likely to throw any light on the blue of the sky![8] Could you get your angle of polarization of 45° by reflection from air or vapour?[9]

Yours very sincerely | John Tyndall.

RS HS/392a

1. *[1863]*: the year is given by reference to Tyndall working on a book about heat (published in 1863), and that this letter is from the January following an 1862 letter from Herschel.
2. *a note*: see letters 1878 and 1880 (Tyndall's reply to Herschel), *Tyndall Correspondence*, vol. 7.
3. *you discovered in 1841*: this refers to Herschel's work on spectroscopy, published as J. Herschel, 'On the Action of the Rays of the Solar Spectrum on Vegetable Colours, and on Some New Photographic Processes', *Phil. Trans.*, 132 (1842), pp. 181–214.
4. *a book on Heat*: LT noted on the letter, 'published in 1863'. This refers to *Heat as a Mode of Motion.*
5. *your son*: Alexander Stewart Herschel.
6. *we might hear him . . . at the R. I.*: Alexander Herschel did indeed give a Friday Evening Discourse at the RI, 'On Luminous Meteors', on 24 April 1863 (*Roy. Inst. Proc.*, 4 (1862–6), pp. 87–97; see letters 1960, 1980, 1981, 1982, and 1983.
7. *(which it is)*: this point was, of course, very much disputed.
8. *it shares the colour of water . . . of the sky!*: Tyndall significantly changed his views on this point; see *Ascent of John Tyndall*, pp. 245–9.
9. *Could you get your angle . . . from air or vapour?*: see Herschel's reply (letter 1960).

From Mary Helen Dawes [5] January [1863][1] 1959

The Deanery, Hereford. | December 6th[2]

Dear Dr. Tyndall

Very many thanks for your kind note[3] and good wishes. We heartily wish you the same—many years of happiness and usefulness, for in that I think your happiness consists. It is always a very great pleasure to hear from you, we only wish we could see a little more of you; that is one of the reasons I regret our living so great a distance from London. When you can come to us,[4] I need not say how happy we shall be to see you. You must recollect also that we now have a house near Malvern, where you may have a run on the Malvern Hills. The drive from Ledbury[5] to Malvern[6] through Eastnor Park[7] is really beautiful, and I should so like to show it to you.

The Dean[8] is flourishing and working for the rising generation as hard as ever. <u>In that consists his happiness</u>. He joins me in kindest regards and may you be blessed with health and strength to carry you through all your hard work is the earnest wish of

Yours very sincerely | Mary Helen Dawes.

RI MS JT/1/TYP/1/324A
LT Typescript Only

---

1. *[5 January 1863]*: although 6 December is given as the date on the typescript LT noted that the envelope was dated 5 January 1863. Given that this appears to be a reply to Tyndall's New Year's greeting to Dawes in letter 1959 (dated 1 January 1863), we have taken the date of the postmark as the more likely.
2. *December 6th*: a more likely date is early 5 January 1863 (see n. 1).
3. *your kind note*: letter 1957.
4. *When you can come to us*: Tyndall had visited the Dawes the previous August (see letters 1902 and 1922), and the Dawes must have been eager to have him visit again. In letters 2106 and 2163 from the following year, Tyndall shared with Richard Dawes that he was having a difficult time finding the time to visit them.
5. *Ledbury*: a market town in the county of Herefordshire.
6. *Malvern*: a scenic town in the county of Herefordshire.
7. *Eastnor Park*: a 300-acre park overlooking Eastnor Castle with views of the Malvern Hills.
8. *The Dean*: Richard Dawes.

From John Frederick William Herschel 5 January 1863 1960

Collingwood | Jan. 5/63

My dear Sir,

I can have no sort of objection to your printing as much of my note referred to by you[1] as you think proper.

I like your idea of accounting for the blue colour of the air as due to the vapour of a blue liquid (water) if you intend it to be regarded as an absorptive colour.[2] Analogies are not wanting. Liquid Chlorine is yellow. so is the gas. Iodine is a very deep coloured substance & I presume its colour if it could be perceived through an excessively thin edge of a crystal of it would be deep purple. Its vapour is so. The vapour of Bromine (a deep ruddy brown liquid) is also ruddy brown. —I do not at once recollect other liquids which distill ever coloured but I dare say one might call to mind many—and it would be interesting to pursue the analogy. Sulphur heated above its melting point is brown—so is its vapour.

As to the polarization of the sky (having a maximum at 45° incidence but far from complete polarization at that maximum) I do not very well see how this can be specially connected with the vapour of water. The broad fact however that the polar[izatio]n of sky light is most intense when the sky is most completely free from visible haze or cloud—seems to look that way. But after all the question remains in all its obscurity—"How is the polarization of the sky light effected?" and I own that all the accounts I have seen of it are to me most unsatisfactory.[3]

It may very well be that the absorptive power of the vapor "pro quantitate materiae"[4] may be much greater than of its liquid. In nitrous gas (when heated) it is enormous & I should think from the very moderate thickness of chlorine gas & bromine vapour (& the exceedingly small one of Iodine D—) which suffices to show colour, that in these liquids it is so. —If not, I cannot understand how there can be water enough in the air to shew colour. —Take a vapour tension of 0.25 in and suppose 20 feet of water sufficient to shew a sky blue colour. This would require reflexion enough to carry the ray through $\frac{20\times12}{0.25}$ = 960 times the entire thickness of the atmosphere—before reaching the eye; which is incredible.[5]

I have delivered your obliging message to my son[6] who will gladly avail himself of your kind offer—always supposing that circumstances—at present in abeyance should permit him giving a lecture at the R. I.[7] this season.

I remain, my dear Sir, | Yours very truly | J. F. W. Herschel

RI MS JT/1/H/99
RI MS JT/1/TYP/2/521–21a

1. *my note referred to by you*: letter 1958.
2. *regarded as an absorptive colour*: see letter 1958, n. 8.
3. *I own that all the accounts I have seen of it are to me most unsatisfactory*: Herschel here effectively refutes Tyndall's suppositions about the colour of the sky. See letter 1958, n. 9.
4. *"pro quantitate materiae"*: amount of matter (Latin).
5. *which is incredible*: i.e., not credible.
6. *your obliging message to my son*: see letter 1958. Tyndall had offered to assist Alexander Herschel with illustrating the proposed lecture.
7. *the R. I.*: see letter 1958, n. 6.

To Mary Helen Dawes [7 January 1863][1] 1961

My dear Mrs Dawes,

There is a heartiness and kindness in your letters which revives and stimulates me; and now on the spur of the moment I will write to you again.

The Dean[2] some months ago wrote and asked me to vote for a certain friend of his at the Athenaeum.[3] His friend did not get in, and I want you to inform the Dean that I did my duty. I voted for him myself and got Sir William Snow Harris and others to vote for him; still he was left behind, not I believe through any objection to the man himself, but clubs are influenced by curious considerations, a great deal depends on the way in which a candidate is brought forward, on the feeling of the club towards his most prominent supporters etc. I was sorry the Dean's friend did not succeed, but do please tell him that I did my duty.

I wish most earnestly you were nearer to London. I never told you what Miss Guthrie[4] advised me, nay urged me to do some months ago. It was during a lecture here:[5] I said something about going down to Hereford. Go, said she and carry away Mrs Dawes from the Dean and bring her to London. I looked at her timidly, as it is my nature to be timid in such cases. She looked at me with a determination which showed the energy of the soul within her. "Do it!" she repeated, and it was as much as I could do to resist the commands. If she is in Hereford, pray remember me most kindly to her.

Goodbye. | Yours always | John Tyndall.

RI MS JT/1/TYP/1/325
LT Typescript Only

1. *[7th January 1863]*: date suggested by LT.
2. *The Dean*: Richard Dawes.
3. *a certain friend of his at the Athenaeum*: letter missing; 'certain friend' not identified.
4. *Miss Guthrie*: Anne Leonora Guthrie (1807–93), half-sister of Mary Helen Dawes who

lived at 4 Berkeley Street until moving in 1858 to Pool House, Hereford; the daughter of the surgeon George James Guthrie (1785–1856) (N. Denison, *Richard Dawes: Education Pioneer and Dean of Hereford* (King's Somborne: The Somborne & District Society, 2017); 'Deaths', *Morning Post*, 1 April 1893, p. 1).

5. *a lecture here*: not identified.

To Rudolf Clausius 8 January 1863 1962

8th January 1863

My dear Clausius.

I send you by this post two sheets of my book[1] and should feel extremely obliged to you if you would read them over for me. The part relating to Mayer & Joule is delicate[2] and I am anxious to have your opinion on it. I do not think either of them will have any fair reason to complain of what I have said.

In addition to these I will ask you to look at one other lecture—that in which I deal with specific & latent heat.[3]

Would you like a good fellow return me the proofs as soon as you can,—I want to get the book off my hands.

Be sure also to present my kindest remembrances to your wife[4]—I should have been very glad to see her in London along with you when you were here.

Yours ever truly | John Tyndall

RI MS JT/1/T/182

1. *two sheets of my book*: Clausius looked over sections of the manuscript proof for Tyndall's *Heat as a Mode of Motion* (1863), which was based on his 1862 lecture series at the RI, 'Twelve Lectures on Heat', delivered on Thursdays at 3pm from 23 January to 10 April 1862 (RI Guard Book, vol. 3, p. 11). See also letters 1963, 1965, 1966, 1967, 1969, and 1972.
2. *The part relating to Mayer & Joule is delicate*: Tyndall discussed these issues in chapters 2, 3, and (more extensively) 12 in *Heat as a Mode of Motion*. He advocated for Mayer's work as in fact preceding the discovery of the mechanical equivalent of heat, which Joule claimed. Tyndall had been quite explicit about his support for Mayer in lecture, but clearly valued Clausius' opinion on his phrasing here. This was a long-running dispute across the period covered by this volume (see the Introduction for further commentary, as well as *Tyndall Correspondence*, vol. 7 and the forthcoming ninth volume).
3. *that in which I deal with specific & latent heat*: Tyndall dealt with this in chapter 5 of *Heat as a Mode of Motion*.
4. *to your wife*: Adelheid Clausius.

To Rudolf Clausius 15 January 1863 1963

15th. Jan 1865

Dear Clausius,

I have sent off to the Dead Letter office[1] directions to have the proof[2] sent to you. The delay is due to stupidity of the assistant at the post office.

Yesterday I sent you a portion of the Lecture[3] on specific heat[4]—I hope it will fare better.

A very happy new year to you & Mrs. Clausius.[5] I hope the authorities in Zurich have done all that they ought to do.[6]

Goodbye | Yours very sincerely | John Tyndall

RI MS JT/1/T/183

---

1. *Dead Letter office*: the dead letter office is where undeliverable mail is held within the Post Office system. Clearly Tyndall had been notified that his letter to Clausius containing proofs had been undeliverable; he indicated here that he had provided them with the correct address.
2. *the proof*: for *Heat as a Mode of Motion* (see letter 1962, n. 1).
3. *Yesterday I sent you a portion of the Lecture*: letter missing; although see letter 1962.
4. *The Lecture on specific heat*: see letter 1962, n. 3.
5. *Mrs. Clausius*: Adelheid Clausius.
6. *I hope the authorities in Zurich . . . all that they ought to do*: a reference perhaps to Clausius's invitation to a post in Braunschweig, which he did not accept (see letter 1965, n. 8).

From Jean Alponse Favre[1] 15 January 1863 1964

Monsieur

Permettez-moi d'avoir recours à votre obligeance pour vous prier de vouloir bien présenter à l'Alpine Club deux cartes que je viens de publier, l'une est une carte topographique des Parties de la Savoie de Piémont & de la Suisse voisines du Mt. Blanc, l'autre est la même carte coloriée géologiquement qui m'a couté beaucoup de travail. C'est un hommage que je suis heureux de rendre aux hardis [croiseurs] de Montagnes qui [comprenent] le Club—et veuillez si vous en avez l'occasion me rappeler au souvenir de M. Mathews que j'ai rencontré à St Jean de Maurienne peu après qu'il était monté au sommet du Mt Viso.

Je vous adresse en même temps ma carte géologique que je vous prie d'accepter, je suis heureux qu'elle soit dans la Bibliothèque du homme qui a jeté tant de lumière sur l'un des plus singuliers phénomènes de nos Alpes, les glaciers.

Je vous adresse encore deux explications de cette carte l'une pour vous, l'autre pour le Club. Vous verrez, Monsieur, que c'est un travail de longue haleine que je viens de terminer, vous trouverez quelques détails d'envoi par M[r] d'Archiac de l'Institut de Paris dans l'Institut 1862 p. 357 ou dans le Cosmos 7 Novembre 1862. Je me permets de vous donner ces indications dans l'espoir que vous voudrez bien faire que ce travail ne passe pas tout à fait inaperçu dans votre club ou dans quelques-uns de vos journaux scientifiques.

Si je n'ai pas fait dans mes montagnes des exploits semblables à ceux des membres de l'Alpine Club; je vous assure cependant que j'ai fait quelques belles courses, entre autres aux sommets des Aiguilles Rouges (Bibliothèq. Universelle 1848, t VII) c'est vrai que je me suis arrêté à 40 pieds de ce sommet puis j'ai été de Courmayeur par le col du Géant au col qui se trouve entre l'Aiguille du Midi de Chamonix et le Mt Blanc de Tacul, de là je suis redescendu à Courmayeur.

J'ai lu avec beaucoup d'intérêt votre notice sur l'origine de nos vallées des Alpes, mais permettez-moi de vous le dire je n'ai pas complètement partagé votre manière de voir. J'ai même fait quelques observations à ce sujet qui seront publiés avec votre article dans nos Archives je pense en Février. Je regrette que vous n'ayez pas développé l'idée que vous avez *[énoncée]* que les vallées n'étaient pas dans le sens des déchirures qui ont dû se former au moment des soulèvements. Il me semblait au contraire que la régularité des vallées qui sont perpendiculaires ou parallèles aux chaines indique une liaison intime entre elles & les chaines de montagnes ?

Je ne sais, Monsieur, si vous avez publié quelques pages sur l'ancienne extension des glaciers & où je pourrai les trouver, mais si vous n'avez-pas fait de travail à ce sujet, ce serait bien intéressant pour tout le monde de connaitre vos idées à ce sujet et nos Archives seraient bien heureux de l'insérer.

Veuillez, Monsieur, recevoir l'assurance de ma considération très distinguée.

Alph[ons][e] Favre | Prof. de Géologie à l'Acad. de Genève.

Genève le 15 janvier | 1863

Sir,

Allow me to have recourse to your kindness to ask you to present two maps that I just published at the Alpine Club.[2] One is a topographical map of the parts of Savoy,[3] Piedmont[4] & Switzerland next to M[t]. Blanc,[5] the other is the same map, coloured geologically, which took me a lot of work. It is a tribute that I am happy to pay to the hardy mountaineers who comprise the Club—and please, if you have the opportunity, mention me to Mr Mathews,[6] whom I met at St Jean de Maurienne[7] shortly after his assent to the summit of M[t] Viso.[8]

I am also sending you my geological map, which I urge you to accept. I would be so happy for it to be in the library of the man who has shed so much light on one of the most singular phenomena of our Alps, the glaciers.

I am also sending you two explanations of this map,[9] one for you and one for the Club. You will see, sir, that this has been a long-term project that I have just completed, you will find details on delivery from Monsieur d'Archiac of the Institut de Paris[10] in the Institut[11] 1862 p. 357[12] or in the Cosmos, November 1862.[13] I am taking the liberty of sending you these instructions in the hopes that you will ensure that this work does not go completely unnoticed in your club or in some of your scientific journals.

I have not, in my mountains, achieved exploits as impressive as those of the members of the Alpine Club, but I promise you that I have had some good journeys, among them one to the summit of the Aiguilles Rouges[14] (Bibliothèq. Universelle 1848, t VII)[15] it is true that I stopped 40 feet from the peak, then I went from Courmayeur[16] through Col du Géant to the pass that is between the Aiguille du Midi[17] in Chamonix[18] and the Mont Blanc du Tacul,[19] from there I descended back to Courmayeur.

I read with great interest your memorandum on the origin of our Alpine valleys,[20] but allow me to tell you that I did not completely share your view. I have even made a few observations on the subject which will be published with your article in our Archives, in February[21] I expect. I regret that you did not develop the idea that you *[mentioned]*, that the valleys were not in the direction of the crevasses that must have formed at the moment of tectonic uplift. It seemed to me, on the contrary, that the regularity of the valleys which are perpendicular or parallel to the ranges indicates an intimate connection between them & the mountain ranges?

I do not know, Sir, if you have published anything on the former expansion of the glaciers & where I could find it, but if you have not done any work on this subject, everyone would be interested to know your thoughts on the matter, and our Archives would be glad to insert it.

Please, Sir, be assured of my highest regards.

Alph[ons][e] Favre | Prof. of Geology at the Acad. of Geneva.

Geneva, January 15th | 1863

RI MS JT/1/F/19

---

1. *Jean Alphonse Favre*: Jean Alphonse Favre (1815–90), Professor of Geology at the Geneva Colony.
2. *the Alpine Club*: a London-based mountaineering club founded in 1857, which met at Ashley's Hotel.
3. *Savoy*: the Savoy region is in the western Alps. It had been annexed to France in 1860.
4. *Piedmont*: a region in north-west Italy that borders France and Switzerland.

5. *M^t. Blanc*: Mont Blanc is the highest mountain in the Alps.
6. *Mr Mathews*: William Mathews (1828–1901), an English mountaineer, botanist, and surveyor known for his role in establishing the Alpine Club.
7. *St Jean de Maurienne*: Alpine village in south-east France.
8. *Mt Viso*: Mont Viso, the highest peak in the Cottian Alps in Italy.
9. *explanations of this map*: M. A. Favre, 'Explication de la Carte Géologique des Parties de la Savoie, Du Piedmont et de la Suisse Voisines du Mont-Blanc', *Archives des Sciences Physiques et Naturelles*, 15 (1852), pp. 238–70.
10. *Monsieur d'Archiac of the Institut de Paris*: Étienne Jules Adolphe Desmier de Saint Simon, Vocomte d'Archiac (1802–68), a French palaeontologist and geologist, who from 1857 was a member of the French Academy of Sciences and, from 1861, Professor of Palaeontology at the Muséum National d'Histoire Naturelle in Paris.
11. *Institut*: Memoires de L'Académie de Sciences de l'Institut de France.
12. *Institut 1862 p. 357*: there is no reference to this on p. 357, or elsewhere in this volume. It is possible that Favre wrongly recalled the details.
13. *Cosmos, November 1862*: 'M. le vincomte d'Archiac communique à l'Academie une Carte des parties de la Savoie, du Piémont et de la Suisse voisines du mont Blanc, par Alphonse Favre, professor à l'Académie de Genéve', *Cosmos*, 21, no. 2 (1862), pp. 526–8.
14. *Aiguilles rouges*: a range of peaks opposite Mont Blanc.
15. *(Bibliothèque Universelle 1848, t VII)*: A. Favre, 'Recherches géologiques faites dans les environs de Chamounix en Savoie', *Archives de Sciences Physiques et Naturelles*, 7 (1848), pp. 265–87.
16. *Courmayeur*: a town in the Aosta Valley in northern Italy.
17. *Aiguille du Midi*: a peak in the Mont Blanc region.
18. *Chamonix*: an alpine town in the Mont Blanc region of France, already popular with visitors in the 1860s.
19. *Mont Blanc du Tacul*: another high peak in the Mont Blanc region.
20. *your note on the origin of our Alpine Valleys*: see letter 1921, n. 6.
21. *our Archives, in February*: M. L'Abbe Stoppani, 'Sur les Couches à Avicula Contorta du Versant Nord-Ouest des Alpes Principales', *Archives des Sciences Physiques et Naturelles*, 16 (1863), pp. 273–93.

From Rudolf Clausius 17 January 1863 1965

Zürich, 17 Jan. 63

Lieber Tyndall,

Ich habe Ihren Druckbogen gestern Abend erhalten und habe die Lecture V sogleich mit grossem Interesse gelesen. Ich finde die Art der Darstellung sehr schön, und halte auch den Inhalt für richtig mit Ausnahmen von einer Stelle, über die ich mir eine Bemerkung erlaube. Sie sprechen auf S. 142 davon, dass das Wasser eine viel grossere specifische Wärme hat, als das Quecksilber, und ebenso das Eisen eine grössere, als das Blei, und erklären dieses daraus dass im Wasser und Eisen die innere Arbeit bei die Erwärmung grösser ist, als im Quecksilber, und im Eisen. Ich glaube nun zwar auch, dass im Wasser die innere Arbeit bei der Erwärmung grösser ist, als im Quecksilber, aber den Haupterklärungsgrund für die Verschiedenheit der specifischen Wärme suche ich doch in einem anderen Umstande.

Wenn zwei chemisch verschiedene Körper gleiche Temperatur haben, so nehme ich an, dass die einzelnen Atome beider Stoffe gleiche lebendige Kraft, oder wie Sie sagen, gleiche actual oder musical energy haben. Wenn ich nun beide Körper um gleich viel erwärmen will, so muss ich die lebendige Kraft aller Atome um gleich viel vermehren, und dazu brauche ich bei einem Körper um so mehr Wärme je mehr Atome er enthält. Vergleichen wir hiernach zuerst Eisen und Blei, so enthält eine Gewichtseinheit Eisen beinahe 4 mal so viele Atome als eine Gewichtseinheit Blei, und daher gebraucht jene auch beinahe 4 mal so viel Wärme zur Temperaturerhöhung als diese. Die dem Eisen mitgetheilte Wärmemenge ist also nicht deshalb soviel grösser weil in ihm mehr Wärme zu innerer Arbeit verbraucht, oder in potential energy verwandelt wird, sondern deshalb, weil die lebendige Kraft der wirklich vorhandenen Bewegung oder die actual energy in einer Gewichtseinheit Eisen grösser ist als in einer Gewichtseinheit Blei. Dabei ist nicht ausgeschlossen, dass auch die zu innerer Arbeit verbrauchte Wärme im Eisen grösser sein kann, als im Blei, aber der Hauptgrund der verschiedenen specifischen Wärme scheint mir nicht hierin sondern in der verschiednen actual energy zu liegen.

Ähnlich wie mit Eisen u. Blei verhält es sich mit Wasser und Quecksilber, nur dass bei ihnen die Verhältnisse etwas complicirter sind. Die chemischen Aequivalentzahlen verhalten sich wie 9:100, also die Anzahl der Molecüle in einer Gewichtseinheit beider Stoffe wie 100:9. Nun enthält ein Molecül Wasser drei Atome nämlich 2 H und 1 O, und ein Molekül Quecksilber wie ich glaube, 2 Atome, so dass die Anzahl der Atomen in einer Gewichtseinheit beider Stoffe sich verhält wie 300:18 oder ungefähr wie 17:1. Ich glaube daher dass in einer Gewichtseinheit Wasser die lebendige Kraft oder actual energy ungefähr 17 mal so gross ist als in einer Gewichtseinheit Quecksilber. Dazu kommt nun, dass, wie ich glaube, im Wasser auch die zu innerer Arbeit verbrauchte Wärme viel grösser ist als im Quecksilber, und dadurch wird die specifische Wärme bis auf das 30-fache von der des Quecksilbers vermehrt. Die von Ihnen angeführte Umstand findet hier also statt, aber er ist nicht die alleinige Ursache für die Verschiedenheit der spef. Wärmen.

Ich habe dieses hier sehr breit auseinandergesetzt, um mich ganz verständlich zumachen; aber ich glaube dass Sie, wenn Sie meine Ansicht theilen, und danach Ihre Schrift ändern wollen, dieses durch geringe Abhänderungen des Wortlautes thun können.

Im Uebrigen habe ich nur unbedeutende Bemerkungen über einige Ausdrücke zu machen. Sie sagen auf S. 137, Sie wollen nach Rankine die Ausdrücke potential energy und dynamic energy gebrauchen. Soviel ich weiss hat aber Rankine die Ausdrücke potential energy und actual energy vorgeschlagen (Phil. Mag. Series 45, Vol 5, p. 106; und Edinburgh Phil. Journ. New Series, Vol. 2, p. 129)—Den Ausdruck Rhythmic Energy einzuführen, würde ich Ihnen deshalb nicht rathen, weil diese energy doch dieselbe sein würde, wie die, welche schon mit actual oder dynamic energy bezeichnet ist.

Die Stelle auf S. 140: This is a work in comparison with which etc. bis of miles, könnte vielleicht, wenn man dabei an die Ausdehnung eines kleinen Körpers denkt, einen zu grossen Begriff von dieser Arbeit geben, und ich möchte Ihnen daher rathen, zur Vermeidung von Missverständnissen, mit einigen Worten hinzuzufügen, dass Sie dabei an die Ausdehnung grosser Massen denken.

Indem ich Ihnen meine Ansichten mitgetheilt habe, mache ich natürlich keinen Anspruch darauf, dass dieselben entscheidend sein müssen, sondern ich wollte nur Ihre Aufmerksamkeit auf diese Puncte lenken, aber die Entscheidung überlasse ich ganz Ihnen. Den Bogen welcher sich auf Joule und Mayer bezieht, haben Sie wahrscheinlich zurückbehalten, weil durch die Verzögerung auf der Post die Zeit zu kurz geworden war. Ich bedauere dieses sehr; ich hätte es gern durchgelesen, und werde auch Alles, was Sie mir sonst noch schicken, mit dem grössten Vergnügen lesen, und Ihnen offen meine Ansichten sagen.

Ich habe Ihnen noch nicht geschrieben wie die Sache mit meiner Berufung nach Braunschweig sich entwickelt hat, aber ich will heute die Absendung des Druckbogens nicht verzögern, den ich selbst gleich befördern will, und schreibe Ihnen daher nur kurz, dass ich in Zürich bleiben werde.

Mit herzlichen Grüssen von meiner Frau und mir, | Ihr | Clausius

Zurich, 17 Jan. 63

Dear Tyndall,

I received your prints[1] yesterday evening and immediately read lecture V[2] with great interest. I find that the form of presentation is very fine and consider the contents to be correct with the exception of one point that I will allow myself to comment upon. You state on p. 142 that water has a much greater specific heat than mercury in the same way that iron has a greater one than lead, and explain that this is because with water and iron the internal work is greater during heating than with mercury and with iron. I too believe that the internal work of warming is greater in water, than in mercury, but I would look for the main reason for the difference of specific heat in other factors.

When two chemically different bodies have the same temperature, I assume that the individual atoms of both substances have the same living force, or as you say, the same actual or musical energy.[3] If I want to heat both bodies equally, then I must increase the living force of all atoms equally and for that I would need in one body inasmuch more heat as much more atoms it contains. If we first compare iron and lead, a weight unit of iron contains nearly 4 times as many atoms as a weight unit of lead, and therefore the former requires nearly 4 times as much heat to raise the temperature as much as the latter. The amount of heat imparted to the iron is therefore not that much greater because more heat is consumed in internal work, or transformed into potential energy, but because the living force of the actually existing motion or the actual energy in a weight unit of iron is greater than in a weight unit of lead. That does not exclude the possibility that the heat consumed in internal work in iron could be greater than in lead. However, it seems to me that this is not the main reason for the different specific heats, that the main reason is the different actual energy.

Things are similar with water and mercury as with iron and lead, only that here the relationships are somewhat more complicated. The chemical equivalent numbers are in a relationship of 9:100, and the number of molecules in a weight unit of both substances 100:9. Now a molecule of water is made up of three atoms namely 2 H and 1 O, and a molecule of mercury is, I believe, 2 atoms, so that the number of atoms in a weight unit of both substances is in a ratio of 300:18 or about 17:1. I believe for that reason that in a weight unit of water the living force or actual energy is about 17 times as great as in a weight unit of mercury. In addition, I believe that the heat consumed in internal work is much greater in water than in mercury, and through that the specific heat is increased to about 30 times more than in mercury. The factor that you mentioned thus also exists here, but it is not the sole cause of the differences in specific heat.

I have discussed this in some breadth here, in order to make myself fully comprehensible, but I believe that if you share my view and want to change your text accordingly, you could do this through minor modifications in expression.

In regard to the rest I only have insignificant remarks to make about a few expressions. On p. 137 you want to use the expressions potential energy and dynamic energy according to Rankine.[4] As far as I know, however, Rankine suggested the expressions potential energy and actual energy (Phil. Mag. Series 45, Vol 5, p. 106;[5] and Edinburgh Phil. Journ. New Series, Vol. 2, p. 129[6])—I would therefore advise you against introducing the expression Rhythmic Energy, because this energy would be the same one that has already been termed actual or dynamic energy.

The point on p. 140: This is a work in comparison with which etc. up to of miles, could perhaps, if one considered the expansion of a small body, give too great a notion of this work, and I would for that reason recommend that in order to avoid misunderstandings, you add a few words to the effect that you are considering the expansion of larger masses.

By communicating my views to you, I naturally do not insist that they must be decisive, I merely wanted to direct your attention to these points, but I leave the decision entirely up to you. You have probably held on to the sheets on Joule and Mayer,[7] because through the delay at the post office there would have been too little time. I regret this very much; I would have happily given them a reading, and will read everything else that you send me with the greatest pleasure, and give you my open opinion.

I have not yet written to you about how the affairs around my call to Braunschweig[8] developed, but I do not want to delay sending you the prints that I will send off immediately, and only briefly say that I will be remaining in Zurich.

With fond greetings from my wife[9] and myself, | Your Clausius

RI MS JT/1/TYP/7/2248–50
LT Typescript Only

---

1. *your prints*: for *Heat as a Mode of Motion* (see letter 1962, n. 1).
2. *read lecture V*: see letter 1962, n. 3.
3. *same living force, or as you say, the same actual or musical energy*: Tyndall is referring to the phenomena now known as kinetic energy. Tyndall and his contemporaries struggled with how best to conceive and describe their theories, as evidenced by their discussion of terminology. The term 'musical' may be a mis-transcription by LT.
4. *Rankine*: William Rankine.
5. *Phil. Mag. Series 45, Vol 5, p. 106*: W. J. M. Rankine, 'On the General Law of the Transformation of Energy', *Phil. Mag.*, 5 (1853), pp. 106–17.
6. *Edinburgh Phil. Journ. New Series, Vol. 2, p. 129*: W. J. M. Rankine, 'Outline of the Science of Energetics', *Edinburgh New Philosophical Journal*, 2 (1855), pp. 120–41.
7. *the sheets on Joule and Mayer*: see letter 1962, n. 2.
8. *my call to Braunschweig*: Clausius had received, but turned down, a professional position at the University of Braunschweig.
9. *my wife*: Adelheid Clausius.

From Rudolf Clausius — 18 January 1863 — 1966

Zurich, 18 Jan 63

Lieber Tyndall,

Als ich eben den Bogen über die spef. Wärme abgesandt hatte, habe ich die beiden anderen Bogen erhalten und habe sie ebenfalls mit grossem Interesse durchgelesen. Ich finde es ebenfalls sehr schöne und bewundere Ihre

Gewandtheit mit der Sie einen abstrakten schwierigen Gegenstand in eine anziehende leicht verständliche Form zu bringen wissen. Nur bei ein Paar unbedeutenden Puncten habe ich Bemerkungen zu machen.

Auf S. 64 wo Sie davon sprechen dass die Blase durch Wärme aufschwillt, sagen Sie, der Druck der äussern Luft nehme deshalb nicht zu, weil die äussere Luft durch die strahlende Wärme nur sehr wenig erwärmt werde. Das letztere ist zwar ganz richtig, aber ich glaube dass man, zur Vermeidung von Missverständnissen, die Stelle doch anders ausdrücken sollte. Wenn nämlich die äussere Luft auch erwärmt würde, so würde ihr Druck doch nicht zunehmen, weil sie sich frei ausbreiten kann, während die innere Luft durch die Blase in ihrer Ausbreitung beschränkt ist. Sie haben gewiss das Richtige gemeint, aber nach den Worten könnte man glauben, dass durch eine Erwärmung der die Blase umgebenden Luft, ihr Druck in entsprechender Weise vermehrt werden würde.

Auf Seit 71 und noch bestimmter auf Seit 73 sagen Sie, Mayer habe das mech. Aequivalent der Wärme im Jahre 1842 zu 771.4 Fusspfund gefunden. Dieses ist, wie ich glaube, nicht ganz streng richtig. Ich habe zwar die Abhandlung von 1842 nicht zur Hand, aber die Broschüre über die organische Bewegung wo auf S. 15 die Rechnung auch vorkommt. Die Art der ist ganz so, wie Sie angegeben haben, aber Mayer musste für die specifische Wärme der Luft statt der von Regnault gefundenen Zahl 0,2377, oder 0,24 die damals geltende von Delaroche und Berard gefunden, nämlich 0,267 anwenden. Dadurch erhielt er nach französischen Maassen die Zahl 367 statt 423, das giebt nach englischen Maassen (engl. Fuss u. Fahrenheitsgrade) 661 statt 771,4: Obwohl dieses eine Abweichung ist, die nicht Mayer zur Last fällt, sondern ihren Grund in der Unvollkommenheit der damals bekannten Beobachtungsdata hat, so glaube ich doch dass es gut wäre, sie zu erwähnen, um nicht Gelegenheit zu einer Einwendung zu geben.

Auf S. 72 steht unter 2 die Zahl 7001. Diese Zahl ist so gross, dass ich fast vermuthe, dass es ein Druckfehler ist. Ich habe die Abhandlung von Joule nicht zur Hand, aber ich möchte Sie auf die Zahl aufmerksam machen, um sie, wenn es ein Druckfehler ist, zu corrigiren.

Auf Seite 76 sagen Sie beim absoluten Nullpunct: 'in reality the gas deviates etc. before it sinks so low; but if no deviation occurred, -273 C. would mark the absolute zero'. Ich glaube man kann in dieser Beziehung noch weiter gehen, und sagen: welches auch die Abweichungen seien, welche das Gas in der Wirklichkeit bei tiefen Temperaturen von dem obigen Zusammenziehungsgesetze zeigen mag, der Schluss über den abs. Nullpunct wird dadurch nicht geändert, weil er aus dem Verhalten des Gases zwischen 0° u. 273° gezogen ist. Bei 273° ist die Expansionskraft doppelt so gross als bei 0°, u. daraus muss man schliessen dass das Gas bei 273° doppelt soviel Wärme enthält als bei 0°. Daher muss man, wenn die Temperatur ein wirkliches Maass für die in dem Körper vorhandene Wärme oder lebendige Kraft sein soll, von -273° an zählen.

Was Sie über Joule u. Mayer gesagt haben, scheint mir ganz gerecht zu sein. Ich glaube auch, wie Sie, dass beide mit der Beurtheilung zufrieden sein können.

Mit herzlichen Grüssen, | Ihr Clausius

Zurich, 18 Jan 63

Dear Tyndall

Just as I sent out the sheets on specific heat, I received both of the other prints[1] and read them with great interest as well. I find them very fine as well and admire your[2] skill in presenting abstract and difficult topics in a way that is appealing and easy to comprehend. I have only a few comments on some insignificant points.

On p. 64, where you talk about the bladder swelling through heat, you say that the reason the pressure of the external air does not increase is because the external air is only heated very little by radiant heat. The latter is quite correct, but I believe that the point should be expressed differently in order to avoid misunderstandings. Namely, if the external air is also heated, its pressure would still not increase because it can expand freely, while the expansion of the inner air of the bladder is restricted. I am certain that you meant this correctly, but following the words, one might believe that by heating the air surrounding the bladder, its pressure would correspondingly be increased.

On page 71 and with more emphasis on page 73, you say that Mayer determined the mech[anical] equivalent of heat in the year 1842 as 771.4 foot-pounds. This is, I believe, not strictly speaking correct. I do not have the paper from 1842[3] at hand, but the brochure on organic motion,[4] where the calculation can be found on p. 15. The method is exactly as you wrote, but instead of the number Regnault found of 0.2377 or 0.24,[5] Mayer must have used the one that Delaroche and Berard[6] found, which was used at the time, namely 0.267. Thus he derived according to French measurements the number 367 instead of 423, which gives in English measurements (Engl. feet and degrees Fahrenheit) 661 instead of 771.4. Although this deviation that cannot be laid at Mayer's feet, but is based on the incompleteness of known observational data at the time, I still think that it would be worth mentioning, so as not to provide an opportunity for objection.

On p. 72 under 2 is the number 7001. This number is so large, that I almost think it is a printing mistake. I do not have Joule's paper at hand, but I would like to direct your attention to the number, so that it can be corrected if it is a printing mistake.

On page 76 you say about absolute zero: 'in reality the gas deviates etc. before it sinks so low; but if no deviation occurred, -273 C. would mark the absolute zero'. I believe that one can go much further in this regard and say: regardless of the deviations that the gas may in reality exhibit at low temperatures due to the laws of attraction, this does not alter conclusions about the abs. zero point, because they have been drawn from the behavior of the gas between 0° and 273°. At 273° the force of expansion is twice as great as at 0°, and from this one must conclude that the gas holds twice as much heat at 273° as at 0°. Thus, if temperature is a real measure for the amount of heat or living force in a body, one must start calculating from -273°.

What you said about Joule and Mayer seems to me to be quite just.[7] I believe, as you do, that both can be very satisfied with the assessment.

With fond regards, | Your Clausius

RI MS JT/1/TYP/7/2251–2
LT Typescript Only

---

1. *I received both of the other prints*: there is clearly at least one letter missing from this exchange.
2. *Your*: Clausius here used the informal address form (*Ihre*).
3. *the paper from 1842*: J. R. Mayer, 'Bemerkungen über die Kräfte der unbelebten Natur', *Justus Liebigs Annalen der Chemie*, 42 (1842), pp. 233–40.
4. *brochure on organic motion*: see letter 1904, n. 3.
5. *Regnault found of 0.2377 or 0.24*: this measurement can be found in W. J. M. Rankine, 'Mechanical Theory of Heat—Specific Heat of Air', *Phil. Mag.*, 5 (1853), pp. 437–9.
6. *Delaroche and Berad*: François-Étienne Delaroche (1743–1812/13), a Swiss physician and naturalist; and Jacques Étienne Bérard (1789–1869), a French naturalist and chemist. Clausius was referring to F. E. Delaroche and J. E. Bérard, *Mémoire sur la détermination de la chaleur spécifique des différens gaz* (Paris: H. Perronneau, 1813).
7. *What you said about Joule and Mayer seems to me to be quite just*: see letter 1962, n. 2.

From Rudolf Clausius 20 January 1863 1967

Zürich, 20. Jan. 63

Lieber Tyndall,

Ueber diesen Bogen habe ich nur wenige unbedeutende Bemerkungen zu machen.

Auf die erste Partie möchte ich dasselbe anwenden, was ich schon früher über die specifische Wärme gesagt habe, dass sie nicht blos die zu Arbeit verbrauchte Wärme, sondern auch die Vermehrung der wirklich vorhandenen Wärme darstellt.

Die latente Wärme ist, wenn Sie es sagen, nur solche Wärme, die zu Arbeit verbraucht ist. Aber die Arbeit ist nicht blos innere sondern auch äussere. Beim Schmelzen kann man die letztere meistens vernachlässigen, aber bei der Verdampfung nicht. Vielleicht können Sie auf S.148, wo sie die innere Arbeit erwähnen, die äussere auch irgendwo kurz anführen.

Bei der Auflösung von Salpeter, Salz, Zucker, etc. Wasser, wäre es vielleicht zweckmässig, an irgend einer Stelle daran zu erinnern, dass hierbei zwei Wirkungen gleichzeitig stattfinden, das Flüssigmachen des festen Körpers, wozu Wärme verbraucht wird, und die Mischung wodurch Wärme erzeugt wird, und dass die beobachtete Temperaturänderung der Differenz beider Wirkungen ist.

Die Stelle auf S. 157: "and thus superpose upon its present coldness that due to its liquefaction, we should obtain a cold still more intense" möchte ich Ihnen rathen, etwas anders auszudrücken. Wenn ein Körper erst flüssig u. dann luftförmig wird, so wird dazu im Ganzen nur ebensoviel Wärme verbraucht, als wenn es direct aus dem festen in den luftförmigen Zustand übergeht und die Kälte kann daher dadurch nicht intensiver werden. Ich glaube auch, dass der Brei dieselbe Temperatur hat, wie der feste Kohlensäure, nämlich die Siedentemperatur der Kohlensäure -78°. Der Brei ist nur mehr geeignet, andern Körpern Wärme zu entziehen, als die schneeartige Masse, welche ein geringes Leitungsvermögen hat, und dadurch wirkt er stärker abkühlend.

In Bezug auf die Stelle S.157: "which exhibits the most intense cold, which has been observed by man" möchte ich Sie darauf aufmerksam machen, dass Natterer sagt, er habe mit einer Mischung von flüssigem Stickstoffoxydul und Schwefelkohlenstoff eine Temp. von -140°C. erhalten, welche Temp. man mit Kohlensäuere, so viel ich weiss, selbst unter der Luftpumpe nicht erhalten hat. In wie weit die Angabe von Natterer zuverlässig ist, kann ich freilich nicht sagen. Sein Aufsatz steht Annalen der Chemie und Pharmacie LIV S.254. Despretz hat ein Gefäss mit Stickstoffoxydul angewandt, welches in einem weiteren Gefäss mit Kohlensäuerebrei stand, u. das ganze unter die Luftpumpe gebracht, u. behauptet dadurch Alkohol zum Gefrieren gebracht zu haben.

Alle diese Puncte sind für einen populären Vortrag nur von geringer Bedeutung, da es in einem solchen weniger auf die äusserste Genauigkeit, als auf eine leichtfassliche Darstellung ankommt. Ich habe sie nur erwähnt, weil ich glaube, dass es Ihnen lieb ist, wenn ich alles, was mir auffällt, anführe. Sie können dann selbst am besten wählen, was Sie berücksichtigen wollen, und was nicht.

Mit herzlichen Grüssen, | Ihr | Clausius

Zurich, 20. Jan. 63

Dear Tyndall,

I only have a few, insignificant comments to make about these pages.[1]

In the first section I would like to point out what I have mentioned earlier about the specific heat, that it is not simply the heat consumed doing work, but also represents an increase in the actually existing heat.

Latent heat is, as you say it, only the heat that is consumed for work. But the work is not only internal, it is also external. One can usually ignore the latter for melting, but not for evaporation. Perhaps you could note the external work somewhere on p. 148, where you mention the internal.

Regarding the dissolution of gunpowder, salt, sugar etc. in water, it would perhaps be advantageous to recall somewhere, that here two effects are taking place simultaneously, a solid body becoming fluid, which consumes heat, and a mixing which generates heat, and that the observed change in temperature is the difference of both effects.

I would recommend that you express the point on p. 157: "and thus superpose upon its present coldness that due to its liquefaction, we should obtain a cold still more intense" somewhat differently. When a body is first in a fluid and then in a gaseous form, on the whole this consumes as much heat, as when it changes directly from the solid into the gaseous state and the thus cold cannot become more intense. I believe as well that the slurry has the same temperature as carbon dioxide snow, that is, the boiling point of carbonic acid -78°. The slurry is only better suited to extract heat from other bodies than a snow-like mass, which has low conductivity and therefore it has a stronger cooling effect.

In regard to the point on p. 157: "which exhibits the most intense cold, which has been observed by man" I would like to draw your attention to what Natterer[2] said, that he got a temp. of -140° with a mixture of fluid nitrous oxide and carbon disulfide, a temp. that, as far as I know, has not been reached with carbonic acid, not even with an air pump. Whether Natterer's claim is reliable, I of course cannot say. His paper is in the Annalen der Chemie und Pharmacie LIV p. 254.[3] Despretz[4] used a container with nitrous oxide, which was placed in another container of carbonic acid slurry and put the whole thing under the air pump & claimed to have frozen alcohol that way.

All these points are of minor importance for a popular lecture, since there it depends less on being completely exact than on an easily understandable presentation. I only mentioned them, because I believe that you would prefer that I note everything that strikes me. You can best decide yourself, what you want to take into consideration and what not.

With fond regards, | Your | Clausius

RI MS JT/1/TYP/7/2253–4
LT Typescript Only

1. *these pages*: for *Heat as a Mode of Motion* (see letter 1962, n. 1).
2. *Natterer*: Johann August Natterer (1821–1900), an Austrian chemist, physicist, and physician. Clausius was referring to his experiment with gas liquefaction. His diaries and specimen collections were lost in the Hoburg Fire (1848), with the exception of a collection of 60,000 insects (*ADB*).
3. *His paper*: J. A. Natterer, 'Ueber die Verdichtung einiger Gase', *Annalen der Chemie und Pharmacie*, 54 (1845), pp. 254–6.
4. *Despretz*: see letter 1920, n. 10.

From James Hall[1] 24 January 1863 1968

Last week, on the 2nd of January, accompanied by a guide, I made the ascent of Snowdon[2] from the Victoria Hotel at Llanberis.[3] There had been a heavy fall of snow the previous night, which circumstance I hoped would give me an opportunity of repeating your experiments relative to the blue colour of newly-fallen snow in Switzerland;[4] nor was I disappointed. As soon as I arrived at a sufficient altitude for the snow to be one foot thick, viz., 1900 feet from the sea, on making the hole with the baton I was delighted to find the blue colour, but rather pale. As we ascended the mountain, we came into a dense mist with the snow deeper, and the blue colour became darker every step until we reached the summit, when the colour was as dark as that of the firmament. The most beautiful colour, however, was in the natural cracks and holes in the drifted wreaths of snow, caused, I suppose, by the particles of snow not having been compressed on each side the cracks or holes, and the reflections consequently more numerous. The light in the crevices of the snow on the side of the mountain overlooking Glaslyn[5] was exceedingly beautiful. The thermometer on the summit was 28 deg. Fah., and the calculated height by an aneroid 3595 feet above the sea.

'THE BLUE SNOW QUESTION—ASCENT OF SNOWDON', *Hampshire Advertiser*, 24 January 1863, p. 2

1. *James Hall*: not identified. This letter was reprinted in the *Hampshire Advertiser* of 24 January 1863 with the introduction: 'The *Athenaeum* gives the following extract from a letter addressed to Professor Tyndall by Mr. James Hall'.
2. *Snowdon*: the highest mountain in Wales.

3. *Victoria Hotel at Llanberis*: the Victoria Hotel lies at the foot of Snowdon and remains popular with climbers.
4. *your experiments . . . in Switzerland*: this refers to Tyndall's observations in the 1850s on glaciers and mountain passes (see e.g., *Glaciers of the Alps*, pp. 29 and 83, and Tyndall's journal entry of 4 September 1856 (RI MS JT/2/13c/899). Presumably Hall read them in *Glaciers of the Alps* or heard them at a lecture.
5. *Glaslyn*: translated as 'blue lake' in Welsh, Glaslyn is a lake found on the eastern flanks of Snowdon.

To Rudolf Clausius 29 January 1863 1969

29th January 1863

My dear Clausius

Accept my best thanks for the great trouble you have taken in looking over the proofs[1] for me. Most of the parts to which you have referred are exactly those which I had marked in my memorandum book for future correction. With regard to the influence of the atomic number, I almost feared to bring it into the course itself, but my intention was to give an analysis at the end of the lecture of the results of Dulong[2] & Petit,[3] Regnault & Neumann.[4] Your letter[5] has caused me to reconsider the matter, and now I have introduced the subject in the text.

I am glad that you think me fair toward Mayer & Joule:[6] it is my most earnest desire to give neither of them any just cause of complaint against me.

Goodbye dear Clausius | & believe me always | yours most sincerely | John Tyndall

RI MS JT/1/T/186

---

1. *the proofs*: for *Heat as a Mode of Motion* (see letter 1962, n. 1).
2. *Dulong*: Pierre Louis Dulong (1785–1838), a French physicist known for his study of the elasticity of steam and conduction of heat. He is especially noted for his research with Alexis Thérèse Petit, which resulted in the Law of Dulong and Petit (*CDSB*).
3. *Petit*: Alexis Thérèse Petit (1791–1820), a French physicist known for his work on heat and steam engines (see n. 2; *CDSB*).
4. *Neumann*: Franz Ernst Neumann (1798–1895), a German physicist and mathematician best known for Neumann's law, which gave a quantitative expression to Faraday's law of electromagnetic induction (*CDSB*).
5. *Your letter*: letter 1966.
6. *fair toward Mayer & Joule*: see letter 1962, n. 2.

To James Croll [January][1] 1863 1970

Your letter[2] was interesting to me as an illustration of power to seize a definite physical image—the molecules acting as hammers was capital. I have no doubt that anything you send me will interest me.[3]

J. C. Irons, *Autobiographical Sketch of James Croll* (London: Edward Stanford, 1896), p. 104

1. *[January]*: according to the *Autobiographical Sketch of James Croll* (p. 104), this was a response to a letter from Croll in early 1863.
2. *Your letter*: letter missing.
3. *anything you send me will interest me*: in early 1864 Croll sent Tyndall a paper which Tyndall published in the *Phil. Mag.* See letter 2072.

From Charlotte Murchison 5 February 1863 1971

16. Belgrave Square, | S. W. | Feb[rua]ry 5 63.

Dear Mr Tyndall

Pray accept my best thanks for your charming little book[1]—It is indeed of deep interest to me, & my Swiss blood warms by the perusal of it—Unluckily for me, the Swiss Alps not abounding in *[1 word illeg]*, my husbands[2] hammer and delight for work in other regions, in the Austrian alps &c &c. . & only twice in my perilous journeys with him, have I been allowed to revisit the abode of my Ancestors—so your book is doubly welcome to me—

ever y[ou]^rs Mr Tyndall | Y[ou]^rs gratefully | M[adam]^e Murchison—

PS I must not laugh at Millais[3] again for Silver Leaves—

RI MS JT/1/M/147

1. *your charming little book*: *Mountaineering in 1861* (published in 1862).
2. *my husbands*: Roderick Murchison.
3. *Millais*: John Everett Millais (1829–96), an English painter (*ODNB*). The reference is likely to his 1856 painting *Autumn Leaves.*

From Rudolf Clausius [10 February 1863][1] 1972

Zürich, 10 Feb. 63

Lieber Tyndall,

Ich würde schon früher an Sie geschrieben haben, wenn ich nicht in der letzten Zeit in mannichfacher Weise in Anspruch genommen wäre. Sie werden dieses verstehen, wenn ich Ihnen mittheile, dass meine Frau mir vor 8 Tagen ein zweites kleines Töchterchen gebracht hat. Bei einer solchen Veranlassung hat man ausser manchen Geschäften u. Sorgen auch so viel Briefe an die verschiedenen Verwandten zu schreiben, dass alle meine freie Zeit ausgefüllt war. Meine Frau ist übrigens nach Umständen ganz wohl und lässt Sie bestens grüssen, und das Töchterchen ist ein prächtiges gesundes Kind.

Die mir zugeschichten Bogen Ihrer Vorträge habe ich mit vielem Vergnügen gelesen und werde auch das Uebrige wenn es erschienen ist, mit grossem Interesse durchlesen. Mein Aufenthalt in London ist mir noch immer in der angenehmsten Erinnerung, und ich kann Ihnen nicht genug danken für die Freundlichkeit welche Sie mir während der Zeit bewiesen haben. Dass ich nicht nach Cambridge gegangen bin, scheint nach allem, was ich über die Versammlung gehört habe, kein so grosser Verlust für mich gewesen zu sein, als ich dachte. Die Versammlung scheint diesesmal nicht so glänzend ausgefallen zu sein, wie in früheren Jahren. Als ich hierher nach Zürich kam, wurde ich von dem Präsidenten des schweizerischen Schulrathes sehr freundlich aufgenommen, u. er bot mir für den Fall, dass ich hier bleiben wollte, eine solche Gehaltzulage an, dass dadurch mein hiesiges Gehalt höher würde, als dass, welches mir in Braunschweig angeboten war, u. ich entschloss mich daher schnell hierzubleiben. Bei dieser Gelegenheit ist noch ein besonderer Zwischenfall vorgekommen. Nachdem ich mit dem Präsidenten verhandelt hatte und meine Sache entschieden war, ist ein fremder Herr zum Präsidenten gekommen, u. hat im Namen eines unbekannten Dritten gefragt, ob der Bundesrath die gehörigen Mittel anwenden würde, um mich zu halten. Als dieses bejaht war, ist er befriedigt fortgegangen; ist dann aber ein anderen Tage wiedergekommen, und hat gesagt, damit es dem Polytechnikum für ähnliche Fälle wo Professoren ins Ausland gerufen würden, nicht an Mitteln fehle, übersende jener Dritte dem Polytechnikum hiermit 50,000 frs als Geschenk, deren Zinsen bei solchen Gelegenheiten zu Gehaltszulagen verwandt werden sollten. Wer der freigiebige Geber ist weiss man nicht, man vermuthet aber, dass es Escher von der Linth sei.

Mit herzlichen Grüssen an Sie und Hirst | Ihr | Clausius

Zurich, 10 Feb. 63

Dear Tyndall,

I would have written to you[2] sooner, if I had not been occupied in many other ways in recent days. You will understand this, when I tell you that my wife[3] gave me a second little daughter[4] 8 days ago. On such occasions, in addition to some business and worries there are so many letters to be written to diverse relations, so that all of my free time was taken up. My wife is by the way quite well under the circumstances and sends her best regards, and the wee daughter is a spanking healthy child.

I read the prints that you sent me of your lectures[5] and will also read the rest with great interest when it is published. I have fond memories of my sojourn in London, and I cannot thank you enough for the friendship you showed me in this time. From everything that I have heard about the meeting in Cambridge,[6] it seems after all that it was not such a great a loss as I had thought, that I did not go to Cambridge. The meeting seems not to have been as illustrious as in past years. When I returned to Zurich, the president of the Swiss Education Council[7] gave me a very friendly reception, and offered me, for the case that I wanted to stay here, such a raise in salary, that my salary here would be higher that what I was offered in Braunschweig, and I quickly decided to stay here. A remarkable incident occurred in the course of this affair. After I had negotiated with the president and made my decision, an unknown gentleman came to see to the president, and inquired in the name of a third party, whether the Education Council would use the appropriate means to keep me. When this was affirmed, he left satisfied; but came again on another day and said, so that Polytechnic Institute[8] is not lacking in means in similar cases in which a professor has received a call to go abroad, this third party is hereby donating 50,000 fr[anc]s as a gift, and the interest is to be used for salary increases in such situations. No one knows who the generous donor is, but it is surmised that it is Escher von der Linth.[9]

With fond regards to you and Hirst, | Your | Clausius

RI MS JT/1/TYP/7/2255–6
LT Typescript Only

1. *[10 February 1863]*: the date was suggested by LT, which is supported by mentions of the Cambridge meeting of the BAAS and the birth of Clausius's second daughter.
2. *Sie*: Clausius here used the formal address form 'Sie'.
3. *my wife*: Adelheid Clausius.
4. *second little daughter*: Helene Clausius (1863–1919), who married Karl Budde. The first daughter was Mathilde Clausius (1861–1907), who married Friedrich Zimmer.
5. *prints that you sent me of your lectures*: for *Heat as a Mode of Motion* (see letter 1962, n. 1).
6. *meeting in Cambridge*: see letter 1904, n. 9.

7. *president of the Swiss Education Council*: not identified. For information about the Swiss Education Council and secondary education in Switzerland, see M. Arnold, 'IV. —Switzerland', in *Schools Inquiry Commission. Vol. VI. General Reports of Assistant Commissioners. Burgh Schools in Scotland, and Secondary Schools in Foreign Countries* (London: Eyre and Spottiswoode, 1868), pp. 608–22.
8. *Polytechnic Institute*: see letter 1904, n. 7.
9. *Escher von der Linth*: Arnold Escher von der Linth.

To William Frederick Pollock [after 14 February 1863][1] 1973

My dear Pollock,

Would you kindly look over this preface[2] for me—I am [*1 word missing*] doubt about the last paragraph—ought it to stand, or would you advise me to expunge it?

My feelings would lead me to dedicate the book to our Managers,[3] but if a shade of doubt could attach itself to the act I would not do it—advise me here pray.

By the Managers I mean the ruling body of the Institution, not of course excluding any of its officers—but I should object to making a parade of President Vice presidents &c. Perhaps on the whole it is safest to omit it—Ask Mrs Pollock[4] what she thinks.

Yours ever | J. Tyndall

I have got a capital Valentine from Fred.[5] "Break!" "Make!"—with awful explosions as a consequence[6]

RI MS JT/1/TYP/6/2001
LT Typescript Only

---

1. *[after 14 February 1863]*: the date is given by reference to a Valentine and to Tyndall's soon-to-be published book *Heat as a Mode of Motion.*
2. *this preface*: to *Heat as a Mode of Motion.*
3. *dedicate the book to our Managers*: Tyndall in fact dedicated *Heat as a Mode of Motion* to the Members of the RI, rather than the Managers. It is unclear whether this was under Pollock's advisement.
4. *Mrs Pollock*: Juliet Pollock.
5. *Fred*: Frederick Pollock.
6. *"Break!" "Make!"—with awful explosions as a consequence*: in the Valentine, Frederick Pollock was making a reference to Tyndall's conception of fracture and regelation in relation to the structure and motion of glaciers (see *Tyndall Correspondence*, vol. 6). The Valentine is held by the RI (RI MS JT/8/3/2).

From Geraldine Jewsbury [16][1] February 1863 1974

Monday Feb 17–63 | 43 Markham Sq[ua]re | SW

Dear Dr Tyndall

Many thanks for y[ou]r kindness. I found y[ou]r Admission[2] on Tuesday at the R.I. & on Saturday (by way of <u>Valentine</u>[3]) I rec[eive]d an official card of permission; but please try to realise the aggravating <u>crossness</u> of human things—I took ill last week with what the Dr called a feverish cold & I have been strictly shut up in the house ever since—I was on my way to see Mrs Huxley[4] when I received the order to go back home I begged to come to the lecture this morning but he w[oul]d not hear of it—but being your admirer he sympathised with the disappointment w[hic]h was something. He promised I shall be out on Wednesday but I don't feel very like it at this moment still you know that in homeopathy we have no <u>after</u> purgatory from the remedies but are well altogether.

If you sh[oul]d see Mrs Huxley will you give her my love & tell her I am still a prisoner—I hope she got a short note I wrote to explain my not keeping my engagement.

I hope dear Dr Tyndall you keep well, please do, for it makes so many people sorry when you don't!

Y[ou]rs truly | <u>G. E. Jewsbury</u>

Mr John Stores Smith[5] requested me to recall him to the kind remembrance of y[ou]r self & of Mr Hirst—

RI MS JT/1/J/29
RI MS JT/1/TYP/2/656

---

1. *[16]*: although this letter is dated 17 February, it is more likely that it was written on 16 February, a Monday. Tyndall's School of Mines lectures started on Monday, 16 February.
2. *Y[ou]r Admission*: Tyndall had sent Jewsbury a card of admission to his School of Mines lecture, which was on the evening of 16 February.
3. *by way of <u>Valentine</u>*: the 'official card of permission' clearly arrived on 14 February.
4. *Mrs Huxley*: Henrietta Huxley.
5. *John Stores Smith*: John Stores Smith (1828–92), an English writer and managing director of the Sheepbridge Coal and Iron Company neat Chesterfield, Derbyshire. An acquaintance of Thomas Carlyle and Charlotte Brontë, he authored *Barnard: A Modern Romance* (1846), *Mirabeau: A Life History* (1848), and *Social Aspects* (1850) (C. Alexander and M. Smith (eds), *The Oxford Companion to the Brontës: Anniversary Edition* (Oxford: Oxford University Press), p. 476).

To Ellen Lubbock 28 February 1863 1975

The Athenaeum[1] | 28 Feb. 1863.

My dear M^rs^. Lubbock,

No doubt you have been already informed of the success of last night's lecture,[2] but probably your husband[3] will not tell how successful it was. The style throughout was clear & straightforward & the winding up admirable.

I could number on my finger ends the lectures which have been equally successful during the ten years of my connexion with the Royal Institution. Not more than three or four have been equally well received. At the conclusion there was a loud & long continued outburst of applause; this subsided, but the feelings of those present had not quite discharged themselves, & like what we call the residual discharge of electricity, a second general rumble occurred. I am quite persuaded that were the opinions of the audience polled, the universal feeling would be that M^rs^. Lubbock has a right to be proud of her husband, Johnny[4] of his father, & Lady Lubbock[5] of her son.

As for me, who it must not be forgotten, stirred up his mind for this performance, I feel quite bright by the reflected radiance. I am like a little glass bead silvered within & set in the sunshine complacently radiating from a little reflected sun pictured on my own surface. With this very pretty figure I leave you to muse pleasantly, as of course, every wife does, on the solid triumphs achieved by my friend, her husband.

Believe me | Always Yours | John Tyndall.

RI MS JT/1/T/1044

---

1. *The Athenaeum*: see letter 1907, n. 7
2. *last night's lecture*: see letter 1945, n. 2.
3. *your husband*: John Lubbock.
4. *Johnny*: John Birkbeck Lubbock (1858–1929), eldest son of John and Ellen Lubbock (R. Birkbeck, *Family of Lubbock*, p. 36). Later he would inherit the family banking business and run it until 1914.
5. *Lady Lubbock*: Harriet Caroline Lubbock (née Hotham, 1810–73), wife of John William Lubbock (1803–65, married in 1833) and mother of John Lubbock (R. Birkbeck, *Family of Lubbock*, p. 35).

To Ellen Lubbock 4 March 1863 1976

4th. March 1863

My dear Mrs. Lubbock

I sat down to write to you on Monday and thought I should have more quiet on Tuesday. Tuesday came & now it is Wednesday and thus these good intentions with which a certain place (not to be uttered for the world in your hearing) is paved, are put off from day to the sin, by repeated commission, tending more & more to become perennial.

I am glad that I wrote to Jon[1]—very glad—though I did not imagine for a moment that he would report as he did.[2] I think he must have known that he was drawing it a little strong, when he said that he lectured "very badly"[3]—Do you know what 'drawing it strong' means? Lombard street to an orange[4] you don't. I met my friend Pollock[5] yesterday & our conversation for a time was about Lyell's book.[6] Suddenly, spontaneously & without the slightest suggestion on my part he exclaimed—"What a beautiful lecture Lubbock gave us!" Now Pollock is a man whose opinion is not to be despised—A true gentleman, of great learning & highly cultivated taste. "It left nothing to be wished for" he continued—so that my report you see is not a subjective delusion of my own. (Do you know what 'subjective' means? Ten to one you don't)

If I should be fit to appear before Amy[7] on the 15th—will run down in the afternoon—All my vigour has filtered away through my hair. I am very tired. I wish I could lie down & sleep for 48 hours at a stretch. Well I must not complain—I cannot have my strength & use it. I will send "John" a copy of my book[8] tomorrow—it is now off my hands but it has left its mark behind. Goodbye.

Love to all the children.[9] Yours ever | John Tyndall

BL Add MS 49640, ff. 31–2

1. *that I wrote to Jon*: John Lubbock; letter missing.
2. *he would report as he did*: letter missing.
3. *lectured "very badly"*: see letter 1945, n. 2. In letter 1975, Tyndall wrote to Ellen Lubbock that the lecture was, in fact, excellent.
4. *Lombard street to an orange*: a Victorian expression meaning to wager something of great value against something of little value.
5. *my friend Pollock*: William Frederick Pollock.
6. *Lyell's book*: C. Lyell, *Geological Evidences of the Antiquity of Man* (London: John Murray, 1863).
7. *Amy*: Amy Harriet Lubbock (1857–1929), eldest daughter of Ellen and John Lubbock (R. Birkbeck, *Family of Lubbock*, pp. 36).

8. *a copy of my book*: *Heat as a Mode of Motion.*
9. *all the children*: for the Lubbocks' children, see letter 1945, n. 3.

From Michael Faraday 5 March 1863 1977

Royal Institution, | 5 Mar. 1863.

My dear Tyndall,

I cannot wait to see you or even to see the work[1] (the inside of it), before I thank you for it. I mean the lectures:—and I rejoice to see them in this form. Those are best off who saw them with these illustrations and apparatus, and can now see them in the book, for the great subject, in its present state, needed the special advantages of both these modes of treatment. I am very happy to be one of the number of those privileged persons.

Ever affectionately yours, | M. Faraday

Faraday Correspondence, 6:4311

1. *the work*: Tyndall's *Heat as a Mode of Motion* (1863), which was based on his 1862 lecture series at the RI, 'Twelve Lectures on Heat', delivered on Thursdays at 3pm from 23 January to 10 April 1862 (RI Guard Book, vol. 3, p. 11).

From François-Napoléon-Marie Moigno 6 March 1863 1978

M[r] l'abbé Moigno 2 rue d'Erfurt, Eglise St. Germain des Prés.
Demande du cliché inséré dans le Philosophical Magazine.

Mon cher Mr Tyndall

J'espère que vous avez reçu les livraisons de ma nouvelle Revue Les Mondes, et que vous en êtes satisfait. J'espère aussi que vous aurez bien voulu lire mon prologue et que cette lecture vous aura rendu favorable à ma cause. Le bon droit était de mon côté et je crois pouvoir compter sur la sympathie de tous. Au Cosmos j'étais esclave, dans Les Mondes je serai roi. Dans le Cosmos je n'avais pas de place, j'en aurai beaucoup dans Les Mondes. Vous pouvez donc compter que je me ferai bien mieux encore l'écho de vos admirables découvertes et de vos 10 belles leçons. Ma traduction de la fameuse leçon sur La Force a eu un succès immense. Que sera-ce dans Les Mondes. La prochaine livraison contiendrait le résumé de votre dernière leçon, si je ne croyais pas nécessaire de le montrer plus haut et je viens vous en demander les moyens. Ne pourriez vous pas obtenir pour moi de la direction du Philosophical Magazine un exemplaire du bois ou du cliché qui représente la mise en expérience de vos recherches sur la radiation à travers les gaz. Je ferais bien graver cette planche, mais cela couterait très cher, et au début d'un nouveau journal on est bien pauvre. J'ai demandé le même bienfait à Mr Miller pour les figures de la lecture sur l'analyse spectrale, et il me l'a accordée. Ma traduction de son mémoire a fait beaucoup de bruit. Vous ne me refuserez pas non plus le bon vu cliché que j'attends avec impatience pour vous consacrer de longues pages, et de belles pages.

Rappelez au souvenir de l'illustre Faraday et de tous mes amis la pauvre victime de la science dont le zèle n'est nullement ralenti, qui se sent plus jeune que jamais, et qui va commencer une nouvelle et grande campagne à la gloire de la science et des savants. Faites pour moi je vous en conjure un généreux apostolat, obtenez qu'on me suive dans mon nouveau journal, implorez pour moi des abonnements. Je suis déjà très satisfait de la manière dont on répond à mon appel, obtenez aussi que toutes les publications de Royal Institution me parviennent le plus rapidement possibles.

A vous d'estime et d'affection vives et sincères | Votre dévoué serviteur | 6 mars 1863 | 2 rue d'Erfurt l'abbé F. Moigno | Église St. Germain des Prés.

Father Moigno 2 Erfurt Street, Abbey of St.-Germain-des-Prés.
A request for the image that appeared in the Philosophical Magazine.

My dear Mr Tyndall,

I hope that you received the deliveries of my new Review Les Mondes,[1] and that you are satisfied. I also hope that you would kindly read my prologue,[2] and that reading it will convince you to side with me. Legitimacy was on my side and I think that I can count on sympathy from everyone. In the Cosmos I was a slave, in Les Mondes I will be king. In the Cosmos there was no place for me, but in Les Mondes there will be. So you can be sure that I will be better able to disseminate your admirable discoveries and your 10 fine lessons.[3] My translation of the famous lesson on Force[4] was an immense success. How much better will it be in Les Mondes. The next delivery would contain the summary of your last lesson,[5] and if I did not deem it necessary to mention it before, I am now asking you to make it possible. Could you not obtain for me, from the directors of the Philosophical Magazine, a copy of the woodblock or image that represents the experiment of your research on radiation through gases? I would have it engraved myself, but it would be very expensive, and at the start of a new journal one is rather poor. I asked the same favour of Mr Miller[6] for the figures of the lecture on spectral analysis[7] and he granted it. My translation of his article made quite a stir. Surely you would not refuse me the good image that I am eagerly awaiting in order to dedicate many excellent pages to you.

Remind the illustrious Faraday and all my friends of the poor victim of science, whose zeal has not faded, who feels younger than ever, and who will begin a grand new campaign for the glory of science and scientists. I beseech you to proselytize for me, encourage people to follow my new journal, implore people to subscribe. I am already very satisfied with the responses that my appeal has. Please also ensure that all the publications of the Royal Institution reach me as quickly as possible.

Truly yours with deep and sincere affection and esteem. | Your devoted servant | 6 March 1863 | 2 Erfurt Street Father F. Moigno | Abbey of St.-Germain-des-Prés.

RI MS JT/1/M/105

---

1. *my new Review Les Mondes*: in 1862, Moigno left his post as editor of *Cosmos* and in 1863 he founded a new journal, *Les Mondes.*
2. *my prologue*: F. N. M. Moigno, 'Science Pratique et Science Pure', *Les Mondes*, 1 (1863), pp. 1–7.
3. *your 10 fine lessons*: possibly a reference to Tyndall's series of lectures on heat at the RI in 1862, though Moigno would be incorrect in stating '10' as the series was twelve lectures, which were then published as *Heat as a Mode of Motion* in 1863 (see letter 1920, n. 4). Indeed, Moigno had some of Tyndall's work on heat published as J. Tyndall, 'La chaleur considérée comme une forme de mouvement', *Les Mondes*, 3 (1863), pp. 32–42, 66–9, 97–102, 318–24. Moigno's French translation of the entire book appeared as J. Tyndall, *La chaleur considérée come un mode de mouvement*, trans. F. Moigno (Paris: Étienne Giraud, 1864).
4. *My translation of the famous lesson on Force*: Moigno's translation of Tyndall's lecture 'On Force', given at the RI on 6 June 1862 (see J. Tyndall, 'On Force', *Roy. Inst. Proc.*, 3 (1862), pp. 527–36), appeared as J. Tyndall, 'Sur la force', *Cosmos*, 21 (1862), pp. 245–8.
5. *summary of your last lesson*: J. Tyndall, 'Exposé de quelques recherches sur la chaleur rayonnante', *Les Mondes*, 2 (1863), pp. 389–94. This was based on: J. Tyndall, 'On the Absorption and Radiation of Heat by Gaseous Matter. Second Memoir', *Phil. Trans.*, 152 (1862), pp. 59–98; and its republication as: J. Tyndall, 'On the Absorption and Radiation of Heat by Gaseous Matter. —Second Memoir', *Phil. Mag.*, 24 (1862), pp. 270–87, 337–50, 422–36. The image that Moigno was probably referring to, which showed Tyndall's experimental apparatus, appeared on p. 91 of the *Phil. Trans.* and p. 429 of the *Phil. Mag.*
6. *Mr Miller*: William Allen Miller (1817–70), an English chemist and MD who held the chair of Chemistry at King's College London and was treasurer of the RS from 1861–70 (*ODNB*).
7. *lecture on spectral analysis*: W. A. Miller, 'On the Photographic Transparency of various Bodies, and on the Photographic Effects of Metallic and other Spectra obtained by means of the Electric Spark', *Phil. Trans.*, 152 (1862), pp. 861–87. Moigno published a summary of this, without any figures, as W. A. Miller, 'Découveries récentes relatives au spectre lumineaux', *Les Mondes*, 1 (1863), pp. 293–6.

To Rudolf Clausius 10 March 1863 1979

10th March 1863

My dear Clausius

It gives me great pleasure to send you a copy of my Lectures on Heat.[1]

I also send you an abstract of a lecture which I gave[2] here some time ago on the action of the aqueous vapour of the air.

Further, I send you a short paper[3] which refers to your friend W[illia]m Thomson: it will explain itself.

Pray read the note which Thomson has appended to his article & which is meant to show that Mayer operated on false principles. I think it will surprise you.[4]

I confine my remarks to Thomson because he is my real antagonist—of Tait I know nothing.

At the time of the glacier controversy[5] in this country Thomson I learned was unsparing in his censures. My reply always was "if he has any thing to say against me let him do so openly". But he did nothing of the kind. Now however he appears in print but not in an open manner of course my duty is to compel him to state his case against me in the presence of those who are expected to give an opinion on the subject.

It is unpleasant work—causing loss of time & loss of temper—but it will never do to allow Mr. Thomson to take such liberties with impunity.

Kindest regards to Madame Clausius.[6] Goodbye | Yours ever sincerely | John Tyndall

RI MS JT/1/T/187

---

1. *Lectures on Heat*: *Heat as a Mode of Motion* (see letter 1962, n. 1).
2. *a lecture which I gave*: see letter 1948, n. 7.
3. *a short paper*: probably J. Tyndall, 'Remarks on an Article entitled "Energy" in "Good Words"', *Phil. Mag.*, 25 (1863), pp. 220–4. The reference to Thomson is to William Thomson. The article referred to is W. Thomson and P. G. Tait, 'Energy', *Good Words*, 3 (October 1862), pp. 601–7. See also *Ascent of John Tyndall*, pp. 176–7.
4. *I think it will surprise you*: Tyndall was in dispute with Joule, Thomson, and others regarding the priority to the discovery of the mechanical equivalent of heat. Tyndall supported Mayer's claim to priority over that of Joule. The dispute occupied Tyndall for the duration of the period covered by this volume. See the Introduction for further commentary.
5. *the glacier controversy*: during the late 1850s and early 1860s, Tyndall was in dispute with James David Forbes, Thomson, and others over the physics of glacier motion. See the Introductory to this and *Tyndall Correspondence*, vol. 6 and vol. 7 for further commentary.
6. *Madame Clausius*: Adelheid Clausius.

From Alexander Herschel 15 March 1863 1980

Collingwood | Hawkhurst | 1863 Mar 15th

Dear Sir

My father[1] was good enough to give me your kind message[2] offering the use of any apparatus which you could furnish for the illustration of my lecture on shooting stars and Fireballs at the Royal Institution on the 24th April.[3] Can I have the use at that time of a Ruhmkorff coil[4] and one or more large vacuum tubes either of the direct discharge construction or the inductive discharge construction? The last contrivance which shews the current in many different gases in successive cells is very beautiful but I am anxious to shew the action of a magnet upon a luminous current in causing it to deviate or be bodily transported to one side. In your lectures on heat[5] I see the great power of originating a current, which a magnet possesses, in a moving conductor, beautifully exemplified and illustrated by a pyramidal reflector (crowning a revolving piece of silver). If a passing current moves a needle, and a moving needle passes a current, but backwards, and in the opposite direction, I suppose that when motion of a needle underneath a galvanic current is permitted, a loss of power is occasioned in the electric current proportional to the effect produced. The same would happen if the current were moveable and the needle fixed. Would not then a mobile or transportable current such as the luminous currents through rarefied air become not only locomotive but also vacillatory or fitful in intensity in proportion to their mere irregular mobility, in presence of a strong magnet? A steady source of electricity and steady magnetism might thus cause the fickle beams of auroral displays by the different characters of the volumes of air successively subjected to the discharge in the sidelong progress of the current? When most stationary the discharge would be brightest.

In alluding to auroras I wish to throw a beam of light upon a basin of water to represent the 'merry dancers' or quickest and most locomotive flashes, with some fidelity, by reflection upon the ceiling. I think by proper treatment of the basin the truth may be pretty fairly imitated.

I remain dear Sir | yours most sincerely | Alex[ande]r S. Herschel.

RI MS JT/1/H/82
RI MS JT/1/TYP/2/583

---

1. *My father*: John Herschel.
2. *your kind message*: see letter 1958.
3. *my lecture on shooting stars and Fireballs at the Royal Institution on the 24th April*: see letter 1958, n. 6.

4. *Ruhmkorff coil*: an induction coil (*OED*); named after Heinrich Daniel Ruhmkorff (1803–77), a German instrument maker who commercialized the use of it (*ADB*).
5. *your lectures on heat*: see letter 1920, n. 4.

From John Frederick William Herschel 15 March 1863 1981

Collingwood | Mar 15/63

My dear Sir,

Accept my best thanks for the copy of your lectures on the Heat-Motion theory[1] which I rec[eive][d]. a day or two ago & which I shall not merely read but study as I doubt not to find in your clear form of exposition the solution of sundry misgivings which I have never been able quite to get over about the dynamics of the matter to say nothing of the difficulty of conceiving the sort of molecular motions, or the way into which vis viva[2] can be converted into (or embodied in) them.

Several of your experiments (in the little way I have read) appear wonderfully delicate.

I have also to thank you for your Lecture on Radiation through the Earth's atmosphere[3]—your brochure on "The Article headed Energy"[4]—& that on the conformation of the Alps.[5] The view taken of the enormous degrading & grinding power of Continental ice in Lyell's last work[6] struck me exceedingly. How far it is original or new I can not of course say—but put in that form it struck me as quite such a new power in Geology.

I remain | my dear Sir | Yours very truly | J. F. W. Herschel.

PS. I find April 24 has been named for my son[7] to deliver a lecture at the RI. on Meteors[8]—I presume it will be desirable for him to come up to arrange his diagrams &c. in the lecture room a day or two before, I do not know (nor I suppose does he) how these things are managed. Any hints from your own experience would be greatly valued.

RI MS JT/1/H/100
RI MS JT/1/TYP/2/522
RS HS/24/14

1. *your lectures on the Heat-Motion theory*: see letter 1977, n. 1.
2. *vis viva*: the operative force of a moving or acting body, reckoned as equal to the mass of the body multiplied by the square of its velocity (Latin; *OED*).
3. *Lecture on Radiation through the Earth's atmosphere*: see letter 1948, n. 7.
4. *The Article headed Energy*: see letter 1979, n. 3.

5. *conformation of the Alps*: see letter 1921, n. 6.
6. *Lyell's last work*: see letter 1976, n. 6.
7. *my son*: Alexander Herschel. See letter 1980.
8. *lecture at the RI. on Meteors*: see letter 1958, n. 6.

To Alexander Herschel 16 March 1863 1982

Royal Institution of Great Britain | 16th. March 1863

My dear Sir

I had replied to your father's note[1] before I opened yours.[2] I shall be here early on the day of your lecture[3] and we will then have a rehearsal of the experiment you need. I can promise you a first rate Ruhmkorff[4]—gorgeous vacuum tubes, and also a rotating luminous current—I mean rotating under the influence of a magnet. We can do almost anything in the way of optical illustration with an electric lamp, so that on that score your mind may be quite at rest. Seeing the facts you will be able to judge whether they make for or against your theoretic view. If you wish we could fix a day previous to the lecture—say one of the Easter holidays for a preliminary rehearsal. This I think would be the best plan for then if the experiments reveal anything you will be better able make your discourse square with these

Yours very sincerely | John Tyndall

Wellcome, MS. 7777/05

---

1. *your father's note*: letter 1981.
2. *yours*: letter 1980.
3. *your lecture*: see letter 1958, n. 6.
4. *Ruhmkorff*: see letter 1980, n. 4.

To John Frederick William Herschel 16 March 1863 1983

Royal Institution of Great Britain | 16th March 1863.

My dear Sir John,

During the whole of the Friday fixed for Mr Herschel's lecture[1] our lecture room will be at his disposal. Trained assistants shall be here to suspend his diagrams and otherwise execute his commands.

The most common error of our lecturers is the accumulating of more matter than can be made to fit within an hour—Our audience comes wound up for an hour, but it is not wise to push them much beyond this.

If Mr. Herschel can turn our electric lamp to account in the way of illustration it shall be also got ready for him. We can reproduce almost any optical effect by means of it. However it is best that he should not swerve from his own independent course.

Imagine the pressure on every square yard of the bed of a glacier say 1000 feet thick—Such a tool urged forward by the force from behind must exert an enormous rasping action. In fact the glacier rivers of Switzerland are loaded to an extent never observed in non glacial streams. The delta of the Rhone has been ploughed out of the mountains by the Rhone glacier.[2]

I cannot help thinking that such an agency active for ages must profoundly affect the contour of a country.[3]

Yours ever truly | John Tyndall.

RI MS JT/1/TYP/2/523
LT Typescript Only

---

1. *the Friday fixed for Mr. Herschel's lecture*: Alexander Herschel; see letter 1958, n. 6.
2. *Rhone glacier*: see letter 1915, n. 2.
3. *I cannot help . . . contour of a country*: here Tyndall was responding to Herschel's comment from the previous letter (1981): 'The view taken of the enormous degrading & grinding power of Continental ice in Lyell's last work struck me exceedingly. How far it is original or new I can not of course say—but put in that form it struck me as quite such a new power in Geology'.

From William Snow Harris 22 March 1863 1984

6 Windsor Villas | 22nd March /63

My dear Tyndall,

Professor William Thomson has never replied to my challenge to state where or when I ever affirmed that the heating effect of the ordinary electrical discharge was proportional to the quantity of electricity simply—nor has he so far as I know ever substantiated or repeated his statement, but has passed the matter by in silence, neither has Professor Whewell[1] or Mr Joule or any of those to whom I alluded in my letter in the Philosophical Magazine[2] thought it worth while to shield themselves in any way from the charge of misrepresentation I brought against them, which as gentlemen and men of honour I think they were bound to do. I have a great love for science and for scientific men, but I have a still greater regard for character and honourable dealing. I can safely say for myself that if I had unintentionally misrepresented either the gentlemen in question as they have misrepresented me and the facts had been made clear to me I should not have rested in my bed before I had made some amends and acknowledged my error. I am glad to find you read my letter which I think you will admit is a fair appeal to Mr Thomson's sense of justice.

Joule was evidently misled upon the subject, and he might certainly have made amends to me with a very good grace. I confess I am very sorry he did not, because his not having done so has left an unfavourable impression on my mind relative to him.

I quite agree with you that personal canvassing for the Royal Society is very undesirable and should not be ever resorted to, but I think you will charitably discriminate between what may be called a general canvass and a submitting to a few friends the pretentions of an individual to consideration: now I wish you particularly to understand, that, I have never solicited a vote in favor of my friend Phipson,[3] or was that the object of my notice of him to you,[4] I only desired to submit to you certain facts relative to his standing in the philosophical world leaving you to your own judgment upon the question. I do assure you that the very last idea which would enter my mind would be to ask of you to support Dr Phipson in the way of a personal favor to myself.

I really could not have paid so bad a compliment to a person whom I so much respect and regard as yourself. I think you will see upon reflection that it is by no means an unfair step to put a few friends in possession of certain facts bearing on the claims of a candidate for the Royal Society, leaving them to their own unbiased judgment in dealing with the question. I have done this with Dr Phipson with three or four of my friends and no more, which I think cannot be called canvassing in the legitimate sense of that word, if something of this kind be not done by the friends of a candidate, his claims may be possibly passed by in silence, so that I hope you will understand that I by no means wish you to take any step whatever in favor of Dr. Phipson, except upon your dispassionate consideration of his merits; I think I have very good grounds for my recommendation of him—I met with him about five years since in Paris on the occasion of my going to Cherbourg at the request of the Minister of Marine and the Emperor to fit my lightning conductors in the French Frigate "L'Imperieuse"[5] Dr. Phipson was then engaged with Moigno in bringing out "Cosmos" a periodical French Journal.[6] He was an ardent student and much esteemed by Members of the Institute. I had a very high character of him from many influential quarters—his modest unassuming manners and zeal in literary and scientific pursuits had won golden opinions in his favor. I regret to find by your letter[7] that his chemistry does not find great favour with your chemical friends but after all a great deal of that may arise from difference of opinion on some particular subjects and should certainly not militate against him upon consideration of his pretentions as a whole, such discrepancies have arisen before in the case of many deserving men—but I will say no more on this subject, I only beg of you to acquit me of any attempt to influence your decision unfairly relative to the election of Dr Phipson into the Royal Society.

I have read with very great pleasure and interest your paper on "Radiation through the Earth's Atmosphere"[8] it contains a great many observations and remarks of singular value to me and I may perhaps trouble you with a short communication on the subject, there is a good deal certainly of debatable ground in it, more especially as relates to the constitution of aqueous vapour, and the existence of atoms and molecules and the wide spaces between[9]

RI MS JT/1/TYP/2/478–9
LT Typescript Only

---

1. *Whewell*: William Whewell (1794–1866), an English philosopher and historian of science, Master of Trinity College, Cambridge, and author of, among many other books, *History of the Inductive Sciences*, 3 vols (London: John W. Parker, 1837) and *The Philosophy of the Inductive Sciences, Founded Upon Their History*, 2 vols (London: John W. Parker, 1840). In 1833, Whewell coined the term scientist (*ODNB*).
2. *my letter in the Philosophical Magazine*: probably W. S. Harris, 'On Some Special Laws of Electrical Force', *Phil. Mag.*, 14 (1857), pp. 156–9.
3. *Phipson*: Thomas Lamb Phipson (1833–1908), a British astronomer and amateur violonist, known for, among many other papers and books, *Meteors, Aerolites, and Falling Stars* (London: Lovell Reeve, and Co., 1867) (C. J. Bouverie, *The Scientific and Literary Work of Dr. T. L. Phipson, F.C.S.* (London: Wertheimer, Lea & Co., 1884).
4. *my notice of him to you*: letter missing.
5. *French Frigate "L'Imperieuse"*: the French fifth rate Junon class frigate, L'Imperieuse, built in 1787, was captured for the Royal Navy off La Spezia by HMS *Captain* in 1793, and refitted for service as HMS *Imperieuse.* She was renamed HMS *Unite* in 1803, leaving active service in 1815 and was eventually broken up at Chatham in 1858.
6. *"Cosmos" a periodical French Journal*: by the time this letter was written, Moigno had left *Cosmos* and started his own journal, *Les Mondes* (see letters 1999 and 2068).
7. *your letter*: letter missing.
8. *your paper on "Radiation through the Earth's Atmosphere"*: see letter 1948, n. 7.
9. *wide spaces between*: fragment only; letter incomplete.

From James P. Joule 23 March 1863 1985

Old Trafford, Manchester | Mar. 23, 1863

My dear Sir

Many thanks for your kindness in sending me copies of your papers.[1] I am sorry I have nothing of much importance to send by way of return. I send by this post the last no. of Proceedings of our Phil[osophica][1] Society in which I profess to have detected the heat radiated by the moon by means of a sort of wind thermometer.[2] Our Secretary, Mr Baxendell[3] says that Melloni had observed the same thing, but I have not yet been able to refer to his paper.[4]

You will observe the lucubrations of one of our local luminaries Mr Dyer.[5] I was absent from the meeting or would have said something more than what is written in the little note appended,[6] which note I intended as a protest. Mr Dyer & I have disputed on these matters for a good share of 20 years but now he seems to be attacking higher games. I dare say you will not think it worth while to write in reply to his observations. I once invited him to go with me to Niagara and offered to pay our joint expenses if a higher temperature was not found at the bottom than at the top of the falls.

Believe me Dear Sir | Yours very truly | James P. Joule

RI MS JT/1/J/139
RI MS JT/1/TYP/2/663

---

1. *copies of your papers*: Tyndall's letter and its enclosures have not been found.
2. *by means of a sort of wind thermometer*: J. P. Joule, 'On a New and Extremely Sensitive Thermometer', *Proceedings of the Manchester Literary and Philosophical Society*, 3 (1862–4), pp. 73–4 (communicated to the society, not by Joule, on 11 March 1863). Joule's short communication was reprinted in the *Phil. Mag.*, 25 (1863), pp. 320–1. Also reprinted as 'A Sensitive Thermometer', *The Electrician*, 10 April 1863, p. 269, with the following editorial note: 'A few months ago, Professor Tyndall, the popular lecturer at the Royal Institution, London, described experiments which he had made, and which he alleged proved that the moon radiated cold, instead of heat. Dr. Joule's new and sensitive instrument appears to negative, in the most direct manner, the Professor's conclusion'.
3. *Mr Baxendell*: Joseph Baxendell (1815–87), a Manchester-born meteorologist and astronomer who constructed a number of major telescopes around Britain for the Royal Astronomical Society, RS, and other institutions. From 1849 he made a number of significant observations of variable stars which he reported to the RS, and was elected FRS on this account in 1858. It was in this year too, that he joined the Manchester Literary and Philosophical Society, to which he made significant contributions throughout the rest of his career; he was one of the founders of the physical and mathematical section of the society, which saw science come to the fore in the society as a whole. In 1859 he succeeded Henry Halford Jones as astronomer to the corporation of Manchester, and later, in 1871, of Southport (*ODNB*).
4. *his paper*: M. Melloni, 'Sur la Puissance Calorifique de la Lumière de la Lune', *Comptes Rendus*, 22 (1846), pp. 541–4. This was also published as M. Melloni, 'On the Calorific Power of the Light of the Moon', *Electrical Magazine*, 2 (1846), pp. 340–3. Tyndall shared about Melloni's experiment in his lectures on heat (see *Heat as a Mode of Motion*, p. 422).

5. *Mr Dyer*: Joseph Chessborough Dyer (see letter 1914, n. 6). Joule's biographer, Donald S. L. Cardwell, notes that Joule is here transferring his own criticisms onto the 'scientifically insignificant' Dyer, and suggests that this was typical of Joule, who did not relish public disputation (D. S. L. Cardwell, *James Joule: A Biography* (Manchester: Manchester University Press, 1990), pp. 209–10).
6. *the little note appended*: J. C. Dyer, 'Notes on the Action of Heat and Force upon Matter', *Proceedings of the Manchester Literary and Philosophical Society*, 3 (1862–4), pp. 77–81 (communicated to the society on 11 March 1863). This is largely Joule's summary of Dyer's paper.

From William Snow Harris 24 March 1863 1986

6 Windsor Villas | Plymouth | 24 March 1863.

My dear Tyndall,

I fully comprehend the object of your letter of this morning's post.[1] It is consistent and considerate in every respect. I see now I was a little out of order in venturing to ask you to see Dr. Phipson but I did so very innocently.[2] I had no idea of his personally soliciting your patronage[3] but I merely thought it would afford an opportunity of your becoming better acquainted with his pretensions and enable you to put a few questions to him relative to his scientific pursuits and of course of his studies on the Continent and elsewhere as also relative to the medal awarded him by the Dutch Academy of Sciences at Haarlem.[4] All this I thought might well take place without any question arising relative to your support

Let all this however be now done with. We understand each other.

I am glad to learn that you mean to call upon Thomson to retract his statement relative to you.[5] I do not think it worth while in my own case to trouble him further in the matter. I think his own conscience will be a sufficient punishment for him.

I really wish I could one day have the pleasure of seeing you here and showing you my electrical instrument[6] which I consider now quite perfect, and conscientiously believe there are no instruments of their class in Europe so perfect.

I have lately revised and extended my researches relative to Electrical charge, a subject which I spoke to you about in a letter now a long time since[7] and one of very deep interest and have hence been led to draw up a paper on the subject for the Royal Society although without much expectation of its being honoured with a place in the Philosophical Transactions. I fear my electrical views are not considered sufficiently orthodox but I really cannot mould them so as to suit what I consider fallacious theories. Rely on it the time will soon come when the theoretical electricity of the schools must be completely revised I have some new and beautiful experimental results upon my experimental ledger. The following is an extremely curious result which you may rely on as being correct and demonstratable by rigid experiment.

Let A.B. be two spheres attracting each other at given distances, A. being charged to a given intensity B. neutral and free to take on any inductive charge which A can induce in it. How the force between these two spheres varies in the inverse duplicate ratio of the square of the tangent m.n. drawn from the touching point of one of the spheres to the other sphere. There are of course a series of consecutive steps leading to this *[distinction]* I will *[not]* however trouble you a present with details[8]

J. Tyndall Esq. F.R.S. etc.

RI MS JT/1/TYP/2/480
LT Typescript Only

---

1. *your letter of this morning's post*: letter missing.
2. *venturing to ask you to see Dr. Phipson but I did so very innocently*: see letter 1984.
3. *soliciting your patronage*: Phipson had appealed to Tyndall for support in election to the RS.
4. *Dutch Academy of Sciences at Haarlem*: the Hollandsche Maatschappij der Wetenschappen (Dutch Society of Sciences and Humanities) was established in 1752, and remains the oldest scientific society in that country. It was, and is still located in the city of Haarlem.
5. *his statement relative to you*: presumably this refers to Thomsons's statement in W. Thomson and P. G. Tait, 'Energy', *Good Words*, 3 (1862), pp. 601–7. See letter 1979, and the Introduction for further commentary.
6. *my electrical instrument*: not identified.
7. *in a letter now a long time since*: possibly a letter of 21 February 1857 (see letter 1342, *Tyndall Correspondence*, vol. 6).
8. *with details*: fragment only; letter incomplete.

From Geraldine Jewsbury 25 March 1863 1987

43 Markham Square. | S. W. | March 25—1863—

Dear Dr Tyndall—you said you were going to take a holiday & read some novels.[1] I am under a Destiny w[hic]h. obliges me to read them (as a Confectioner eats sweets) in the way of business so I venture to send you a list of novels[2] that well are worth reading. May they fulfil their mission & amuse you! Also I venture to beg y[ou]r acceptance of a copy of the last novel I wrote[3]—novels are scarcely worth the room they take up on shelves—but you can at any time turn it over to a friend going to the Antipodes or to live in the back woods. In both which conditions people are thankful for very small mercies in the reading way!—

I am sorry the Lectures[4] are coming to a close for tho I c[oul]d. not stand an examination still I am sure one takes up a great deal of instruction w[hic]h. goes to "the strengthening & refreshing" of our inner life tho' we may not be able to give the tale of it in words just as food nourishes the body one does not know how. —When you have an hour, not laid out before it comes,—I sh[oul]d. be so glad if you w[oul]d come & see me—evening of course you sh[oul]d have some tea—a most charming woman[5] wants to make y[ou]r acquaintance you warm her heart by y[ou]r kindness to her son[6]—now at Woolwich[7] where he thinks of your Lectures to w[hic]h he now cannot come—

Don't plague y[ou]r self to answer this note I know the bother of writing notes when busy—but accept my grateful thanks for letting me come to y[ou]r lectures—(w[hic]h gratitude has an eye to future favours of a similar kind

& I am dear Dr Tyndall | y[ou]rs very truly & respectfully | Geraldine E. Jewsbury

RI MS JT/1/J/30
RI MS JT/1/TYP/2/657

1. *you said . . . read some novels*: letter missing.
2. *a list of novels*: enclosure missing.
3. *the last novel I wrote*: G. E. Jewsbury, *Right or Wrong*, 2 vols (London : Hurst and Blackett, 1859).
4. *the Lectures*: this refers to Tyndall's lecture series at the Government School of Mines and Science Applied to the Arts, established in 1851 (and from 1863 known as the Royal School of Mines). Tyndall had held the Chair of Natural Philosophy there since 1859. Jewsbury clearly attended the 1863 series of thirty lectures on 'Physics', which began on 16 February (*Ascent of John Tyndall*, p. 173).
5. *a most charming woman*: not identified.
6. *her son*: not identified.
7. *Woolwich*: the unidentified son would have been enrolled at the Royal Military Academy at Woolwich.

From George Hector Tyndale 27 March 1863 1988

Philadelphia 27th March 1863

My dear John

Before going back to my command[1] I must say a few words of remembrance to yourself and our good friend Tom Hirst. Your reply to my letter,[2] written when very ill yet, came duly as did a short while after, your relation to the Tindalls of Wexford[3]—quite interesting to all of us. Tom's good, kind letter to me came too. My heart warms to him for his goodness. He is the first Englishman who seems to recognise <u>our</u> nationality—he says "your Fatherland". That is the word: We are <u>a people</u> & must be so acknowledged. But now of myself.

I had thought of resuming duty shortly after I wrote you last, but the disease which was in me for some time previous to the time of my wound, has been very slow in yielding—in fact not yet gone. Typhoid fever with diarrhea had nearly used me up. I am better now however, and though not well shall go to my command in a day or two. Last December I believe the President[4] sent my name to the Senate for promotion to a Brigadier Generalship—this was unknown to me until January and was therefore unsolicited by me. I have not yet received the Commission as all parties at the Government are hurried so, but suppose it will come after more important things are done. Until that coming I still am the Lieut[enant]: Col[onel]: of the 28th Reg[imen]t: P[ennsylvani]a: Vol[unteer]s:; but presume that it will come. It is no folly which make me dread the responsibility of the office, no mockery. But I do dread it and feel ashamed that I should take a trust involving life and health of thousands of brave men. Yet I <u>do</u> accept it—perhaps from perversity, perhaps from vanity, ambition or other unknown organic motive.

I have sent you two papers, one with a speech of Wendell Phillips[5] the great Abolition orator, in which he uses my name to point a moral, and while he gets the point he wants, he misstates facts concerning my position as Brig[adier]: Gen[era]l: &c. I did burn part of Harper's Ferry[6] as a military necessity—and certainly Mr. Phillips does not hold the circumstances to be more curious—providential—than I do. Certainly my post then and all circumstances connected are strange. Why should it have happened that out of the tens of thousands of officers in the Service I the only one with any connection with John Brown[7] should have accidentally been posted then in command in the position of Colonel <u>at least</u>—I was then only a Major? It was wholly accidental as my small relation to the raid or rather the body of John Brown was unknown to every officer above me. The other paper makes a complimentary notice of my appointment. I was ashamed to send you some other papers which praised too much.

I send you some photographs, of which you can give one or other to Tom Hirst. Please hand the enclosed letter to him. We are all well Mrs Tyndale[8] sends kind remembrances to Mr. Hirst & yourself. Her only brother[9] was wounded at Fredericksburg[10] but he has recovered & returned to his regiment, promoted to the Adjutancy of his regiment.

Affectionately yours | Hector Tyndale

I shall be glad to hear as often as you can spare time of yourself Tom Hirst & Debus. | H. T.

RI MS JT/1/T/53

---

1. *my command*: this refers to Tyndale's military service during the American Civil War; see below in letter.
2. *Your reply to my letter*: see letters 1946 and 1953.
3. *your relation to the Tindalls of Wexford*: see letter 1956.
4. *President*: Abraham Lincoln.
5. *speech of Wendell Phillips*: Wendell Phillips (1811–84), an American attorney (he graduated from Harvard Law School) and abolitionist. Unlike many who opposed slavery rather than argue that slavery was unconstitutional, he attacked the constitution as endorsing the inhumane practice. He was also an outspoken advocate of female suffrage. The speech referred to was given at Plymouth Church, Brooklyn (New York) on 3 February 1863 (see 'Our Future', *New York Times*, 4 February 1863, p. 8).
6. *Harper's Ferry*: see letter 1912, n. 4.
7. *John Brown*: John Brown (1800–59), an American abolitionist most noted for his role in the raid on Harper's Ferry on 16 October 1859 which most historians agree escalated national tension and led to secession and later the American Civil War (1861–5) (*ANB*).
8. *Mrs Tyndale*: Julia Tyndale.
9. *Her only brother*: not identified.
10. *Fredericksburg*: the Battle of Fredericksburg was fought between December 11–15, 1862 in Fredericksburg, Virginia.

To Juliet Pollock 28 March 1863 1989

To the Moon[1]
1863, Feb. 14th.

Say does the crimson of the drooping rose
When soft it falls upon delighted eyes
Close up those eyes against the glorious sun
Which gave all flowers their odours and their bloom?
Or does the song of lark and nightingale
Mingling at dawn along the Devon shore
Make the full heart less fitted to enjoy

The grander music of the gleaming sea?
Is it not rather so, that where a love
So large as that which fills my soul for thee
Unlocks the doors, the smaller loves of earth
Troop in without disturbance to the great?
Dismiss thy fear; retract thy strong reproach
And bend thy beauty o'er me as of yore.
Nor Bromine richly brown, nor Chlorine green—
Nor Aqueous Vapour which the praying earth
Swings from her censers underneath thy beams,
Has ever caused my love to swerve from thee.
These are but melodies of minor note
Which mingle with that grander holier strain
My soul for ever sendeth to that heaven
Where thou dost reign, the Queen of all the stars.
Dissolve those clouds, unpucker that fair brow,
Nor let thy lover for a moment deem
The shock of worlds could move thy steadfastness
Thou'rt bright once more—come nearer then my love,—
Still nearer—stoop—a little lower—there!
I kiss thy silver cheek, Goodnight! Goodnight!

———

Thus nobly mated we shall love through time—
Our time—and send the memory of our love
To other times; a torch to kindle trust
To burn up doubt, and give the sinking hearts
Of men reliance on the force of love.
I knew thee by thy eyes (J's)
Hamlet.[2]

———

The queenly Moon commands the plastic Sea
Which rolls around the world its silvered brine
And thus on Sunday evening drawn by thee
I'll roll from woodless 'woods'[3] to 59.[4]

J. T. | March 28.

RI MS JT/1/TYP/6/2002–3
LT Typescript Only

1. *the Moon*: Tyndall had compared Juliet Pollock to the Moon. See letter 2107.
2. *Hamlet*: the quote has not been identified.
3. *woodless 'woods'*: probably a reference to Tyndall's residence in London: 14 Waverley Place, St. Johns Wood.
4. *59*: a reference to 59 Montagu Square in Marylebone, London, where the Pollocks had their primary residence.

From François-Napoléon-Marie Moigno 28 March 1863 1990

Mon cher Mr Tyndal

Je vous dirai bientôt les clichés que je désire avoir, et à qui vous devrez les remettre. Aujourd'hui je réponds à votre question relativement à Mr Seguin.

Mr Seguin a d'abord énoncé la théorie de la convertibilité de la chaleur en force dans une lettre à Sir John Herschel.

Il a développé cette théorie dans son ouvrage sur les chemins de fer publié en 1837 et 1838 j'ai dit à Mr Tremblay de vous l'envoyer immédiatement.

Il a calculé pour la première fois l'équivalent mécanique de la chaleur après Joule en 1847. Sa note est dans les comptes rendus de l'Académie tome 25 p. 420.

Dans une leçon faite le 25 janvier 1856 à Royal Institution Mr Grove a dit en effet que calculées d'après les *[coefficients]* de Regnault les expériences de Mr Seguin donneraient 1666 au lieu de 500 Cosmos tom 8 p. 224.

Tout dernièrement Mr Seguin a écrit à Mr Joule une lettre imprimée dans le Mecanic's Magazine et les Chimical notes? où vous trouverez des renseignements utiles.

Vous avez vu que j'ai déjà traduit votre préface. A quelle condition votre éditeur accorderait-t-il le droit de reproduction en français et donnerait-il des clichés de vos bois? Seriez-vous bien que cette traduction fut bientôt faite?

Vous ne m'avez pas dit si vous avez lu mon prologue. Je tiens beaucoup à ce que vous le lisiez.

Avez-vous été assez bon pour recommander mes Mondes à vos confrères les physiciens de Royal Society; Wheatstone, Grove, Gassiot, Miller, General Sabine, etc. Faites pour moi je vous en prie un *[peu]* de propagande. Je mérite, je crois de réussir.

Mes respects profonds et respectueux à Mr Faraday. Votre humble et dévoué serviteur. | l'abbé F. Moigno | 28 Mars 1863

My dear Mr Tyndal,

Soon I will tell you about the images that I would like to have, and to whom you should give them. Today I am answering your question about Mr Seguin.

Mr Seguin first stated the theory of converting heat into energy in a letter to Sir John Herschel.[1]

He developed this theory in his book on railways[2] published in 1837 and 1838; I told Mr Tremblay[3] to send it to you immediately.

He calculated for the first time the mechanical equivalent of heat after Joule in 1847. His note is in the reports of the Academy, volume 25 p. 420.[4]

In a lesson presented on January 25th 1856 at the Royal Institution,[5] Mr Grove said, in effect, that according to the calculations of Regnault's[6] *[coefficients]*, Mr Seguin's experiments would give 1666 instead of 500, in Cosmos vol. 8 p. 224.[7]

Most Recently, Mr Seguin wrote a letter to Mr Joule, which was printed in the Mechanic's Magazine[8] and the Chemical News,[9] where you will find useful information.

You saw that I have already translated your preface.[10] Under what conditions will your publisher grant the right of reproduction in French, and will he provide copies of your woodblock prints? Are you content to have the translation finished soon?

You did not tell me if you read my prologue.[11] I am very much counting on you to read it.

Have you been kind enough as to recommend my Mondes to your physics colleagues in the Royal Society; Wheatstone, Grove, Gassiot, Miller,[12] General Sabine, etc. I beg you to do a little bit of propaganda for me. I deserve it, I believe I will succeed.

My profound and respectful wishes to Mr Faraday. Your humble and devoted servant. |

Father F. Moigno | 28 March 1863

RI MS JT/1/M/106

1. *letter to Sir John Herschel*: M. Seguin, 'Letter to Sir J. Herschel:—Observations on the effects of heat and of motion', *Edinburgh Philosophical Journal*, 10 (1824), pp. 280–3. The letter itself is dated 12 September 1822.
2. *his book on railways*: Marc Séguin, *De L'Influence des Chemins de Fer et de l'art de les tracer et de les construire* (Paris: Carilian-Gery et Dalmont, 1839).
3. *Mr Tremblay*: a publisher in Paris, notably of *Cosmos*, especially after Moigno was no longer an editor.
4. *reports of the Academy, volume 25 p. 420*: M. Sequin, 'Note à l'appui de l'opinion émise par M. Joule, sur l'identité du movement et du calorique', *Comptes Rendus*, 25 (1847), pp. 420–2.
5. *In a lesson . . . at the Royal Institution*: W. R. Grove, 'Inferences from the Negation of Perpetual Motion', *Roy. Inst. Proc.*, 2 (1854–6), pp. 152–9.
6. *Regnault's*: Henri Victor Regnault.
7. *Cosmos vol. 8 p. 224*: W. Grove, 'Conséquences de la Négation de la Possibilité du Mouvement Perpétuel', *Cosmos*, 8 (1856), pp. 219–24; the mention of '500' and '1666' is on p. 224.
8. *Mechanic's Magazine*: 'Séguin to Joule', *Mechanics Magazine*, 30 January 1863, pp. 78–9.
9. *Chemical News*: 'Séguin to Joule', *Chemical News*, 10 January 1863, pp. 21–2.
10. *your preface*: the preface to *Heat as a Mode of Motion.*

11. *my prologue*: see letter 1978, n. 2. Moigno had urged Tyndall to read this prologue in letter 1979.
12. *Miller*: see letter 1978, n. 6.

To Alexander J. Ellis[1] 30 March 1863 1991

Copies of Letters. | 1. Prof. Tyndall to Mr Ellis.
Royal Institution 30th March 1863

Dear Sir—I have no hesitation in stating that the work of Professor Helmholtz to which you refer[2] is excellent,—incomparably superior, in my opinion, to anything hitherto written on the same subject. —It is also written so clearly as to be quite intelligible to the general reader. I have had some thoughts myself of urging its translation. There cannot be a doubt that it would be a most important addition to our scientific literature.

Yours very truly | John Tyndall.

Professor Helmholtz I may add is a man of the very highest scientific eminence. His name alone is a guarantee of the superior character of any work to which it is appended.

J. T.

Archiv der Berlin-Brandenburgischen Akademie der Wissenschaften 131

---

1. *Alexander J. Ellis*: Alexander J. Ellis (1814–90), an English philologist and mathematician, later the translator of Helmholtz's work *On the Sensations of Tone as a Physiological Basis for the Theory of Music* (London: Longmans, Green, & Co., 1875).
2. *to which you refer*: letter missing. The reference is probably to Helmholtz's *Die Lehre von den Tonempfindungen als Physiologische Grundlage für die Theorie der Musik* (Braunschweig: F. Vieweg und Sohn, 1863). Ellis went on to translate the 1870 third German edition into English in 1875 with Helmholtz's approval (see n. 1).

To Dr. Lawson[1] [March 1863][2] 1992

Royal Institution of Great Britain | Saturday

My dear Sir.

The Royal Society Council grant the use of the plate, but they require that the copies should be struck of by

Mr Basire[3] | 7 Princes Street | Lincoln's Inn Fields

the Society's engraver.

The plate however will be in the hands of Messrs Longman[4] till about

Friday next—after that M[r] Basire would print you as many as you require. You of course will arrange your terms with him.

You speak of a "block", but I suppose you mean the copper plate at the end of my book on Heat. On this point you will please be precise on writing to M[r] Basire.

Yours truly | John Tyndall

D[r] Lawson

Wellcome, MS. 7777/81–81a

1. *Dr. Lawson*: not identified.
2. *[March 1863]*: dated in relation to content. The line 'The plate however will be in the hands of Mess[rs] Longman till about Friday next' implies that Tyndall's book *Heat as a Mode of Motion* was in the process of being published (published 4 March 1863).
3. *M[r] Basire*: James Basire III (1796–1869), a London engraver, like his father and grandfather before him, who did the engravings for figures that were printed in the RS's journal *Phil. Trans.* (*ODNB*).
4. *Mess[rs] Longman*: the brothers Thomas Longman (1804–79) and William Longman (1813–77) of Longman, Green, Longman, Roberts, & Green, publishers of Tyndall's *Heat as a Mode of Motion.*

To Thomas Archer Hirst [3 April 1863][1] 1993

My dear Tom

It is Good Friday and the post is starting. I have just time to say that I should hardly like writing directly to any member of the Council of Military Education[2] and chiefly for this reason: when I resigned my examinership[3] I wrote strongly recommending you,[4] but they took no notice of my recommendation. One does not like to have such things repeated.

Of course all my pen can do in the way of testimony shall be done with all my heart. Just think what I can do and let me know.

Yours ever | John Tyndall

RI MS JT/1/HTYP/523
LT Typescript Only

1. *[3 April 1863]*: the date is given by relation to letter 1994, and by reference to Good Friday, which in 1863 fell on 3 April.
2. *Council of Military Education*: from 1856–7, Tyndall wrote and graded examinations in scientific topics for commissioned officers of the Royal Artillery and Royal Engineers at

the Royal Military Academy at Woolwich. He stopped being an examiner when the War Office decreased the pay in April 1857 (see letters 1375 and 1376, *Tyndall Correspondence*, vol. 6). In January 1858, he again became an examiner for commissioned officers of the Royal Artillery and Royal Engineers at the Royal Military Academies at both Woolwich and Sandhurst for the newly-established Council of Military Education, which advised the British military on matters of efficiency and civic profession, but resigned again in 1861 (see n. 3). For more on Tyndall's and others' roles as examiners for military education, see J. Elwick, 'Economies of Scales: Evolutionary Naturalists and the Victorian Examination System', in G. Dawson and B. Lightman (eds), *Victorian Scientific Naturalism: Community, Identity, Continuity* (Chicago: University of Chicago Press, 2014), pp. 131–56.

3. *when I resigned my examinership*: Tyndall resigned his examinership in 1861 over the issue of his delegating the grading of papers from an exam on 22 January 1861 to his friend Edward Frankland which led to a delay in his turning them in (Journal, RI MS JT/2/13c/1198).
4. *recommending you*: Tyndall recommended Hirst for the position of assistant Professor of Mathematics at the Royal Military Academy at Woolwich (see letter 1994).

To Joseph Edward Addison[1] 7 April 1863 1994

[copied to Thomas Hirst][2]
Clarendon Hotel Chale | Isle of Wight 7th Ap. 1863

My dear Col[onel]. Addison.

I trust to your good nature to pardon my present act if, judged from my position, it should seem an unwarranted liberty. It has come to my knowledge that the Council of Military Education will probably appoint an assistant Professor of Mathematics in the Academy of Woolwich,[3] and I thought on receiving the intelligence that I might without presumption introduce to your notice a gentleman who is certainly eminently qualified for the post. This is Mr. Thomas Archer Hirst, at present head mathematical master in University College School. He is a Fellow of the Royal Society and has been educated under some of the most eminent mathematicians of the Continent. He is a pupil of Derichlet's,[4] of Steiner,[5] Louisville[6] & Chasles.[7] His own position as a mathematician may be inferred from the original investigations which he has published in the Philosophical Transactions, the Philosophical Magazine, the Comptes Rendus,[8] Crelle's Journal,[9] & the Journal of Tortolini.[10] He has an excellent knowledge of physics; he is perfectly good tempered, and I am acquainted with no man more likely to live in peace & harmony with his colleagues. He has had considerable experience as a teacher, and wherever he has taught he has invariably won the respect & confidence of his pupils. Of his personal character it belongs perhaps to me to speak <more> than to any other man, for although he is esteemed by all who know him, there are few so well acquainted as I am with his private excellence. I know he will do his duty with wholeheartedness: <and> that should he be appointed to Woolwich he will justify by his deeds all that I have here written regarding him.

Believe me | Most sincerely yours | John Tyndall

Approximately correct | J. T. | Frankland has written a strong letter to Sir Fred[eric]k. Abbott.[11]

RI MS JT/1/T/657

---

1. *Joseph Edward Addison*: Joseph Edward Addison (1821–90), a British Army officer and member of the Council of Military Education. He had been promoted to colonel in 1861 (*Modern English Biography*).
2. *[copied to Thomas Hirst]*: comments at the end of the letter suggest that this is a copy of a letter to Addison that Tyndall forwarded to Hirst. See letter 1993.
3. *Academy of Woolwich*: the Royal Military Academy at Woolwich.
4. *Derichlet*: Johann Peter Gustav Lejeune Dirichlet (1805–59), a German mathematician (*ADB*).
5. *Steiner*: possibly Jakob Steiner (1796–1863), a Swiss mathematician (*HLS*).
6. *Louisville*: possibly Joseph Liouville (1809–82), a French mathematician (*CDSB*).
7. *Chasles*: Michel Floréal Chasles (1793–1880), a French mathematician known for his work in geometry (*CDSB*). His work *Traité des Sections Coniques, faisant suite au Traité de Géométrie Supérieure* was published in 1865 (Paris: Gauthier-Villars).
8. *Compte Rendus*: the common name for the French journal *Comptes rendus de l'Académie des Sciences.*
9. *Crelle's Journal*: the common name for the *Journal für die reine und angewandte Mathematik*, founded in 1826 by the German mathematician August Leopold Crelle (1780–1855).
10. *Journal of Tortolini*: *Annali di scienze matematiche e fische*, a journal founded in 1850 by the Italian mathematician Barnaba Tortolini (1808–74).
11. *Sir Fred[eric]k. Abbott*: Frederick K. Abbott (1805–92), a British Army officer and engineer of the East India Company. In 1859, he was appointed to the Royal Commission on the Defense of the United Kingdom which embarked on a program to fortify the nation's dockyards. He later served on a number of royal commissions (*ODNB*).

From Mary Somerville[1] 9 April 1863 1995

Spezia[2] 9th April 1863

Dear Dr Tyndall

What a wonderful branch of science you have created by your experiments on the absorption and radiation of radiant heat,[3] I cannot tell you how much they have interested me. The vast superiority in absorptive power of the compound gases & vapours over that of dry air and the elementary gases, and the enormous increase produced by tension, gives an entirely new view of the energy of the compound molecules of aëriform matter. The discovery of the compound nature of perfumes[4] seemed to be the very limit of molecular action, but even that is surpassed by Dynamic absorption and radiation the whole of which is truly marvelous, for although the quantities of vapour are minute beyond imagination, the experiments give the conclusion that they and their molecules have as true an existence as the objects around us. I shall only add that you make it clear that one might live very comfortably in Mercury if his atmosphere had enough of aqueous vapour

Accept my best thanks and | Believe me truly yours | Mary Somerville

RI MS JT/1/S/80

---

1. *Mary Somerville*: Mary Fairfax Somerville (1780–1872), a Scottish science writer and polymath, mathematician, and astronomer. Somerville, who was largely self-taught was, alongside Caroline Herschel, nominated as one of the first female members of the Royal Astronomical Society. She had married in 1804 and was widowed in 1807. Her status as a widow gave her a limited financial independence as well as more social freedom than she would have otherwise enjoyed, which allowed her to seek out teachers and other scholars of mathematics. In 1812 she remarried, and her husband William Somerville, an army doctor, was supportive of her studies. She made connections among the mathematical and scientific community in London, and, following a visit in 1817–8, also on the continent. Her first published work appeared in 1821, but it was her work popularizing Laplace's five-volume *Mécanique céleste* (1798–1827) describing his nebular hypothesis which really brought her to notice. She consulted with John Herschel, Augustus De Morgan, and Charles Babbage during the preparation of this work. Her second book *On the Connection of the Physical Sciences* (1834) brought her into contact with Faraday, Lyell, Whewell, Ampère, and Becquerel. It went through nine editions in her lifetime, and appeared in American, French, Italian and other languages. Somerville was recognised and received awards by many scientific organizations. When John Stuart Mill organized a petition to Parliament to give women the right to vote, Somerville was among the first to sign (*ODNB*).
2. *Spezia*: La Spezia is a city on the west coast of northern Italy.
3. *your experiments on the absorption and radiation of radiant heat*: by this time Tyndall had published several papers on the absorption and radiation of heat, which Tyndall may have sent her copies of: J. Tyndall, 'On the Absorption and Radiation of Heat by Gases and Vapours, and on the Physical Connexion of Radiation, Absorption, and Conduction', *Phil. Trans.*, 151 (1861), pp. 1–36; J. Tyndall, 'On the Absorption and Radiation of Heat

by Gaseous Matter. Second Memoir', *Phil. Trans.*, 152 (1862), pp. 59–98; and J. Tyndall, 'On the Relation of Radiant Heat to Aqueous Vapour', *Phil. Trans.*, 153 (1863), pp. 1–12.

4. *perfumes*: Somerville is referring to Tyndall's second memoir (the 1862 paper in n. 3), in which he discussed the measurement of scents.

To François-Napoléon-Marie Moigno 10 April 1863 1996

10th. April 1863

My Dear Mons. l'Abbé.

I was obliged to quit London to get a little rest & for the last 10 days I have been staying in the Isle of Wight. I wrote to Messrs. Longman[1] regarding the reproduction of my book:[2] they reply to me[3] in the following words:—"You may inform your Paris correspondent that we shall sell a set of the woodcuts to your work on heat for £25 to be paid for in cash on delivery of the casts to his agent in London".

With regard to the book itself I think no difficulty w[oul][d]. stand in the way of an arrangement. I think I could induce Mr. Longman to agree to any reasonable arrangement. My idea is that you should pay nothing for the book until after your profits had reached a certain sum—say £50. After that you would perhaps be willing to allow the publisher a share in the profits.

If any other proposition should appear to you more reasonable pray let me know. I shall do what I can to arrange matters.

Many thanks for the information you have given me regarding M. Séguin.[4] I shall return to London on Saturday next, and shall be glad to hear from you.

Yours most truly | John Tyndall

Staatsbibliothek zu Berlin (Stabi) Slg Darmstaedter 1855 Tyndall Be. 1–58, 40–1

1. *I wrote to Messrs. Longman*: letter missing; see letter 1992, n. 4.
2. *reproduction of my book*: Tyndall's *Heat as a Mode of Motion* (published in March 1863) appeared in French as J. Tyndall, *La chaleur considérée come un mode de mouvement*, trans. F. Moigno (Paris: Étienne Giraud, 1864).
3. *they reply to me*: letter missing.
4. *M. Séguin*: see letter 1990.

To Mary Somerville[1] 13 April 1863 1997

Royal Institution 13th. Ap. 1863

Dear Mrs. Somerville

The same post brought me your note,[2] and also the intelligence that a cousin of mine[3] for whom I have a great affection has been behaving so gallantly in America that he has been made a brigadier General. I do not know which pleased me most—your letter, or the intelligence of my cousin's good fortune. I think it extremely kind of you to write to me as you have done; and there is certainly no gentleman philosopher now existing whose letter could give me so much pleasure as that which yours has given.

I am myself sometimes astounded at the action of these attenuated gases & vapours on radiant heat. It seems to me to open up new ideas as to the qualities & powers of matter; and it is extremely satisfactory to me to find that you take an interest in the subject.

Wishing you health & strength to pursue your labours through many coming years

Believe me | Most sincerely yours | John Tyndall

Oxford Bodleian Dep c.372 (MST-1)

1. *Mary Somerville*: see letter 1995, n. 1.
2. *your note*: letter 1995.
3. *a cousin of mine*: George Hector Tyndale; see letter 1988.

From Rudolf Clausius [16 April 1863][1] 1998

Zürich, 16. April 63

Lieber Tyndall,

Sie haben Sich gewiss schon gewundert, dass ich mit der Zusendung Ihres Buches und auch Ihren freundlichen Briefen solange mit der Antwort gewartet habe. Der Grund lag aber nur darin, dass ich Ihr Buch gern erst ganz lesen wollte, bevor ich schrieb, und ich nachher noch durch einige nothwendige Geschäfte in Anspruch genommen wurde. Ich habe Ihr Buch mit dem grössten Vergnügen gelesen. Die ganze Art der Darstellung ist ausserordentlich lebendig und anschaulich u. zugleich so fesselnd, dass, wenn man angefangen hat, es zu lesen, man auch bis zu Ende lesen muss. Sehr lieb war es mir auch, Ihre eigenen schönen Untersuchungen über die strahlende Wärme, darin noch einmal in übersichtlicher Zusammenstellung zu lesen.

Die Resultate, die Sie gefunden haben, sind ebenso überraschend als interessant u. wichtig, und werden gewiss dazu dienen neue Aufschlusse über die Constitution der zusammengesetzten Stoffe zu geben. Ich danke Ihnen sehr für die Zusendung Ihres Buches.

Was Ihren Streit mit Thomson anbetrifft, so müssen Sie Sich dadurch nicht zu sehr aufregen lassen. Thomson hat durch seine unberufenen und oft ganz ungerechtfertigten Einmischungen in Prioritätsfragen schon mit so vielen Leuten Streit angefangen, dass er schon allgemein dafür bekannt ist. In Bezug auf Mayer kann ich dasselbe wiederholen,was ich Ihnen früher geschrieben habe. Sie werden Sich erinnern, dass ich in meinem ersten Briefe über Mayer schrieb, es schiene mir, dass er nicht ganz klare Ideen über die mechanischen Principien habe. Dieses lag daran, dass ich damals nur die erste Abhandlung von Mayer kannte, und diese ist wirklich etwas unklar geschrieben. In meinem folgenden Briefe, wo ich Ihnen mittheilte, dass ich in den späteren Schriften von Mayer viel mehr gefunden hätte als ich erwartet hatte, fügte ich, wenn ich mich recht erinnere, hinzu es schiene mir, dass Mayer durch seine weiteren Studien über den Gegenstand auch eine schärfere Auffassung der mechanischen Grundbegriffe gewonnen habe.

Ich bin aber nicht sicher, ob Mayer, als er seine erste Abhandlung schrieb, wirklich ungenaue Ansichten über gewisse mechanische Grössen hatte, oder ob nur seine Ausdrucksweise ungewöhnlich und unbehülflich war, was bei einem Arzte der zum ersten Male über solche Gegenstände schrieb, nicht zu verwundern wäre. Wenn jemand bei seinen Schlüssen zu richtigen Resultaten kommt, so muss man damit, ihm unrichtige Prämissen zuzuschreiben, sehr vorsichtig sein, und besonders bei genialen Männern, welche keine Gelegenheit hatten, regelrechte mathematische Studien zu machen, kommt es nicht selten vor, dass sie sich zwar nicht klar ausdrucken können, aber doch das Richtige meinen. In seinen späteren Schriften drückt sich Mayer viel schärfer und bestimmter aus, woraus ich die oben erwähnte Vermuthung geschöpft habe, dass er in der Zwischenzeit bedeutende mechanische Studien gemacht habe. Sie selbst haben diesen Unterschied zwischen der ersten Arbeit und den späteren Arbeiten von Mayer auch deutlich angesprochen auf S. 71 Ihres Buches, wo Sie sagen: „Had this first paper stood alone Mayer's relation etc." In dieser Beziehung ist es auch um Mayer's Verdienste zur richtigen Anerkennung in England zu bringen, gewiss von aller zweckmässigsten, auch die spätern Schriften dem englischen wissenschaftlichen Publikum zugänglich zu machen, was, wie ich aus dem neusten Hefte des Phil. Mag. gesehen habe, jetzt mit der Dynamik des Himmels geschieht.

Nun, lieber Tyndall, leben Sie recht wohl, grüssen Sie Hirst von mir, und seien Sie selbst auf das freundlichste von meiner Frau u. mir gegrüsst.

Ihr Clausius

Zurich, 16. April 63

Dear Tyndall,

You must have wondered, that I waited so long to respond to your friendly letters[2] and the book you sent.[3] The reason is that I first wanted to finish reading your book, before I replied, and afterwards I was taken up with some pressing business. I read your book with great pleasure. The entire mode of presentation is unusually vivid and clear and at the same time so captivating, that when one has begun to read it, it must be read through to the end. I appreciated too, being able to read about your own beautiful studies on radiant heat again in a well laid out summary. The results you found are as surprising as they are interesting and important, and will certainly lead to new conclusions about the constitution of compound substances. I thank you very much for sending me your book.

In regard to your dispute with Thomson, you must not let it upset you. Thomson has started arguments with so many people through his uncalled for and often completely unjustified interference in priority questions that he is well known for this. In regard to Mayer, I can repeat what I wrote to you earlier. You will recall that I wrote in my first letter about Mayer,[4] it seemed to me that he did not have very clear ideas about mechanical principles. The reason for this was that I was only familiar with Mayer's first paper[5] at the time, and it really is somewhat unclearly formulated. In my following letters,[6] in which I told you that I had found much more than I expected in Mayer's later papers,[7] I added, if I remember correctly, that is seemed to me that Mayer had attained a sharper conception of fundamental mechanical concepts through his further studies of the matter.

I am not sure whether Mayer really had unclear notions about particular mechanical quantities when he wrote his first paper or whether his form of expression was unusual and clumsy, which would not be surprising for a physician writing about such matters for the first time. If someone gets the right results through his conclusions, one must be very cautious about ascribing false premises to him, and especially with brilliant men who never had the opportunity for proper mathematical study, it is not uncommon that they cannot express themselves clearly but mean the right thing. In his later papers, Mayer expressed himself much more precisely and definitively, and that is why I speculated as previously mentioned that he had made significant studies of mechanics in the interim. You yourself clearly pointed to the difference between the first paper and the later papers from Mayer on p. 71 of your book,[8] where you say: "Had this first paper stood alone Mayer's relation etc." In this regard, in order to draw the right attention to Mayer's achievements in England, it is probably most advantageous to make the later papers accessible to the English scientific public, as I see has now been done in the latest issue of the Phil. Mag. with the Dynamik des Himmels.[9]

Now, dear Tyndall, a very good bye, give Hirst my greetings, and the most friendly ones from my wife and myself to you.

Your Clausius.

RI MS JT/1/TYP/7/2257–8
LT Typescript Only

---

1. *[16 April 1863]*: date suggested by LT.
2. *your friendly letters*: besides letter 1979, the extant correspondence does not include any other letters from Tyndall to Clausius following Clausius's letter of 10 February 1863 (1972).
3. *book you sent*: *Heat as a Mode of Motion* (published in March 1863).
4. *my first letter about Mayer*: see letter 1892, *Tyndall Correspondence*, vol. 7. After reading more of Mayer's work, Clausius agreed that Mayer deserved more recognition from the scientific community than he had received.
5. *Mayer's first paper*: see letter 1904, n. 3.
6. *my following letters*: see letter 1896, *Tyndall Correspondence*, vol. 7. In letters 1904 and 1906 of this volume, Clausius made brief remarks about Mayer deserving more recognition.
7. *Mayer's later papers*: J. R. Mayer, *Die organische Bewegung in ihrem Zusammenhang mit dem Stoffwechsel. Ein Beitrag zur Naturkunde* (Heilbronn: Verlag der C. Drechler'schen Buchhandlung, 1845); J. R. Mayer, 'Sur la transformation de la force vive en chaleur et réciproquement', *Comptes Rendus*, 27 (1848), pp. 385–7; and J. R. Mayer, *Bemerkungen über das mechanische Äquivalent der Wärme* (Heilbronn: Verlag von Johann Ulrich Landherr, 1851).
8. *your book*: see n. 4.
9. *Phil. Mag. with the Dynamik des Himmels*: J. R. Mayer, 'On Celestial Dynamics', *Phil. Mag.*, 25 (1863), pp. 241–8 . This paper was published in three parts. Clausius was referring to the first part published in the April issue, while the rest was published in the May and June issues: J. R. Mayer, 'On Celestial Dynamics', *Phil. Mag.*, vol. 25 (1863), pp. 387–409, 417–28.

From François-Napoléon-Marie Moigno 17 April 1863 1999

Avril 17 | 1863

Mon cher Mr Tyndall

Je suis tout à fait décidé à faire la traduction de votre charmant volume, et à en entreprendre la publication à mes risques et périls.

J'accepte vos conditions et celles de votre éditeur. Nous paierons 25 livres la collection de tous les clichés ou gravures. Dès que nos bénéfices auront atteint 50 livres, nous partagerons par tiers avec vous pour le reste; c'est à dire qu'au delà de 1200 francs, un tiers des bénéfices sera pour vous, un tiers pour mon libraire, un tiers pour moi. Nous désirons ardemment qu'en nous accordant le droit de reproduction pour la France vous nous l'accordiez aussi pour la Belgique, sans cela il nous serait difficile de faire nos frais.

Ne tardez pas à m'écrire que nous sommes d'accord, car nous allons nous mettre à l'œuvre immédiatement.

Avez vous lu ma petite note sur l'équivalent mécanique de la chaleur. Pourquoi n'avez vous pas dit au moins un mot du thermo-générateur de Mrs Beaumont et Mayer, c'est le meilleur appareil pour mesurer l'équivalent mécanique.

La traduction de votre leçon sur la radiation de l'atmosphère publiée dans Les Mondes a fait une très grande sensation.

Il me tarde que votre volume soit imprimé.

A vous d'estime et d'affection sincère | l'abbé F. Moigno

N'oubliez pas Les Mondes qui vous seront si dévoués; recommandez les bien, tentez qu'on en dise un mot dans le Philosophical Magazine, comme on l'a fait dans les Monthly Notices.

J'ai besoin qu'on m'encourage, car c'est une entreprise presqu'au-dessus des forces d'un homme.

Que je serais heureux si Mr Faraday me disait un petit mot gracieux. Appelez son attention sur la dernière livraison.

April 17 | 1863

My dear M[r]. Tyndall,

I am quite determined to translate your charming volume,[1] and to undertake the publication at my own risk.

I accept your conditions and those of your editor.[2] We will pay 25 livres for all the pictures and engravings. As soon as our earnings will reach 50 livres, we will share the rest with you by thirds. That is to say that after 1200 francs, a third of the profit will go to you, a third to my bookseller, and a third to me. We fervently hope that having given us the right of reproduction in France, you will allow us the same in Belgium, without which it would be difficult to cover our expenses.

Do not delay in writing to me[3] that you agree, as we will get to work immediately.

Did you read my small note[4] on the mechanical equivalent of heat? Why have you not said a word about the thermo-generator of M[r] Beaumont and Mayer?[5] It is the best tool for measuring the mechanical equivalent.

The translation[6] of your lesson on the radiation of the atmosphere[7] published in Les Mondes was an absolute sensation.

I eagerly await the printing of your volume.

Yours sincerely, with respect and affection | Father F. Moigno

Do not forget Les Mondes that will be so loyal to you; recommend it widely, try to have it mentioned in the Philosophical Magazine, as was done in the Monthly Notices.[8]

I need some support, since this undertaking is almost beyond the power of one man.

I would be so glad if Mr Faraday could send me a kind word. Bring his attention to the last published issue.

RI MS JT/1/M/140

1. *your charming volume*: Moigno translated Tyndall's *Heat as a Mode of Motion* into French, appearing as J. Tyndall, *La chaleur considérée come un mode de mouvement*, trans. F. Moigno (Paris: Étienne Giraud, 1864).
2. *your conditions and those of your editor*: see letter 1996.
3. *in writing to me*: if Tyndall wrote a reply to this letter, it is missing.
4. *my small note*: F. N. M. Moigno, 'Mécanique appliquée sur l'équivalent mécanique de la chaleur', *Les Mondes*, 2 (1863), pp. 134–9.
5. *thermo-generator of Beaumont and Mayer*: Jean-Baptiste Armand Louis Léonce Élie de Beaumont (1798–1874), a French geologist (*Encyclopædia Britannica*, 11th edn, 29 vols (New York: Encyclopædia Britannica Company, 1911), vol. 9, pp. 272–3); and probably Julius Mayer. See 'Apparat zur Erzeugung von Wärme durch Reibung und ohne Brennmaterial', *Schweizerische Polytechnische Zeitschrift*, 1 (1856), pp. 30–1.
6. *translation*: J. Tyndall, 'Du rayonnement à travers l'atmosphère terrestre', *Les Mondes*, 1 (1863), pp. 230–4.
7. *your lesson on the radiation of the atmosphere*: see letter 1948, n. 7.
8. *Monthly Notices*: *The Monthly Notices of the Royal Astronomical Society*, edited at this time by Arthur Cayley (see letter 2138, n. 4).

From Gustav Magnus 28 April 1863 2000

Berlin 28 April 63

Verehrter Freund

Vor einigen Tagen von Paris zurückgekehrt fand ich Ihren freundlichen Brief vom 20 v. M. so wie Ihr Werk on Heat. Entschuldigen Sie daß ich Ihnen für beides nicht früher gedankt habe. Gestern habe ich nun auch die officielle Anzeige, daß mich die R. Society zum Foreign Member erwählt hat, durch Prof. Miller erhalten. Ich gestehe daß ich die Ehre sehr hoch veranschlage, und daß ich Ihnen für Ihre Bemühungen in dieser Sache ganz besonders verbunden bin. Ich werde in den nächsten Tagen meinen Dank an Prof. Miller senden, der sein officielles Schreiben noch mit einigen sehr herzlichen Worten begleitet hat für die ich ihm noch besonders dankbar bin.

Meine Reise nach Paris ist sehr plötzlich zu Stande gekommen, da Prof. Riess sich unerwartet entschloß mich zu begleiten, was die schnelle Entscheidung herbeiführte. Ich hegte im Stillen die Hoffnung Sie in Paris zu finden, und wäre unsere Abreise weniger unvorbereitet erfolgt, so würde ich Sie aufgefordert haben dorthin zu kommen, um, ähnlich wie vor 10 Jahren, in Ihrer Gesellschaft einige Zeit dort zuzubringen. Wir hätten dann auch über unsere wissenschaftliche Differenz plaudern können. Ich habe mich während des vergangenen Winters noch mit dieser Sache beschäftigt. Der Versuch ohne Steinsalzplatten, den sie so gut waren mir zu zeigen, und den Sie in Ihrer Abhandlung beschrieben haben, war so wenig in Uebereinstimmung mit dem was ich gefunden, daß ich glaubte ihn wiederholen zu müssen. Ich habe ähnliche Ausschläge erhalten wie Sie; aber ich habe niemals eine größere Erwärmung gefunden wenn trockne Luft durch die Röhre ging, sondern stets mit feuchter. Ich habe mich dann überzeugt daß die Luft aus der Röhre bis zur Säule sich bewegt, und daß dadurch der Ausschlag des Galvanometers hervorgebracht wird; ohne Zweifel indem aus der feuchten Luft der Wasserdampf absorbirt wird und Wärme frei wird, dagegen bei Anwendung von trockner Luft, die auf der Säule vorhandene Feuchtigkeit verdunstet. Was besonders beweist daß die Diathermansie nichts mit der Sache zu thun hat, ist der Umstand daß derselbe Erfolg eintritt wenn man die Wärmequellen ganz fortläßt, und daß der Ausschlag des Galvanometers außerordentlich groß wird, wenn man die Luft direkt in den Reflector der Säule bläst.

Ich habe zwar in der kleinen Note die ich in der Academie gelesen, meine Methode gegen den Tadel verwahren müssen den Sie dagegen ausgesprochen haben, allein Sie werden finden daß alles vollkommen sachlich gehalten ist. So bald die Note gedruckt ist, was nun wohl, nachdem ich zurück bin, bald geschehn sein wird, so theile ich Ihnen dieselbe mit.

Ihr dankbar ergebener | G Magnus.

Meine Frau, Tochter und Paul lassen vielmals grüssen.

Berlin 28 April 63

Cherished friend

Returned from Paris a few days ago, I found your kind letter of the 20th[1] of last month as well as your work on Heat.[2] Please excuse me that I did not thank you earlier for either. Yesterday I also finally received, through Prof. Miller,[3] the official announcement that the R. Society[4] had elected me as a Foreign Member.[5] I confess that I estimate this honour very highly and that I am especially indebted to you for your efforts in this matter. In the next days I will write my thanks to Prof. Miller, who added some very kind words to the official note for which I am especially grateful.

My journey to Paris came about very suddenly, as Prof. Riess[6] unexpectedly decided to accompany me, which led to the quick decision. I privately hoped to meet you in Paris, and if our departure had not happened with such little preparation, I would have asked you to go there so that I could spend some time in your company, as 10 years ago. We could have also chatted about our scientific differences. During the last winter, I kept working on that subject. The experiment without rock salt plates, which you were good enough to show me and which you described in your article, agreed so little with my findings that I thought I had to repeat it. I obtained deflections similar to yours; but I never found a greater temperature increase when dry air flowed through the tube; but always with moist air. I then assured myself that the air from the tube moved up to the column and that this caused the deflection of the galvanometer, doubtless because steam from the moist air is absorbed and heat is released. On the contrary, when using dry air, the moisture present in the column evaporates. The proof that diathermancy plays no role is that one obtains the same result even if one omits altogether the sources of heat and that the deflection in the galvanometer is especially strong when the air blows directly into the reflector of the column.

Although I had to defend my method against your objections in the short note that I read in the Academy,[7] you will find that everything is dealt with entirely objectively. I will inform you as soon as the note is printed, which now that I have returned should happen soon.

Your gratefully devoted | G Magnus.

My wife, daughter and Paul send you their greetings.

RI MS JT/1/M/29

1. *your kind letter of the 20th*: letter missing.
2. *your work on Heat*: *Heat as a Mode of Motion.*
3. *Prof. Miller*: see letter 1943, n. 3.
4. *R. Society*: the RS.
5. *elected me as a Foreign Member*: see 'Anniversary Meeting', *Roy. Soc. Proc.*, 12 (1863), p. 22.
6. *Prof. Riess*: Peter Theophil-Pierre Riess (1804–83), a German physicist known for his work in electrical physics, particularly friction electricity, but also in the field of electromagnetic induction (*CDSB*).
7. *the Academy*: the Royal Prussian Academy of Sciences (Königlich-Preußische Akademie der Wissenschaften), or 'Berlin Academy', founded in 1700.

To Edward Livingston Youmans 29 April 1863 2001

MY DEAR SIR: As soon as I received the letter[1] with which you kindly favored me some months ago, I communicated at once with Mr. Longman[2] and requested him to forward you the separate sheets of my work on Heat[3] according as they appeared. I intended to accompany the sheets with a letter which should express my desire to leave the management with the Messrs. Appleton[4] entirely in your hands, but I have been so knocked about-sometimes so ill, sometimes so hard worked, and sometimes engaged so far away from London that I have delayed thus far to write to you. My friend Spencer[5] called to see me a few days ago, and from him I had the great gratification of learning that the book has interested you. Indeed, he read portions of letters from Mrs. Youmans[6] and yourself which gave me very great pleasure. Since the appearance of the work I have had communications from many of my eminent Continental friends regarding it, and they, I am happy to say, concur in your opinion. A French translation[7] of it has already been commenced. I can assure you that I have spared no labor to render a difficult subject intelligible, and it gives me great pleasure to find that I have, at least in some measure, succeeded.

I am now giving a short course of lectures on Sound,[8] at the Royal Institution. If I have time I may throw them into a readable form. I have for some time entertained the idea of publishing my lectures gradually and of afterward collecting them and fusing them into a book on general physics. But the time necessary to the proper accomplishment of the task deters me almost from undertaking it. However, it may perhaps be executed by slow degrees.

Mr. Huxley informs me that you are thinking of bringing out his work[9] also. I am glad to hear this, for it is an extremely able production. Indeed, there are parts of it which in point of writing power have scarcely ever been excelled. . . .

Good-by, my dear sir; accept my best thanks for the trouble you have taken in my behalf, and believe me | Most sincerely yours, | JOHN TYNDALL.

E. A. Youmans, 'Tyndall and his American Trip', *Popular Science Monthly*, 44 (1894), pp. 502–14, on pp. 505–6

1. *the letter*: letter missing.
2. *Mr. Longman*: letter missing; either Thomas Longman (1804–79) or William Longman (1813–77) of Longman, Green, Longman, Roberts, & Green, publishers of Tyndall's *Heat as a Mode of Motion.*
3. *my work on Heat*: *Heat as a Mode of Motion.*

4. *Messrs. Appleton*: Appleton's was an American publishing company based in New York City, founded by Daniel Appleton (1785–1849). At this time the company was under the leadership of his son William Henry Appleton (1814–99). The publishers issued the works of many popular scientists such as Herbert Spencer, Thomas Huxley, and Charles Darwin. They published the American edition of *Heat as a Mode of Motion* under a slightly modified title: *Heat Considered as a Mode of Motion* (New York: D. Appleton & Company, 1869).
5. *My friend Spencer*: Herbert Spencer.
6. *Mrs. Youmans*: Catherine Newton Lee Youmans (*c.* 1819–94). They were married in 1861 and she often served as an editor for Youmans (*ANB*).
7. *French translation*: see letters 1978, 1990, and 1999.
8. *a short course of lectures on Sound*: in 1862 Tyndall delivered a series of lectures at the RI on sound, and these were later developed and published as J. Tyndall, *Sound: A Course of Eight Lectures* (London: Longmans, Green, and Co., 1867), and reprinted by D. Appleton and Co. in 1871.
9. *his work*: T. H. Huxley, *Evidence as to Man's Place in Nature* (London: Williams and Norgate, 1863). This was reprinted by D. Appleton and Co. in 1863.

To William Thomson [April][1] 1863 2002

Remarks on the Dynamical Theory of Heat. | *By* JOHN TYNDALL, *F.R.S.* | *To* WILLIAM THOMSON, *Esq., F.R.S., Professor of Natural Philosophy in the University of Glasgow.*

SIR,

THE article in 'Good Words'[2] which drew from me the brief remonstrance published in the March Number of the Philosophical Magazine[3] bore your name, which preceded, though not by right of the alphabet, that of Professor Tait. From this gentleman's communication to the Philosophical Magazine of last month,[4] I learn that it was in compliance with requests made to you that the article in 'Good Words' was written. Indeed, irrespective of these considerations, in any article of the kind where your name is associated with that of Prof. Tait, you are sure to be regarded as the 'principal'. You are the older and more famous man, and it is your behaviour in this controversy, and not that of your colleague, which will interest the scientific world. I trust, therefore, Prof. Tait will see that simple chivalry makes it my duty to decline entering into any contest with him at present; and seeing this, he will, I doubt not, have the grace and modesty to stand aside and allow you and me to settle this affair[5] between ourselves.

At your request, we are informed, your friend has replied to my "Remarks" both for you and for himself. Doubtless his reply has passed through your hands and received your assent. I may therefore, I hope, regard it as *your* reply, and deal with it accordingly. You open thus:—"I think it right at starting to call Prof. Tyndall's attention to the fact that, in the Philosophical Magazine (1862, vol. xxiv. P. 65), he has published the following words:—'I do not think a greater disservice could be done to a man of science than to overstate his claims: such overstatement is sure to recoil to the disadvantage of him in whose interest it is made.'" There is surely nothing in these words of which an honest man need be ashamed. They are, I submit, sensible words; and were they not *my* words, I should on various grounds strongly recommend them to your own particular attention. But besides reviving my memory of these words, you remind me "that any unpleasant results which may follow from the course which he [Prof. Tyndall] has pursued are, by his own acknowledgement, to be laid to his charge."

I hope you will forgive me if I now remind you that, in conducting this correspondence, it is extremely desirable that each of us should endeavor to make his meaning clear. This 'reminder' of yours is by no means clear. As a deduction it would be illogical, as an assertion it would be untrue. If in such a simple matter your writing is made difficult of comprehension, how am I to seize your meaning when you come to deal with a really difficult subject? Let each of us try then to render his meaning unmistakable; but this once done, let neither of us seek to divert the words of the other from their plain and obvious sense. This course will shorten controversy, and make the world feel that we mean fairly by each other. In your last communication you write thus:—"A journal which contains in nearly every Number a scientific paper by yourself [Sir David Brewster,[6] to whom Prof. Tait addresses his communication], Forbes, Herschel,[7] or Piazzi Smyth,[8] can surely not be regarded as unsuitable for a paper on 'Energy.'" Did I say it was? Did I not rather state that it was both laudable and desirable that men in your high positions should *instruct* the readers of 'Good Words;' and I may here add, what I am known to have stated a hundred times, that in my opinion it would be an incalculable boon to English society if our best men could be prevailed upon to write our scientific manuals, and to teach the public through our leading magazines. But I also said, and I think rightly, that 'Good Words' is not the place to give vent to insinuations prejudicial to the character of a scientific fellow-worker. And inasmuch as I know that the character you impugn is of a quality to defy all your attacks, if only openly made, it is only natural that I should invite you to make your charges against it in the presence of a jury competent to understand your evidence.

With regard to the publication of my lecture on Force,[9] the course pursued was this: the abstract of the lecture was published, as all such abstracts are, in the 'Proceedings of the Royal Institution,'[10] and it was transferred thence to the Philosophical Magazine.[11] With its further publication I had nothing whatever to do. Until you informed me of the fact, I did not know that the lecture had been honoured by admission into the 'Illustrated London News,' [12] the 'Engineer,' [13] and 'Macmillan's Magazine.' [14] The gentleman[15] whom you denominate my 'pupil' is quite unknown to me. But even had I written my abstract expressly for any of these journals, there would, I submit, be nothing unworthy or inconsistent in the act; for the lecture contains imputations upon no man's character. My object in giving the lecture was not to defame anybody, but to raise a noble and a suffering man[16] to the position which I believed his labours entitled him to occupy, and from which I am persuaded your efforts will be unavailing to remove him.

In proof of your acquaintance with Mayer, you refer me to a paper of yours in the Edinburgh Transactions,[17] in which I am informed you give him "the full credit that his scientific claims can possibly be admitted to deserve." This is an assertion made without knowledge. Regarding Mayer's labours you knew, I assume, what everybody knew—the contents of his first paper,[18] which, able and important as it is, is in reality nothing more than a preliminary note*.[19] Everybody was acquainted with this paper, because it was published in Liebig's *Annalen*:[20] and had Mayer's subsequent memoirs been published in the same way, I do not think you would ever have ventured to provoke this controversy. The contributions from which, in connexion with his first one, Mayer's merit as a philosopher is to be inferred, were published as separate essays at Heilbronn.[21] To these essays you never refer, and from your silence I charitably inferred your ignorance of their contents. This, however, does not satisfy you. And yet if I assume that you are acquainted with these memoirs, *how am I to reconcile your knowledge with your silence?*

You have thought it necessary to refresh my memory on another point which I should willingly have allowed to remain at rest. "Prof. Tyndall," you say, "is most unfortunate in the possession of a mental bias which often prevents him (as, for instance, in the case of Rendu and Glacier-motion) from recognizing the fact that the claims of individuals whom he supposes to have been wronged*[22] have, before his intervention, been fully ventilated, discussed, and settled by the general award of scientific men." This language, as applied to Mayer, can only spring from ignorance; and as regards Rendu, it would be wise in you to remember that your opinions on this matter are not those of the world. Even among your own countrymen there are persons too just and too free to be warped by a misnamed patriotism, who, in common with the great majority of its readers in other lands, consider that in my work upon the Alps the claims of Rendu have been for the first time fairly set forth.[23] No man, however, can be less persuaded on his own infallibility than I am of mine, and I clearly see the possibility of my having greatly erred in my writings upon glaciers. But I am willing at all events to bring my deeds to the light, so that, whether good or evil, they may be made manifest. I have now to request you to refer me to the passages from my writings on Rendu and Glacier-motion which establish the unfortunate mental bias you have ascribed to me.

It may shorten our correspondence, by making reflexion the precursor of expression, if I here assure you that it is my fixed intention not to permit random assertions to stand in this discussion. What you have the hardihood to affirm, you certainly must have the goodness to prove or the manliness to retract. I ask your proof, then, of the assertion you have made regarding Rendu and Glacier-motion.

After disposing of the matters to which I have had thus far occasion to refer, apparently much to your own satisfaction, you put the following pointed question:—"Does Professor Tyndall know that Mayer's paper has *no claims or novelty or correctness at all*, saving this, that by a lucky chance he got an approximation to a true result from *an utterly false analogy*"? I am able with the utmost sincerity of heart to avow that I did *not* know anything of the kind, and that you are the first man who has furnished me with this information. I should doubtless stand abashed and confounded at my convicted ignorance were it not that I find myself surrounded by an extremely respectable company who are actually as ill-informed as myself. Permit me to illustrate my meaning. In a discourse delivered at Königsberg[24] on the 7th of February, 1854, Prof. Helmholtz analyses that relationship of natural forces which he, in 1847, called the Conservation of Force—a phrase to which you object as inconvenient and erroneous, but which means the same thing as the phrase 'Conservation of Energy' which you adopt. This conservation Helmholtz showed to follow directly from the negation of a perpetual motion; and it is with reference to this great principle that he expresses himself in the following words*:[25]—"The first man who correctly perceived and rightly enunciated the general law of nature (*das allgemeine Naturgesetz*) which we are here considering, was a German physician, J. R. Mayer, of Heilbronn, in the year 1842. A little later (in 1843) a Dane named Colding[26] presented a memoir[27] to the Academy of Copenhagen, in which the same law was enunciated . . . . . In England Joule began about the same time to make experiments on the same subject." You say that you "never intended to hint that Prof. Tyndall could have meant to put Mayer forward as having any claims to this great generalization." Bad as I am, you would not think of charging me with this grotesque folly. And still for nine years Helmholtz, who was a master in this field before you ever entered it, has stood self-convicted of this very absurdity.

Let me now refer you to another of my companions in ignorance. On the 7th and 21st of February, 1862, two lectures on the Mechanical Theory of Heat[28] were given by M. Verdet before the Chemical Society of Paris†.[29] This philosopher, to whom everybody has hitherto ascribed a thorough knowledge of the literature of science, expresses himself on the subject in hand in the following words:—"Finally, I will terminate this first part of my historic review by calling to mind that Mr. Séguin, in a work published in 1839,[30] and devoted more to political economy than to physics, has presented considerations regarding the steam-engine which closely resemble those by which I endeavoured, in our first lecture, to make you comprehend the transformation of heat into mechanical power."

I should deem it probable that M. Verdet knows as much about the labours of M. Ségun[31] as you do, and he certainly knows more about those of Mayer. But he does not see in the former the annihilation of the latter. He proceeds thus:—"I now come to the researchers which from 1842 to 1849 definitely founded the science. These researchers are the exclusive work of three men, who without concert, and even without knowing each other, arrived simultaneously in almost the same manner at the same ideas. The priority in the order of publication belongs without any doubt to the German physician Jules Robert Mayer, whose name has occurred so often in these lectures; and it is interesting to know that it was by reflecting on certain observations in his medical practice that he perceived the necessity of an equivalence between work and heat..... He perceived in the act of respiration the origin of the motive power of animals; and the comparison of animals with thermic engines afterwards suggested to him the important principle with which his name will be connected for ever. . . . . We also find," continues M. Verdet, "in the same memoir (Liebig's *Annalen*, 1842) a first determination of the mechanical equivalent of heat, deduced from the properties of gases, which is perfectly exact in principle*."[32] You see then that I am not the only person who has failed to arrive at your remarkable conclusion that "Mayer's paper has *no claims to novelty or correctness at all.*"

After referring to M. Colding, M. Verdet passes on to Mr. Joule. "The third inventor," he says, "of whom it remains for me to speak is Mr. Joule, who perhaps has done most for the demonstration of the principle and for its final adoption. His first investigation, published only in 1843, is incontestably posterior by some months to the first publications of Mayer and Colding†.[33] All this is certainly very much a variance with your statement in 'Good Woods.' "Curiously enough," you there say, "although similar coincidences are common, while Joule was pursuing and publishing his investigations, there appeared in Germany a paper by Mayer, of Heilbronn. Its title is 'Bemerkungen über die Kräfte der unbelebten Natur,' and its date is 1842." Would you have the goodness to point out a single word regarding the mechanical value of heat, or the mechanical equivalent of heat in any of Mr. Joule's writings prior to the month of December 1843? If no such word occurs, if Helmholtz and Verdet be correct in assigning this date to the earliest experiments of Mr. Joule on the 'Mechanical Value of Heat,' how am I to characterize your statement that it was while Joule was pursuing and publishing his investigations that Mayer's paper, dated May 1842, appeared? If you should urge that these investigations *led* Mr. Joule to his experiments on the mechanical equivalent of heat, you simply repeat what I have myself stated in the very lecture which has been the object of your unwarrantable attack.

With reference to the manner in which I have met your insinuations in my brief communication to the Philosophical Magazine, we have the following remarks:'—"I am unwilling to enter upon matters of a more personal character, but it is impossible to pass over the fact that, in answer to an expostulation from Joule (having reference to his lecture), Prof. Tyndall referred (Phil. Mag. 1862, second half year, p. 173) to statements which he had made in a course of lectures which, he now tells us, were completed *before* he acquired those views of Mayer's claims to which Joule so naturally objected, and that he now replies to Prof. Thomson and me, not by showing that his lecture, to which we referred, was free from the objections which we urged against it, but by showing that these objections could not be urged against a certain statement which he quotes from a work, not published, but promised for publication." This reads smart; but my frame of mind disqualifies me from meeting it with mere smartness. In replying to it I do not forget that my aim must not be to satisfy *you* as to the uprightness of my course of action, but to enable the scientific public to judge between your actions and mine. On my return from Switzerland in 1862, I became acquainted with the "expostulation" to which you refer; and I also found waiting for me a private letter from Mr. Joule,[34] reminding me of what he had done in connexion with this subject. In one of my morning lectures given six months previously, I had recognized in the strongest terms the claims of Mr. Joule on the very points to which, in his private note, he had directed my attention. In my answer to his "expostulation" I quoted this lecture, not expecting that Mr. Joule ought to be satisfied with a statement made six months before, but for the purpose of publicly renewing the value of that statement by resubscription to it. This was done in the following words addressed to Mr. Joule:—"Such has been my language regarding you, and to it I still adhere: I trust you find nothing in it which indicates a desire on my part to question your claim to the honour of being the experimental demonstrator of the equivalence of heat and work." It was in the presence of this statement that you ventured to insinuate in 'Good Words' your charge of depreciation and suppression; and it is in the presence of this statement, requoted for your especial information in the March Number of the Philosophical Magazine, that you now dare to reiterate this charge. The quotation from my book will be recognized by the court before which you and I now stand, as springing from the same spirit of goodwill towards Mr.

Joule, and the same high appreciation of his services which have marked all that I have said and written regarding him. It is not only *'now'* that I entertain those feelings, as you intimate in your concluding paragraph, they have always been mine; and they will continue to be mine despite an advocacy on your part so unwise and so unwarranted as to be well calculated to produce a reaction against the man who is unfortunate enough to find in you the assertor of his claims.

Your article in 'Good Words' was professedly written to counteract the errors and absurdities of previous writers. "We have aimed," you say, "at preparing an article which shall at all events be *accurate* as far as human knowledge now reaches," or as you more cuttingly express it in reference to my poor lecture, you "seized the opportunity of distributing among its ['Good Words'] 120,000 readers a corrective to the erroneous information which you saw stealing upon them through the medium of *popular* journals." Well, let us inquire a little into the quality of the "corrective." In putting Mr. Joule forward as the founder of the dynamical theory of heat, you write thus:—"In 1843 he (Mr. Joule) published the results of a well-planned and executed series of experiments, by which he ascertained that a pound of water is raised one degree Fahrenheit in temperature by 772 foot-pounds of mechanical work done upon it. In other words, if a pound of water fall from a height of 772 feet, and the kinetic energy thus acquired in the form of ordinary motion be entirely transformed into the kinetic energy of heat, the water will be one degree hotter than before its fall." You here prove yourself to be as ill-informed regarding the labours of Joule as you are regarding those of Mayer. It was in 1849, and not in 1843, that, Mr. Joule proved the mechanical equivalent of heat to be 772 foot-pounds. His determinations of the mechanical equivalent of heat, published in 1843, varied from 1040 to 587 foot-pounds. They were so discordant that nobody attached any value to them. It was with reference to Mr. Joule's earlier experiments that Helmholtz expressed himself thus, in his celebrated essay 'Uber die Erhaltung der Kraft,' published in 1847:—"Joule has examined the quantities of heat generated by the friction of water in narrow tubes, and in a vessel in which the water was set in motion by a kind of turbine. He has found in the first case that the heat which raises 1 lb. of water 1°C. would lift 452 kilogrammes to the height of 1 metre; in the second case he found the weight to be 521 kilogrammes. Nevertheless his methods of experiment correspond too little to the difficulties of the investigation to allow of these results having the least claim to accuracy'*[35] (*als dass diese Resultate irgendwie auf Genauigkeit Anspruch machen könnten*). I translated this paper myself for the 'Scientific Memoirs†,'[36] and was careful to append to this passage a note to the effect that Helmholtz when he wrote thus was only acquainted with the earlier experiments of Mr. Joule. Judged on broad and liberal grounds, Mr. Joule is worthy of such praise as rarely falls to the lot of a philosopher. He is, in my opinion, the true experimental demonstrator of the dynamical theory of heat. But to write as you write regarding his first experiments is simply to betray a want of acquaintance with the requirements of refined experimental inquiry.

You state in 'Good Words' that Mayer's hypothesis is at best only "*approximately*" true for air. Yes; but the approximation is so close that it cost Mr. Joule six years of labour, according to his own methods, to arrive at the degree of accuracy attainable by the method of Mayer. You coldly deny to Mayer's first paper all claim to novelty or correctness; and yet a comparison of that paper[‡37] with the first part of your "corrective" in 'Good Words' will show that the ideas which you have there enunciated are the ideas of Mayer. The very illustration which you employ, of stirring water in a basin to produce heat, is an experiment of Mayer's mentioned in this first paper. In the subsequent portion of the article referred to, you come to what you denominate "the grandest question of all." You put it thus:—"Whence do we immediately derive all those stores of potential energy which we employ as fuel or food? What produces the potential energy of a loaf or a beefsteak? What supplies the coal and the water-power without which our factories must stop?" These "grandest questions of all' were asked and answered by Dr. Mayer seventeen years before you wrote your article; and yet you never mention his name[§].[38] You proceed, "Whence does the sun produce the energy which he so continuously and liberally distributes?" You then consider the various hypotheses framed to account for the permanence of solar emission, and develop the meteoric theory of the sun's heat. Mayer did the same fourteen years before you wrote this article in 'Good Words,' and six years before you published your earliest paper on the subject in the Edinburgh Transactions.[39] There is not an idea of any originality in the whole of that paper that is not to be found in the memoirs of Mayer; and yet you do not give him an iota of credit in this article of yours in 'Good Words,' the accuracy of which you have so trumpeted forth. Mayer, moreover, went further than you did, and showed that the heat produced by tidal friction must be generated at the expense of the earth's rotation. This is but an episode in Mayer's 'Essay on Celestial Dynamics,'[40] but it is eminently illustrative of the clear glance which this hard-working physician had obtained into the system of nature. Here was a phenomenon which had been for centuries the subject of observation, measurement, and calculation; still no astronomer, no mathematician, no natural philosopher perceived its inevitable mechanical effect. But Mayer, following out his own principles, saw and enunciated that the motion of the tides, indefinitely continued, must finally stop the earth's rotation. Name, if you can the "mathematician or naturalist, from Galileo[41] to Davy," who ever "elaborated" these things. You cannot do so: and while science lives the name of Mayer will be associated with these questions. In the presence of such facts, it ill becomes you to talk to me of suppression and depreciation. You may send your statements into the world labelled 'Good Words,' but the world before which you and I now stand will see that the "trade mark" is incorrect; that if to be pitiful, if to be courteous, if to cherish that charity which thinketh no evil, be the marks of goodness, these utterances of your are *not* good words, but the reverse. Judged of by the facts, and apart from your own uninformed convictions, they are not even words of truth.

In your last communication you anticipated "unpleasant results," and, doubtless, taking it for granted that the unpleasantness was to be all on my side, you were good, enough to inform me (still by anticipation) that I had only myself to blame. Certainly it is in the highest degree unpleasant to me to be compelled to write as I have here written. It wastes that commodity which to me is the most precious of all—my time, and it stirs up feelings between you and me which I would make any sacrifice consistent with self-respect to annul. But the responsibility is yours. No taint of personality is to be found in anything that I have ever written regarding you; even when facts were against you I held my peace. When I wrote my brief and temperate remonstrance to the Philosophical Magazine, all that I have stated in this letter was known to me. I then knew, as well as I do now, that your treatment of Mayer could not bear the light of criticism; still my remonstrance breathed no syllable of accusation. My desire was to clear myself, and not to criminate you, trusting that the translations of Mayer's words now appearing in the Philosophical Magazine would in due time set public opinion right regarding him. I hoped that good sense and gentlemanly feeling wood so far predominate as to induce you to withdraw an accusation which, though vaguely uttered, was offensive. Had you done so, this discussion would have perished in the bud, and I should never have written a sentence which could be personally disagreeable to you. But instead of withdrawing your accusation, you repeat it, and thus compel me in self-defence to lay a true statement of the case before the scientific public, and invoke its judgment between us.

I have the honour to be, Sir, | Your obedient Servant, | JOHN TYNDALL.

Phil. Mag., 25 (1863), pp. 368–78

---

1. *[April]*: the month is given as this letter was published in the May 1863 issue of the *Phil. Mag.*, and an article referred to as from the *Phil. Mag.* 'last month' was published in the March issue (see n. 4).
2. *article in 'Good Words'*: see letter 1986, n. 5.
3. *March Number of the Philosophical Magazine*: see letter 1979, n. 3.

4. *From this gentleman's communication to the Philosophical Magazine of last month*: P. G. Tait, "Remarks on a Paper on 'Energy' in 'Good Words'. *Phil. Mag.*, 25 (1863), pp. 263–6. Roland Jackson notes that Thompson and Tait had used the *Good Words* essay to reach a popular audience; the magazine had a readership of 120,000 at that time. In the essay they claimed Joule as the founder of the modern dynamical theory of heat and criticised Tyndall—although without naming him directly, for championing Mayer (*Ascent of John Tyndall*, pp. 176–7).
5. *to settle this affair*: Thomson and Tyndall were feuding over claims of discovery in thermodynamics, namely the credit due to Mayer.
6. *David Brewster*: David Brewster was one of the editors of the *Phil. Mag.*
7. *Herschel*: John Herschel.
8. *Piazzi Smyth*: Charles Piazzi Smyth (1819–1900), an English astronomer who served as Astronomer Royal for Scotland from 1846–88. He is known for many innovations in astronomy, namely in spectroscopy, and his pyramidological (a term of his own making) and meteorological studies of the Great Pyramid of Giza (*ODNB*).
9. *lecture on Force*: see letter 1905, n. 3.
10. *Proceedings of the Royal Institution*: J. Tyndall, 'On Force', *Roy. Inst. Proc.*, 3 (1858–62), pp. 527–36.
11. *Philosophical Magazine*: J. Tyndall. 'On Force', *Phil. Mag.*, 24 (1862), pp. 57–65.
12. *Illustrated London News*: not found.
13. *Engineer*: not found.
14. *Macmillan's Magazine*: J. Tyndall, 'The Indestructibility of Force', *Macmillan's Magazine*, 6 (1862), pp. 337–44.
15. *The gentleman*: not identified.
16. *to raise a noble and a suffering man*: a reference to Mayer, who had struggled with mental health in the wake of the death of three of his children between 1845–8 (*CDSB*).
17. *Edinburgh Transactions*: W. Thomson, 'On the Dynamical Theory of Heat; with numerical results deduced from Mr. Joule's "equivalent of a thermal unit," and M. Regnault's "observations on steam", *Transactions of the Royal Society of Edinburgh*, 20 (1853), pp. 261–88.
18. *his first paper*: J. R. Mayer, 'Bermerkungen uber die Kräfte der unbelebten Natur', *Annalen der Chemie und Pharmacie*, 42 (1842), pp. 233–40.
19. The following footnote appeared here in the original: '* "To secure what I had discovered against eventualities, I put together its most important portions in a short paper, which I sent to Liebig in the spring of 1842, with a request that it might be published in the *Annalen der Chemie und Pharmacie.* It is to be found in vol. xlii. Page 233 of the *Annalen* under the title 'Bemerkungen über die Kräfte der unbelebten Natur.'" (Mayer, *Bemerkungen über das mechanische Equivalent der Wärme.* Heilbronn and Leipzig, 1851.)'.
20. *Liebig's Annalen*: Liebig's *Annalen der Chemie und Pharmacie.*
21. *Heilbronn*: this refers to two of Mayer's early works; see letter 1904, n. 3 and n. 4.
22. The following footnote appeared here in the original: '* I did not suppose Mayer to be wronged: I supposed simply that he had wronged himself by his method of publication'.

23. *in my work upon the Alps the claims of Rendu have been for the first time fairly set forth*: in chapter 14 of *Glaciers of the Alps*, 'Rendu's Theory', Tyndall described Rendu's work on glacial motion, his early hypotheses and publications (notably Rendu's 'Théorie des Glaciers de la Savoie' published in 1841). Tyndall had championed the priority of Rendu against the claims of Forbes, suggesting that Forbes had not given due credit to Rendu's published work. See N. K. L. Kaalund, 'A frosty disagreement: John Tyndall, James David Forbes, and the early formation of the X-Club', *Annals of Science*, 74 (2017), pp. 282–98.
24. *a discourse delivered at Königsberg*: this 7 February 1854 lecture, 'Ueber die Wechselwirkung der Naturkräfte und die darauf bezüglichen Ermittlungen der Physik', was later published as H. Helmholtz, 'On the Interaction of Natural Forces', *Phil. Mag.*, 11 (1856), pp. 489–518.
25. The following footnote appeared here in the original: '* Ueber die Wechselwirkung der Naturkräfte. Königsberg. Phil. Mag. vol. xi. p. 489'.
26. *Colding*: Ludwig A. Colding (1815–88), a Danish civil engineer and physicist known for his work on the conservation of energy, simultaneously and separately from Joule and Mayer (*CDSB*).
27. *memoir*: Colding's memoir 'Nogle saetninger om kraefterne' was read to the Royal Danish Academy of Sciences and Letters in 1843 and published as L. Colding, *Nogle saetninger om kraefterne* (Copenhagen: Lunos, 1856). See also [L.] A. Colding, 'On the History of the Principle of the Conservation of Energy', *Phil. Mag.*, 27 (1864), pp. 56–64.
28. *two lectures on the Mechanical Theory of Heat*: M. Verdet, *Exposé de la Théorie Mécanique de la Chaleur, presente a la Societe Chimique de Paris, Le 7 et le 21 Février 1862* (Paris: Imprimerie de Ch. Lahure, 1862).
29. The following footnote appeared here in the original: '† *Exposé de la Théorie Mécanique de la Chaleur*, par M. Verdet'.
30. *work published in 1839*: see letter 1990, n. 2.
31. *Ségun*: the journal misspelled 'Séguin'.
32. The following footnote appeared here in the original: '* Page 116'.
33. The following footnote appeared here in the original: '† Though the history of the subject compels him to write thus, M. Verdet recognizes in the most explicit manner the merits of Mr. Joule. After giving an account of Joule's experiments, he thus addresses his audience:— "Vous ne refuserez pas à voir dans le travail aujourd'hui classique de M. Joule le prevue expérimentale de l'exactitude des nouveaux principes." (P. 20.)'.
34. *a private letter from Mr. Joule*: letter 1914.
35. The following footnote appeared here in the original: '* See also Verdet, pp. 91 & 165'.
36. The following footnote appeared here in the original: '† 1853'.
37. The following footnote appeared here in the original: '‡ A translation of it is published in the Philosophical Magazine, vol. xxiv. Ser. 4, p. 371'.
38. The following footnote appeared here in the original: '§ "Ces idées introduite pour la première fois dans la science en 1845, par Jules Robert Mayer, font faire à la physiologie générale un progrès assurément égal au progress qui est résulté, vers la fin du siècle dernier, des découvertes de Lavoisier et de Senebier sur la respiration."—Verdet, p. 101'.

39. *Edinburgh Transactions*: W. Thomson, 'An Account of Carnot's Theory of the Motive Power of Heat, with numerical results Deduced from Regnault's Experiments on Steam', *Transaction of the Royal Society of Edinburgh*, 16 (1849), pp. 541–74.
40. *Mayer's 'Essay on Celestial Dynamics'*: see letter 1904, n. 4.
41. *Galileo*: Galileo Galilei (1564–1642), an Italian mathematician and astronomer known for his controversial publications and for his formative role in the transformation of observational science (*CDSB*).

From Geraldine Jewsbury 1 May 1863 2003

43 Markham Square | Kings Road Chelsea | S. W. May 1—1863

Dear Dr Tyndall—I have received a message for you w[hic]h. it w[oul]d. be worse than embezzlement on my part not to give up to you—so I send it you[1] in the original hand writing that you may have it genuine.

Whose were the fairy fingers[2] that illustrated y[ou]r lecture the other day?[3]—I envied them the good luck of being selected to do any thing to give you pleasure & make you some return for all the good you do to so many others—I hope you are pretty well—

in haste y[ou]rs always faithfully | G. E Jewsbury

never mind returning the enclosure nor answering this—

RI MS JT/1/J/31
RI MS JT/1/TYP/2/658

1. *I send it you*: enclosure missing.
2. *Whose were the fairy fingers*: not identified, but clearly female! Possibly Henrietta Mary Adair. See letter 2009, n. 1.
3. *y[ou]r lecture the other day*: see letter 1987, n. 4.

To the editor of the *Athenaeum* 2 May 1863 2004

Royal Institution, April 27, 1863.

I trust you will permit me to make a few remarks in connexion with the brief review of my Lectures on Heat which appeared in your last number.[1] It is quite possible that my mode of treating the subject may be open to even graver criticism[2] than that which your reviewer has bestowed upon it; but I should be extremely sorry if an alleged personal defect on my part should stand between the readers of the *Athenaeum* and a generalization which they ought to be among the earliest to understand and appreciate. The dynamical theory of heat, forming as it does the most important part of the theory of the interaction of natural forces, is deemed by the most competent authorities of this age fit to stand, in point of importance, beside the Newtonian theory of gravitation. It is held, without misgiving, by the leading natural philosophers of all the countries of Europe. Had the theory still to be discussed, I should never have presumed to bring it before the public as I have done. But proofs of it of the most varied and conclusive character have been accumulating for the last twenty years, until they have at last disarmed all opposition. This theory must not only be the future guide of the practical engineer in his applications of heat as a motive power, but in its more purely intellectual bearings the theory must profoundly affect the whole course of philosophic thought and inquiry. It, moreover, opens up views of vital phenomena, and of the organization of the material universe, which cannot be regarded with indifference by any thinking man; and it is for these reasons that I have endeavoured, as far as in me lay, to divest the subject, without injuring it, of technical difficulties, and thus to render it accessible to the intellectual public of England.

The passage[3] which your reviewer has done me the honour to quote is the most highly coloured in the book. It reads, indeed, more like romance than science; and nothing is more natural than that it should stagger those who have not closely followed the developments of modern physics. It may be, and probably is, open to the charge of unnecessary iteration, but certainly no coolness or calmness on my part could add to the substantial truth of the statements therein contained. When, for example, I say that "every fire that burns and every lamp that glows dispenses light, and heat which originally belonged to the sun," I mean to express a *fact*, and not a figure of speech. There heat of every fire in London is as much a part of the sun's heat as if every glowing coal had been taken by tongs from the body of the sun and put into our grates; and the sun's heat has been diminished by the exact quantity emitted by our fires. This seems very wonderful, but it is only a small part of the wonders which the dynamical theory of heat lays open to our view. The lifting of a weight from the earth is a commonplace act; but it is an act of the same mechanical quality as that performed by the sun which he lifts the oxygen from the carbon of carbonic acid gas, and permits the latter to store itself in the holes, branches, and leaves of our trees. It is the solar force thus invested in the vegetables, that afterwards becomes the source of all animal power. Muscular force is simply that sun's force transmuted. The helmsman stands at the wheel, but he cannot add one jot, or tittle to the motion of the ship. He directs, but cannot create. Equally ineffectual is human volition, or the organic processes of the animal body, to generate one unit of mechanical force. All such force is primarily derived from the sun. These are the latest verities of science, and, read in their light, the passage quoted by your reviewer will not appear extravagant. I may add, that these considerations are confined to the twelfth lecture of the course, and that the work embraces the relationship of the dynamical theory to all the phenomena of heat. It is, moreover, so written that the reader who wishes it may also make himself acquainted with the old theory. Indeed, the old theory comes out with greater vividness by its contrast with the new one.

John Tyndall

J. Tyndall, 'Heat as a Mode of Motion', *Athenaeum*, 2 May 1863, p. 585.

---

1. *which appeared in your last number*: the anonymous review of *Heat as a Mode of Motion* was published in the *Athenaeum*, 25 April 1863, p. 549. All reviews in the *Athenaeum* were published without authorial attribution.
2. *even graver criticism*: the reviewer had criticized Tyndall's work for displaying 'in many portions of these lectures too much of the spirit of an advocate, and sometimes there is evidence of special pleading which is unbecoming the philosopher … Readers of the book will not expect to find the calmness of Herschel or Faraday; for Prof. Tyndall has his own individualities as a writer and thinker, which lend to his work their character and charm' (see n. 3 for reference).
3. *The passage*: in the lengthy quoted passage, taken from Lecture XII, Tyndall asserted, in quite poetic language, that all energy finds its source in the sun (p. 549).

To John Couch Adams[1] 8 May 1863 2005

8th May. /63

My dear Adams.

I express no formal sentiment when I congratulate you, and wish you and your wife a long continuance of this world's best happiness. Indeed your example has raised a kind of vague tending towards imitativeness in myself—an incipient desire to complete what is at present but half a life.

I heard yesterday that you were at Capel Curig[2]—Were I at present in that lovely region, I should be happy to offer my services as guide to Mrs. Ada[ms][3] through the "craggy chaos" of Snowdonia. I am sure I should take good care of her both on her account and yours.

Believe me. | ever yours sincerely | John Tyndall

Wellcome, MS. 7777/06

---

1. *Adams*: recipient identified by reference to marriage; John Couch Adams married Eliza Bruce in 1863 (W. G. Adams (ed), *The Scientific Papers of John Couch Adams*, 2 vols (Cambridge: Cambridge University Press, 1896), vol. 1, p. xxxv).

2. *Capel Curig*: a small village in the heart of Snowdonia, western Wales.
3. *M^rs^. Ada[ms]*: Eliza Adams (née Bruce, 1827–1919). In the original letter, the 'ms' in 'Adams' appears to be covered in residue from having been previously glued into a volume.

To Rudolf Clausius 8 May 1863 2006

8th May. 1863

My dear Clausius.

It gave me very great pleasure to receive your last letter,[1] and to learn from it that my little book[2] had interested you. There is no man in Europe whose opinion on such a matter I prize more highly than yours, and hence the gratification which your letter gave me.

I have delayed writing to you until I could send you my reply to the attack of Prof. William Thomson & Prof. Tait.[3] I can assure you that I should be rejoiced to live in peace and friendship with those men. Nothing grieves me more than that two men of nearly the same age like Thomson and myself, engaged in pursuits which ought to make us friendly towards each other, should be found in hostile conflict. It is an awful waste of time and good temper. But there is no knowing how far this habit of imputing unworthy motives will proceed if it be not stopped in time. Ever since the publication of my work upon the Alps[4] Thomson has been unsparing in his censures, but has never had the manliness or frankness to attack me openly. Even in his article in 'Good Words'[5] his attack, though manifest, is more like an insinuation than a fair open manly proceeding. His putting Tait forward[6] & keeping in the background himself is quite characteristic, but he has for once mistaken his man.

His treatment of poor Mayer[7] you will I think agree with me in regarding as perfectly inexcusable. I should be willing to grant him the excuse of ignorance of Mayer's merit, but such an excuse he refuses to employ and hence lays himself open to the charge of treating Mayer unjustly.

Believe me I am most anxious to live in peace with Thomson and every body else. A man's life is too short to have any part of it devoted to such controversy,[8] but with some people forbearance is interpreted as timidity. It is very hard to deal generously with people who mistake your generosity for fear. It is now more than two years since Prof. Forbes wrote a pamphlet[9] against my book on glaciers, but I was content to allow him to have the last word, though it has been in my power all this time to answer him successfully. For this forbearance on my part I am now taunted by his countrymen Thomson & Tait.

If you had time I wish you would look over the chapter '<u>Moteurs</u>' in Seguin's work "Sur 1'Influense des Chemins de Fer" (1839).[10] He has uttered many truths, but I think his publications show him to be a very different man from Mayer. He is a man with a certain amount of insight; mixed with a large amount of confusion & error. Goodbye—kindest regards to your wife[11] & believe me

always yours, sincerely, | John Tyndall

RI MS JT/1/T/233

---

1. *your last letter*: letter 1998.
2. *my little book*: *Heat as a Mode of Motion.*
3. *the attack of Prof. William Thomson & Prof. Tait*: see letter 1986, n. 5. For Tyndall's reply, see letter 1979, n. 3. Also see letter 2002 of this volume. Roland Jackson provides further commentary on this debate in *Ascent of John Tyndall*, pp. 176–7.
4. *my work upon the Alps*: see letter 2002, n. 23.
5. *his article in 'Good Words'*: see letter 1986, n. 5.
6. *his putting Tait forward*: P. G. Tait, 'Reply to Professor Tyndall's Remarks on a paper on "Energy" in *Good Words*,' *Phil. Mag.*, 25 (1863), pp. 263–6. Also, although the original paper was co-authored, Tyndall clearly saw Thomson and not Tait as the driving force behind the essay, and said as much to Thomson (see letter 2002).
7. *His treatment of poor Mayer*: Tyndall refers here to Thompson's charge that Mayer's paper had 'no claims to novelty or correctness' (see letter 2002).
8. *A man's life is too short . . . controversy*: this is somewhat ironic since Tyndall's career was marked by a number of such controversies, many of which were of his own making. See the Introduction to this volume for further commentary.
9. *Prof. Forbes wrote a pamphlet*: J. D. Forbes, *Reply to Professor Tyndall's Remarks in his work, 'On the Glaciers of the Alps,' relating to Rendu's 'Théorie des Glaciers'* (Edinburgh: Adam and Charles Black, 1860).
10. *Seguin's work*: see letter 1990, n. 2.
11. *your wife*: Adelheid Clausius.

From Samuel Haughton[1] 9 May 1863 2007

9 May 1863.

My dear Sir,

I forward you with this an account of some novel experiments on Muscular Action,[2] which may lead to further speculations in physiology

I have read your book on Heat[3] with great pleasure, and your controversy with Thomson[4] with still greater pleasure—as I feel satisfied you have only done simple justice to Mayer and Joule—Thomson's own share in the Mechanical Theory of Heat is considered here to be of the minutest kind.

Yours faithfully | Sam[uel]l. Haughton

John Tyndall Esq.

RI MS JT/1/TYP/2/488
LT Typescript Only

1. *Samuel Haughton*: Samuel Haughton (1821–97), an Irish mathematician, geologist, and physiologist, born in Carlow. From 1851 he held the chair of geology at Trinity College, Dublin, and from 1859, despite his chair and fellowship, enrolled as a medical student, until 1862. He subsequently published extensively on comparative anatomy, dissected large animals at the Dublin zoo, and, in 1866, proposed that criminals condemned to hang should be subject to a longer drop than was then current (death would be instantaneous and more humane, he argued) (*ODNB*).
2. *an account of some novel experiments on Muscular Action*: S. Haughton, *Outlines of a New Theory of Muscular Action* (London: Williams and Norgate, 1863).
3. *your book on Heat*: *Heat as a Mode of Motion.*
4. *your controversy with Thomson*: see letters 1998, 2002, 2004, and 2006, as well as the Introduction essay for further commentary.

From William Frederick Pollock 10 May 1863 2008

59 Montagu Square[1] | 10th May | 1863

My dear Tyndall

Not until yesterday did I see the Philosophical Magazine for May and read your last letter to Prof[esso]r Thomson[2]—It seems to me, if you will let me say so, most admirable, and I must be allowed to congratulate you upon the entire success of your reply to a most unmerited attack—In substance, style, tone, & temper I can conceive nothing better—Juliet[3] much wishes to see it, and if you can lend her a copy of the magazine for that purpose it will much oblige her—

With kindest regards—| Yours ever truly | W. F. Pollock

RI MS JT/1/P/232
RI MS JT/1/TYP/6/2005

1. *59 Montagu Square*: see letter 1954, n. 2.
2. *and read your last letter to Prof Thomson*: letter 2002.
3. *Juliet*: Juliet Pollock.

From Mary Adair[1] [19 May 1863][2] 2009

Heatherton Park[3] | Tuesday

Dear M^r^. Tyndall,

The book[4] arrived this morning, and I am quite delighted with all I have had time to read of it. I am so very much obliged to you for your kindness—and shall hope to read it with Miss Hamilton[5] during a visit I am soon going to pay them in Wales—It seems a "grand" work, and I shall value it much.

Believe me | Very truly yours | H Mary Adair

RI MS JT/1/A/8

1. *Mary Adair*: Henrietta Mary Adair (1843–1921), the youngest daughter of Alexander Adair (1791–1863) and Harriet Eliza Atkinson (1811–78). She attended Tyndall's lectures at the RI and they enjoyed conversing about scientific matters. Tyndall would later fall in love with Mary and eventually propose to her in late 1869, which she declined (*Ascent of John Tyndall*, p. 257).
2. *[19 May 1863]*: this date is written on the letter but does not appear to be in Adair's or Tyndall's hand; perhaps added by LT. 19 May 1863 was indeed a Tuesday.
3. *Heatherton Park*: Heatherton Park Estate, Somerset, had been purchased by (Hugh) William Adair (1754–1844) in 1807 from the Gunston family. At his death it passed, with Colehayes Park (Devon), to his second surviving son, Alexander Adair (1791–1863).
4. *The book*: probably *Heat as a Mode of Motion*, which was published in March 1863.
5. *Miss Hamilton*: not identified. In his journal entry for 28 November 1869, Tyndall wrote that 'Years and years ago I remember her friend Miss Hamilton sending me down to Ventnor a draft of Miss Adair's notes of my lecture' (RI MS JT/2/13c/1357). Like Adair, this woman likely took a similar interest in Tyndall's lectures.

From George Hector Tyndale 20 May 1863 2010

Philadelphia May 20th 1863

My dear John

In my idleness here I should sooner have acknowledged yours of the 12th April.[1] Just now, however, Mrs Tyndale[2] hands me a letter to send you[3] and this brings me to the writing. I am much obliged for your kindness and interest in me. Your pride makes me stronger, though I feel it is too high for me—this is really so. In fact I have done much less than very many others, some of whom have been deservedly well named, others not known. It is, however, natural for me, your kinsman, to speak of my own doings (even if it be vanity) as of more interest than another's to you. And so with my wife. Indeed when a man is housed up or confined to a small society he naturally magnifies his own little circle of events. However I am sincerely glad if anything I've done has given you prideful pleasure.

After sending my last I was sorry to have said anything of Mr Debus, at least to have said what I did,[4] but, in truth, I cannot, and do not desire to, speak calmly of this War, it's causes and it's results. It has been more than my life to me, and to all real Americans it is the same. Without claiming more than common nature I can say that I think nothing of men, individual men, comparatively, but wholly of man. My social condition, my advancement, my blood, my individuality is nothing (even to as selfish a man as myself) compared with this greatest and best of Governments. One which has ameliorated every other and which in such a country as this has blessed and cheered millions of the Juggernaut-crushed. And we too have our Juggernauts, but they are lighter and fewer than the older ones. I claim for this country a basis of the results of science and of faith in religious liberty.

Feeling this I cannot accept anything for myself if it be in the road, as an obstacle, to a love for this country and its government.

Please make my kind regards to Mr Debus, who as an earnest & truly religious seeker I greatly respect. Give my kindest love to "Tom" Hirst and tell him we, Mrs Tyndale and myself, often speak of him with the kindest remembrances. It may again happen that we shall meet, as if I get out of this war alive, I may visit Europe again, but if not, then one of the pleasantest of kindly feelings to the Earth will be to the genial, kindly face and heart and the earnest, honest soul of Tom Hirst—may his shadow never be less.

I am still awaiting orders and to be assigned to a command, but our armies are new and large and it requires time to place individuals.

We are all much interested in hearing of our distant relatives on the other side the ocean. Perhaps some day, should life continue and fortune smile on me we may see each other.

Mrs Mitchell[5] is well as are all my brothers and sisters.[6] Not promising quick replies, I shall always be glad to get a letter from you.

Aff[ectionate]ly yours as ever | Hector Tyndale

RI MS JT/1/T/56
RI MS JT/1/TYP/5/1655–6

1. *yours of the 12th April*: letter missing.
2. *Mrs Tyndale*: Julia Tyndale.
3. *a letter to send you*: letter missing.
4. *After sending my last I was sorry to have said anything of Mr Debus, at least to have said what*

I did: letter missing; this does not appear to be the previous letter from Tyndale in this volume (1988).

5. *Mrs Mitchell*: Elizabeth Mitchell.
6. *all my brothers and sisters*: besides Elizabeth Mitchell, other sisters of Tyndale are unidentified. For his brothers, see letter 1946, n. 19.

From Juliet Pollock [28 May 1863][1] 2011

59 Montagu Square[2] | Thursday M.

Trusted and good friend

Can & will you send me a pass for my cousin Miss Herries[3] for Max Müller[4] for tomorrow afternoon? Don't trouble yourself to do more than post the pass to Miss Herries.

80 Westbourne Terrace

Yours most really | Juliet Pollock

RI MS JT/1/P/215

---

1. *[28 May 1863]*: the date is given by relation to Max Müller's Friday Evening Discourse at the RI, 'On the Vedas, or the Sacred Books of the Hindus', delivered on 29 May 1863 (*Roy. Inst. Proc.*, 4 (1862–6), pp. 135–40). William Frederick Pollock noted hosting Müller in May 1863 (*Remembrances*, vol. 2, p. 112–3), and this letter by Juliet being written on a Thursday with mention of 'tomorrow afternoon' gives a date of 28 May 1863. It is important to note that Müller had also delivered a series of lectures on language at the RI from February to May 1863, which were published the following year as *Lectures on the Science of Language at the Royal Institution of Great Britain delivered in February, March, April and May 1863. Second Series* (London: Longman, Green, Longman, Roberts & Green, 1864). Thus, the referred to lecture could have been part of this series, and the date of 28 May 1863 would still hold.
2. *59 Montagu Square*: see letter 1954, n. 2.
3. *Miss Herries*: possibly Isabella Herries, a cousin of Juliet Pollock (see letter 1550, n. 11, *Tyndall Correspondence*, vol. 6)
4. *Max Müller*: Friedrich Max Müller (1823–1900), a German-born philologist and Orientalist, who lived and studied in Britain for most of his life. He was one of the founders of the western academic field of Indian studies and the discipline of comparative religion (*ODNB*).

From George Gabriel Stokes 28 May 1863 2012

Lensfield Cottage Cambridge | 28 May 1863

My dear Tyndall,

I return you your paper[1] by book post, partly to let you see a few trifling alterations, but mainly with reference to a point in p 11.

"Precipitate the act of condensation" p. 8 struck me as very odd immediately after the use of precipitation in its technical (i.e. meteorological) sense. On first reading I expected "water" or "vapour" instead of "the act" to follow precipitation.

Now for p. 11, there is no reference given. I referred to Leslie's's[2] book on heat,[3] but it is weary work to hunt out a single sentence in a book without an index or even table of contents. So I was unable to judge from the context what Leslie's views might have been. But it is to me perfectly inconceivable that Leslie should have supposed that the vapour acted by its pressure. I feel pretty sure either that "pressure" was a misprint for "presence", or, which is more probable, that as we frequently speak of tension (a most unhappy substitute for pressure) of aqueous vapour as a measure of, or rather mode of expressing, its quantity, so here Leslie used pressure as tantamount to quantity.

An eminent man is not to be lightly charged with a bit of utter nonsense. If the context supports the charge, well; but I think the reference should be given to enable the reader to judge [Hereafter LT Transcript Only] for himself. But if my suggestion seems to you correct or probably so, the passage should be altered.

Thank you for your article in the Phil. Mag.[4]

Yours very truly | G. G. Stokes

Proof by book post.

RI MS JT/1/S/232
RI MS JT/1/TYP/4/1387

---

1. *your paper*: J. Tyndall, '"On the Relation of Radiant Heat to Aqueous Vapour', *Phil. Trans.*, 153 (1863), pp. 1–12.
2. *Leslie's*: John Leslie (1766–1832), a Scottish mathematician, physicist, and professor of natural philosophy at Edinburgh University (*ODNB*).
3. *Leslie's book on heat*: J. Leslie, *An Experimental Inquiry into the Nature and Propagation of Heat* (Edinburgh: J. Mawman, 1804).
4. *your article in the Phil. Mag*: see letter 1948, n. 7.

From Julius Robert Mayer 31 May 1863 2013

Hochgeehrtester Herr!

Kaum weiss ich die Worte des Dankes und der Verehrung zu finden, um die Gefühle auszudrücken, welche meine Brust bewegen. Schon unter dem 15 Juni v. J. hat mir Herr Prof. Clausius in Zürich die höchst angenehme Mittheilung von der Rede gemacht, welche Sie im Hörsaale der Royal Institution am 6 Juni 1862 gehalten haben, und welche sich u.a. auch im Scientific American von New York abgedruckt findet. Wenn Sie aber, hochverehrtester Herr, schon hier mir ein, meine je im Stillen gehegten kühnsten Hoffnungen übersteigendes Lob ertheilen, so machen mich vollends Ihre Mittheilungen aus dem Philosophical Magazine, welche Sie die Güte hatten, mir selbst zu übermachen, bestürzt, und zwar um so mehr, als ich mich längst an Kundgebungen ganz entgegengesetzter Art habe gewöhnen müssen. —

Die Prioritätsfrage über das mechanische Aequivalent der Wärme hielt ich schon durch mein letztes Schreiben an die Pariser Akademie (Comptes rendus XXIX, 534. Jahrg: 1849) erledigt, da mein berühmter Concurrent, Herr Joule, meines Wissens nicht mehr darauf geantwortet hat. Auf dieses Document habe ich mich auch in einem Schreiben an die Wiener Akademie berufen, das in den Sitzungsberichten Jahrg: 1851, Bd vi, Hft 5. S. 601. abgedruckt ist. Gewiss ist es aber nie meine Absicht gewesen, die ausserordentlichen und auch überall anerkannten Leistungen des grossen Physikers zu Manchester irgend wie verkleinern zu wollen; vielmehr habe ich denselben ganz in der Weise, wie Sie hochverehrtester Herr Sich so treffend ausgedrückt haben, immer nur als meinen sehr geschätzten und auch bevorzugten Mitarbeiter auf dem nemlichen Gebiete des Denkens und nie als Gegner betrachtet und gerne gestehe ich, dass ohne dessen vortreffliche experimental-physikalische Arbeiten die ganze Lehre von der Erhaltung der Kraft, oder wie ich mich auch ausdrücken möchte, von der physikalischen Stöchiometrie nicht das schöne Gedeihen hätte finden können, dessen sich dieselbe heute erfreut. Auch ist der Name "Joule" so viel ich weiss in Deutschland so hoch gefeiert, als er es nur selbst in England seyn kann. Die Herren Professoren Thomson und Tait aber anlangend, so schulde ich denselben für ihre Kundgebung in „Good Words" insoferne indirect viel Dank, als sie dadurch Ihnen zu den "Remarks" Veranlassung gegeben haben. Im Übrigen würde ich aber, an Herrn Joule's Stelle, auf derartige fromme Bundesgenossen gerne Verzicht leisten.

Der Inhalt der übrigen Zusendungen—nämlich die Abhandlung über die Ausstrahlung des Erdkörpers, über die Wärme Absorption und Radiation der Gase und über die Conformation der Alpen—welche Sie mir zu übermachen die Güte hatten, war mir theilweise schon durch deutsche Übersetzungen bekannt. Es sind dieser Abhandlungen welche ebensowohl durch ihre tief eingreifende Wichtigkeit, wie durch den grossen Scharfsinn und die musterhafte Gründlichkeit, welche dieselben überall verrathen, Bewunderung erregen. Für meinen Theil muss ich aber leider gestehen, dass meine Kräfte zu schwach sind, um solche classische Bearbeitungen an und für sich schwieriger Gegenstände nach ihrem vollen Umfange würdigen zu können.

Ihre Reisen in die Schweiz von denen Ihre Arbeiten so schöne, bleibende Zeugnisse geben, werden Sie hoffentlich bald wiederholen. Zu besonderer Ehre würde ich es mir dann schätzen, wenn sie die Güte hätten, mir einen Ort zur Zusammenkunft zu bezeichnen, wo es mir vergönnt wäre, auch die persönliche Bekanntschaft eines Mannes zu machen, dem ich mich für immer zum höchsten Danke verpflichtet fühle.

Schliesslich habe ich noch Ihre verzeihende Nachsicht in Anspruch zu nehmen, dass ich mir in deutscher Sprache an Sie zu schreiben erlaubt habe. Der Grund hiervon ist, dass mir leider das Englische sehr wenig geläufig ist, während Sie, wie ich aus Ihren Arbeiten ersehe, meiner Muttersprache vollkommen mächtig sind.

Genehmigen Euer Hochwohlgeboren den Ausdruck der höchsten Verehrung, mit welcher ich verharre

Dero | gehorsamster Diener | Dr. J. R. Mayer

Heilbronn (Königr. Württemberg) | 31 May 1863.

Esteemed Sir![1]

I hardly know how to express the gratitude and admiration I feel. On June 15 last year in Zurich, Prof. Clausius conveyed to me the most pleasant intelligence of the lecture you held on June 6, 1862[2] in the rooms of the Royal Institution, and which has been printed, among other places, in Scientific American of New York.[3] But if the hopes I ventured to nourish in stillness were exceeded, most esteemed Sir, by the praise you offered me here, your communications to the Philosophical Magazine,[4] which you were kind enough to convey to me personally,[5] affect me deeply and even more so, as I have long become accustomed to enunciations of a quite opposed nature. —

I considered the question of priority as to the mechanical equivalent of heat to be settled by my note to the Parisian Academy (Comptes rendus XXIX, 534. Year: 1849),[6] since my celebrated rival, Mr. Joule, did not to my knowledge respond to me. I also referred to this document in a note to the Viennese Academy,[7] which is printed in the meeting reports, year 1851, vol. vi, issue 5, p. 601. Certainly, however, it was never my intention to diminish in any way the extraordinary and universally recognised accomplishments of the great physicist from Manchester; quite the opposite—and as you, esteemed Sir, yourself so aptly expressed—I have always seen him as my esteemed and favoured colleague in this domain of thought, and never as an opponent, and I gladly avow that without his outstanding work in experimental physics the whole doctrine of the conservation of force or, as I would like to express it, of physical stoichiometry, would not be flourishing, as it is today. The name "Joule", moreover, is, as far as I know, as highly praised in Germany as it is in England. In relation to Professors Thomson and Tait, however, I am indirectly grateful to them as their pronouncement in "Good Words"[8] gave you a reason to write your "Remarks."[9] In general, however, I would, in Mr. Joule's position, gladly do without such pious allies.

I was already in part familiar with the contents of the other mailings—namely the paper on the terrestrial radiation, on the absorption and radiation of heat in gases, and on the conformation of the Alps[10]—that you were kind enough to send me, through German translations. These papers arouse admiration through the deep import, the ingenuity, and the exemplary rigour that they reveal. For my part I must unfortunately confess that my faculties are too weak to do full justice to such accomplished analyses of matters of such inherent complexity.

I hope that you will be able to soon repeat your travels to Switzerland, of which your writings offer such beautiful and enduring memories. I would feel deeply honoured if you were so kind to suggest a place that we could meet, and where I could enjoy the opportunity to make the personal acquaintance of the man to whom I shall always feel deeply indebted.

Finally, I must appeal to your forgiving forbearance that I have allowed myself to write to you in German. The reason is that I am unfortunately not very familiar with the English language, while you, as I can see from your work, have complete command of my native tongue.

Please accept, your Excellence, the expression of my highest admiration with which I remain,

Your | most humble servant | Dr. J. R. Mayer

Heilbronn (Kingdom of Württemberg) | 31 May 1863.

RI MS JT/1/M/80
RI MS JT/1/TYP/7/2510–1

---

1. Tyndall provided a translation of most of this letter in J. Tyndall, 'Remarks on Professor Tait's last letter to Sir David Brewster', *Phil. Mag.*, 26 (1863), pp. 65–7, on pp. 66–7.
2. *lecture you held on June 6, 1862*. see letter 1905, n. 3.
3. *Scientific American of New York*: J. Tyndall, 'On Force—Laws of Motion', *Scientific American*, 7 (1862), p. 100.
4. *your communications to the Philosophical Magazine*: these would most likely have included his letter to William Thomson that was published in the *Phil. Mag.* (letter 2002); and J. Tyndall, 'Remarks on an Article entitled "Energy" in 'Good Words'', *Phil. Mag.*, 25 (1863), pp. 220–4.

5. *to me personally*: letter missing.
6. *my note to the Parisian Academy*: J. R. Mayer, 'Réclamation de priorité contre M. Joule, relativement à la loi de l'équivalence du calorique', *Comptes Rendus*, 29 (1849), p. 534.
7. *note to the Viennese Academy*: *Sitzungsberichte der kaiserlichen Akademie der Wissenschaften: Mathematisch-naturwissenschaftliche Classe*, 6 (1851), pp. 601–2.
8. *their pronouncement in "Good Words"*: see letter 1979, n. 3.
9. *your "Remarks"*: see n. 4 above.
10. *the paper on the terrestrial radiation, on the absorption and radiation of heat in gases, and on the conformation of the Alps*: see letters 1948, n. 7; 1941, n. 2; and 1921, n. 6, respectively.

To George Gabriel Stokes 8 June 1863 2014

Royal Institution of Great Britain | 8th June 1863

My dear Stokes,

You were kind enough last year to inform me[1] that you do not think the fact of my giving an account of an investigation at the Royal Institution previous to the handing in of the paper to the Royal Society would be any obstacle in the way of the acceptance of the paper.

I am this year in circumstances similar to those which environed me last year. I have a paper[2] almost ready & by a tremendous shuffle I can have it in your hand by Thursday[3]—but to have to do this will press me terribly hard Would it matter if I lectured on the subject on Friday ev[eni]ng next[4] before putting the paper in your hands, or must you have it on Thursday?

Yours very truly | John Tyndall

CUL SC—add 7656

1. *to inform me*: this does not appear to be in any correspondence from Stokes to Tyndall from 1862, so possibly this occurred in person.
2. *a paper*: J. Tyndall, 'On the Absorption and Radiation of Heat by Gaseous and Liquid Matter. Fourth Memoir', *Phil. Trans.*, 154 (1864), pp. 201–25 (received at the RS on 18 June 1863).
3. *Thursday*: 11 June 1863.
4. *Friday ev[eni]ng next*: Friday, 12 June 1863 when Tyndall delivered his Friday Evening Discourse at the RI, 'An Account of Some Researches on Radiant Heat' (*Roy. Soc. Proc.*, 4 (1862–6), p. 146–50).

From Henry Bence Jones 9 June 1863 2015

My dear friend

Since our conversation I have been thinking how I can help you & Frankland without hurting you.

It would please me much if you and he would give me the wording of the objects of the fund[1] unless indeed no wording whatever will do & then I shall prefer avoiding offence to you to helping science & the whole thing must stop.

My intention was something like the following subject to your & Franklands approval

To enlarge the objects of the Holland fund[2] as that Donation fund for the promotion of Scientific Research at the Royal Institution.

1st by providing scientific instruments

2nd by procuring the best assistants

3rd by enabling the professors to give as much of their time as possible to research in the laboratory

Will you and Dr Frankland alter this in any way you please

I oblige | *[1–2 words illeg]* | H Bence Jones | June 9 1863

RI MS JT/1/J/53

1. *the objects of the fund*: originally Bence Jones had intended a fund to support the researches of Tyndall and Frankland by increasing funding for their research and to pay their assistants. The result was the 'Holland Fund' (see n. 2; also see *Ascent of John Tyndall*, pp. 180–1).
2. *Holland fund*: the Holland Fund, or Donation Fund, established to promote experimental researches by the RI, was named after Henry Holland (1788–1873), first baronet, an English physician and travel writer who would serve as the president of the RI from 1865–73 (*ODNB*). Holland donated £40 each year from 1859, but did not succeed in attracting a large sum.

To Julius Robert Mayer 17 June 1863 2016

Royal Institution | Albemarle Street London | 17th. June 1863

My dear Sir,

It gave me the greatest pleasure to receive a letter from you,[1] and still greater pleasure to find that what I have written and spoken regarding you has given you satisfaction.

My task has not been a difficult one. From the moment when I first became acquainted with your writings I knew that your reputation was secure. And though I have encountered an amount and a kind of opposition which I did not anticipate,[2] still with so good a cause I should not fear to encounter far greater.

It is a great delight to me to be the means of setting before the people of England the claims of a man to whom we owe so much. And depend upon it as far as England is concerned your reputation will triumph over all antagonism. With regard to Mr Joule no man esteems him more highly than I do & I knew that you would share this feeling. You ought never to be put in opposition to each other,[3] you ought both to stand side by side like brothers, and no man ought to attempt to put the claims of one above those of the other. This I think will be the course which the future historian of science will pursue.

It would give me great delight to make your personal acquaintance and should I, during the next vacation, or any subsequent vacation come near you I shall certainly do myself the pleasure of finding you out.

Your works are appearing in the Philosophical Magazine. Knowing their value I induced the Editor to publish translations of them. Three of them have already appeared and the only one that remains is "Die Organische Bewegung" which will also appear in due time.[4]

Let me express to you again the great pleasure which your letter has given me, and wish you health and happiness to enjoy your well won honours.

Yours most sincerely | John Tyndall

RI MS JT/1/T/1064
LT Typescript Only

---

1. *a letter from you*: letter 2013.
2. *opposition which I did not anticipate*: this is a reference to disputes over Mayer's work, which spurred controversy between Peter Tait, William Thomson, and John Tyndall. See the Introduction of this volume for further commentary.
3. *You ought never to be put in opposition to each other*: Tyndall had characterized William Thompson as doing exactly this; see letters 2004 and 2006.
4. *Your works . . . in due time*: J. R. Mayer, 'Remarks on the Forces of inorganic Nature (1842)', *Phil. Mag.*, 24 (1862): pp. 371–7; J. R. Mayer, 'On Celestial Dynamics (1848)', *Phil. Mag.*, 25 (1863), pp. 241–8, 387–409, 417–28; J. R. Mayer, 'Remarks on the Mechanical Equivalent of Heat (1851)', *Phil. Mag.*, (1863), pp. 493–522. Mayer's lengthy essay, *Die Organische Bewegung in ihren Zusammenhange mit dem Stoffwechsel* (Heilbronn: C. Drechsler, 1845) had not been translated at this point in time, nor would it be published in the *Phil. Mag.* at all (in letter 2102 to Hirst, dated 31 May 1864, Tyndall noted that Mayer's essay was so far 'unpublished in England'). The essay did appear in a later German edition of Mayer's work (see letter 1904, n. 3).

From Edward Frankland 19/20 June 1863 2017

MY DEAR TYNDALL,—At your request I have made a number of experiments on the comparative transcalency of common air, and of air deprived of its moisture by contact with monohydrated sulphuric acid. The apparatus which I used was that described by you in the Philosophical Transactions for December 1862*.[1] It was exclusively under my own control; and I arranged the details of manipulation in such a manner as appeared to me best calculated to eliminate all sources of error. My mode of operating was as follows:—The brass tube (AB, fig. 2 [see antè, p. 34]), open at both ends, formed the conduit for a portion of the thermal radiation from the source (S). These heat-rays, after passing through the tube, traversed several inches of intervening air-space before entering the cone of the thermo-electric pile, where they produced their effect, in opposition to that arising from another constant source of heat affecting the opposite face of the pile (the compensating cube). The differential action was indicated as usual by a delicate galvanometer. These arrangements being once for all made, I was able by means of an air-pump to introduce at pleasure into the tube either the ordinary air of the laboratory, dry air, or air rendered moist by passage over extensive surfaces of wet glass. At the commencement of the experiments the tube was of course full of the common air of the laboratory; the needle of the galvanometer marked 42°, and remained steady for a quarter of an hour within a degree of that point. I now interposed in the path of the rays entering the brass tube a sheet of tin-plate; the needle at once bounded from 42° up to 90°. It was thus evident that any obstruction to the passage of the rays of heat through the tube, or, in other words, any *cooling* of that face of the pile which was turned towards the tube, would be indicated by an increased deflection of the needle on the same side of zero, which I will call the—side, whilst a *heating* of the same face of the pile would be attended by a diminished deflection, or even by a passage of the needle to the opposite or + side of zero. The following are the results which I observed:—

Tube filled with	Permanent deflection.
Common air	-42°
Air dried by contact with monohydrated sulphuric acid and introduced gently into the open tube.	+13
Common air which had spontaneously displaced the foregoing dry air	-43
Common air gently drawn in by air-pump . .	-43.3
Common air gently blown in from caoutchouc[2] bag.	-45

Same air gently blown in from caoutchouc bag, but dried by passage over sulphuric acid .	+6
Air from same bag but not dried.	-46.5
Air of laboratory	-42
Air dried and introduced as before.	+14
Air dried as before with sulphuric acid, but after—wards passed over fragments of glass moistended with water	-46
Common air gently drawn in by air-pump . .	-42.5

At the conclusion of the experiments I found that the deflection due to the total radiation was 86°.2.

I also saw the following experiments made by yourself when the ends of the brass tube were closed by plates of rock-salt:—

	Permanent deflection of needle (°)
Tube filled with dry air. . . .	+7
After exhaustion of tube. . . .	-3
After admission of laboratory air .	-42

Rain was falling at the time these last determinations were made, and the air was very moist. On removing the plates of rock-salt from the tube they appeared to be quite dry; and after being breathed upon, the film of moisture soon disappeared and they recovered their previous lustre. I ought perhaps to mention that these experiments are not selected, they are the only ones I have made upon the subject, and they were performed in the sequence given above: after a very careful scrutiny I have been unable to detect any source of fallacy in them, and they therefore appear to me to prove conclusively that obscure radiant heat passes much more readily through dry than through moist air.

In conclusion, I cannot but express my surprise and admiration at the precision and sharpness of the indications of your apparatus. Without having actually worked with it I should not have thought it possible to obtain these qualities in so high a degree in determinations of such extreme delicacy, and which are so well known to be exposed to numerous sources of derangement.

Believe me, | Yours very truly, | E. Frankland. | Royal Institution, | June 19, 1863.

P.S. Since writing the foregoing letter, I have repeated the experiments there recorded without any source of heat at either end of the pile, in order to ascertain whether the introduction and withdrawal of dry air at all affected the galvanometer. The tube was first full of the common air of the laboratory, and the needle remained steadily at +12.5 for a quarter of an hour. A current of moistened air was now drawn through the tube for ten minutes in precisely the same manner as when the two sources of heat were employed, the needle being closely watched during the whole time. It oscillated between +12 and +13, but never passed these limits on either side. The current was not interrupted and the needle closely watched for five minutes: it remained perfectly steady at 12.5. A current of dried air was now conveyed through the tube for five minutes, the needle oscillated as before between 12° and 13°.

Thus far I operated on the air exactly as in the experiments recorded in the foregoing letter. I then quadrupled the velocity of the current through the tube, introducing in the first place dry air: the needle in a first experiment moved 6 degrees in the direction of cold; but on repeating the experiment with both dry and moist air no effect whatever was produced. I now removed the tube and delivered a gentle current of dry air into the cone of the pile, immediately the needle moved 90° in the direction of *cold;* the current was continued uninterruptedly for ten minutes, during which time the needle gradually returned to nearly its original position. The current of dry air being now stopped, the needle moved 40° in the direction of *heat,* returning again gradually and slowly to its normal position. The same temporary deflection for heat was also produced in an exalted degree when the dry current was immediately succeeded by a moist one.

These supplementary experiments lead me to the following conclusions:—

1st. The gentle currents of air which were caused to flow through the tube in the experiments detailed in my letter did not in any way disturb the results of those experiments, neither would they have done so in any material degree even had their velocity been quadrupled.

2nd. The impact of air dried than that previously in contact with the pile cools that face of the instrument with which it comes in contact, whilst the like impact of moister air produces the opposite effect.

3rd. It is, however, impossible to confound the effects obtained in the above experiments on transcalency with those produced by the impact of dray and moist air upon the face of the pile, because in the first place the former are *permanent,* whilst the latter are essentially *transitory;* and in the second place the deflections due to the impact of dry or moist air against the face of the pile are always *in the opposite direction* to those obtained by the interposition of the same kind of air in the path of radiant heat. Thus, if the heat-rays falling upon one face of the pile be made to traverse dry air, the needle will move in the direction of heat, but if the apparatus be so arranged as to cause the dry air to impinge upon the face of the pile, the effect due to the greater transcalency of the dry air would be at first more or less neutralized, or even altogether overborne, by the cooling influence due to evaporation at the surface of the pile so brought into contact with dry air. —E. F.

Royal Institution, | June 20, 1863.

J. Tyndall, 'On the Passage of Radiant Heat through Dry and Humid Air', *Phil. Mag.*, 26 (1863), pp. 44–54, on pp. 46–9

1. *Philosophical Transactions for December 1862**: the following footnote appeared here in the original: '* [And also in the present Number of the Philosophical Magazine. —J. T.]'. Frankland was referring to the following: J. Tyndall, 'On Dynamic Radiation and Absorption of Heat by Gaseous Matter. Second Memoir', *Phil. Trans.*, 152 (1862), pp. 59–98. Tyndall noted that the description of the apparatus was also published in: J. Tyndall, 'On the Relation of Radiant Heat to Aqueous Vapour', *Phil. Mag.*, 26 (1863): pp. 30–44.
2. *caoutchouc*: India-rubber, or Gum Elastic; the milky resinous juice of certain trees in South America, the East Indies, and elsewhere, which coagulates on exposure to the air, and becomes highly elastic, and is waterproof; it is now a most important substance in arts and manufactures (*OED*).

From Edwin Wyndham-Quin[1] 22 June [1863][2] 2018

June 22[d]

My dear Mr Tyndall

Pray do not think me very changeable when I say that I find I must give up the very great enjoyment I had anticipated in going with you to Switzerland: about the greatest which would have been offered me: but the principal reason is (what I had quite overlooked when I saw you) that my Mother[3] is going over to Adare[4] in the beginning of July, & will only remain a couple of months with us: I am required for some time to be a fortnight about the end of August in Kerry[5] with friends who are coming from England in purpose to pay us a visit; & therefore were I to be away 3–6 weeks more I feel it would not do, in fact I should scarcely be at Adare any part of my Mothers visit. Pray excuse these details, but I wish to tell you the exact state of the case. I am <u>very</u> disap[pointe][d]., & I cannot tell you my chum Mr Tyndall how much I feel your kindness in inviting me to be your companion—should we both be alive & well next year, & that it would be convenient by you, I would make every arrangement to accompany you, & probably be able to bring my Daughter[6] with me—but Lady D[7] felt that it would not do for her to be away from Adare next month. My Mother is old & perhaps may never go over again. —I certainly never before made such a sacrifice of enjoyment to duty—

Believe me to be | Yours most sincerely | <u>Dunraven</u>

RI MS JT/1/D/192

1. *Edwin Wyndham-Quin*: Edwin Wyndham-Quin (1812–71), 3rd Earl of Dunraven and Mount-Earl in the Irish Peerage, and First Baron Kenry in the United Kingdom. As Viscount Adare he served as Member of Parliament for Glamorganshire (1837–51). He had a deep interest in both science and spiritualism, studied astronomy for a number of years, and was a notable archeologist and founding member of the Irish Archeological Society. Wyndham-Quin had converted to Roman Catholicism in 1855 and became an ardent advocate of safeguarding religious education in Ireland (*ODNB*).
2. *[1863]*: the year is suggested by Anna-Maria Hajba, consultant archivist at the Special Collections and Archives Department of Glucksman Library at the University of Limerick. The diary of Dunraven's mother Caroline Wyndham-Quin (see n. 3), reveals that she was in Adare from 13 July to 18 September 1863, and during her stay Edwin made two solo trips to County Kerry, from 18 to 28 July and 29 August to 9 September, respectively. Another, but less likely year is 1856, when Caroline was in Adare from 5 July to 8 September and visited County Kerry with Edwin between 31 July and 9 August.
3. *Mother*: Caroline Wyndham-Quin (1790–1870), Countess of Dunraven, the daughter and heiress of Thomas Wyndham of Dunraven Castle, Glamorganshire. She inherited from her father property in Gloucestershire, as well as the Wyndham estate in Glamorganshire (O. Clarke, 'Divine Providence and Resignation: The Role of Religion in the Management of the Emotions of the Anglo-Irish Countess of Dunraven, Caroline Wyndham-Quin (1790–1870)', in W. Ruberg and K. Steenbergh (eds), *Sexed Sentiments: Interdisciplinary Perspectives on Gender and Emotion* (Amsterdam/New York: Rodopi, 2011), pp. 71–91).
4. *Adare*: a village in County Limerick, Ireland, long highly regarded for its scenic beauty. Adare Manor was also the family seat of the Earl of Dunraven and Mount-Earl.
5. *Kerry*: County Kerry in the southwest of Ireland.
6. *Daughter*: not identified; Wyndham-Quin had five daughters.
7. *Lady D*: Augusta Charlotte Wyndham-Quin (née Goold, 1810–66), Lady Dunraven, the third daughter of Thomas Goold of Rossbrien, Dromadda and Athea, a Master in the Irish Court of Chancery.

To George Hector Tyndale 9 July 1863 2019

9th July 1863.

My dear Hector

In the midst of your present perturbation[1] you will I fear have but little attention to give to this letter. But as I am going off to Switzerland and shall not for two months think of writing letters I wish to say a word to you before I go.

For the last couple of weeks I have had staying with me the two sons of Dr

John Tyndall of Gorey Co. Wexford.[2] They are nice gentlemanly lads, & one of them has creditably taken out his diploma for surgery & midwifery at the Royal College of Surgeons in Ireland. I am trying to get berths for them, but as yet my outlook is very vague.

In a late number of the Wardes newspaper[3] I found under the head of the 'Wexford Archers' a number of names among which were M^r^. Tyndall[4] and the Misses Tyndall.[5] I have learned that this M^r^. Tyndall is a man of substance living at a place called Oakland near Enniscourthy Co. Wexford.[6] I have written to him,[7] & expect a reply from him by Monday next.

I have also had two letters from a namesake in Holland[8] who writes anxious to learn whether we are relatives. His father was named Joseph Tyndall, and he was Judge of Surinam when it belonged to the English. He appears to have been a sugar planter in Surinam. His title & address are

The Hon. J. F. Tyndall de Veer | *[Suchtenberg]* House | near Breda | Holland.

The name de Veer was added to his by Royal letters patent—it is his wife's name. He appears to consider it certain that we are related. His arms bear the three corn sheaves which you find on the arms of Baron de Tyndale[9] in Tyndale's[10] testament. He strongly disputes the correctness of the e at the end, and avows that the martyr was a Tyndall.

They have been getting up the monument to him of late.[11]

If I had time I should like much to take a sweep into Ireland & see those people at home. The old original farm is still occupied by a Joseph Tyndall.[12] It lies, as you mentioned when you were here between Arklow & Wicklow.

And my dear Hector goodbye and with kindest remembrances to M^rs^. Tyndale[13]

Believe me | always affec[tiona]t[el]y yours | John Tyndall

RI MS JT/1/T/1447
RI MS JT/1/TYP/4/1657

---

1. *your present perturbation*: the American Civil War. See letters 1946, 1988, and 2010.
2. *Dr John Tyndall*: John Tyndall (of Gorey). In an earlier letter to Tyndale (1956), Tyndall wrote that this John Tyndall had 'three sons, one of whom has passed all his examinations as a surgeon, and the two others he intends for mercantile pursuits'.
3. *In a late number of the Wardes newspaper*: not identified.
4. *M^r^. Tyndall*: not identified.
5. *Misses Tyndall*: not identified.
6. *Oakland near Enniscourthy Co. Wexford*: in the mid-nineteenth century, Oakland was a small village near the town of Enniscourthy, in County Wexford on the south-eastern coast of Ireland.

7. *I have written to him*: letter missing.
8. *two letters from a namesake in Holland*: not identified; letters missing.
9. *Baron de Tyndale*: the Barony of Tyndale or Tindale refers to the Baron of South Tynedale in the county of Northumberland dating back to the eleventh century.
10. *Tyndale's*: William Tyndale.
11. *the monument to him of late*: presumably the monument at Nibley Knoll in Gloucestershire (see *Ascent of John Tyndall*, p. 256).
12. *Joseph Tyndall*: not identified.
13. *M^rs^. Tyndale*: Julia Tyndale.

To John William Draper[1] 15 July 1863 2020

The Athenaeum | Royal Institution 15th. July | 1863

My dear Sir,

I had the pleasure of receiving two days ago your new & important work.[2]

Thus far I have had barely time to make myself acquainted with its general character—a character which requires courage as well as ability—on the part of its author.

I trust the time is coming when that word 'courage' will not need to be applied to any undertaking which consists in the casual utterance of a man's conviction; but that each will be permitted, without fear of persecution to speak out the thought that is in him.

It is only the real sceptics as to the power of truth that deprecate this consummation.

May you & such as you continue to make their numbers less.

Believe me dear Sir, | Very sincerely from, | John Tyndall

Library of Congress, John William Draper Family Papers

---

1. *John William Draper*: John William Draper (1811–82), an English philosopher, physician, chemist, historian, and photographer, who, from 1832 made his home in the United States. He made significant improvements in photographic techniques. Draper's lecture 'On the Intellectual Development of Europe, considered with reference to the views of Mr. Darwin and others, that the progression of organisms is determined by law' provided the backdrop to the debate between Thomas Huxley and Samuel Wilberforce at the Oxford meeting of the BAAS in 1860 (*ODNB*). He went on to write *The History of the Conflict between Religion and Science* (New York: D. Appleton and Company, 1874), a major contribution to establishing what later became known as the 'conflict thesis' regarding the relationship between science and religion. On the 'conflict thesis', see B. Lightman (ed), *Rethinking History, Science, and Religion: An Exploration of Conflict and the Complexity*

Principle (Pittsburgh: University of Pittsburgh Press, 2019); B. Lightman, 'The Victorians: Tyndall and Draper', in J. Hardin, R. L. Numbers, and R. A. Binzley (eds), *The Warfare between Science and Religion: The Idea That Wouldn't Die* (Baltimore: John Hopkins University Press, 2018), pp. 65–83; and J. C. Ungureanu, *Science, Religion, and the Protestant Tradition: Retracing the Origins of Conflict* (Pittsburgh: University of Pittsburgh Press, 2019).

2. *your new & important work*: J. W. Draper, *History of the Intellectual Development of Europe* (New York: Harper and Brothers, 1863), in which Draper applied the methods of science to writing history.

To [Julius Haast][1] 15 July 1863 2021

15th. July 1863

My dear Sir.

I have received your letter and the newspaper which accompanied it,[2] contrary the account of your last adventurous expedition.[3] I read your account with great interest, and I inferred from it that you were adapted in a special degree for the exploration of strange lands. When all your wanderings & observations are put together in a connected form they will I doubt not be read with very great [in]terest in England.

I am just on the point of starting for the Alps, where I purpose spending the next five or six weeks. I never tire of the mountains. They are to me objects of perennial pleasure, and I feel besides that each year they make me a new man and send me back to London fitted for London work. Were I several years younger I should entertain the thought of going over to New Zealand for I certainly should like to see those mountains of which you give so animated a description.

Right or wrong the action of glaciers has had an importance ascribed to it of late years which though it has excited strong opposition has nevertheless many able supporters. I suppose you have seen my friend Ramsay's paper[4] on the origins of the Lakes of Switzerland. Ramsay ascribes them to the scooping action of the glaciers of former days. The idea is a hardy one, but it is certainly deserving of attention & the arguments which may be adduced in its favour are by no means easy to refute.

Is it your intention to publish a connected and complete work on the topography of New Zealand? I certainly think that out of such material as you are now collecting a most interesting work might be produced.

Wishing you health and strength to pursue your arduous labours.

believe me dear Sir | most truly yours | John Tyndall

Alexander Turnbull Library, National Library of New Zealand, MS-Papers Turnbull 0037–143

1. *[Johann Franz Julian von Haast]*: probably Julius Haast, based on the content of the letter and New Zealand archive address.
2. *your letter and the newspaper which accompanied it*: letter and enclosure missing.
3. *your last adventurous expedition*: in early 1863, Haast led an expedition to search for an overland route to the west coast of the South Island of New Zealand. See J. Haast, *Report on the Geological Exploration of the West Coast* (Christchurch: Provincial Government of the Province of Canterbury, 1865).
4. *Ramsay's paper*: A. C. Ramsay, 'On the Glacial Origin of Certain Lakes in Switzerland, the Black Forest, Great Britain, Sweden, North America, and Elsewhere', *Quarterly Journal of the Geological Society*, 18 (1862), pp. 185–205.

From Julius Robert Mayer 16 July 1863 2022

Heilbronn

Hochverehrtester Herr Professor!

Durch Ihr gütiges Schreiben vom 17 v. M. haben Sie mich aufs Neue sehr erfreut und zu weiterem grossen Danke verpflichtet. Da ich in London wo ich ohnedies auch Verwandte besitze, die ich einmal besuchen wollte, noch nie gewesen, und diese Weltstadt gerne einmal sehen möchte, so erlaube ich mir an Sie die Frage zu richten, ob ich, wenn ich in allernächster Zeit dorthin komme, die Ehre haben könnte, Ihnen meine Aufwartung zu machen. Es würde mir den höchsten Genuss gewähren, mit einem solchen Manne einige Stunden über wissenschaftliche Gegenstände sprechen zu können. Bei dieser Gelegenheit würde ich dann wohl auch einen seit vielen Jahren gehegten Plan zur Ausführung bringen und Herrn Joule in Manchester aufsuchen.

Genehmigen Sie den Ausdruck meiner vollkommensten Verehrung mit welcher ich verharre.

Ihr | ergebenster Diener | Dr. J. R. Mayer | Heilbronn | 16 Juli 1863.

Most respected Professor!

You have delighted me again with your kind letter of the 17th[1] of last month for which I am very much obliged. As I have never been to London, where I have some relatives notwithstanding whom I would like to visit, and would very much like to see this metropolis, I allow myself to ask you kindly whether, when I come there shortly, I might have the honour of paying you my respects.[2] It would give me great pleasure to discuss scientific matters with such a man for some hours.

On this occasion, I would also realise a plan that I have had for many years and visit Mr Joule in Manchester.

Please accept the expression of my most perfect admiration with which I remain

Your | most humble servant | Dr. J. R. Mayer | Heilbronn | 16 July 1863.

RI MS JT/1/M/81
RI MS JT/1/TYP/7/2512

1. *your kind letter of the 17th*: letter 2016.
2. *of paying you my respects*: in letters 2024 and 2041, Tyndall and Mayer further discussed the possibility of visiting London, 'before next spring" Mayer had hoped, but it does not appear that such a trip came to fruition.

To Henry Bence Jones 30 July 1863 2023

30th. July. 1863.

My dear Dr. Jones

I am here in the midst of the mountains once more. I left London on the 17th accompanied by Mr. Sclater of the Zoological Society.[1] He left me yesterday to return to his wife[2] and being alone to day I will spend a quarter of an hour in writing to you.

Our first day upon the Glaciers was a very long grind. We were on ice and snow for more than 12 hours, and had a little mountain work afterwards. The sun was hot and blazed up against our faces so that I have not yet recovered the broiling which I then endured. That is to say my lips and face still suffer though my general health is good. With the strongest desire to keep out of danger I was near coming to grief the first day, as I found myself alone in an unknown region, and had to perform some ugly work to extricate myself. We crossed the Oberaarjoch[3]—a very fine pass—a few days ago, but its magnificence was obscured by clouds and before we reached our hotel[4] we were assailed by a thunderstorm which drenched us through and through.

Since then we have been dawdling about here, breathing the mountain air, eating the mountain mutton and partaking in moderation of the mountain brandy. Yesterday was splendid, but with this exception the weather has not been good. Last night this house was in a state of commotion regarding a party of two ladies and a gentleman who were lost on the mountains. I summoned the forces of the place and dispatched them in various directions; taking a man with myself I struck into the route which seemed most likely to lead to the wanderers. We had a horrid scramble; along water courses, over torrents, shooting down slippery grass slopes and shouting as loud as our lungs could yell. We found them at length, but could not get them home. So we forced open a chalet for their reception, made a pine fire and left them to rest on the hay which fortunately the place contained.

I will stay here for a few days in the hope of my guide Bennen turning up. Afterwards I will go to the Engadin[5] for a week or two, then to London. I must I think pay a flying visit to Ireland,[6] but I hope by the end of August to be once more at work.

I have had a letter from Dr. Mayer[7] in which he informs me that he intends to visit England. I hope he will come when I am there. I have urged him to visit Newcastle.[8] If you meet him there any services you render him will be done to a true Philosopher.

Yours ever | John Tyndall

RI MS JT/1/TYP/3/792
LT Typescript Only

1. *Mr. Sclater*: Philip Lutley Sclater (1829–1913), an English lawyer and zoologist, known especially for his works in ornithology. He was secretary of the Zoological Society of London from 1860–1902 (*ODNB*).
2. *wife*: not identified.
3. *Oberaarjoch*: see letter 2025, n. 6.
4. *hotel*: see letter 2025, n. 4.
5. *Engadin*: see letter 2025, n. 7.
6. *flying visit to Ireland*: see letter 2025, n. 8.
7. *a letter from Dr. Mayer*: letter 2013.
8. *I have urged him to visit Newcastle*: see letter 2024, n. 3.

To Julius Robert Mayer 30 July 1863 2024

Switzerland | 30th. July 1863

My dear Sir,

Your letter[1] has reached me amid the mountains of Switzerland. I can assure you it would give me extreme pleasure to meet you in London[2] and I should be most happy to do all in my power to render your visit there agreeable to you.

But I shall not be in London until the end of August. I hope your arrangements will enable you to come when I am there. Mr. Joule I doubt not will be delighted to see you.

I am very glad that you have resolved to visit England. The British Association meets this year at Newcastle upon Tyne.[3] Could you not go there? You would meet a great number of the scientific men of England all of whom would be happy to make your acquaintance—The association commences on the 26th of August & continues its sittings for a week.

You are now perfectly well known in England. It was only this morning that I received a letter from a scientific friend[4] thanking me in the warmest manner for a copy of the translations of your "Abhandlung Zur Dynamik des Himmels"[5] and expressing the very highest opinion of its merits.

I had 100 separate copies of the translation printed & distributed.

I did not send you a copy of my last letter to Prof. Thomson[6] on the relative claims of Mr. Joule & yourself. I trust when you read the letter that you will not find that I have permitted your reputation to suffer.

Believe me dear Sir | most sinc[ere]ly yours | John Tyndall

Dr. J. R. Mayer &c. &c. | Heilbronn

Stadtarchiv Heilbronn, 155 Tyndall, John (London) mit RM und Mayer, Emma
04 1863 Juli 30 Schweiz; Brief Tyndall to Mayer
[Fasc. 32 d; Englisch; Druck: V 12, S. 116]

1. *Your letter*: letter 2022.
2. *to meet you in London*: in letter 2022, Mayer had mentioned his desire to visit with Tyndall during a proposed trip to England.
3. *The British Association meets this year at Newcastle upon Tyne*: the BAAS held its thirty-third annual meeting at Newcastle Upon Tyne from 26 August to 2 September 1863 (*Brit. Assoc. Rep. 1863*); 'Analysis of Attendances at the Annual Meetings, 1831–1906', *Brit. Assoc. Rep. 1907*, p. lxxxviii). Mayer did not make the trip.
4. *letter from a scientific friend*: letter missing.
5. *Abhandlung Zur Dynamik des Himmels*: see letter 1904, n. 4.
6. *Prof. Thomson*: William Thomson. The reference is probably to letter 2002.

To Thomas Archer Hirst 3 August 1863[1] 2025

Aeggischorn[2] July 29th 1863.

My dear Tom

I reached Switzerland without accident.[3] My first day's journey was an awfully long grind over three successive glaciers with a bright sun white snow and no veil. My lips have not yet recovered the broiling which they then endured. At the close of the day I found myself alone and in difficulty. I had to get down some steep glaciated rocks to the Grimsel Hotel.[4] The rocks flanking the pass not those of the pass itself, and twice I very nearly came to grief.

From the Grimsel[5] I crossed the Oberaarjoch[6] to this place which I reached in a thunder storm. With one day's exception the weather has been bad. To day the valley is filled with cloud, and a smoky drapery obscures all the summits. I have however moved about a little each day and find my health getting firmer. Indeed I seem quite fit for work again but to render this fitness permanent I must of course remain longer. In this barbarous weather it is hard to decide on a plan of operation. If it clears I will start for the Engadin[7] in a few days and spend a week or so there. But I do not yet see my way clear about Germany. I must make a flying visit to Ireland to see the Tyndalls there.[8] I have had a letter from the man who lives near Enniscorthy,[9] who merely knows the old tradition, but promises to exert himself to procure me more definite information. This trip to Ireland will occupy me more than a week, and I almost think that I cannot accomplish both it and Germany. In fact if this weather should continue I will return direct to England from this place as there is no satisfaction in living amid clouds.

On the whole Tom the time we could spend together seems to me to be necessarily so short that I hardly think it would be worth while making use of it. Nevertheless if you desire it I am ready to meet you about the 16th at Basle[10] and to run on with you to Marburg.[11] When you receive this pray write to me at the same time to Grindelwald[12] and to this place.

3rd Aug.

We have had a fearful night's rain, and a day or two will be required to sweep the snow of the glacier. The weather defeats all plans. Let us put off our journey to Germany until we can do it pleasantly. I shall give up the Engadin, remain at the Oberland[13] for some days and then return to England and Ireland

Yours affectionately | John.

RI MS JT/1/HTYP/523
LT Typescript Only

---

1. *3 August 1863*: Tyndall began this letter on 29 July 1863, and finished it on 3 August 1863.
2. *Aeggischorn*: see letter 1907, n. 6.
3. *I reached Switzerland without accident*: in a letter to Henry Bence Jones (2023), Tyndall noted that he left London for Switzerland on 17 July.
4. *Grimsel Hotel*: the Grimsel Hotel is located at 2,165 metres above sea level on the Grimsel Pass.
5. *Grimsel*: the Grimsel Pass, a mountain pass crossing the Bernese Alps.
6. *the Oberaarjoch*: a high mountain pass across the eastern Bernese Alps, connecting the Fiescher Glacier with the Oberaar Glacier in central southern Switzerland.

7. *Engadin*: a long Alpine valley in southeastern Switzerland.
8. *flying visit to Ireland to see the Tyndalls there*: in a letter to Hector Tyndale (2019), Tyndall had mentioned that if he had the time he 'should like much to take a sweep into Ireland & see those people at home'. He indeed ventured toward Ireland following the BAAS meeting at Newcastle Upon Tyne (see letter 2024, n. 3), but, as he wrote to Juliet Pollock in letter 2031, 'when I reached Stockport the thought of the storm so sickened me that I swerved from my route to Chester & came on to London'.
9. *I have had a letter from the man who lives near Enniscorthy*: letter missing; 'man' not identified. Enniscorthy is a town in County Wexford, Ireland.
10. *Basle*: see letter 1909, n. 4.
11. *Marburg*: see letter 1940, n. 2.
12. *Grindelwald*: an alpine town in the Swiss canton of Bern.
13. *Oberland*: the geographically higher region of the Swiss canton of Bern.

From George Hector Tyndale 6 August 1863 2026

Head Quarters 1st Brigade, 3 Division, | 11th Army Corps U.S. Army | Near Warrenters Junction V[1] | August 6th 1863

My dear John

A few days ago I got yours of the 9th July[2]—before you went to Switzerland[3]—and reply early as there may be no chance again to write to you, as our positions may vary often.

Since I wrote you,[4] the Union Armies have been universally victorious, and had the battle of Gettysburg[5] (stupid names people give places to show their stupidity) been followed up promptly, the rebellion would have been near crushed—as it is, it will draw out, like a scotched snake, it's vile, painful form for some time longer. At the time of that battle I was in command of a brigade on "the Peninsula"—made by the York and James Rivers near Richmond V[irgini]a. But as all supposed that the decisive battle of the war would come off on the Potomac, to which Lee (rebel General in Chief)[6] had retreated, I was sent up with some regiments, whose time of enlistment had nearly expired, to take a part in the expected actions. But Lee escaped and is now in this State (Virginia), and relations between the Armies are about as before his foolish attempt on Pennsylvania—except that the rebels have been badly defeated with a loss of about forty thousand (40,000) men.

In the South West (the States of Tennessee, Mississippi, Alabama, Arkansas and Louisiana) we have swept everything clean. The great American Artery, the Mississippi, and all it's grand tributaries, (in all perhaps over ten thousand navigable miles (10,000 miles) is now in our possession again. This cuts the rebel states in two equal, geographical parts, and hastens the day of peace. The Rebels make their strongest fight in this state—the most populous, richest and strongest of them all. This is a very strong defensive State, large rivers and mountain ranges traverse it, and it's face of hills and dense woods make it strong to defend, hard to hold. But by the blessing of God we shall conquer and restore peace through this pivot land and, I should like to trust too, through the World. But this, so easily written, will be difficult to do and bloody. The soldiers on both sides fight desperately and all are now old soldiers, even in the European sense. Over two years of incessant fighting in campaigns has inured all to great sufferings and privations.

I read your account of the Tyndall's with much pleasure. When you visit Ireland[7] I shall be glad to hear of your opinions on them all—together with such, if any, points of resemblance you may find with ourselves—ie whether they are like our people in Carlow, in London and in America? And so, of all places, you find relations in Holland! Tyndall de Veer[8] (now Tynedale, or dahl de Vere Norse, Saxe-Norman I could have understood but de Veer and in Holland!) Strange. I didn't give the /pasty/ stock so much credit for colonizing and propagating as Nature seems to have done. Had I told my Father[9] that he had relatives, distant even as the "old Gardener," in Holland, he would have laughed—and replied only by the lucus a non lucendo![10] But you see what it is to have pushed to the front, my Cousin! You may even expect to see a fair vrow or fraulein[11] turning both cheeks to be kissed by the new found bachelor Cousin, who has climbed the Alps and made a footprint on the beaten hardness of English literature! Think of it, your tap roots (like Emerson's[12] vine "reaching under the Andes to the Cape"[13]) going over-ground (a /bull/ palpably) to Gorey[14] and through all the extensive Irish lands, to America to Australia, even unto Holland! What a pity none of 'em are rich bachelors or maidens without nearer kin! Well John, it may turn out well with you. As for me, and our family here, we shall never see any of them, probably, and they will never hear of us. But if God wills it, we shall know all and see all some brighter day. When I heard from Mrs Tyndale[15] she was well—it was only two days since. My brothers and sisters[16] are well. Give my kind regards to Hirst ask him to write and remember me to Debus, of whom I have a most kindly memory.

When you can write I shall be glad to read & learn that you are well. Affectionately your cousin Hector Tyndale

Address as usual to Philadelphia, as changes are rapid and frequent.

H. T.

RI MS JT/1/T/57
RI MS JT/1/TYP/5/1659–60

---

1. *V*: Virginia.
2. *yours of the 9th July*: letter 2019.

3. *to Switzerland*: Tyndall left London for Switzerland on 17 July 1863.
4. *Since I wrote you*: letter 2010.
5. *Gettysburg*: the deadliest battle of the American Civil War, the Battle of Gettysburg from 1–3 July 1853 was fought in and around the town of Gettysburg, Pennsylvania. It is considered a turning point in the war because the northern advance by the Confederates was halted.
6. *Lee (rebel General in Chief)*: Robert E. Lee (1807–70), commander of the Confederate States Army (1862–5) during the American Civil War (1861–5). He would surrender to Ulysses S. Grant at Appomattox Court House on 9 April 1865 (*ANB*).
7. *visit Ireland*: see letter 2025, n. 8.
8. *Tyndall de Veer*: see letter 2019.
9. *my Father*: Robinson Tyndale (1775–1842), who had emigrated to Philadelphia from Leighlinbridge around 1804 with his brother William and founded Tyndale & Company, a glass and ceramics import business. For the children of Robinson and his wife Sarah Tyndale (née Thorne, 1792–1859), Hector's siblings, see letter 1946, n. 9 and n. 19.
10. *lucus a non lucendo!*: an illogical assertion. Derived from the Latin *lūcus ā nōn lūcendō*, which literally means 'a grove from not shining', echoing the assertion made by the Roman rhetorician Quintilian (ad *c.* 35–*c.* 96) in his *Institutio Oratoria* that the Latin *lūcus* ('grove') derived from *lūceō* ('I shine') because groves were dark places.
11. *vrow or fraulein*: 'vrow' is not identified. 'Fraulein' is German for 'woman'.
12. *Emerson*: Ralph Waldo Emerson (1803–82), an American essayist and poet (*ANB*).
13. *"reaching under the Andes to the Cape"*: from Emerson's 1847 poem 'Bacchus' (lines 3–4): 'Or grew on vine whose taproots reaching through | Under the Andes to the Cape' (R. W. Emerson, *Poems* (London: Chapman, Brothers, 1847), p. 149).
14. *Gorey*: a town in County Wexford, Ireland.
15. *Mrs Tyndale*: Julia Tyndale.
16. *My brothers and sisters*: see letter 1946, n. 9 and n. 19.

To Juliet Pollock [21 August 1863][1] 2027

My dear Mrs Pollock,

We have had two such jolly days! On Wednesday[2] the anger of the weather rained itself all away, and on Thursday[3] we started from Windermere[4] and made our way along the magnificent ridge of Fairfield[5] purposing to go to Patterdale,[6] from Grisdale tarn.[7] Dr. John Davy[8] rather puzzled us at Ambleside[9] by pointing out a ridge which we did not pursue as Fairfield. But we were certainly right and he was wrong.

As we came near the end of Fairfield the summit of Helvellyn[10] loomed in the distance and we at once resolved to assail the 'mighty' mount. We did not expect to have to dip so deep into Grisdale but no matter, on we went nothing daunted but awfully hungry, especially Huxley who has a much higher capacity for food than I have. We saw some newspapers twirling down the mountain and inferred the existence of picnic parties at the top. We were right, and a most hospitable party of 3 ladies and a gentleman shared their bread and cheese with us and made the summit joyful. I wanted to make the acquaintance of 'Striding Edge,'[11] but the wind was a little high and the people whom I asked about it drew long faces and said that on such a day it would be madness to venture upon the 'edge'—of course I knew what this madness meant. So I sent Huxley down Swirling Edge,[12] and I scampered over the jagged crags of Striding Edge myself. Certainly a man might break his neck there but there is not the least necessity for it if the traveller knows what he is about. In fact I would undertake to conduct you along Striding Edge without allowing you to strike your foot rudely against a stone. We put up at the Water Hotel.[13] How lovely!

There we met the Dean of Westminster[14] and two charming daughters.[15] I had known the Dean only in contest at the Athenaeum[16] & I feared he had reason to believe me to all good words reprobate. I heard he wanted to make his way to Lodore,[17] and it was not without some misgiving as to the results that I proposed guiding them over the intervening country. He at once closed with the proposal, and we had such a jolly day. The deep organ tone of his voice as from time to time he ejaculated a piece of poetry was in thorough keeping with the grandeur of the mountains. We made a new pass which we called the Dean's Pass and the terrors of which Huxley afterwards illustrated in the visitors book at Thirlmere.[18] We crossed two chains of mountains. Were long entangled knee deep in heather, with here and there a stretch of bog, but we hit our point correctly and we left them at the Lodore hill yesterday. The girls were charming, and acquitted themselves in a manner that did credit to the English race. The Dean himself moreover behaved most manfully and on the whole we made as jolly a party as ever crossed those mountains

J. T.

RI MS JT/1/T/1171
RI MS JT/1/TYP/6/2128–9

---

1. *[21 August 1863]*: the handwritten letter is undated, but '21st. Aug' was added to it; the typescript includes '21 August'. Tyndall went on this trip while *en route* to the BAAS meeting in Newcastle Upon Tyne (26 August–2 September 1863), and Huxley, who was mourning the recent death of his brother George Knight Huxley, joined Tyndall and then returned to London while Tyndall made his way to the meeting (*Ascent of John Tyndall*, p. 231; A. Desmond, *Huxley: From Devil's Disciple to Evolution's High Priest* (Reading, MA: Perseus Books, 1994), p. 321–2).
2. *Wednesday*: i.e. 19 August 1863.

3. *Thursday*: 20 August 1863.
4. *Windermere*: name of a town and the largest lake in the Lake District.
5. *Fairfield*: a mountain in the Lake District, above Grisedale tarn.
6. *Patterdale*: a village in the Lake District, in the Ullswater valley.
7. *Grisdale tarn*: Grisedale tarn, a tarn in the Lake District.
8. *D^r. John Davy*: John Davy (1790–1868), a Cornish doctor, amateur chemist, and brother of the noted chemist Humphry Davy. He lived in Ambleside during the 1860s where he died in 1868 (*ODNB*).
9. *Ambleside*: a town in the Lake District.
10. *Helvellyn*: a mountain in the Lake District, north of the town of Ambleside; it is the highest peak in the Helvellyn range.
11. *'Striding Edge'*: a high ridge that requires climbers to scramble to the top of Helvellyn.
12. *Swirling Edge*: Swirral Edge, a ridge on the opposite side of Helvellyn from Striding Edge, with a less intense descent than Striding Edge.
13. *Water Hotel*: possibly the Ullswater Hotel in Patterdale (H. I. Jenkinson, *Practical Guide to the English Lake District*, 4th edn (London: Edward Stanford, 1875), pp. 28–9).
14. *Dean of Westminster*: Richard Chenevix Trench (1807–86), an Anglican archbishop and poet, and Dean of Westminster from 1856–64 (*ODNB*).
15. *two charming daughters*: along with eight sons, Trench had six daughters: Melisina Mary Chenevix Trench (1834–1918), Edith Chenevix Trench (1844–1942), Helen Emily Chenevix Trench (1846–1935), Frances Harriet Chenevix Trench (1847–1941), and Rose Juliet Chenevix Trench (1848–1902). It is not clear which two are referred to here.
16. *Athenaeum*: see letter 1907, n. 7.
17. *Lodore*: a high cascade and waterfall in Cumbria which flows down into Borrowdale Valley.
18. *Thirlmere*: a lake, and now reservoir, in the Lake District, below Helvellyn.

From Thomas Archer Hirst 26 August 1863 2028

Ponte | Engädin[1] | Aug[u]^st 26th/63

My dear John.

I arrived at S^t. Moritz[2] more than a week ago, and made, daily, enquiries for you at Samaden[3] and Pontresina.[4] On Sunday last I received your letter,[5] written from the Eggischorn[6] and forwarded to me by Mrs Tunstall,[7] and to my great disappointment learned that our projected tour together was impossible this year.[8] If you had but persevered a little longer, and come to the Engädin[9] you would have been rewarded with magnificent weather. I was preparing to leave this neighbourhood on Monday when I learned that precisely on that day the Naturforschende Gesellschaft[10] would meet in Samaden. This induced me to prolong my visit and I am glad I did so for I have spent three very pleasant days with the Swiss Naturforscher.[11] Clausius was here and Wartmann.[12] The former and his friend Dedikind,[13] the mathematician, have been my companions in the three excursions that the Society have made. Your absence was greatly regretted by all—Clausius Wartmann Studer,[14] Desor,[15] Martins[16] Escher von der Linth and all others to whom I was introduced. Your theory of the formation of Glaciers[17] was attacked, I hear, in the Geological Section; probably by Studer for from his conversation I gathered that he was not a convert of yours—I wish you had been here to defend yourself. I will give you further particulars when I return. I am just going to walk over the Albula pass to Chur[18] whence I shall proceed northwards through Frankfort & Marburg to Hamburg perhaps. I am at a loss what to do with myself however, and unless I come to a speedy decision I shall simply return home to scold you. I am well and gaining strength daily. I finished my dictionary[19] before I started, and please God when I return I will occupy myself more profitably.

Mrs Tunstall tells me you are looking well. I trust you may not have cause to regret your too hasty return and with love remain as ever Dear John

Yours affectionately | T A Hirst

[Envelope] Prof. Tyndall F.R.S. | Royal Institution | Albemarle Street | London

RI MS JT/1/H/249

1. *Ponte | Engädin*: Pontresina, a small town about 3 miles to the east of St. Moritz, in the Engadine, a long Alpine Valley in southeastern Switzerland.
2. *S[t]. Moritz*: an alpine resort town in southeastern Switzerland in the Engadine.
3. *Samaden*: a small town in the Maloja Region of southeastern Switzerland around 3 miles to the northeast of St. Moritz, also in the Engadine.
4. *Pontresina*: see n. 1.
5. *I received your letter*: letter 2025.
6. *Eggischorn*: see letter 1907, n. 6.
7. *Mrs Tunstall*: Tyndall and Hirst's housekeeper at 14 Waverley Place in London.
8. *our projected tour together was impossible this year*: in letter 2025, Tyndall wrote that the weather had been 'bad' and 'defeats all plans', and thus cancelled his plans to climb in the region.
9. *Engädin*: see n. 1.
10. *Naturforschende Gesellschaft*: see letter 1906, n. 6.
11. *Naturforscher*: naturalists (German).
12. *Wartmann*: Elie Wartmann (1817–86), a Swiss physicist and Professor of Natural Philosophy at the Academy of Geneva (*HLS*).

13. *Dedikind*: Julius Wilhelm Richard Dedekind (1831–1916), a German mathematician who made important contributions to abstract algebra (particularly ring theory), algebraic number theory, and the definition of the real numbers (*CDSB*).
14. *Studer*: Bernhard Studer (1794–1887), a Swiss geologist, mountaineer, and professor of geology at the University of Bern. He was elected foreign member of the Geological Society of London and the president of the Geological Survey of Switzerland (*HLS*).
15. *Desor*: Pierre Jean Édouard Desor (1811–82), a German-Swiss geologist and naturalist (*HLS*).
16. *Martins*: Charles Frédéric Martins (1806–89), a French physician, botanist, and geologist. He held several professorships at the Faculté de Médecine, Paris, the Sorbonne, and Montpellier (see 'Charles Frédéric Martins', *Annals of Botany*, 3 (1889–91), pp. 480–2).
17. *Your theory of the formation of Glaciers*: after Tyndall's first alpine explorations in the late 1850s, he proposed a theory of glacial formation counter to that proposed by James Forbes. Tyndall described the motion and structure as due to fracture and regelation of the ice, rather than viscosity, as the glacier also slid under gravity (see *Glaciers of the Alps*; and the introductions to *Tyndall Correspondence*, vol. 6 and vol. 7).
18. *Albula pass to Chur*: the Albula Pass is an alpine pass in eastern Switzerland in the Central Eastern Alps. Chur, one of the oldest towns in Switzerland, lies in the Grisonian Rhines Valley, where the Rhine River turns north.
19. *I finished my dictionary*: in his journal entry for 12 October 1862, Hirst recorded that he began working on articles for W. T. Brande and G. W. Cox (eds), *Dictionary of Arts, Literature and Science*, 3rd ed, 3 vols (London: Longman, Brown, Green, and Longmans, 1866) which he finished in 1863 (Brock and MacLeod, *Hirst Journals*, p. 1624).

From George Biddell Airy 27 August 1863 2029

Royal Observatory Greenwich | S E | 1863 August 27

My dear Sir,

I have to thank you for copies of several papers, relating to the question of Radiation through Vapour[1] and the attack in Good Words.[2]

The effect of aqueous vapour on radiation is, in my opinion, most important for meteorology. If in addition to this we could have information on the rate at which aqueous vapour disseminates itself through the air, I think that we should be in a way for beginning Meteorology—that collection of useless observations to which I cannot at present give the name of science—I was much interested in the account of your interview with Magnus.

In the matter of the Good Words animadversion you had good grounds for complaint.

I wish that I could acknowledge your kindness in a corresponding manner. But my occupations lead to little more in the first instance than very dull tabular results, which however I hope are of standard value in their kind. I have been longing for a Phil. Trans. paper[3] to send you, but hitherto have no copies, I must try to get some.

I am, my dear Sir, | Yours very truly | G B Airy
Professor Tyndall

RI MS JT/1/A/37
RI MS JT/1/TYP/1/27

---

1. *question of Radiation through Vapour*: see letter 2012, n. 1.
2. *the attack in Good Words*: Thomson and Tait critiqued Tyndall's claims about Mayer's contribution to energy physics in *Good Words* (see letter 1986, n. 5), and Tyndall's lengthy reply, published in the *Phil. Mag.*, is reproduced in this volume as letter 2002.
3. *a Phil. Trans. paper*: probably G. B. Airy, 'On the Diurnal Inequalities of Terrestrial Magnetism, as deduced from observations made at the Royal Observatory, Greenwich, from 1841 to 1857', *Phil. Trans.*, 153 (1863), pp. 309–16.

From Thomas E. Jenkins[1] 28 August 1863 2030

Louisville, Ky, Aug 28th 1863 | Prof Tyndall | Royal Institution,

Dear Sir,

I have been looking for a report of your last spring lectures on Sound;[2] but have been disappointed in finding one in either of this scientific periodicals which I receive, viz, the Philosophical Magazine and the Chemical News.[3]

As professor of Physics in the University of the Public Schools of this city I announced to my class and to a number of citizens who are interested in science that I would repeat your experiments illustrative of the subject of sound which you were at the time making for your auditory in London, provided I could get a report of them.

Last year I repeated your lecture on light as reported in the Chem. News[4] and I think that nothing of that kind had given to much satisfaction here, and we were anticipating an unusual scientific treat, in the event of obtaining your discourses on Acoustics.[5]

Not being able to find a report in any of the scientific periodicals to which I have access, I must beg you to excuse me for addressing you direct; but I am hoping that some other publication may contain it, and in that case I shall esteem it as an extraordinary favor if you will furnish me with one.

I would refer you to MS David Taylor & Sons,[6] Mark Lane, | Dr Francis,[7] Phil Mag, | Dr Dewitt,[8] Hertford St Mayfair, or Mr Crookes,[9] Ed Chem News,

I am yours most res[pect]f[ul]ly | Tho[ma]s. E. Jenkins

P.S. For the purposes of illustration we have all the important pieces of optical and acoustic apparatus as made by MM. Duboscq & Koenig[10] of Paris.

RI MS JT/1/J/26

1. *Thomas E. Jenkins*: Thomas E. Jenkins (dates unknown), professor of physics in the University of the Public Schools in Louisville, Kentucky, and possibly also a pharmacist and member of the American Pharmaceutical Association and Medico-Chirurgical Society.
2. *spring lectures on Sound*: see letter 2001, n. 8.
3. *Chemical News*: *The Chemical News and Journal of Physical Science*, a British periodical dedicated to topics in chemistry and physics, edited by William Crookes (see n. 9).
4. *as reported in the Chem. News*: J. Tyndall, 'Notes: A Course of Six Lectures on Light', *The Chemical News*, 7 (1862), pp. 20–6.
5. *your discourses on Acoustics*: see n. 2.
6. *David Taylor & Sons*: David Taylor established the firm in 1792 and it was managed by his sons until 1880. It was located first at 46 Cross Street until 1851, then 76 Mark Lane until 1865, and finally 61 Mark Lane until 1880 (H. Taylor (ed), *A sketch, left by David Taylor, of his own life between 1759 & 1792* (London, 1880), p. 15).
7. *Dr Francis*: William Francis.
8. *Dr Dewitt*: not identified.
9. *Crookes*: William Crookes (1832–1919), a British chemist, physicist, and editor who studied organic chemistry at the Royal College of Chemistry where he became assistant to August Wilhelm von Hoffman. It was through Hoffman that he met Faraday at the RI, who convinced him to change the focus of his work to optical physics. From 1855 he lectured in chemistry at Chester Diocesan Training College, but following a significant inheritance upon the death of his father, he set up his own physics laboratory. In 1859 he founded and edited *Chemical News*, and was an enthusiastic supporter of the methods of spectral analysis proposed by Bunsen and Kirchhoff. Indeed, it was through the application of their methods in his own work that he discovered, and named, thallium. He was elected FRS in 1863 (*ODNB*).
10. *MM. Duboscq & Koenig*: Louis Jules Duboscq (1817–86), a French instrument maker, and Rudolph Koenig (1832–1901), a German physicist, both operated instrument making firms in Paris, specializing in optical and acoustic equipment, respectively (P. Brenni, 'Nineteenth-century French scientific instrument makers, XIII: Soleil, Duboscq, and their successors', *Bulletin of the Scientific Instrument Society*, 51 (1996), pp. 7–16; P. Brenni, 'Nineteenth-century French scientific instrument makers, VI. The triumph of experimental acoustics: Albert Marloye (1795–1874) and Rudolph Koenig (1832–1901)', *Bulletin of the Scientific Instrument Society*, 44 (1995), pp. 13–7).

To Juliet Pollock [3 September 1863][1] 2031

Tuesday morning

My dear Mrs Pollock,

I sent you a few lines from Keswick[2] just before starting with Mr Huxley over Derwentwater[3] & through Borrodale.[4] While passing I thought it just possible that I might see Mrs Spedding.[5] I went to Greta Bank[6] but the hour was too early. We had a most pleasant day in Borrodale—the colouring is so tender & beautiful and the former so fine. Nor had I until this visit any adequate idea of the magnitude of the ancient glacier action of the region. We crossed Borrodale to Buttermere[7] by Honister Crag[8] and next day went to Scale Force.[9] Instead of returning to Buttermere and crossing Scarf Gap[10] we went up the mountain and worked round Red Pike,[11] mostly in the midst of driving fog, which to me is detestable. I have no bump of locality and am therefore bewildered by a fog. Nevertheless we crossed the mountain and dropped down its craggy sides into Ennerdale.[12] Thence over Black Sail to Wastwater.[13] Here we stayed two days—for the most part wet ones—I climbed the pillar mountains and endeavoured to climb the pillar rock,[14] but the driving rain & cloud & the wet slippy crags were too much for me. You might break your neck very respectably on the Pillar Rock. I hardly ever was amid such a commotion of the elements, such gloom and such wild pelting rain as on the summit of this mountain. Had I not luckily borrowed a compass I do not know how I should have escaped from it. I should certainly have found myself in the wrong valley after my descent. We found a party of gallant young Cambridge men at Wastdale head[15] & amid drenching rain and intermittent gleams of sun crossed by Borofell to Langdale.[16] There are beautiful snatches of scenery on the way. Huxley quitted me at Windermere[17] and came to London greatly improved in health, while I went on to Newcastle[18] purposing to spend two days there—But the entertainment at Armstrong's[19] was so regal and the company so excellent that I was induced to remain the whole week. Thence Dr. Bence Jones and myself went on to Alnwick Castle[20] and remained three days there. It is a magnificent place, but for a bird of the mountain like myself a less stately lodging would be more suitable. The Duke and Duchess[21] however were exceedingly kind—the Duchess especially so—She has always been genial towards me. From Alnwick we went to Chillingham to Lord Tankerville's[22]—Sir Wm. Armstrong,[23] Professor Owen & Mr Gould[24] were of our party. I stalked the wild castle with Owen & had a capital view of the herd. They smelt us and became uneasy, and finally set off at full gallop like buffaloes over a prairie. Zigzagging up a hill so as to give us a famous view of their motions. The same evening I tried deerstalking, but though I might have shot a couple of does the fine antlered bucks kept away from me. We spent the best of two very joyous days at Chillingham—They are wonderfully frank and pleasant people. Bence Jones and myself returned to York, and there parted company, he going to London and I intending to go to Ireland.[25] It happened to be on the day of that wild gale, so when I reached Stockport[26] the thought of the storm so sickened me that I swerved from my route to Chester[27] & came on to London.

I am now hard at work—horribly entangled in a magnetic question through which I do not yet see my way. I was yesterday informed that you were at Freshwater[28] and thither I send this. Give my love to your husband[29] and the boys[30] & believe me

Yours as ever | John Tyndall

I intend to rush down to Abury[31] next Sunday to see the Druidical remains. I should very much like to take a rush down to Plumbleys[32] while you are in the neighbourhood. A walk over that "noble down" would be refreshing—

RI MS JT/1/T/1183
RI MS JT/1/TYP/6/2126–7

---

1. *[3 September 1863]*: the date is given by the reference to Newcastle as Tyndall attended the BAAS meeting at Newcastle Upon Tyne, which began on 26 August 1863. The Tuesday following the meeting was 3 September 1863. Tyndall had already written to Juliet Pollock before the meeting about where he had been exploring (see letter 2027).
2. *a few lines from Keswick*: letter missing. Keswick is a market town and civil parish in the Lake District.
3. *Derwentwater*: one of the largest lakes in the Lake District just south of Keswick.
4. *Borrodale*: or Borrowdale, a valley and civil parish in the Lake District.
5. *Mrs Spedding*: probably Sarah Spedding (see letter 1936, n. 16).
6. *Greta Bank*: see letter 1936, n. 4.
7. *Buttermere*: a lake and village in the Lake District.
8. *[Honister] Crag*: a high crag in the Lake District, above Honister Pass.
9. *Scale Force*: near Buttermere, the highest waterfall in the Lake District.
10. *Scarf Gap*: Scarth Gap, a pass between Buttermere and Ennerdale.
11. *Red Pike*: a 2,476 ft mountain in the High Stile range in the western Lake District, which separates Ennerdale from the valley of Buttermere and Crummock Water.
12. *Ennerdale*: a scenic and somewhat remote valley in the Lake District.
13. *over Black Sail to Wastwater*: Black Sail is a mountain pass separating Wasdale from Buttermere; Wastwater is a ribbon lake formed by glacial erosion and is the deepest of the lakes in the region.
14. *pillar mountains and endeavoured to climb the pillar rock*: Pillar is a mountain in the Lake District, above Ennerdale, and Pillar Rock is a large rocky outcrop on its northern side. By this date, it had not been climbed very often.

15. *Wastdale head*: Wasdale Head, the area in upper Wasdale including a small scattered hamlet in the Lake District, surrounded by some of the country's highest mountains.
16. *Borofell to Langdale*: probably Bowfell, one of the mountains above the Langdale valley.
17. *Windermere*: see letter 2027, n. 4.
18. *I went on to Newcastle*: see n. 1.
19. *Armstrong's*: William Armstrong.
20. *Alnwick Castle*: a castle and stately home in Alnwick in the English county of Northumberland. It is the seat of the Duke of Northumberland.
21. *Duke and Duchess*: Algernon Percy (1792–1865), 4th Duke of Northumberland and Eleanor Percy (1820–1911), Duchess of Northumberland.
22. *to Chillingham to Lord Tankerville's*: Charles Bennet (1810–99), 6th Earl of Tankerville, a Conservative politician and peer. He and Tyndall were good friends. Chillingham Castle is a medieval castle in the village of Chillingham in the northern part of Northumberland, England.
23. *Sir W^m^. Armstrong*: William Armstrong.
24. *Mr Gould*: John Gould (1804–81), a British ornithologist and publisher (*ODNB*).
25. *intending to go to Ireland*: see letter 2025, n. 8.
26. *Stockport*: a town near Manchester, *en route* from where he was to the west coast of England where he had planned to travel over to Ireland, probably from Holyhead.
27. *Chester*: a town southwest of Stockport.
28. *Freshwater*: a large village on the western coast of the Isle of Wight.
29. *your husband*: William Frederick Pollock.
30. *the boys*: Frederick, Walter, and Maurice Pollock.
31. *Abury*: or Avebury, a Neolithic henge monument in Wiltshire in southwest England.
32. *Plumbleys*: either a reference to the Plumley family who lived in Newport on the Isle of Wight, or Plumbly's Freshwater Bay Hotel in the village of Freshwater Bay, Isle of Wight (near the town of Freshwater where Pollock was staying).

To Elizabeth Dawson Steuart 8 September 1863 2032

8th. Sep. 1863.

My dear Friend,

On my return from the North of England last night[1] I found your friendly letter.[2] It would give me great delight to pay you a visit and I hope to do so some day, but I must I regret to say postpone it for the present. I have been wandering about so long that the calls upon me in London have at length become peremptory.

I have been spending a most pleasant week with my friend Sir William Armstrong at Newcastle, afterwards I spent two or three days at Alnwick Castle,[3] the residence of the Duke of Northumberland.[4] Thence I went to

Chillingham Castle[5] the seat of the Earl of Tankerville[6] and spent a short but joyous time there—yesterday morning I started [south] and came last night to London.

I have been in the Country of the Tyndall's—but they were scattered abroad from Tyndale during the wars of the Roses. The Duke once said to me that I ought to belong to Northumberland as my name indicated relationship to the valley of the Tyne.[7] The only trace of the name that I alighted on was on a sign over a grocer's shop in a little country village.

I have been lately in correspondence with Mr Robert Tyndale[8] of Oaklands near New Ross.[9] Some day I hope to pay you all a visit.

I shall keep my eyes open regarding George and John[10]

Yours ever | John Tyndall.

RI MS JT/1/TYP/10/334
LT Typescript Only

1. *the North of England last night*: Tyndall had attended the annual meeting of the BAAS, which was hosted in Newcastle Upon Tyne (see letter 2024, n. 3).
2. *your friendly letter*: letter missing.
3. *Alnwick Castle*: see letter 2031, n. 20.
4. *the Duke of Northumberland*: see letter 2031, n. 21.
5. *Chillingham Castle*: see letter 2031, n. 22.
6. *the Earl of Tankerville*: see letter 2031, n. 22.
7. *valley of the Tyne*: the valley through which the river Tyne flows.
8. *Mr Robert Tyndale*: Robert Tyndale.
9. *Oaklands near New Ross*: Oaklands is a small village just outside the town of New Ross in Country Wexford, Ireland.
10. *George and John*: possibly two of the three sons of John Tyndall (of Gorey), who had stayed with Tyndall from late June into July 1863 (see letter 2019, n. 2). Tyndall was trying to secure positions for them, although he was not optimistic about being successful in this endeavour.

From Robert Bunsen — 9 September 1863 — 2033

Heidelberg

Mein theuerster Freund.

Erst jetzt, wo ich aus der Schweitz zurückgekehrt bin, ist Ihr schönes Werk in meine Hände gelangt für deßen freundliches [Senden] ich Ihnen meinen herzlichsten Dank sage. Ich lese es so eben und schöpfe mit jeder neuen Zeile neues Vergnügen und neue Belehrung daraus. Ihre Erwiederung auf Thomsons Angriffe ist wahrhaft vernichtend für ihn. Möge er sich eine Lehre daraus nehmen und möge er aufhören, ganz unberechtigter Weise in versteckten Seitenhieben die Arbeiten Anderer zu verdächtigen, wie er es in der letzten Zeit so häufig gethan hat!

Ich war in der Schweitz und hoffte Ihnen dort zu begegnen, bin aber leider gleich im Anfang der Reise auf einer höchst unerfreulichen Weise hierher zurückgetrieben indem ich unter dem Kamm des Surnenpaßes mir das Fußgelenk derart verstauchte, daß ich nur mit äußerster Anstrengung noch das Reußthal erreichen konnte. Jetzt bin ich schon so weit wieder beßer, daß ich mit Vorsicht wieder ausgehen kann.

Kirchhoff ist schon seit 4 Wochen in Königsberg bei seinen Schwiegereltern und wird wohl erst in 14 Tagen mit seiner Familie zurückkommen; sonst würde ich Ihnen Grüße von ihm zu bestellen haben.

In herzlicher aufrichtigster Freundschaft | Ihr | RW Bunsen

Heidelberg den 9[n] Sept. 1863.

My dearest friend.

Only now that I have returned from Switzerland has your splendid work[1] come into my hands, and I thank you from the heart for kindly sending it. I am reading it now and draw fresh pleasure and new instruction from every new line. Your response to Thomson's attacks[2] is truly devastating for him. May he take a lesson from it and stop throwing suspicion at the works of others by making completely unfounded shrouded gibes, as he has so often done of late!

I was in Switzerland and hoped to meet you there, but unfortunately was forced back in an extremely unpleasant way at the very the beginning of the journey. Below the ridge of the Surenen Pass[3] I sprained my ankle so severely that only with the utmost effort did I manage to reach the Reuss Valley.[4] I am doing better now so that I can go out again with care.

Kirchhoff has been at his in-laws' in Königsberg[5] for 4 weeks now and will probably only return with his family in 14 days; otherwise, I would have greetings from him to pass on to you.

With warm and most sincere friendship | Your | R W Bunsen

Heidelberg, 9 Sept. 1863.

RI MS JT/1/B/149

1. *your splendid work*: *Heat as a Mode of Motion.*
2. *Your response to Thomson's attacks*: Thomson and Tait critiqued Tyndall's claims about Mayer's contribution to energy physics in *Good Words* (see letter 1986, n. 5), and Tyndall's lengthy reply, published in the *Phil. Mag.*, is reproduced in this volume as letter 2002.

3. *Surenen Pass*: a high mountain pass across the Urner Alps in the Swiss canton of Uri.
4. *Reuss Valley*: the valley of the Reuss river in central Switzerland.
5. *Königsberg*: a former German city that is now Kaliningrad, Russia.

From Charles P. Hooker[1] 12 September 1863 2034

Leigh House | September 12th. 1863.

My dear Dr. Tyndall,

I had a letter from Papa[2] the other day and he told me your direction. I like school very much indeed. How do you get on with your lectures? We have got a field which we often go to play in when it is fine. There are a great many caterpillars. I have begun to learn Latin. When are you going to Switzerland?

I send my love to you | From your affectionate little friend | Charles P. Hooker

RI MS JT/1/TYP/8/2562
LT Typescript Only

1. *Charles P. Hooker*: Charles Paget Hooker (1855–1933), second son of Joseph Dalton and Frances Hooker. He went into medical practice and later married Sophie Hayes Evans-Lombe (1857–1931).
2. *Papa*: Joseph Dalton Hooker.

From Rudolf Clausius [19th September 1863][1] 2035

Zürich, 19. Sept. 63

Lieber Tyndall!

Ich muss mich grosser Nachlässigkeit anklagen, dass ich Ihren letzten freundlichen Brief so lange unbeantwortet gelassen habe, besonders wegen der darin enthaltenen Aufforderung, Seguin's „Sur l'Influence des chemins de fer" anzusehen. Der Grund weshalb ich Ihnen darüber nicht geschrieben habe, lag nur darin, dass ich mir das Werk hier nicht beschaffen konnte. Ich glaube, Sie haben sehr gut daran gethan, die betreffende Stelle von Verdet im Phil. Mag. abdrucken zu lassen, denn Verdet ist der Mann, dem man die beste Kenntniss der französischen Autoren und ausserdem das grosste Wohlwollen für dieselben zutrauen kann. Was Ihren Streit mit Thomson und Tait anbetrifft, so habe ich freilich, wie ich offen gestehen muss, Ihren Artikel im Phil. Mag. etwas scharf gefunden, aber die Art, wie die beiden Herrn darauf geantwortet haben, hat mir durchaus missfallen, besonders das Benehmen von Thomson, der Sie erst angegriffen hat, und dann, da Sie sich gegen den Angriff vertheidigen, sich vornehm von dem Streite fern zu halten sucht. Ein solches Benehmen finde auch ich unmännlich, und es wird gewiss auch von Anderen nicht gebilligt. Ich habe mit Interesse gehört dass Sie in diesem Herbste wieder der Session der Brit. Assoc. beigewohnt haben, und dass auch Mayer dort gewesen ist, und ich bin gespannt darauf, zu erfahren ob dort ein Zusammentreffen mit Thomson, Joule u. anderen bei dieser Sache interessirten Personen stattgefunden hat.

Mit Erstaunen hörte ich in diesem Sommer von Hofmann dass Sie in der Schweiz sein, wovon ich gar nichts gewusst hatte, und noch mehr war ich überrascht, in Samaden Hirst zu treffen, und zu erfahren, dass Sie beabsichtigt hatten, auch dorthin zu kommen. Sie glauben nicht, wie sehr ich es bedauert haben, dass Sie diesen Plan nicht ausgeführt haben, und wie sehr ich mich gefreut haben würde, dort mit Ihnen zusammenzutreffen. Wir haben dort einige sehr angenehme Tage verlebt, wobei ich besonders mit Hirst und einen anderen Freunde, einem Mathematiker Dedekind zusammen war. Im nächsten Jahre wird die schweizerische naturforschende Gesellschaft ihre Sitzung in Zürich halten, und ich hoffe dass Sie und Hirst ihr dann beiwohnen werden, und in dieser Hoffnung möchte ich Sie schon jetzt einladen, dann mit Hirst bei uns zu logieren, womit Sie meiner Frau und mir einen grosse Freude machen werden.

Vor einigen Tagen habe ich auch dass Vergnügen gehabt Ihren Collegen Frankland kennen zu lernen, der mir sehr gefallen hat.

Mit herzlichen Grüssen an Sie und Hirst,
Ihr
Clausius

Zurich, 19. Sept. 63

Dear Tyndall,

I must charge myself with great negligence that I have left your last friendly letter[2] unanswered for so long, most especially because of the request it contained to have a look at Seguin's "Sur l'influence des chemins de fer".[3] The reason I did not write to you was simply because I could not get hold of that work here. I think you have done very well in having the relevant passage from Verdet published in the Phil. Mag.,[4] since Verdet is the man who can be trusted to have the best knowledge of French authors and in addition the greatest goodwill for them. In regard to your dispute with Thomson and Tait, to be free, I must admit that I found your essay in the Phil. Mag. somewhat harsh, but the manner in which both gentlemen answered it[5] completely displeased me, especially the way Thomson behaved, first attacking you and then, since you are defending yourself against the attack, trying to aristocratically distance himself from the dispute. I find such a behavior unmanly and it will certainly not be condoned by others. I was interested to hear that you attended the meeting of the Brit[ish] Assoc[iation][6] again this year and that Mayer was also there, and I am curious to hear whether a meeting took place with Thomson, Joule and other persons interested in the matter.

I was surprised to hear from Hofmann[7] this summer that you were in Switzerland, which I had not known at all, and I was even more surprised to meet Hirst in Samaden[8] and to hear that you intended to come there as well. You will not believe how much I regretted that you did not carry out these plans and how much I would have enjoyed meeting you there. We spent some very pleasant days there, whereby I spent time particularly with Hirst and another friend, a mathematician Dedekind.[9] Next year the Society of Natural Science[10] will hold its meeting in Zurich, and I hope that you and Hirst will attend, and it is with this hope that I would like to invite you and Hirst now to lodge with us, which would make my wife[11] and myself very happy.

A few days ago I had the pleasure of meeting your colleague Frankland, who I liked very much.

With best regards to you and Hirst,
Your Clausius

RI MS JT/1/TYP/7/2259–60
LT Typescript Only

1. *[19th September 1863]*: date suggested by LT.
2. *your last friendly letter*: letter 2006.
3. *Seguin's "Sur l'influence des chemins de fer"*: see letter 1990, n. 2.
4. *published in the Phil Mag*: see letter 2002.
5. *answered it*: P. G. Tait, 'On the Conservation of Energy' (letter to David Brewster), *Phil. Mag.*, 25 (1863), pp. 429–31; and W, Thomson, 'Note on Professor Tyndall's "Remarks on the Dynamical Theory of Heat"', *Phil. Mag.*, 25 (1863), p. 429.
6. *meeting of the Brit[ish] Assoc[iation]*: see letter 2024, n. 3.
7. *Hofmann*: August Wilhelm von Hofmann (1818–92), a German chemist and co-founder of the German Chemical Society (*ADB*).
8. *Samaden*: see letter 2028, n. 3.
9. *Dedekind*: see letter 2028, n. 13.
10. *Society of Natural Science*: see letter 1906, n. 6.
11. *my wife*: Adelheid Clausius.

From Michael Faraday 19 September 1863 2036

Sat. 19th. Sept. 1863.

My Dear Tyndall

I was to have asked you from my wife[1] whether next Thursday we shall see you at 3 o'clock (dinner) or not until 6 o'clock (tea &c). My only way of satisfying her will be to ask you to send her a single line saying which.

Ever yours | M. Faraday

Faraday Correspondence, 6:4362

1. *my wife*: Sarah Faraday.

From Joseph Plateau[1] 21 September 1863 2037

Gand, 21 7bre 1863

Mon Cher Monsieur Tyndall

J'ai l'honneur de vous adresser trois exemplaires de mon dernier mémoire; j'espère que vous voudrez bien accepter l'un d'eux, et offrir le second, de ma part, à la Société Royale, et le troisième à l'Institution Royale.

Dans ce travail, j'ai résolu une question qui a dû se présenter déjà, je pense, à l'esprit des physiciens, mais qui n'avait jamais été traitée, à ma connaissance du moins. Malheureusement j'arrive à une solution absolument négative; Toute fois cette solution me paraît assez *[curieuse]* en elle-même pour exciter l'attention des savants.

Agréez, Mon Cher M. Tyndall, l'assurance de mes sentiments d'affection. | J. Plateau.

Ghent, 21 September 1863

My dear Mr Tyndall,

I have the honour of sending you three copies of my latest article;[2] I hope that you would kindly accept one of them, offer the second, on my behalf, to the Royal Society, and the third to the Royal Institution.

In this work I resolved an issue that physicists, I believe, must have already been aware of, but which has never been dealt with, at least to my knowledge. Unfortunately, I arrived at an utterly negative solution; nevertheless that solution seemed to me *[curious]* enough in itself to merit the attention of scientists.

Please accept, my dear Mr. Tyndall, the assurance of my affectionate regards. | J. Plateau.

RI MS JT/1/P/96

1. *Joseph Plateau*: Joseph Plateau (1801–83), a Belgian physicist. Born in Brussels, Plateau studied at the University of Liège, where he graduated in 1829 with a doctorate degree in mathematics and physics. In 1832, Plateau invented a device which could give the illusion of a moving image, which he called the phenakistoscope. In 1835, he was appointed professor of experimental physics at the University of Ghent. Plateau is best known for his experimental work in which he investigated what is now known as the 'Plateau Problem'; from 1840, he conducted experiments to demonstrate that a minimal surface could be formed from a soap film, made of a solution of soapy water and glycerine, which was stretched on a wire frame (*CDSB*).
2. *my latest article*: J. Plateau, 'Réponse aux observations présentées par M. Chevreul', *Comptes Rendus*, 57 (1863): p. 1029–32.

From Joseph Plateau[1] 30 September 1863 2038

Gand, 30 Septembre 1863

Mon cher Monsieur Tyndall,

J'ai l'honneur de vous adresser des exemplaires de trois notes dont deux publiées par moi et une par mon fils; ces exemplaires sont en triple. Veuillez, pour chaque note, accepter le premier, et offrir le second à la Société Royale et le troisième à l'Institution Royale.

Depuis nos anciennes relations, j'ai continué l'étude du liquide glycérique et c'est seulement à présent que je suis sur le point de l'avoir terminée; la chose n'était pas facile, je vous assure.

Agréez, mon cher M. Tyndall, l'assurance de tous mes sentiments d'affection | J. Plateau

Ghent, 30 September 1863

My dear Mr Tyndall,

I have the honour of enclosing copies of three notes,[2] two of which were published by me and one by my son.[3] Those copies are in triplicate. For each note, please accept the first, offer the second to the Royal Society, and the third to the Royal Institution.

Since we last communicated, I have continued the study of the glycerol liquid and it is only now that I am at the point of completion. The thing was not *[easy]*, I assure you.

Please accept, my dear Mr. Tyndall, the assurance of my warmest regards. | J. Plateau

RI MS JT/1/P/95

1. *Joseph Plateau*: see letter 2037, n. 1.
2. *three notes*: not identified. Both Plateau and his son published numerous papers between 1842 and 1868 that he assembled in 1873 in his massive work, J. Plateau, *Statique experimentale et theorique des liquides soumis aux seules forces moleculaires*, 2 vols (Gand et Leipzig: F. Clemm, 1873). The one by his son is possibly: F. Plateau, 'Sur un mode particulier de production de bulles de savon', *Bulletins de l'Académie royale des sciences et belles-lettres de Bruxelles*, 13 (1862), pp. 286–8.
3. *by my son*: Félix Auguste Joseph Plateau (1841–1911), who after graduating from the University of Ghent became a notable physician. He is most well-known for his researches in anatomy and zoology (*CDSB*).

From Alexandre-Eugène Béguyer de Chancourtois[1] 14 October 1863 2039

Le Havre | 14 octobre 63

Cher Monsieur

En revenant à Paris l'année passée, après vous avoir vu à Londres, je m'étais mis en devoir de vous envoyer les quelques détails concernant les glaces du Groenland sur lesquels je désirais appeler votre attention. Mais ma lettre a été interrompue par l'obligation de me rendre auprès d'une personne dont la santé me donnait de vives inquiétudes et depuis, entrainé par le courant des labeurs quotidiens je n'ai su trouver une bonne occasion de la reprendre. Dans ces derniers temps j'espérais que quelque <*1–2 words missing*> d'été vous amènerait à Paris <*2–3 words missing*><. Puisque> je ne peux plus guère compter sur cette bonne chance, que j'ai peut-être d'ailleurs manquée depuis que je suis en vacance ici, je veux au moins, avant la rentrée, régler mon arriéré envers vous. C'est pourquoi je vous adresse mes croquis avec le commencement de mon ancienne lettre à titre de légende, vous priant bien de ne pas même accuser réception de cet envoi dont le principal but aujourd'hui ne peut être que de me rappeler à votre souvenir et de vous demander de ne pas oublier, lorsque vous passez par Paris, le visiteur qui a conservé, lui, une si bonne impression de votre excellent accueil et qui se dit toujours votre tout dévoué serviteur.

EB de Chancourtois

Mes com[pliments pour] votre dernier livre que je me suis procuré avec empressement seraient maintenant bien tardifs. Il m'a vivement interessé et j'en attends la traduction pour le lire à fond comme il le mérite. Vous êtes bien heureux, Monsieur, de mener de front l'experience et la théorie.

Veuillez me rappeler au bon souvenir de monsieur Hirst.

Le Havre | 14 October 63

Dear Sir,

Following my return to Paris last year, after meeting you in London, I was determined to send you a few details about the ice floes of Greenland, to which I wanted to draw your attention. But I was forced to interrupt my letter when I had to visit a person whose health gave me serious concern, and since then, caught up by daily chores, I could not find a good opportunity to get back to it. As of late, I was hoping that some summer <*1–2 words missing*> would bring you to Paris <*2–3 words missing*><. Since> I can hardly count on this, and may have even missed it since I have been on vacation here, I want at least to settle my debt to you before my return. This is why I am sending you my sketches[2] and the beginning of my old letter[3] for caption purposes, asking you not to even acknowledge receipt of this consignment, the main goal of which today is simply to send you my regards, and ask that you don't forget, when you travel through Paris, the visitor who was left with such a good impression of your excellent hospitality, and who remains your devoted servant.

EB de Chancourtois

I apologize for the delay in sending you my congratulations for your last book,[4] which I hastened to obtain. I was greatly interested in it and am eagerly waiting for the translation to read it as thoroughly as it deserves. You are very fortunate, Sir, to be able to do justice to both the experimentation and the theory.

Please send my regards to Mr. Hirst.

RI MS JT/1/C/19

---

1. *Alexandre-Eugène Béguyer de Chancourtois*: Alexandre-Eugène Béguyer de Chancourtois (1820–86), a French mineralogist and geologist, who is best known for his arrangement of the chemical elements by atomic weight in 1862. Despite publishing prior to Dmitri Mendeleev, his work was largely ignored by chemists because it was written for geologists. A celebrated member of the École Polytechnique, he is also noted for his work on mining inspection and safety (*CDSB*).
2. *my sketches*: sketches missing.
3. *my old letter*: letter missing.
4. *your last book*: *Heat as a Mode of Motion.*

From Louis Soret 26 October 1863 2040

Heidelberg 26 Octobre 1863

Cher Monsieur

Depuis que j'ai eu le dit plaisir de vous rencontrer à Bâle, j'ai toujours voulu vous envoyer quelques notes pour le Philosophical Magazine sur mes dernières expériences sur l'ozone. Il serait trop long de vous raconter toutes les occupations qui m'en ont empêché et les motifs qui ont retardé ma publication dans les Archives. J'aurais voulu pouvoir vous envoyer un extrait en anglais, j'ai essayé de le faire mais je suis si bien habitué a écrire dans votre langue et l'étude de l'allemand m'a si bien fait oublier le peu d'anglais que je savais qu'il me faudrait un temps considérable pour en venir à bout. Si vous croyez que mes expériences en soient dignes je vous serais bien reconnaissant de faire faire la traduction à Londres. Si vous voulez un extrait très court vous pouvez faire traduire ce que les Comptes rendus ont publié (5 octobre); si vous voulez la mémoire in extenso, il se trouve dans les Archives de Septembre. Je vous en envoie un exemplaire sous bandes ainsi qu'un exemplaire de ma précédente note que vous avez déja eu l'obligeance de faire connaitre aux lecteurs de Phil. Mag.) Enfin, si cela vous convenait mieux on pourrait prendre un intermédiaire, et traduire l'exemplaire que je vous envoie sous ce pli, et dans lequel j'ai rayé de nombreux passages. L'extrait reste suffisamment complet quoique fort abrégé.

Dans le cas où vous n'avez pas d'autre objection à l'insertion d'un de ces trois articles, que la difficulté de trouver un traducteur, faites le moi savoir en m'indiquant celui des trois que vous préférerez. Je tacherai de le traduire ou de le faire traduire.

Dans le cas où vous feriez faire la traduction du mémoire complet, je vous prierai de faire corriger une erreur; à la page 14 on a mis une note, commençant par ces mots: « On sait d'après les belles expériences de MM. H. St Claire Deville etc » cette note se rapporte à la page 17, 9eme alinea. La note qui devait être mise à la page 14 est destinée à indiquer qu'on n'a pu faire une analyse correspondant à une expérience sur l'action de la potasse. Les deux notes sont relatives telles qu'elles doivent l'être, dans l'exemplaire abrégé que je vous envoie ci-joint.

Vous avez vu que le N° de Sept. des Archives a donné une traduction des mémoires de Magnus et un extrait de l'un de vos mémoires sur le diathermanie.

J'ai regretté que l'on n'y ait pas joint un extrait de votre réponse à Mr Magnus (inséré The Philos. Mag. July 1863). J'avais indiqué à mes collaborateurs de Genève l'intérêt qu'il y aurait à tenir nos lecteurs au courant de cette discussion; il paraît que je n'ai pas été très bien compris ou que le Philosophical Mag. n'était pas arrivé à Genève quand l'article a été rédigé. J'en écrirai un mot à Mr Prevost pour voir comment on pourrait arranger la chose. Je pense que votre nouveau mémoire annoncé, sur l'identité d'action d'une couche liquide et de la vapeur des mêmes liquides en fournira l'occasion. Je n'ai pas encore vu le mémoire, j'ignore même s'il a paru.

Je resterai probablement encore 15 jours ici, puis j'irais à Paris pour quelques temps (mon adresse ici est toujours la même, 22 Neue Anlage, et à Paris ce sera: chez MM. Hentsch Lütscher et Cie).

MM. Bunsen, Kirchhoff, Helmholtz sont bien, je regrette bien de devoir les quitter bientôt.

Recevez, cher Monsieur, l'assurance de mes sentiments les plus affectueux. | L. Soret

Heidelberg 26 October 1863

Dear Sir

Since I have had the pleasure of meeting you in Basel,[1] I have wanted to send you some notes for the Philosophical Magazine about my last experiments on ozone. It would take too long to list the tasks that prevented me from doing so and the reasons why my publication in the Archives[2] was delayed. I wish I could have sent you an extract in English, I tried to do so but I am so used to writing in your language and the study of German made me forget the little bit of English I knew, to the point that I would need a considerable amount of time to achieve it. If you think my experiments are worthy, I would be grateful if you would have them translated in London. If you want a very short excerpt, you could have translated what the Comptes rendus published (5 October);[3] if you want the article in extenso,[4] it can be found in September's Archives.[5] I am sending you a copy of it as well as a copy of my previous note kind enough to share with the readers of the Phil. Mag.[6] Finally, if it is more convenient for you, we could use an intermediary to translate the copy that I send to you here enclosed,[7] and in which I have crossed out many passages. The excerpt remains complete enough, albeit significantly shortened.

If you have no further objections to the inclusion of one of these three articles, except for the difficulty of finding a translator, please let me know by telling me which of the three you prefer, and I will try to translate it or to get it translated.

If you would have the entire article translated, I would ask you to correct a mistake; on page 14[8] there was a note that started with these words: "We know, according to the excellent experiments of Messrs H. St Claire Deville[9] and". This note relates to page 17, 9th paragraph. The note that should have been on page

14 should have indicated that we could not do an analysis corresponding to an experiment on the action of lye. These two notes are laid out as they should be in the abridged copy that I have attached to this letter.

You saw that the September issue of the Archives had a translation of Magnus' memoirs[10] and an excerpt of your memoirs on diathermancy.[11]

I regret that they did not attach an excerpt of your reply to Mr Magnus[12] published in the Philos[ophical]. Mag[azine]. of July 1863. I told my colleagues in Geneva that our readers would find it interesting to keep abreast of this discussion. It seems that I was not well understood, or that the Philosophical Mag[azine]. had not arrived in Geneva when the article was written. I wrote a word to Mr Prevost[13] to see how we can fix this. I think we will have the opportunity with your newly announced article on the action of a layer of liquid and of the vapour of the same liquids.[14] I have not yet seen the article; I am not even aware if it has been published.

I will most likely stay here for another 15 days, and then I will go to Paris for some time (my address here is still the same, 22 Neue Anlage, and in Paris it will be: at Messrs. Hentsch Lütscher and Co.)[15].

Messrs. Bunsen, Kirchhoff, and Helmholtz are well; I regret that I have to leave them soon.

Please accept, dear Sir, my warmest regards. | L. Soret

RI MS JT/1/S/87

1. *Basel*: see letter 1909, n. 4.
2. *Archives*: J. L. Soret, 'Sur les Relations Volumétriques de l'Ozone', *Archives des Sciences Physiques et Naturelles*, 18 (1863): pp. 65–82.
3. *what the Comptes rendus published (5 October)*: M. J.-L. Soret, 'Sur la production de l'ozone par l'électrolyse et sur la nature de ce corps', *Comptes Rendus*, 56 (1863), pp. 390–3.
4. *in extenso*: in full (Latin).
5. *found in September's Archives*: see n. 2.
6. *my previous note that you were kind enough to share with the readers of the Phil. Mag.*: M. Soret, 'Experiments on Ozone. In a letter to Professor Tyndall', *Phil. Mag.*, 25 (1863), pp. 208–10.
7. *here enclosed* : enclosure missing.
8. *on page 14*: here Soret was referring to the fourteenth page of the article cited in n. 2, which was p. 78. He perhaps was consulting an off print of the article without the original pagination.
9. *H. St Claire Deville*: Henri Étienne Sainte-Claire Deville (1818–81), a French chemist who worked at the Chemistry Laboratory, Ecole Normale Supérior, Paris. He developed a process for producing aluminium (*CDSB*). In the letter, Soret's quoted footnote implied other persons besides Sainte-Claire Deville, and the article shows these to be 'Troost et de M. Bineau' (p. 78).

10. *September issue of the Archives had a translation of Magnus' memoirs*: G. Magnus, 'Sur le passage de la chaleur rayonnante à travers l'air humide et sur les propriétés hygroscopiques du sel gemme', *Archives des Sciences Physiques et Naturelles*, 15 (1862), pp. 21–5.
11. *excerpt of one of your memoirs on diathermancy*: J. Tyndall, 'Remarques sur des recherches récentes relatives à la chaleur rayonnante', *Archives des Sciences Physiques et Naturelles*, 15 (1862), pp. 26–40.
12. *your reply to M^r. Magnus*: probably J. Tyndall, 'On the Passage of Radiant Heat through Dry and Humid Air', *Phil. Mag.*, 26 (1863), p. 44–54. Tyndall and Magnus were in a friendly dispute over the absorption of radiant heat; indeed, the two met at the RI when Magnus was in London visiting the International Exhibition, and Tyndall demonstrated his findings in Magnus's presence. See also H. G. Magnus, 'On the Diathermancy of Dry and Moist Air', *Phil. Mag.*, 26 (1863), p. 21–30; and J. Tyndall, 'On the Relation of Radiant Heat to Aqueous Vapour', *Phil. Mag.*, 26 (1863), p. 30–44.
13. *M^r. Prevost*: not identified.
14. *your newly announced article on the action of a layer of liquid and of the vapour of the same liquids*: see letter 2014, n. 2.
15. *Messrs. Hentsch Lütscher and Co.*: Hentsch, Lütscher & Co. was a banking firm in Paris.

From Julius Robert Mayer 31 October 1863 2041

Heilbronn

Hochverehrtester Herr Professor!

Sehr gerne hätte ich Ihrer ehrenden Einladung die Versammlung in New-Castle zu besuchen entsprochen, wenn nicht meine zu geringe Kenntniss der englischen Sprache, der ich übrigens neuerdings nach Kräften abzuhelfen suche, mich davon abgehalten hätte; ich zog es deshalb vor, der beinahe gleichzeitig stattfindenden Naturforscher Versammlung in Stettin anzuwohnen. Dort traf ich unter anderen unsern gemeinschaftlichen Freund Knoblauch, der mir vieles von Ihnen erzählte. Auch Dove war anwesend. Auf der Rückreise wurde ich in Berlin bei Poggendorff durch den Anblick Ihrer Photographie erfreut, welche sich neben der meines Freundes Schönbein, welcher aber leider in Stettin fehlte, befindet. Poggendorff hat mir nebst Knoblauch die besten Grüsse an Sie aufgegeben. —Die projectirte Reise nach London betreffend, so hoffe ich dieselbe bis nächstes Frühjahr ausführen zu können, denn es wäre für mich vom höchsten Werthe, ihre persönliche Bekanntschaft zu machen, und dabei Ihre Ansicht und Ihren Rath über verschiedene Puncte zu hören. —

Von meinem Freund Reusch, Professor der Physik in Tübingen, wurde mir vor kurzem Ihr berühmtes Werk „Heat considered as a mode of motion" zugeschickt, das ich gegenwärtig mit grossem Interesse studire. Ohne Zweifel habe ich es auch diesem Werke zu verdanken, dass mir von der neugegründeten naturwissenschaftlichen Fakultät Tübingen vorgestern ein Doctor Diplom unter den schmeichelhaftesten Ausdrücken zugesendet wurde. Ich könnte überhaupt nicht enden, wollte ich Ihnen die Anerkennungen alle aufzählen, die mir seit dem 6 Juni 1862 zu Theil geworden sind,—Anerkennungen, die mich in soferne schüchtern machen, als ich wohl fühle, dass meine schwachen Kräfte nicht hinreichen, das Werk durchzuführen, das ich mir in früheren Jahren als Lebensaufgabe bezeichnet habe. Dass ich mich aber Ihnen, hochverehrtester Herr, ewig zu höchstem Danke verpflichtet fühle, dies bedarf wohl kaum einer Wiederholung!

Genehmigen Sie die Versicherung meiner ausgezeichnetsten Verehrung mit welcher ich verharre

Ihr | ergebenster Diener | J. R. Mayer

Heilbronn, | 31 October 1863

Most respected Professor!

I would gladly have accepted your flattering invitation to visit the meeting in Newcastle,[1] were I not hindered by my poor knowledge of the English language, an issue that I am now trying to address to the best of my ability. Accordingly, I preferred to attend the Congress of Natural Researchers[2] which took place almost concurrently in Stettin.[3] There, I met, among others, our mutual friend Knoblauch who told me a great deal about you. Dove was also present. On the journey back, at Poggendorff's in Berlin, I was pleased to see your picture placed beside that of my friend Schönbein,[4] who unfortunately was missing in Stettin. Poggendorff, as well as Knoblauch, asked me to send you kind greetings—Regarding the projected journey to London, I hope to be able to execute it before next spring, as I would greatly value meeting you personally and hearing your views and advice on various matters. —

My friend Reusch, professor of physics in Tübingen, recently sent me your famous work "Heat considered as a mode of motion"[5] which I am currently studying with great interest. Without a doubt, I also owe it to this work that the newly-founded faculty of natural sciences in Tübingen sent me a doctoral diploma the day before yesterday with the most flattering remarks. If I were to list all the acknowledgements I have received since 6 June 1862,[6] I would not come to an end—acknowledgements that make me bashful because I am aware that feeble powers are insufficient to complete the works that in former years I set myself as a lifelong task. Nevertheless, most respected Sir, it is not necessary to repeat that I am eternally grateful to you![7]

Please accept the assurance of my greatest admiration with which I remain

Your | most humble servant | J. R. Mayer

Heilbronn | 31 October 1863

RI MS JT/1/M/82
RI MS JT/1/TYP/7/2513–4

1. *meeting in Newcastle*: see letter 2024, n. 3.
2. *Congress of Natural Researchers*: the Versammlung Deutscher Naturforscher und Ärtze, or Society of German Natural Scientists and Physicians, met in Stettin in 1863.
3. *Stettin*: a town near the German-Polish border. Currently part of Poland (Szczecin), at the time it belonged to Prussia.
4. *Schönbein*: Christian Friedrich Schönbein (1799–1868), a Swiss-German chemist known for his work on the fuel cell and discovery of ozone (*HLS*).
5. *"Heat considered as a mode of motion"*: *Heat as a Mode of Motion.*
6. *6 June 1862*: see letter 2013, n. 2.
7. *I am eternally grateful to you*: Tyndall had done a lot to advance the priority of Mayer's work in relation to the conservation of energy; this is a theme of many of the letters in this volume, see for example, note 6 above, but also letters 2002, 2016, and 2024. See the Introduction for further commentary on Tyndall's defence of Mayer's claims.

To Walter White[1] 13 November [1863][2] 2042

Royal Institution of Great Britain | 13th. November

My dear Mr White.

Dr. Frankland is not Fullerian Professor. Prof. Faraday still holds the office.

Dr. Frankland's title is Professor of Chemistry in the Royal Institution & that of your humble servant is Professor of Natural Philosophy in the Royal Institution—There is no Fullerian Professorship of Natural Philosophy.

To the minds rather than to the eyes of my friends I wish to be commended. But if you wish for ornamentation here is some.

John Tyndall member of the Academies & Societies of Holland Geneva, Göttingen, Zurich, Halle, Marburg, Breslau, Upsala, la Societé Philomathique de Paris,[3] Cam. Phil. Soc.[4] Physical & Philosophical Societies of Berlin &c. Professor of Natural Philosophy in the Royal Institution & in the Royal School of Mines.[5]

Will that satisfy you? If not wait a few years & you shall have more.

Please omit the Ph.D.[6] | Yours truly | John Tyndall

Have you succeeded in finding the drawing[7] that I sent to Prof. Stokes some months ago? Mr Marshall[8] is Fullerian Prof. of Physiology

RS MS/769/111

1. *Walter White*: probably Walter White (1811–93), an English travel writer who served as librarian (1844–61) and then assistant secretary (1861–85) of the RS. He was also a prolific contributor to the periodical *Chambers's Journal* (*ODNB*).
2. *[1863]*: although indicated 1863 this may perhaps relate to 1864 prior to the presentation of the Rumford Medal to Tyndall, though the reason for the reference to Frankland is not clear.
3. *la Societé Philomathique de Paris*: the Philomatic Society of Paris, founded in 1788.
4. *Cam. Phil. Soc*: the Cambridge Philosophical Society, Cambridge's oldest scientific society, was founded in 1818 by the geologist Adam Sedgwick (1785–1873) and botanist John Stevens Henslow (1796–1861). See S. Gibson, *The Spirit of Inquiry: How One Extraordinary Society Shaped Modern Science* (Oxford: Oxford University Press, 2019).
5. *Royal School of Mines*: see letter 1987, n. 4.
6. *Please omit the Ph.D*: it is not clear why Tyndall would make this request.
7. *the drawing*: letter missing and drawing not identified.
8. *M[r] Marshall*: John Marshall (1818–91), an English surgeon and anatomist. He was Fullerian Professor of Physiology at the RI from 1862–5 (*ODNB*).

To Michael Faraday 23 November 1863 2043

Royal Institution, London

Assistance in the Physical Laboratory

I beg to recommend to the managers that Mr. William Barrett be appointed Assistant in the Physical laboratory at a salary of 50 pounds a year.

Mr. Barrett has been my assistant for a year, at no expense to the Royal Institution, and I can confidently speak of him as worthy in the highest degree of the post to which I recommend him.

In addition to Mr. Barrett the aid of a second trained assistant for a portion of the year would be necessary. 50, or at most 60 pounds, would cover his claim.

The sum total therefore required at present for assistance in the physical laboratory would be 100 or 110 pounds a year.

At the recommendation of their professor at the time I hope the managers will think it right to advance Mr. Barrett's Salary to 60 pounds next year. The expenses of the original investigations which I have presented during the last ten years have been defrayed in small parts by myself, but mainly by funds voted to me, on my personal application, by a committee of the Royal Society. In framing their new decisions the Managers I suppose will not calculate upon this foreign aid, but proceed on the assumption that the Researches conducted at the Institution shall be carried out at the expense of the Institution.

John Tyndall | 20th November | 1863

Endorsed by Faraday: Sent to Mr. Faraday Nov[embe]r. 20, 1863

Faraday Correspondence, 6:4409
RI MS CG/3/1/7

From Karl Hermann Knoblauch 26 November 1863 2044

Halle, November 28th | 1863.

My dear friend,

I take the liberty to send you 4 copies of my last inquiry,[1] relative to the transmission of radiant heat through rock-salt etc. and the diffusion of rays, for you, my dear friend; our friend Hirst; Prof. Faraday and the London Royal Society.

It is a long time since I have nothing heard of you. I hope you are well and are satisfied with your experiments.[2]

My cousin[3] who had the pleasure to make your acquaintance last year, has since last summer left London and is at present in New York.

As I have little hope to come over once more to your beautiful country, where I have spent so happy days, especially by your friendship. I think that you, who come over to the continent regularly every summer, will call upon Halle one day, take a view of our new established Institution for Natural Philosophy at our University[4] and allow me to return your hospitality.

You will make the acquaintance of Miss Knoblauch[5] and my little boys, John, Randolphe and Oscar and we will remember the good old Marburg-time.[6]

With my best compliments to Prof. Faraday, Dr. Hirst and who else may remember me in London.

Ever yours most truly | H Knoblauch

[in verso] To Dr John Tyndall | Professor | Royal Institution, | London

RI MS JT/1/K/23
RI MS JT/1/TYP/7/2508

---

1. *my last inquiry*: K. H. Knoblauch, 'Ueber den Durchang der strahlenden Wärme durch polirtes, mattes und berusstes Steinsalz, und über die Diffusion der Wärmestrahlen', *Poggend. Annal.*, 120 (1863), pp. 177–286.
2. *with your experiments*: probably a reference to the work Tyndall was doing that would be published as J. Tyndall, 'On the Passage of Radiant Heat through Dry and Humid Air', *Phil. Mag.*, 25 (1863), pp. 44–54.
3. *My cousin*: not identified.
4. *our University*: the University of Halle, where he had been since 1853.
5. *Miss Knoblauch*: Bertha Marie Emilie Knoblauch (née Winter, 1835–87) (Jan Mende, Stadtmuseum Berlin, personal communication).
6. *good old Marburg-time*: see letter 1940, n. 2.

To Julius Haast 27 November 1863 2045

My dear Sir

I am extremely obliged to you for the map[1] which you have been kind enough to send me,[2] and on which I readily found the position with which you have been pleased to honour me. It also gratified me much to receive your photograph which is now enrolled in company with some of my best friends. I wish I had an equally good one of myself to send you in return, but the fact is I do not possess the repose necessary to the production of a good photograph. Such as it is however it gives me pleasure to send you mine.

At a meeting of the philosophical club a month ago[3] my friend Dr. Hooker showed us some sketches taken by your colleague Dr. Hector,[4] and I was much pleased to hear Hooker express in the presence of the Club the high opinion which he entertained of your labours in New Zealand.

I am now working at subjects which would hardly interest you, but if I knew how to send books and papers to you I would in this way inform you as to what I am doing from time to time. I published in the spring of this year a work on heat[5] which I should like to present to you & parts of which might excite your interest.

I trust you will long continue to enjoy the health and possess the energy necessary for labours such as yours—goodbye

Yours very sincerely | John Tyndall

Dr. Julius Haast | &c &c &c | Royal Institution 27th. November 1863

Alexander Turnbull Library, National Library of New Zealand, MS-Papers Turnbull 0037–143

---

1. *the map*: this refers to Haast's 'Map of the Southern Alps in the Province of Canterbury (New Zealand)', in which he named one of the peaks Mt. Tyndall (there are also peaks named for Darwin, Hooker, Forbes, Stokes, and Ramsay, and glaciers for Faraday, Huxley, Lyell, and Agassiz). Haast later published the map in: J. Haast, 'Notes to accompany the Topographical Map of the Southern Alps, in the Province of Canterbury, New Zealand', *Journal of the Royal Geographical Society*, 40 (1870), pp. 433–41; the map precedes the article).
2. *to send me*: letter missing.
3. *meeting of philosophical club a month ago*: the Philosophical Club of the RS, founded in 1847, was a rather informal society but one that brought together prominent men of science to dine and enjoy each other's company. The club had the somewhat peculiar requirement that at least 35 of its stipulated 47 members should reside within a ten-minute walk of the London General Post Office. T. G. Bonney notes in his study of the club that it

arose out of dissatisfaction with the direction of the RS, and with its method of election of Fellows in particular—namely, that the chief requirement for Fellows seemed to be that they were 'well-born, well-dressed, and moderately learned [in science]' (T. G. Bonney, *Annals of the Philosophical Club of the Royal Society, written from its Minute Book* (London, Macmillan & Co., 1919), p. v). It is notable that Bonney's work was published in response to Archibald Geike's *Annals of the Royal Society Club. The Record of a London Dining-Club in the Eighteenth and Nineteenth Centuries* (London: Macmillan & Co., 1917). During the 29 October 1863 meeting, Hooker 'exhibited photographs of drawings by Dr. Hector, made in the South Island of New Zealand, at comparatively low levels above the sea, which showed the effect of glacial action and much resembled similar phenomena in the Himalayas' (Bonney, *Annals of the Philosophical Club of the Royal Society*, p. 165). Tyndall had been elected a member of this club on 10 May 1855 (Journal, RI MS JT/2/13c/745).

4. *D^r. Hector*: James Hector (1834–1907), a Scottish geologist and naturalist, known for his lengthy political and scientific careers in New Zealand. One of his major contributions to New Zealand geological studies was a survey of the Otago region on the southern island, known as 'Hector's Map' (*ODNB;* S. Nathan, 'James Hector and the First Geological Maps of New Zealand', *Journal of the Royal Society of New Zealand*, 44 (2014), pp. 88–100).
5. *a work on heat*: *Heat as a Mode of Motion.*

From [Alexandre-Eugène Béguyer de Chancourtois][1] 28 November *[1863]*[2] 2046

Paris 28 9bre

Cher monsieur

Voici les figures dont je vous ai parlé—Je demande toute votre indulgence pour celle du glacier d'Assakak. J'ai voulu tendre mon papier pour perfectionner le coloriage qui le faisait goder d'une manière très gênante mais au premier coup de pinceau, appliqué heureusement dans la moraine, il s'en fait une terrible déchirure et j'ai dû ne pas toucher au glacier de peur d'y déterminer des crevasses. Au surplus, avec le profil mis en marge, le trait calqué sur un dessin du docteur Rink vous fera bien comprendre les points principaux de la structure. Les veines sont régulièrement développées en a—on les voit aussi un peu en b. Cependant il est évident que la glace inférieure est bien moins laminée que celle des couches supérieures. La présence des pierres enchâssées en c y est aussi remarquable. J'avais compris que le docteur Rink me donnait ce dessin comme spécimen de la coupe terminale résultant du détachement d'un Eisberg mis à flot au fond d'un des fiords qui atteignent le grand champ de neige du Groenland lequel ne s'élève sur ses bords qu'à une hauteur de 600 mètres . . .

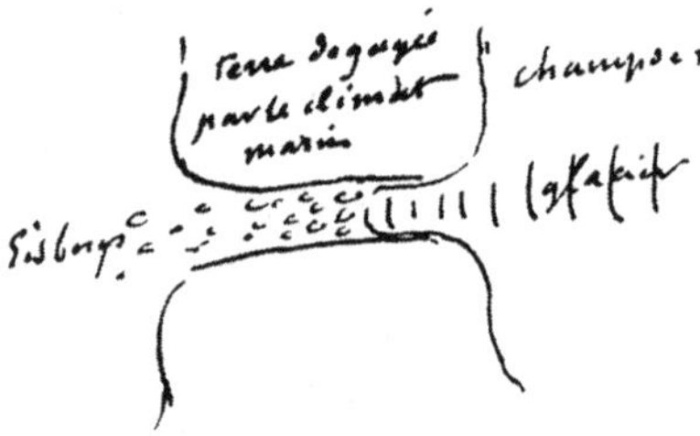

Dans de tels points <on> trouve les conditions les plus différentes de celles de nos glaciers de montagnes. La surface limite des neiges perpétuelles est là plutôt verticale qu'horizontale. L'étendue indéfinie du réservoir de neige compense la faiblesse de la pente sur laquelle s'opèrent les transformations qui semblent devoir être accomplies dans un parcours relativement restreint etc etc. Malheureusement je vois en me reportant à la carte du docteur Rink que le glacier d'Assakak est un petit glacier descendant d'un champ de neige limité de la presqu'île de Noursoak et situé à une hauteur de près de 2000 mètres. C'est-à-dire qu'il est à peu près dans les conditions de nos glaciers—Je ne suis même pas sûr qu'il atteigne la mer car il y en a deux très voisins du point marqué Assakak et l'un d'eux se termine dans la vallée—je crois cependant toujours que le dessin représente la cassure produite par la mer, que le prolongement de la moraine au dessous est un effet de perspective et que l'Esquimaux ne peut servir d'échelle. Quoiqu'il en soit il y a une ressemblance évidente entre la disposition générale de la coupe et celle que l'on observe dans les plus grands Eisberg comme vous pouvez en juger par les croquis que j'ai calqués sur mon carnet de voyage—on voit dans le premier et le dernier une couche plus résistante de glace feuilletée—surmontant des couches de glace grenue (au moins à en juger par sa blancheur mate)—il semble que l'on voit une tranche de la coupe d'Assakak inclinée à environ 450—dans le second croquis pris de plus près sur le reste d'un Eisberg sans doute primitivement plus gros, mais évidement brisé, on voit la structure feuilletée rendue manifeste au plus haut point par le creusement des feuillets tendres. L'ouvrage du docteur Rink qui contient une carte mais pas de figures et traite de toute l'histoire naturelle et climatérique et économique du Groenland septentrional est intitulé De Danske Handelsdistricter i Nordgronland—il a été publié à Copenhague en 1852 et 1855. Ce sont deux minces brochures—mais d'une grande valeur à en juger par la partie minéralogique que j'ai seule essayée de comprendre. Le docteur Rink retournait au Groenland en 1856 pour faire l'étude des établissements du sud, il est probable qu'il a fait de nouvelles publications.

Paris 28 November

Dear Sir,

Here are the sketches I told you about.[3] I request your complete leniency for that of the glacier of Assakak.[4] I wanted to stretch my paper to perfect the colouring which crinkled it in a very embarrassing way, but the first brush stroke, fortunately applied to the moraine, tore it badly and I had to refrain from touching up the glacier in fear of introducing crevasses there.[5] Furthermore, in addition to the profile placed in the margin, the sketch modelled on one of Doctor Rink's[6] drawings will help you understand the main elements of the structure. The veins are developed consistently in a—they are also slightly visible in b. However, it is obvious that the lower ice is much less foliated than that of the upper layers. The presence of stones embedded in c is also remarkable. As I understood it, Dr. Rink gave me this drawing as a specimen of the final section resulting from the detachment of an iceberg set afloat in one of the fjords that extend to the great snowfield in Greenland, which only rises on its edges up to a height of 600 meters . . .

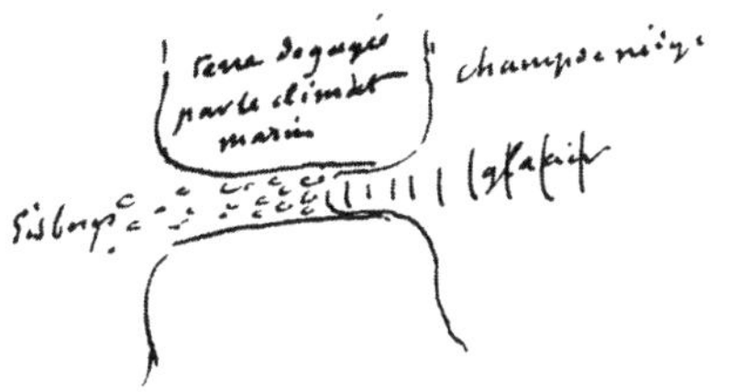

It is in such places that we find the conditions most dissimilar to those of our mountain glaciers. The boundary of the surface of perpetual snow is vertical there, rather than horizontal. The indefinite extent of the snow reservoir compensates for the weakness of the slope, on which take place the transformations that seem to happen over a relatively limited route etc. etc. Unfortunately, when I refer to Dr. Rink's map, I see that the glacier of Assakak is a small glacier descending from a localised snowfield in the peninsula[7] of Noursoak[8] and situated at the height of approximately 2,000 metres—in other words, the conditions are fairly similar to those in our own glaciers—I am not even certain that it reaches the sea, since there are two glaciers very close to the spot indicating Assakak, and one of them ends in the valley. However, I still believe that the drawing represents the fracture made by the sea, that the extension of the moraine below is a trick of perspective, and that the Eskimos cannot be used for scale. In any case, there is an obvious similarity between the general layout of the section and the layout observed in the largest icebergs, as you can see from the sketches that I copied in my travel journal. The first and the last show a more resistant layer of foliated ice—topping off layers of grainy ice (at least if we judge it by its matte whiteness).

It is a view of a slice of the Assakak section tilted at about 450—in the second sketch, drawn closer, we see the remains of an iceberg which was undoubtedly once larger, but is now obviously broken. The foliated structure can be seen at the highest point by the sinking of the soft layers. Dr. Rink's book, which contains a map but no figures and which covers the entire natural, economic, and climatic history of northern Greenland, is entitled: De Danske Handelsdistricter i Nordgronland[9]—it was published in Copenhagen in 1852 and 1855. It comes as two thin brochures, but is of great value, judging solely by the mineralogical part, the only part which I tried to understand. Dr. Rink returned to Greenland in 1856 to study the southern colonies and it is likely that he has produced new publications.

RI MS JT/1/C/20

1. *[Alexandre-Eugène Béguyer de Chancourtois]*: the author is probably Alexandre-Eugène Béguyer de Chancourtois (see letter 2039, n. 1).
2. [*1863*]: the year is given by relation to letter 2039.
3. *sketches I told you about*: see letter 2039.
4. *glacier of Assakak*: the glacier is located in the Uummannaq Fjord in western Greenland.
5. *made a terrible tear . . . in fear of introducing crevasses there*: wordplay in French between the act of tearing the paper and that of representing a fake crevasse in a glacier.
6. *Doctor Rink*: Hinrich (Henrik) Johannes Rink (1819–93), a Danish geologist and a pioneer of glaciology, known for his exploration of Greenland. Along with his wife Nathalie Sophia Nielsine Caroline Rink née Møller (1836–1909), he founded Greenland's first newspaper *Atuagagdliutit* (R. Brown, 'Obituary: Dr. Hendrik Rink', *The Geographical Journal*, 2 (1894), pp. 65–7).
7. *a localised snowfield in the peninsula*: while there is no distinction in English, the French differentiate between the 'Presqu'île' (used here) and 'Peninsule'—the isthmus connecting the 'Presqu'île' to the mainland is not as large as that of a 'Peninsule'.
8. *Noursoak*: Four Islands Point, North Greenland.
9. *De Danske Handelsdistricter i Nordgronland*: H. Rink, *De dankse Handelsdistricter I Nordgrenland deres geographiske Beskaffenhed og productive Ervervskilder*, 2 vols (Kjøbenhavn [Copenhagen]: Andr. Fred. Host, 1852, 1855).

From Henry Bence Jones 29 November 1863 2047

London

My dear Tyndall

This note you sent me with your prospectus[1] made me fear that you were ill; and as soon as I had a moment to spare I went to Mr Faraday with it to talk to him about you when I met you in the hall and heard from yourself that you were better. When I went up stairs I soon found from Mr Faraday's manner that the expression in your note had a deeper foundation than your health.

He would not tell me what had passed between you.[2] He absolutely refused & said it must come from you & that he had begged you to come to me for he was sure that you were ill.

Please do please come to me and bring me the letter you shewed me the other day & would not give me then & talk to me as you would to your best friend.

I am so grieved to think that all I have been trying to do for the good of scientific research[3] should bring trouble upon you and distress upon Mr Faraday; the two men of all in the world I wish to serve & help

Ever your sincere friend | H. Bence Jones | Nov 29 1863

RI MS JT/1/J/54
RI MS JT/1/TYP/2/685

---

1. *note you sent me with your prospectus*: letter missing; prospectus not identified.
2. *what had passed between you*: as Roland Jackson notes in his *Ascent of John Tyndall*, there had been some 'uncharacteristically unkind words from Faraday, perhaps over the appointment of laboratory assistants to Tyndall and Faraday. Tyndall had asked Faraday if the Institution could fund his assistant, William Barrett, whom he had been paying out of his own pocket' (p. 180). See letter 2043.
3. *for the good of scientific research*: Bence Jones had suggested expanding the Holland Fund (see letter 2015, n. 2).

To Rudolf Clausius 3 December 1863 2048

London | 3rd <u>December</u> 1863

My dear Clausius

Your last kind letter[1] has remained a long time unanswered. It would give me great delight to be your guest next year at the meeting in Zürich,[2] but when I quit London I am so worn down—so utterly tired of science that I usually make my escape into the solitudes of the mountains. If however I go to Switzerland next year I should make an effort to visit you, but I am thinking of going to Norway.

I sent a little note to the Philosophical Magazine of November[3] regarding Laplace's correction of the theoretic velocity of sound. I cannot help thinking that what I have said has some foundation and that men have relied too much on the applicability of such experiments as those executed by Dulong[4] in determining the specific heat of gases. For olefiant gas he found the ratio of the two specific heats to be 1.240. Now I should be disposed to conclude that if the gas demanded any interior work the ratio of the two specific heats, instead of being less than that found in the case of air, ought to be greater. I think that there is the very highest a priori probability against the concreteness of the number 1.240 as expressing the ratio of the two specific heats. Allowing air & olefiant gas to perform the same amount of exterior work the quantity of heat necessary to the performance of this work would in the two cases, if Dulong be right, be as 0.421:0.240; but inasmuch as air demands no interior work, and utilises in exterior work the entire excess of heat imparted to it, there must, if Dulong be right, exist in olefiant gas some magnitude of heat which it develops when expanding, and which enables a less amount of heat communicated from without to perform the same amount of work as that performed by expanding air:—This, to my mind is absurd.

I continue my experiments on Radiant Heat; the subject is exceedingly difficult; but I am attaining gradually a mastery over it: & I trust after some time that we shall know more of molecular physics than we do at present.

Give my kindest regards to Madame Clausius—& believe me

Yours ever | John Tyndall

RI MS JT/1/T/188

1. *Your last kind letter*: letter 2035.
2. *the meeting in Zurich*: the Naturforschende Gesellschaft, or Swiss Society of Natural Science, met in Zürich in August 1864. See letter 1906, n. 6 and letter 2130, n. 3.
3. *Philosophical Magazine of November*: J. Tyndall, 'Note on Laplace's Correction for the Theoretic Velocity of Sound', *Phil. Mag.*, 26 (1863), pp. 384–7.
4. *Dulong*: Pierre Louis Dulong (1785–1838), a French physicist and chemist. Amongst other researches, Dulong worked on the conduction of heat (*CDSB*). See P. L. Dulong, 'Recherches sur la chaleur spécifique des fluids élastiques', *Annales de chimie et de physique*, 41 (1829), pp. 113–59.

From Rudolf Clausius [6 December 1863][1] 2049

Zürich, den 6. Dezember 63

Lieber Tyndall,

Ich habe sehr bedauert, aus Ihrem lieben Briefe zu entnehmen, dass Sie uns zu unserer Naturforscher-Versammlung vielleicht nicht durch Ihren Besuch erfreuen werden. Da Sie indessen noch nicht bestimmt abgesagt haben, so dürfen wir vielleicht doch noch hoffen, Sie und Hirst hier zu sehen. Jedenfalls können Sie versichert sein dass Sie uns zu jeder Zeit durch Ihren Besuch eine grosse Freude machen werden.

Was Ihre Ansichten über die specif. Wärme der Gase anbetrifft, so kann ich denen nach meiner Auffassung der Sache nicht ganz beistimmen. Nach der mechanischen Wärmetheorie ist nicht vorauszusetzen, dass das Verhältniss der beiden specifischen Wärmen für alle Gase gleich sei, sondern die Grösse, welche für alle Gase angenähert gleich sein muss, ist eine andere, welche ich in meiner ersten Abhandlung über die mech. Wärmetheorie (Pogg. Ann. Bd 79 S. 394) bestimmt habe.

Denken wir uns die specif. Wärme nicht für die Gewichtseinheit, sondern für die <u>Volumeneinheit</u> bestimmt. Wir können z.B. von jedem Gase ein Cubikmeter unter der Temp. 0° und dem Drucke von 1 Atm. betrachten. Für diese Gasmenge sei die specif. Wärme bei const. Volumen mit *[1 word illeg]* bezeichnet, d. h. die Grösse bedeutet die Wärmemenge welche diese Quantität Gas bedarf, um 1° Grad erwärmt zu werden, wenn dabei das Volumen constant erhalten wird. Entsprechend heisse für dieselbe Quantität Gas die specif. Wärme bei const. Drucke dann ist nach Gleichung 11 (Pogg. Ann. LXXIX S.394)

$$y' - y = H . \frac{p_0}{a + t_0}$$

worin die Temperature und der Druck ist, unter denen die Volumeneinheit genommen ist. Also wenn die Temp. 0° und der Druck einer Atm. gewählt ist, so ist zu setzen Ferner H ist = 273 und H ist das calorische Aequivalent für die Einheit der Arbeit, also H = $\frac{1}{423{,}55}$. Es kommt somit:

$$y' = y = \frac{1}{423{,}55} \cdot \frac{10333}{273} = 0{,}0894$$

Diese Differenz, ist es, welche für alle Gase angenähert gleich sein muss, und welche, wenn alle Gase vollkommene Gase wären, bei denen gar keine innere Arbeit vorkäme, ganz gleich sein müsste.

Betrachten wir z.B. atm. Luft. Für <u>ein Kilogramm</u> ist die spec. Wärme bei constantem Drucke 0,2377. Die spec. Wärme bei constantem Volumen ist, wenn man das Verhältniss 1,421 annimmt $\frac{0{,}2377}{1{,}421}$ = 0,1673 oder wenn man das Verhältniss 1,410 annimmt $\frac{0{,}2377}{1{,}410}$ = 0,1686. Will man die spec. Wärme für <u>ein Cubikmeter</u> haben, so muss man die vorigen Zahlen mit 0,7733 dividiren, weil ein Kilogramm Luft bei 0° und unter dem Drucke von 1 Atm. 0,7733 Cubikmeter einnimmt. Dadurch kommt:

Specif. Wärme bei const. Druck	0,3075
" " " " Volumen	0,2164
Oder	0,2181

Daher

~~Dieser~~ $y' - y \;=\; 0{,}3075 - 0{,}2164 \;=\; 0{,}0911$

oder $= 0{,}3075 - 0{,}2181 = 0{,}0894$

Die letztere Zahl stimmt ganz mit der obigen. Die erstere 0,0911 weicht etwas ab, woraus man schliessen muss, entweder das Verhältniss 1,421 nicht ganz genau und 1,410 genauer ist, oder dass in den andern bei den Berechnungen angewandten Zahlen Ungenauigkeiten vorkommen. Aber selbst diese Abweichung zwischen der Zahl 0,0911 und der obigen 0,0894 ist so gering dass sie kein Misstrauen gegen die Theorie erwecken kann.

Wir wollen nun die Gleichung, = 0,0894 weiter anwenden. Sie giebt,

$$\frac{y'}{y} - 1 = \frac{0{,}0894}{y}$$

Der Bruch $\frac{y'}{y}$ ist das Verhältniss der beiden specifischen Wärmen, dann für dieses Verhältniss ist es gleichgülitg auf welche Quantität Gas man die specifischen Wärmen bezieht. Aus dieser Gleichung folgt dass man nicht erwarten darf, dass die Differenz $\frac{y'}{y}-1$ für all Gase gleich sei. Sondern sie muss um so kleiner sein, je grösser die auf die Raumeinheit bezogene specifische Wärme ist. Bei den einfachen Gasen, und solchen zusammengesetzten, die bei der Verbindung keine Volumverminderung erlitten haben, ist gleich, und daher auch die Differenz $\frac{y'}{y}-1$ und somit auch der Bruch $\frac{y'}{y}$ gleich.

Bei solchen zusammengesetzen Gasen dagegen, welche bei der Verbindung eine Volumenverminderung erlitten haben, ist grösser und daher muss $\frac{y'}{y}-1$ kleiner sein als bei den einfachen Gasen. Da nun ölbildendes Gas ein solches zusammengesetztes Gas ist, so musste man im Voraus erwarten, dass bei ihm die Differenz $\frac{y'}{y}-1$ kleiner sei als bei den einfachen Gasen, also kleiner als 0,421 oder 0,410. Die von Dulong gefundene Zahl 0,240 kann daher in dieser Beziehung nicht überraschen, ja ich halte die Zahl eher noch für zu gross als für zu klein.

Sollten Sie mit den vorstehenden Auseinandersetzungen übereinstimmen, so könnten Sie vielleicht einen Nachtrag zu Ihrem Aufsatze im Novemberhefte des Phil Mag schreiben, in welchem Sie ihr dortigen Betrachtungen in entsprechender Weise abändern. Eine Berichtigung von der Hand des Autors selbst kann am leichtesten die etwa sonst entstehenden Discussionen abschneiden. Ich habe Ihnen deshalb meine Ansicht so vollständig mitgetheilt und bin so offen gewesen, wie es unter Freunden sein muss. Durch die immer weitere Fortsetzung Ihrer Versuche über die strahlende Wärme erwerben Sie sich ein grosses Verdienst um die Wissenschaft, denn schon Ihre bis jetzt gefundenen Resultate sind von ausserordentlicher Wichtigkeit. | Herzlich grüssend | Ihr | Clausius

Zurich, 6 December 63

Dear Tyndall,

I was very sorry to gather from your last kind letter,[2] that you might not grant us the pleasure of your presence at our next meeting of natural scientists.[3] Since you have not yet definitively declined, perhaps we may still hope to see you and Hirst here. In any case, you may be assured that a visit from you at any time would give us great pleasure.

In regard to your views on the specific heat of gases, my conception of the matter does not completely agree with them. According to the mechanical theory of heat, one cannot assume that the relationship of both specific heats is the same for all gases, but in contrast the quantity that must be approximately the same for all gases is the one that I determined in my first paper on the mech[anical] theory of heat[4] (Pogg. Ann. Vol. 79 p. 394).

Let us consider that the specif[ic] heat is determined not by the unit of weight but by the unit of volume. We can, for example, take one cubic meter of any gas under the temp. 0° and the pressure of 1 atm. For this amount of gas, the specif[ic] heat at constant volume is indicated by *[1 word illeg]*, that is the quantity means the amount of heat that this quantity of gas needs to be heated by 1° when the volume is kept constant. Correspondingly, this means for the same quantity of gas the specif[ic] heat at const. pressure is according to equation 11 (Pogg. Ann. LXXIX p. 394)

$$y' - y = H \,.\, \frac{p_0}{a + t_0}$$

in which is the temperature is the pressure on the unit of volume. Thus when the temp. 0° and the pressure of one atm[osphere] is selected, one must set In addition H is = 273 and H is the caloric equivalent for the unit of work, thus H = $\frac{1}{423{,}55}$. Thus:

$$y' = y = \frac{1}{423{,}55} \cdot \frac{10333}{273} = 0{,}0894$$

This difference, is the one that all gases must approach, and which would have to be completely equivalent, if all gases were perfect gases, in which no internal work took place.

Take for example atm[osphere] air. For <u>one kilogram</u> the spec[ific] heat at constant pressure is 0.2377. The spec[ific] heat at constant volume is, if one take the ratio 1.421, $\frac{0{,}2377}{1{,}421}$ = 0.1673 or if one takes the ratio 1.410, it is $\frac{0{,}2377}{1{,}410}$ = 0,1686. If one wants the spec[ific] heat for one <u>cubic meter</u>, one must divide the previous numbers by 0.7733, because one kilogram of air at 0° and under the pressure of 1 atm[opshere] takes up 0.7733 cubic meters. Thus one gets:

Specific Heat at constant Pressure		0,3075
" " " "	Volume	0,2164
	Equals	0,2181

Thus:

~~This~~ $y' - y = 0.3075 - 0.2164 \quad = \quad 0.0911$

or $= 0.3075 - 0.2181 \quad = \quad 0.0894$

The latter number completely agrees with the above. Since the first 0.0911 deviates somewhat, one must conclude either that the ratio 1.421 is not completely exact and 1.410 is more exact or that there are imprecisions in the other numbers used in the calculations. But even this deviation between the number 0.0911 and the above 0.0894 is so small that it cannot arouse mistrust against the theory.

Let us continue with the equation = 0.0894. It gives

$$\frac{y'}{y} - 1 = \frac{0{,}0894}{y}$$

The fraction $\frac{y'}{y}$ is the ratio of both specific heats, since for this ratio it makes no difference what quantity of gas the specific heat refers to. It follows from this equation that one cannot expect the difference $\frac{y'}{y}-1$ to be the same for all gases. Instead, the greater the specific heat in relation to the spatial unit, the smaller it must be. In simple gases and compound ones that did not undergo a reduction in volume through the bonding, is the same, thus the difference $\frac{y'}{y}-1$ is the same, and with it the fraction $\frac{y'}{y}$.

In such compound gases, however, where bonding causes a reduction in volume, is greater and therefore $\frac{y'}{y}-1$ must be smaller than in the simple gases. Since olefiant gas is such a compound gas, one must in advance expect that the difference $\frac{y'}{y}-1$ will be smaller in it than in simple ones, that is, smaller than 0.421 or 0.410. Dulong's number[5] 0.240 is not surprising in this regard, nonetheless I consider the figure more likely to be too great rather than too small.

Should you agree with the preceding explanations, perhaps you could write an addendum[6] to your paper[7] in the November issue of the Phil Mag, in which you modify the observations you made in it accordingly. A correction from the pen of the author is the easiest way to cut off discussions that would arise otherwise. I communicated my view so thoroughly to you for that reason and am as forthright as one must be between friends. You are doing a great service to science through the continuation of your experiments on radiant heat, and the results you have found thus far are of extraordinary importance.

With best regards, | Your Clausius

RI MS JT/1/TYP/7/2261–3
LT Typescript Only

1. *[6 December 1863]*: date suggested by LT.
2. *your last kind letter*: letter 2048.
3. *our next meeting of natural scientists*: see letter 2048, n. 2.
4. *my first paper on the mech[anical] theory of heat.*: R. Clausius, 'Ueber die bewegende Kraft der Wärme und die Gesetze, welches ich daraus für die Wärmelehre selbst ableiten lassen', *Poggend. Annal.*, 79 (1850), pp. 368–97. This was later published as: R. Clausius, 'On the Application of the Mechanical Theory of Heat to the Steam-engine,' *Phil. Mag.*, 12 (1856), pp. 241–65.
5. *Dulong's number*: see letter 2048, n. 4.
6. *addendum*: J. Tyndall, 'Note on Laplace's Correction for the Theoretic Velocity of Sound', *Phil. Mag.*, 26 (1863), pp. 384–7. See also letter 2057.
7. *your paper*: J. Tyndall, 'On the Relation of Radiant Heat to Aqueous Vapour', *Phil. Mag.*, 26 (1863), pp. 30–44.

To George Gabriel Stokes 7 December 1863 2050

7th Dec. 1863

My dear Stokes,

Though I cannot say that I have read the paper of Messrs. Mattheissen & Vogt[1] very critically still the inspection that I have made of it and the faith that I have in the trustworthiness of its authors, induces me to recommend it for publication in the Philosophical Transactions.

The subject of the paper is important, and the determinations contained in it will be of permanent value to scientific progress.

Yours very truly | John Tyndall
1863 | Matthiessen | by | Tyndall[2]

RS RR/5/149 Tyndall 7 December 1863

1. *paper of Messrs. Mattheissen & Vogt*: Augustus Matthiessen (1831–70), a British chemist and physicist, obtained his PhD at Giessen and later worked with Bunsen in Heidelburg, and, by the time of this letter, was lecturer on Chemistry at St. Mary's Hospital (*ODNB*). Karl [Carl] Christoph Vogt (1817–95), a German scientist, philosopher, and politician who lived in Switzerland (*HLS*). The paper referred to is: A. Matthiessen and C. Vogt, 'On the influence of temperature on the electric conducting-power of alloys', *Phil. Trans.*, 154 (1864), pp. 167–200.
2. *1863 | Matthiessen | by | Tyndall*: written either on an envelope or reverse side of letter.

To William Ginty 16 December 1863 2051

Royal Institution. | 16th. December 1863.

My dear Ginty,

You referred to the Jungfrau[1] in your last letter to me,[2] and when it reached England I happened to be in the vicinity of that noble mountain. Soon after I received it I found myself one glorious morning at 7 o'clock upon the summit of the majestic peak.

The death of Mr Edmondson[3] I trusted would not make much difference in the constitution of Queenwood School.[4] It seemed to be the intention of Mrs Edmondson to keep the old staff together, and thus to maintain the place on its former footing. Mrs Edmondson however has since died and at Christmas the school will pass into other hands.

Informed of this I went down there on Sunday to make enquiries regarding the probable future of the School. Many of the old masters are going away; but the new principal who visited the place some time ago appears to have made a favourable impression. I think he would have acted more wisely had he endeavoured to retain the men to whose exertions the prosperity of Queenwood was really due. Still I am not clear at present whether it will be advisable to remove your boys to another establishment.

In the event of their being removed one of the masters, Mr Wright,[5] has opened a school at Southsea,[6] a very healthy locality. He is a thoroughly respectable man, and his wife[7] is an excellent woman. In his house I think your boys would be well taken care of in case you thought it well to remove them from Queenwood. Mr Wright you are probably aware was the surveyor and teacher of drawing at Queenwood.

I am tremendously hard worked at present; otherwise I would send you a longer letter. Give my kind regards to Mrs Ginty.[8] Yesterday Col[onel] Wynne[9] called to see me. He has been called over by the Government in connexion with the fortifications of Corfu. He looks well—the same upright fellow morally and physically as when we first served under him.

Believe me | yours ever truly | John Tyndall.

RI MS JT/1/TYP/11/3643
LT Typescript Only

---

1. *the Jungfrau*: see letter 1915, n. 7.
2. *your last letter to me*: letter missing.
3. *Mr Edmondson*: George Edmondson (1798–1863), headmaster of Queenwood College in Hampshire from 1847 until his death. Edmondson, a Quaker, had attended Ackworth

School in Yorkshire. From 1817 to 1824 Edmondson lived in Russia, tutoring the children of his mentor, Daniel Wheeler. After returning to England he opened a school at Blackburn in 1825, then one near Preston, before becoming the head teacher at Queenwood College when it opened in 1847. Under his influence science and technological subjects were emphasized (*ODNB*).

4. *Queenwood School*: Queenwood College, a boys' school near Stockbridge, Hampshire, England, opened by George Edmondson in 1847 (the site was previously Queenwood Farm). Queenwood was an experimental school dedicated to science teaching, and Tyndall and Hirst both taught mathematics for surveying there, Tyndall from 1847–8 and again from 1851–3, and Hirst from 1853–6. Edward Frankland had been employed by the school (for the 1847–8 school year), as well as Heinrich Debus (from 1851–67). Although Edmondson had hired Tyndall, Hirst, Frankland, and Debus, they all became critical of his management of Queenwood. The school remained operational until 1896 (see W. H. Brock, 'Queenwood College Revisited,' article XVII in *Science for All: Studies in the History of Victorian Science and Education* (Aldershot: Ashgate Variorum, 1996).
5. *Mr Wright*: Richard Pears Wright (1829–92), an English mathematician and educator who taught engineering and surveying at Queenwood College from 1851–63. From 1870 until his death in 1892 he was a teacher of mathematics at University College School, London (J. Hodge (ed), *Lady Tennyson's Journal* (Charlottesville: University of Virginia Press, 1981) p. 226, n. 21).
6. *Southsea*: coastal resort near Portsmouth on the south coast of England.
7. *wife*: Fanny Wright (née Bellamy, dates unknown), married Richard Pears Wright in July 1852 at Saint Martin-in-the-Fields, London. In 1855 she gave birth to Alice Maud Wright, who died in January 1857. Tyndall attended the child's funeral (see letter 1324, n. 3, *Tyndall Correspondence*, vol. 6).
8. *Mrs Ginty*: Margaret Ginty.
9. *Col. Wynne*: George Wynne.

To Maurice Holzmann 16 December 1863 2052

16th December 1863

Dear Sir

In reply to the first of the two questions which you have done me the honour of addressing to me[1] I have to state that though my aim is to make the Christmas lectures[2] an organic whole, still the appliances which we possess at the Royal Institution will enable me to illustrate the last two lectures of the Course in a manner which, I think, would render them agreeable and instructive to His Royal Highness the Prince Arthur.[3]

It is not in the amount of knowledge conveyed by such lectures that their chief value consists, but in the intelligent interest which they excite in the works of Nature;—an interest which often abides long after its exciting cause has passed away. My opinion therefore is that even the last two lectures of the Course would not be without value to the Prince.

With regard to the second question I would beg to state that at all times my duty and my delight will go hand in hand in imparting to any member of the Royal Family such knowledge as it is in my power to communicate. If I were permitted to make a remark as to the best way of carrying out the wishes of Her Majesty,[4] I would say that I do not think a formal lecture the most suitable means of securing the interest and instruction of their Royal Highnesses the Princesses, for, as you rightly observe, the apparatus necessary to illustrate such a lecture could hardly be introduced into a drawing room. Instead of a lecture I should therefore propose a conversation on such a series of experiments as I could by simple appliances execute in the presence of their Royal Highnesses.[5] My desire indeed would be to employ the humblest means and to show their Royal Highnesses the wonders which underlie common things, and the wealth of interest which Nature possesses for those who appeal to her aright.

The arrangement of my duties at the Royal Institution gives me a little breathing time from the 9th to the 16th of January inclusive. On any one day or on any two days of this period I shall be prepared to execute to the best of my ability her Majesty's commands.

Believe me | dear Sir | very faithfully yours | John Tyndall

M. Holzmann Esq[ui]r[e]. | &c &c &c

RI MS JT/2/11 (in loose plastic sleeve)

---

1. *of addressing to me*: letter missing.
2. *Christmas lectures*: the RI's annual juvenile Christmas Lectures began in 1827 and still continue. Tyndall delivered the Christmas Lectures a dozen times over his time with the RI, his first in 1861 (see F. A. J. L. James (ed.), *Christmas at the Royal Institution: An Anthology of Lectures* (Hackensack, NJ: World Scientific, 2007)). This season's lectures, his second time, was on 'Electricity at Rest and Electricity in Motion', and the lectures were given on 26, 29, and 31 December 1863, and 2, 5, and 7 January 1864 (*Roy. Inst. Proc.*, 4 (1862–6), pp. 157; RI Guard Book, vol. 3, p. 29). The reference here is to the lectures on 5 and 7 January.
3. *Prince Arthur*: Arthur William Patrick Albert (1850–1942), Prince Arthur, Duke of Connaught and Strathearn, the seventh child and third son of Queen Victoria and Prince Albert (*ODNB*).
4. *Her Majesty*: Queen Victoria.
5. *their Royal Highnesses*: following the end of his Christmas lectures on 'Electricity at Rest and Electricity in Motion,' Tyndall lectured at Osborne House on 30 and 31 January 1864

to Queen Victoria and the Princesses Victoria Adelaide Mary Louisa (1840–1901), Alice Maud Mary (1843–78), Helena Augusta Victoria (1846–1923), Louise Caroline Alberta (1848–1939), and Beatrice Mary Victoria Feodore (1857–1944) (*ODNB*; *Ascent of John Tyndall*, p. 181). Osborne House was the royal residence in East Cowes, on the Isle of Wight. The house, designed by Prince Albert, served as a summer home and rural retreat. Thomas Hirst noted in his journal entry for Sunday, 31 January 1864, that Tyndall 'left for Osborne' on Friday (Brock and Macleod, *Hirst Journals*, p. 1663).

To John Murray [before 21 December 1863][1] 2053

Copies wanted for gifts[2]
Royal Institution of Great Britain | Tuesday

My dear Mr Murray.

I want to present to the Duchess of Northumberland[3] & to another noble lady[4] two copies of the "Glaciers of the Alps".[5] Do you happen to have suitable copies beside you?, & if not would you kindly have two bound for me?

[in verso, in another hand] Yours very sincerely | John Tyndall
1863 | Tyndall, J.

National Library of Scotland, John Murray Archive, Ms. 41214

---

1. *[before 21 December 1863]*: while no month or day are given in this letter, it would appear to precede letter 2054, dated 21 December 1863, in which Tyndall received thanks from the Duchess for sending the book.
2. *Copies wanted for gifts*: these words are written at the top of the letter, either in John Murray's hand or someone else from his publishing firm.
3. *Duchess of Northumberland*: see letter 2054, n. 1. In that letter, Eleanor Percy sent Tyndall thanks for sending a copy of the book.
4. *another noble lady*: not identified.
5. *Glaciers of the Alps*: *Glaciers of the Alps*.

From Eleanor Percy[1] 21 December 1863 2054

Dec. 21st 1863.

Dear Sir

Pray accept my best thanks for the two gifts[2] you have so kindly sent me; I shall value them very highly; your photograph is placed in my book side by side with Professor Faraday, these portraits reminding me most agreeably of the many pleasant evenings I have passed at the Royal Institution. For your interesting book[3] I am also very grateful; it makes me long to have you as my Pioneer over some of the easiest Passes of enchanting Switzerland.

The Duke[4] unites in kind wishes for this season with | yours truly | E Northumberland.

I enclose our photographs, both from pictures by Grant,[5] for your acceptance. | Professor Tyndall.

RI MS JT/2/13c/1263
LT Typescript Only

---

1. *Eleanor Percy*: Eleanor Percy (1820–1911), Duchess of Northumberland.
2. *the two gifts*: his photograph and book (see n. 3).
3. *your interesting book*: *Glaciers of the Alps.* See letter 2053.
4. *The Duke*: Algernon Percy (1792–1865), 4th Duke of Northumberland. He had been President of the RI since 1842.
5. *Grant*: Francis Grant (1803–78), a Scottish portrait painter who painted many notable Victorians, including Queen Victoria and a number of prominent aristocrats and politicians (*ODNB*).

To Maurice Holzmann 23 December 1863 2055

23rd December 1863

My dear Sir

On my return from Wiltshire[1] this day I received your letter[2] informing me of Her Majesty's desire that His Royal Highness Prince Arthur should be present at the two concluding lectures of our Christmas Course. I trust and hope that those two lectures will not be without interest to the Prince.

You are quite right in understanding that the Course I ventured to recommend[3] would not consist in a succession of questions and answers. My idea was this:—The subject of our Christmas Lectures is Electricity,[4] and I thought it might be agreeable to Her Majesty,[5] and interesting to Their Royal Highnesses the Princesses, if for the first afternoon's proceedings I made a selection from the experiments /embraced/ in the first four lectures of the course. These experiments would be best seen if Their Royal Highnesses were tolerably close to the table at which the experiments would be made. My remarks on the experiments would therefore be naturally conversational in tone, and not in the manner of a lecture room.

I thought of devoting a second afternoon to the subject of sound, selecting such experiments as I thought most likely to interest Their Royal Highnesses. I trust however that I may rely on your kindness to advise me if any other course should seem to you more likely to be agreeable to Her Majesty. Were I to choose something connected with Light a certain experimental splendour would be attainable, which does not belong to other subjects; but to deal with Light I should require an electric lamp & it would be necessary to operate in a room capable of being darkened. The amount of apparatus required in the room would be very small, but a Voltaic battery would have to be placed outside, from which wires would be conveyed to the Electric lamp. If this could be arranged I should be equally ready to converse upon Light and Sound as upon Electricity and sound.

I shall be quite ready to attend on the two days proposed by Her Majesty.[6]

Your practical knowledge of Scientific matters will enable you to give me valuable advice, & I should be much obliged for your friendly counsel.

Believe me | dear Sir | yours very sincerely | John Tyndall

M. Holzmann Esq[ui]r[e] | &c &c &c

American Philosophical Society

1. *Wiltshire*: a county in southwest England.
2. *your letter*: letter missing, although see letter 2052.
3. *the Course I ventured to recommend*: see letter 2052, n. 5.
4. *subject of our Christmas Lectures is Electricity*: see letter 2052, n. 2.
5. *Her Majesty*: Queen Victoria.
6. *two days proposed by Her Majesty*: 30 and 31 January 1864 (see letter 2052, n. 5).

From Maurice Holzmann 25 December 1863 2056

OSBORNE | Dec[em]b[e]r. 25. 1863.

Dear Sir,

I have no doubt that the course you propose to have in the lectures at Osborne[1] will be the best to make them interesting and instructive. From the lectures at the Royal Institution I know very well what beautiful experiments may be introduced when the subject treated of is in any way connected with Light; but unfortunately there would be no possibility of finding here a room capable of being darkened. It will be sufficient to say that there is no gas in the rooms.

The Queen proposes to have the lectures in the evening from 6 to 7 o'clock, and her Majesty very probably would be present herself with the Princesses,[2] Prince Leopold[3] and the Ladies and Gentlemen of the Household.

As the Queen is very anxious that the Princesses should know something more of natural sciences than may be learnt from books alone, and especially as the late Prince Consort[4] took so great an interest in the cultivation of science, her Majesty will very likely have a continuation of the lectures at a later time. It is to be hoped that the Court may be then at Windsor or London, that the arrangements for the lectures might be made easier.

Believe me, | dear Sir, | very sincerely yours | M. Holzmann

John Tyndall, Esq., F.R.S.

RI MS JT/2/11/308

1. *course you propose to have in the lectures at Osborne*: see letter 2052, n. 5.
2. *Princesses*: see letter 2052, n. 5.
3. *Prince Leopold*: Leopold George Duncan Albert (1853–84), the eighth child and youngest son of Queen Victoria and Prince Albert (*ODNB*).
4. *the late Prince Consort*: Francis Albert Augustus Charles Emmanuel (1819–1861), Prince Albert, husband of Queen Victoria (*ODNB*).

To the Editors of the *Phil. Mag.* [December 1863][1] 2057

IV. *On* Laplace's *Correction for the Velocity of Sound.* | *By* JOHN TYNDALL, *F.R.S.*

To the Editors of the Philosophical Magazine and Journal.

GENTLEMEN,

IN the last November Number of the Philosophical Magazine I published a brief "Note on Laplace's Correction for the Theoretic Velocity of Sound,"[2] the main point of which was to show that the extraordinary radiant and absorbent power which recent experiments have proved to be possessed by various gases and vapours, would, in all probability, affect the correction to be applied to the velocity of sound through such gases and vapours. The portion of the note which refers to this point, resting as it does on sure experiments, will, I think, always retain its force and pertinence: this portion terminates with the words "nor could a correct ratio be deduced from the velocity of sound in the medium."

The remaining portion of the note I would ask permission to cancel for the present.[3] I have recently made some experiments with a view to testing the possible influence of radiation; and if the continuance of these should lead to any definite result, they shall in due time be communicated to the Philosophical Magazine.

I am, Gentlemen, | Your obedient Servant, | JOHN TYNDALL.

J. Tyndall, 'On Laplace's Correction for the Velocity of Sound,' *Phil. Mag.*, 27 (1864), p. 41

1. *[December 1863]*: the month and year are given based on Tyndall's mention of the November 1863 issue and that this letter to the editor was published in the January 1864 issue.
2. *I published a brief "Note on Laplace's Correction for the Theoretic Velocity of Sound,"*: J. Tyndall, 'Note on Laplace's Correction for the Theoretic Velocity of Sound', *Phil. Mag.*, 26 (1863), pp. 384–7.
3. *The remaining portion . . . cancel for the present*: here Tyndall was addressing points that Rudolf Clausius had suggested he tackle in his essay (see letter 2048).

To Maurice Pollock [December 1863][1] 2058

Dearest Moss

You are a wonderful engineer: The Admiralty shall know all about it.

Yours aff[ectionate]ly. | John Tyndall

Admit | My Right Honourable friend[2] | Maurice Pollock | to the | Juvenile Lectures | 1863–4 | John Tyndall.

RI MS JT/1/TYP/6/2006
LT Typescript Only

1. *[December 1863]*: the month and year are given based on the reference to Tyndall's Christmas lectures (see letter 2052, n. 2).
2. *Admit | My Right Honourable friend*: this is in effect a ticket for Maurice Pollock to attend Tyndall's Christmas lectures at the RI.

To [George Gabriel Stokes][1] [December][2] 1863 2059

My dear Sir,

I return M^r^. Matthiessen's[3] paper*[4]; and am of opinion that it is worthy of being published in the Philosophical Transactions.

Yours very faithfully | John Tyndall

* I presume this should be Messrs. Matthiessen & Vogt? GS[5]

1863 | Matthiessen | & Vogt | by Tyndall[6]

RS RR/5/144 Tyndall undated 1863

1. *[George Gabriel Stokes]*: the recipient is given based on content. Tyndall had first communicated to Stokes about the paper referred to here in letter 2050.
2. *[December]*: the month is given as this 1863 letter follows letter 2050 to Stokes, dated 7 December.
3. *M^{r}. Matthiessen*: see letter 2050, n. 1.
4. *paper*: see letter 2050, n. 1.
5. *I presume this . . . GS*: this note was added in a different hand, signed 'GS', presumably George Gabriel Stokes.
6. *1863 | Matthiessen | & Vogt | by Tyndall*: written either on an envelope or reverse side of letter.

1864

To Hermann von Helmholtz [late 1863/early 1864][1] 2060

I[2] had the pleasure of making the acquaintance of your friend Fraulein Becker,[3] and also of her charming pupil Miss Mary Strutt,[4] at Lord Belper's[5] last August. I was struck exceedingly by the energy of your friend, who seemed to be a great favorite with the family.

It would be a great pleasure to me to make the acquaintance of Mrs Helmholtz:[6] and if you do not bring her to England with you,[7] I shall make a pilgrimage to Heidelberg to see you and her.

With best wishes | Believe me ever yours truly | John Tyndall

I have closely studied the first part of the Ton-Empfindungen[8] lately—It is admirable.

RI MS JT/1/T/503
LT Typescript Only

---

1. *[late 1863/early 1864]*: the date is given based on content: *Tonempfindungen* was published in 1863, and Helmholtz visited England in April 1864 (see n. 7).
2. Fragment only; letter incomplete.
3. *Fraulein Becker*: not identified; possibly the wife of Ernst Becker.
4. *Miss Mary Strutt*: daughter (dates unknown) of Edward Strutt (see n. 5).
5. *Lord Belper*: Edward Strutt (1801–80), 1st Baron Belper, a British liberal politician. In 1856 he was raised to the peerage as Baron Belper, of Belper, in the County of Derby (*ODNB*).
6. *Mrs Helmholtz*: Anna Helmholtz (née Mohl, 1834–99), a Berlin salonnière and translator of scientific works. She married Herman Helmholtz in 1861. With Clara Wiedemann she translated Tyndall's books on heat and sound into German (*ADB*).
7. *bring her to England with you*: Anna did not join him for the trip to London. Helmholtz left Heidelberg on 11 March 1864, and in the month before his RI lectures, stayed with and visited various acquaintances, attended meetings of the RS, met politicians and diplomats, enjoyed London cultural life, and made a visit to Oxford, among many other activities (D. Cahan: *Helmholtz: A Life in Science* (Chicago: University of Chicago Press, 2018), pp. 293–305). After Easter he delivered a series of six lectures at the RI, 'On the Natural

Law of Conservation of Energy', on 5, 7, 12, 14, 19, and 21 April 1864 (RI Guard Book, vol. 3, pp. 105–6; *Roy. Inst. Proc.*, 4 (1862–66), p. 157), and delivered the Croonian Lecture for the RS and Royal College of Physicians on 14 April (H. Helmholtz, 'On the Normal Motions of the Human Eye in Relation to Binocular Vision', *Roy. Soc. Proc.*, 13 (1863–64), pp. 186–199).

8. *Ton-Empfindungen*: H. Helmholtz, *Die Lehre von den Tonempfindungen als physiologische Grundlage für die Theorie der Musik* (Braunschweig: Druck und Verlag von Friedrich Vieweg und Sohn, 1863). This would be later published in English as *On the Sensations of Tone as a Physiological Basis for the Theory of Music* (London: Longmans, Green, and Co., 1875), translated by Alexander J. Ellis.

To Annette Roche 2 January 1864 2061

John Tyndall | 2nd. January. 1864.

My dear Miss Roche

I rise every morning before day dawn, and work by lamplight in order to cope with the heavy duties which I have at present to perform. But my first act this morning must be to express to you the deep sorrow with which I have heard of the untimely loss of your nephew.[1] Believe me I sympathise with you; I wish I could in any way lighten your grief, but save the sharing of it, this, I fear, is beyond my power. There seems a fatality in these things, and looking back upon my own life, and the "hairbreath escapes" which I have had, makes me wonder that I am here to deplore the loss of another. The inevitable however must be borne, and the best way of meeting such calamity is by the performance of our duties; not to brood, or sorrow too much over what cannot be revoked; but to do the duties which still remain to us. Thus met, time soon softens sorrow and often leaves us all the better for having had to endure it.

But I am a poor hand in dealing with matters so tender as the present one—all I can say is that I partake of your sorrow, and would do a great deal to lessen your affliction in the least degree.

Believe me dear Miss Roche | Yours most truly | John Tyndall

RI MS JT/1/TYP/11/3852–3
LT Typescript Only

1. *your nephew*: not identified.

From Joseph Gill 4 January 1864 2062

XVI. *On the Dynamical Theory of Heat.* | By JOSEPH GILL, Esq. | *To John Tyndall, Esq., F.R.S. &c.*

SIR,

THE kind courtesy with which you received my letter "On the Dynamical Theory of Heat," inserted in the Philosophical Magazine for August last,[1] encourages me to hope that you will be so good as to give to the following remarks, in continuation of the subject, an opportunity of meeting the notice of scientific men through the same channel. My only object is to elicit truth. I have neither time nor inclination nor ability for scientific controversy, and I long shrunk back from the idea of making public the results of my researches on a subject to which I have given much attention during a large portion of the term of a medium lifetime; but at length my conviction of the truth of some of the glimpses of light which have been afforded me in these intricate researches forced itself into utterance; and my anxiety to see the difficulties which have baffled me removed by abler hands has caused me to persevere in the effort to induce some one of the eminent men who have distinguished themselves in the development of the new science of thermo-dynamics to reconsider these knotty points, and clear away if possible, once for all the doubts and obscurity which still cloud some parts of the subject.

Since the date of my last letter I have read your excellent "Lectures on Heat as a mode of Motion,"[2] and I feel that the highest praise is due to the ability with which the subject has been divested of technical and mathematical intricacies and brought within the grasp of mind of every earnest reader of average intellect. By thus exhibiting the results of deep and involved researches in "that simple form in which great truths present themselves to those who thoroughly understand them," an inestimable boon is conferred on numerous readers possessing a knowledge of physics quite sufficient to handle subjects fairly within the range of physical reasoning, though wanting the mathematical knowledge requisite for treating these subjects on the higher scientific plane to which their full theoretical development more properly belongs. It was well remarked by Dr. Lardner,[3] that "the phenomena of heat all admit of being explained without the aid of abstruse reasoning, technical language, or mathematical symbols. The subject abounds in examples of the most felicitous processes of induction, from which the general reader may obtain a view of that beautiful logic, the light of which Bacon[4] first let in on the obscurity in which he found physics involved." I may truly say that after the perusal of your book I understand the subject better than I did before; but still some of my doubts are unresolved, as you will perceive from the following statements.

Although a fixed equivalence undoubtedly exists between the heat generated by friction and the energy expended in producing the friction, *experiment proves* that elastic fluids suffer no diminution of heat in the act of producing work, but only a fall of temperature. I have long perceived the difficulties of admitting this hypothesis, as it involves the apparent paradox of a continual investment of energy in the production of molecular motion, of the reconversion of which back again into energy no account can be given. But the results of numerous and varied experiments leave me no alternative, and I must repeat my conviction that heat *runs down* but does not disappear in the production of work. Even admitting that quantitative heat disappears in the process, still many reasons might be given for doubting that any fixed equivalence exists between the heat disappearing and the work done; but for the present I will confine myself to the simple consideration that, with the disappearance of equal quantities of heat in the expansion of air of constant initial density, different quantities of work should be produced according to the mode of conducting the process.

Imagine a cylinder of 2 cubic feet capacity to contain 1 cubic foot of air of atmospheric pressure at the temperature of 32°F. confined by a piston resting at half stroke, supposed to be without weight or friction. Let the piston be fixed in this position, and heat applied to the air until its temperature is raised 490° above the initial point, and let the quantity of heat required to produce this effect be represented by 100. In this process the air is simply heated, and although statical tension is developed, no dynamical effect is produced, it is merely a preparation for work—a transfer of heat from one substance to another if the heat is taken either ready made, or produced chemically, or a change of some form of energy into heat by friction; but, as long as no work is done by the least air, things are considered to remain in a state of balance or equilibrium.

Suppose the air to be again at 32°, and the piston free to move when any disturbing cause should be applied. If heat be gradually communicated to the confined air it will expand, raising the piston and with it the superincumbent column of the atmosphere, which may be taken in round numbers at 15 lbs. per square inch. When 490 degrees of temperature have been given to the confined air, its volume will be doubled, and if the piston be 1 square foot area, it will have moved upwards through 1 foot; and it would be said that in its ascent it had performed the work of 144 square inches x 15 lbs. = 2160 lbs. raised 1 foot. In this process 1 cubic foot of air has been heated 490 degrees, as in the former case, but the circumstances are different. In the former case the air was confined to constant volume, and tension was induced, but no work was done. In the present case the air was free to expand, no tension was induced, and work was performed; and as heat was the only energy employed in the operation, we should naturally expect an excess of heat to be used equivalent to the work done. In fact we find that 142 of heat has been thrown into the air in this case, while to heat the air to the same degree under constant volume 100 sufficed; it is therefore said that the difference, represented by 42, has done the work.

The work assumed to be done in the case under consideration, by raising the column of the atmosphere resting on the piston, is not available, it cannot be detached from the apparatus—a circumstance which might induce doubt as to its reality. If it possesses the entity which it appears to have, it should not be impossible to detach it from the matrix where it exists, and to exhibit it in some other form, thus removing the suspicion of imperceived coefficients acting in the phenomena. By altering the circumstances, the work which exists as potential energy in the raised atmospheric column ought, if all real, to be convertible into an equivalent of actual energy, or work done in another shape. By withdrawing the heat from the air in the cylinder, the piston will descend with a power corresponding to the difference of pressure between the total atmospheric pressure outside and the diminished pressure inside; and this power will be equal to half the atmospheric pressure at the beginning of the piston's stroke, diminishing to nothing at the end, when the confined air will, as a first, occupy 1 cubic foot at atmospheric pressure and 32° temperature. The amount of power obtained in this operation will be less than a quarter of an atmosphere on the area of the piston moved through 1 foot. For simplicity of reasoning we may suppose it to be one quarter of an atmosphere.

In the working of a theoretically perfect thermic engine, *all the heat consumed*, or which permanently disappears, is supposed to be converted into work in the proportion of Joule's equivalent of 772 foot-pounds for every unit of heat, neither more nor less. By means of a *regenerator* (as used in Stirling's and Ericsson's air-engine[5]) of theoretically perfect action, all the heat withdrawn from the air as above described could be stored up in such a way as to be transferred back again to the same air under the same circumstances as before, and the operation could be repeated over and over again indefinitely, with the production of available work, provided that at each stroke of the piston a fresh quantity of heat represented by 42 be thrown into the apparatus. The heat consumed by 1000 strokes of the piston would be 1000 times 42, and the work produced would be 1000 quarters of an atmosphere on the area of piston raised 1 foot.

In the former case 1000 repetitions of the operation, with fresh air and fresh heat each time, would equally consume 1000 times 42 of heat, the 100 of heat remaining unchanged at the end of each operation; but if it is assumed that in each cycle of heating and cooling, expanding and contracting the confined air, work is actually done equal to raising the atmosphere column on the piston through 1 foot (*which is the apparent effect*), we must allow that the work done by 1000 repetitions of the operation would be 1000 atmospheres on the piston raised 1 foot. How is this apparent discrepancy to be accounted for?

Another serious difficulty in the dynamical theory is the question of the specific heat of air. Regnault's experiments appear to prove that the specific heat of air is constant at different densities. This result is in accordance with the dynamical theory, and in fact the theory could not stand unless this assumption were admitted. My own experiments have given very different results; but as it might be thought presumptuous to place these experiments in opposition to those of such a high authority as Regnault, I beg to submit the following reasoning to candid scrutiny.

Imagine a vacuum to be made in a cylinder by drawing up its piston by a descending weight, in which operation the work done by the descent of the weight gives an equivalent amount of potential energy in the tendency possessed by the atmospheric column raised by the piston to fall back into the vacuum; and so far things are in a state of equilibrium. Imagine further a second cylinder, twice the size of the former one, full of atmospheric pressure air, which should be compressed into half its original volume by some exterior force applied allowing the developed heat to pass out during the compression. In this operation work has been expended, and heat has been produced, which heat, according to the dynamical theory, would be considered equivalent to the work expended, as by allowing the air to expand back under moderated resistance, restoring to it the heat it had lost, the work done in the compression would be recovered. Now suppose a communication to be made between the two cylinders—the air will rush into the vacuum, and as the contents of the two vessels now filled by the expanded air are equal to the volume which the air originally occupied, it is evident that the pressure will also be equal to what it was before the compression, namely atmospheric pressure. It is clear that no power can be obtained from the expanded air, as it is in equilibrium with the surrounding atmosphere; but the energy which might have been availed of by allowing the atmosphere to fall into the vacuum, and thus produce work, has disappeared. It is impossible to conceive of the annihilation of energy, and we must therefore suppose that it has taken some other shape in the confined air. The dynamical theory does not allow the existence of energy in elastic fluids under any other form than that of molecular motion or heat, consequently there should be more heat in the expanded air; but as the experiments of Gay-Lussac[6] and Joule have proved that in the expansion of air into a vacuum the temperature of the mass remains constant, we are forced to admit that the specific heat of the expanded air is greater than that of the air in a compressed state: in other words, the susceptibility of air to the influence of heat as affecting the thermometer increases with increase of density.

According to the dynamical theory, a common non-condensing steam-engine, working without expansion, utilizes only about 1/40 of the theoretical power of the steam. By the addition of a condenser this effect is doubled, or the engine gives about 1/20 of useful effect. Such were the deductions published by Regnault, conveying to abstract theorists the idea that the steam-engine is one of the most imperfect of prime movers. Some of the results obtained by M. Hirn[7] from the practical working of large steam engines were as high as 1/6, none lower than 1/10; and 1/8 may be taken as a possible degree of practical efficiency—better certainly than Regnault's deductions, but still showing that the steam-engine under the most favourable circumstances is so far from perfection as to throw away 7/8 of the theoretical power of the steam. However, the marked discrepancy between the theoretical deductions and the practical results showed that something was wrong in the theoretical calculation; and the excess of work obtained in Hirn's results has been accounted for by supposing that some of the steam condenses in the act of expansion in the working of the engine, and that this condensation represents the heat converted into work beyond the theoretical calculation which supposed all the steam to pass into the condenser in the vaporous state. This explanation involves the idea that steam at different densities contains a variable quantity of heat in proportion to the work it can perform; and this inference obtains some slight support from the results of experiment, showing that the total heat of steam increases with increase of density, but by no means enough to account for the greater amount of work it can perform by expansion.

One of M. Hirn's experiments with an engine cutting off the steam at 1/3 of the stroke showed at the same time a production of work and a creation of heat. In another of his experiments steam was thrown into a sheet-iron vessel immersed in cold water, and it was found that the heat thrown out by the condensing steam was more than that due to its constitution as deduced from calculations of acknowledged correctness. Moreover the fact that steam becomes superheated in the act of free tumultuous expansion is not disputed, though it involves the idea of some other form of energy existing in the steam besides molecular motion. The dynamical theory necessitates the assumption that steam is condensed during expansion in the production of work, as theoretically shown by Clausius and Rankine; and another of Hirn's experiments would seem to corroborate this assumption. A copper tube upward of 6 feet long and 6 inches diameter, with strong glass ends, was put in communication, by means of stopcocks, with a steam-boiler at one end and with the atmosphere at the other. The tube full of high-pressure steam was perfectly transparent; but the moment the cock communicating with the atmosphere was opened, and part of the steam rushing out allowed the remainder to expand, a cloud was formed in the tube: the steam became quite opake, showing, as it is supposed, the condensation of the steam which should accompany expansion.

On this experiment, I would remark that a film of moisture must exist on the interior surface of the tube, though at the same time the contained mass of steam may be dry, or free from particles of water mechanically suspended in it. The temperature of the copper must be somewhat lower than that of the high-pressure steam, but much higher than the temperatures of the steam at low pressure; consequently, on the sudden lowering of pressure in the tube when the steam is allowed to blow off, an instantaneous "flashing" into steam of the water on the now comparatively hot metal must ensue, with a large amount of *priming*, or dispersion of the water in the shape of fine spray, probably quite sufficient to account for the opacity observed. The experiment should be made with the tube enclosed in a case with glass heads, filled with steam of the same temperature as the steam inside, to prevent the deposition of moisture on the interior surface of the tube. This precaution would in a degree obviate a serious source of error, though still it should be borne in mind that glass has an attractive power for moisture which prevents its free evaporation at temperatures considerably higher than the regular boiling point under any given pressure.

I must be allowed to express my opinion that the experiments of M. Hirn are of great practical value, and I submit that the importance of the facts which they furnish is not depreciated by the circumstance that the experimenter, apparently borne down by scientific authority, recently presented to the Academy of Sciences an interpretation of his results in accordance with the dynamical theory. It should also be borne in mind that the results of Hirn's experiments coincide in the main with those of Séguin made more than twenty five years ago, when his reasoning on the mode of action of the steam-engine pointed out the necessity of supposing a *conversion of heat into work*, of which, nevertheless, he could find no appreciable indication in the results of the numerous experiments made by him with this express purpose, and conducted with the attention and accuracy suggested by his great scientific knowledge and superior practical skill in thermo-dynamics as then known.

In my experiments above alluded to on the specific heat of air at different densities, I found that certain proportions of surface in the apparatus relative to the masses of air, their condition, and the times occupied by their passage through the calorimeter, were essential to correct results; and I venture to think that some concealed source of error must have existed in Regnault's experiments on this subject, undetected even by his extraordinary ability in similar researches.

I would also repeat that my long practical acquaintance with the general phenomena of steam, as well as the results of direct experiment, have tended to convince me that the total heat of steam is a constant quantity. I believe that this important point is again engaging the attention of physicists; and it may be well to remark that in the condensation of high-pressure steam there is a tendency to produce heat from the effects of tumultuous expansion, which may seriously affect the results, showing, as in some of Hirn's experiments, more heat than is due to the constitution of the steam. In the cooling of gases under ordinary circumstances there is no change of state, and the contraction of the volume should result from a gradual shrinking of the sphere of action of each particle in the mass, a process which suggests no idea of tumultuous motion or persuasive action of the particles. In saturated vapours each particle must possess the precise specific amount of heat requisite for its existence in the vaporous state; if heat be added, the vapour becomes a gas; if heat be withdrawn, some of the particles must collapse into liquid. This collapse should evidently be sudden and each particle in assuming the liquid state should, for the moment, leave an almost perfect vacuum in the space it occupied in contact with the cold walls of the condenser. The void is instantly occupied by a fresh particle of vapour, which in its turn collapses, and so the process continues as long as vapour remains to be condensed. It may be conceived that the particles of vapour thus rapidly precipitated into a vacuous space, bounded on the outside by the hard metallic walls of the condenser, must strike with violence against these surfaces, and generate heat by this molecular battering.

It may be replied that the internal work done by the remaining particles of vapour in forcing their fellows down the precipice of the vacuum should cause an amount of chilling of the mass by an actual disappearance of heat equal to that generated by the percussive action of the particles against the sides of the condenser. To this I would answer that, until good account can be given of the actual excess of work-producing power which high-pressure steam is allowed to possess as compared with steam of low pressure, weight for weight, and which is *apparently lost* by the quiet condensation of the steam in the vessel which contains it, we may suppose some unacknowledged form of energy to exist in the steam distinct from mere molecular motion, though always acting in cooperation with it, and more than sufficient to cause the violent fall of the vapour particles against the walls of the condenser, thus generating heat in excess of what is fairly due to the constitution of the steam.

The very nature of the steam suggests the idea that, as far as quantitative heat is concerned, each individual particles should have a fixed specific temperature the general temperature of any given mass resulting simply from the number of particles in a given space, or, in other words, the distance of the particle from each other, and consequently the amplitude of their molecular motions. Thus it is certainly reasonable to conceive of a fall of temperature in expanding steam without any idea of absolute loss of heat. However this may be, I would beg to impress on the attention of physicists who may be making further experimental researches on this interesting point, the importance of guarding against all probable causes of error which can be foreseen, keeping in mind the results obtained by Seguin and Hirn, and also the experiments of Siemens, in which high-pressure steam seemed to *superheat itself* by tumultuous expansion.

I remain, Sir, | Your most obedient Servant, | JOSEPH GILL | Palermo, January 4, 1864.

Phil. Mag., 27 (1864), pp. 84–91

1. *Philosophical Magazine for August*: J. Tyndall, 'On the Dynamical Theory of Heat', *Phil. Mag.*, 26 (1863), pp. 109–16.
2. *"Lectures on Heat as a mode of Motion"*: *Heat as a Mode of Motion.*
3. *Dr. Lardner*: Dionysius Lardner (1793–1859), an Irish scientific writer (*ODNB*). D. Lardner, 'Dr. Lardner on Heat', *Mechanics Magazine*, 18 (1833), pp. 365–8.
4. *Bacon*: Francis Bacon (1561–1626), a prominent British philosopher of science and author of *Novum Organum* (1620) (*ODNB*).
5. *Stirling's and Ericsson's air-engine*: a hot air engine developed by Reverend Robert Stirling (1790–1878), a Scottish inventor (*ODNB*); simultaneously John Ericsson (1803–89), a Swedish-American inventor (*ANB*), developed a similar engine. A regenerative heat exchanger, or more commonly a regenerator, is a type of heat exchanger where heat from the hot fluid is intermittently stored in a thermal storage medium before it is transferred to the cold fluid.
6. *Gay-Lussac*: Joseph Louis Gay-Lussac (1778–1850), a French chemist and physicist, known for his discovery that water is made of two parts hydrogen and one part oxygen (*CDSB*).
7. *M. Hirn*: Gustave-Adolphe Hirn (1815–90), a French physicist, astronomer, mathematician, and engineer known for his measurements of the mechanical equivalent of heat and applied science in the practical development of steam engines (*CDSB*).

From Juliet Pollock 14 January 1864 2063

AUDACTER ET STRENUE[1]

Are you good or bad, and how is it that you have disappeared from our world without a word or a sign, and what are you about and what is Bromine about and how are you, and may I wish you a happy new year, and many more to come for use and enjoyment? And will you answer me and will you care for what I have to tell? Know then that we have all been laid up with colds and that poor Moss[2] coughs to day still so much that he is not fit to leave home (although not ill) and that therefore I remain at home while my husband[3] and elder sons[4] go to S^t^. Julians[5] & the good Sarah Spedding[6] stays with me: if Moss is pretty well on Friday I shall go down with her to S^t^. Julians just for two days leaving him in Polly's[7] care in order that my Aunt may not be altogether disappointed and that Sarah S. may have still the pleasure of a little outing. Now Sir—does all that signify any thing to you, or are you indifferent; as I am not yet evaporated into a gas—and we are all mere common solid bodies?

Charades in active preparation! Will you play a ruffian who has a part of dumb show & who will be inserted in the bills as <u>A ruffian</u>—<u>by a ticket of leave man</u>? Nights of performance | 13th. & 14th. Jan[uar]^y^. —

So! M^r^. Savory[8] is going to give us four lectures on life & death.[9]—what can he tell us of that, that we long to know? Can he say why one goes & another stays, why one career happy & useful is suddenly cut short, & another drivelling & wretched is prolonged to the extremest verge?—no—he cannot—he certainly cannot & therefore I suppose will only deal with the thing physically. But why do I talk so much to you who don't care—just as if the needle would deflect for me?

Some people talk of the value of a solid friendship but not you!—If I were but vapourish instead of /being/ | still most truly your Juliet Pollock

RI MS JT/1/P/223

1. *AUDACTER ET STRENUE*: see letter 1936, n. 2.
2. *Moss*: a nickname for Maurice Pollock.
3. *my husband*: William Frederick Pollock.
4. *elder sons*: Frederick and Walter Pollock.
5. *S^t^. Julians*: see letter 1954, n. 14.
6. *Sarah Spedding*: see letter 1936, n. 16.
7. *Polly*: probably a governess for the Pollock's son Maurice, then aged 6 years old (born 28 March 1857).
8. *M^r^. Savory*: William Scovell Savory (1826–95), a British surgeon and former Hunterian Professor of Anatomy and Physiology (1859–61) (*ODNB*).
9. *four lectures on life & death*: Savory had delivered four lectures 'On Life and Death' at the RI in the spring of 1863 (*Roy. Inst. Proc.*, 3 (1858–62), p. 542). These were soon after published as W. Savory, *On Life and Death: Four Lectures delivered at the Royal Institution of Great Britain* (London: Smith, Elder, & Co., 1863).

To Hermann Helmholtz 16 January 1864 2064

Royal Institution | 16th. January 1864.

My dear Helmholtz,

The reading of your excellent work on "Tonempfindungen"[1] has excited in me the desire to have a syren[2] such as you have there described. Sauerwald[3] I think is the maker of it. I shall feel exceedingly obliged to you if you would have the kindness to order the instrument for me: I think the cost of it is about 60 thalers, but I should be willing to go to a little further expense if you thought it desirable.

I am also writing to Kirchhoff[4] to ask him to be good enough to order a Spectroscope for me. Both instruments are for the Royal School of Mines;[5] and the Government requires of us that we should have our accounts sent in before the end of March. If therefore Sauerwald cannot have the instrument constructed by the end of March he might send me his bill (quittert) and then I could send in the bill and allow him a little more time for the construction of the Syren.

We are all looking forward with pleasant anticipation to your course of lectures.[6] If I can do anything to help you it will give me the greatest pleasure to do so.

Believe me | Yours most sincerely | John Tyndall

RI MS JT/1/T/481
LT Typescript Only

1. *"Tonempfindungen"*: see letter 2060, n. 8.
2. *syren*: a siren, or syren (as Tyndall often spelled it); an acoustical instrument (invented by Cagniard de la Tour in 1819) for producing musical tones and used in numbering the vibrations in any note (*OED*). The specific siren Tyndall referred to here was a double siren, composed of two of Dove's polyphonic sirens with discs attached to a common axis. On each of the two discs of the sirens are four rows of holes, which can be either blown separately or together.
3. *Sauerwald*: Ferdinand Sauerwald, a German scientific instrument maker whose workshop was located in Berlin. He constructed apparatuses for Gustav Magnus, John Tyndall, Emil du Bois-Reymond, and Hermann Helmholtz, among others (D. Cahan: *Helmholtz: A Life in Science* (Chicago: University of Chicago Press, 2018), pp. 186, 264).
4. *Kirchhoff*: letter missing; see letters 2065 and 2070.
5. *Royal School of Mines*: see letter 1987, n. 4.
6. *your course of lectures*: see letter 2060, n. 7.

From Gustav Kirchhoff 19 January 1864 2065

Heidelberg den 19ten Jan. 1864.

Verehrter Freund!

es ist mir ein grosses Vergnügen für Sie hier einen Auftrag ausrichten zu können. Dennoch habe ich bis jetzt noch nicht Ihren Spektral-Apparat bestellt; derjenige den ich zu meinen Beobachtungen benutzt habe, hat 900 francs gekostet, also fast das Doppelte von dem Preise, den Sie in Ihrem Briefe mir bezeichnen; ich weiss nicht, ob Ihnen das nicht zu viel sein wird. Wenn Sie hierüber mich mit einer Zeile aufklären wollen, so will ich augenblicklich die Bestellung ausführen. Ich denke mich dabei an Steinheil in München zu wenden, der meinen Apparat gemacht hat, und dessen Prismen, wie ich glaube, noch immer unübertroffen sind. Dass der Apparat im März fertig sein wird, bezweifle ich sehr; aber die Rechnung kann Ihnen Steinheil gewiss dann zusenden.

Noch möchte ich fragen, ob die Rechnung ausgestellt sein soll „for the Government-School of Mines"?

Mit herzlichen Grüssen | Ihr | GKirchhoff

Heidelberg, 19th Jan. 1864.

Honoured friend!

It is a great pleasure for me to place an order for you here. Nevertheless, I have not yet commissioned your spectral-apparatus; the one that I used for my observations cost 900 francs; that is, almost twice the price that you indicated in your letter;[1] I am not sure whether this will not be too much to you. If you would confirm this in a note to me, I will instantly place the order. I intend to contact Steinheil in Munich,[2] who made my apparatus and whose prisms, I believe, are still unsurpassed. I very much doubt that the apparatus will be ready in March; but Steinheil will certainly be able to send you the invoice then.

Further, I would like to ask whether the invoice should be issued as "for the Government-School of Mines"?

With kind regards | Your | G Kirchhoff

RI MS JT/1/K/8

1. *your letter*: letter missing, but see letter 2064.
2. *Steinheil in Munich*: Carl August von Steinheil (1801–70), a German physicist, inventor, engineer, and astronomer. In 1854, he founded C. A. Steinheil & Söhne in Munich, an optical-astronomical equipment company (*CDSB*). Henry Bence Jones described

Steinheil's apparatus as one 'for producing the spectrum of different substances' that 'gave a new method by which the action of magnetic poles upon light could be tried' (H. Bence Jones, *The Life and Letters of Faraday*, 2 vols (London: Longmans, Green, and Co., 1870), vol. 1, pp. 449–50).

From William Snow Harris 25 January 1864 2066

6 Windsor Villas | Plymouth | January 25th 1864

My dear Tyndall,

Allow me to place in your hands, a copy of my paper on Intensity[1] lately printed in the last number of the Phil. Mag. which I trust may prove of some little interest to you. I am more than ever satisfied that very erroneous views have been entertained relative to Electrical Force, and in whatever that force consists, we have no evidence whatever of its being susceptible of change in its elementary state; thus gravity is the same in a feather, as in a lump of lead, it is not more intense in the one, than in the other.

I am more confirmed in my views since reading the interesting and instructive review of your lectures, on the "Dynamical Theory of Heat" printed in the last number of the Edinburgh review.[2] If you have a few minutes at any time to spare I should like you to cast your eye over the little geometrical diagram I have given in my paper, in illustration of the operation of the Leyden Jar.

The advances in Philosophical research will soon oblige the Continental philosophers to abandon the idea of Electric fluids, susceptible of change in what they call Intensity, and such like

I am about to send another paper to the R. S.[3]

Believe me to be | My dear Tyndall | ever your sincere friend | W. Snow Harris

In a former letter when adverting to your little misunderstanding with Thomson relative to your exp[eriments] &c. concerning the development of heat by motion, I called attention to <u>Rennie</u> papers[4] which I have since found and enclose in my paper on Intensity thinking it may be of interest to you.

W. S. H.

Should my life be spared[5] I shall hope to see you this way in the course of the ensuing summer.

RI MS JT/1/TYP/2/481
LT Typescript Only

1. *my paper on Intensity*: W. S. Harris, 'On the Correct Interpretation of the Electrical Terms Intensity and Tension', *Phil. Mag.*, 26 (1863), pp. 504–15.

2. *the "Dynamical Theory of Heat . . . the Edinburgh review*: 'Review of Tyndall's *Heat Considered as a Mode of Motion*, Joule's *On the Mechanical Equivalent of Heat*, Mayer *On Celestial Dynamics*', *Edinburgh Review*, 119 (1864), pp. 1–13.
3. *another paper to the R. S*: possibly W. S. Harris, 'Further inquiries concerning the laws and operation of electrical force', *Roy. Soc. Proc.*, 13 (1864), pp. 364–75 (received by the RS on 8 June 1864).
4. *former letter when adverting . . . attention to Rennie papers*: letter missing. There appears to be no other letter from Snow Harris in the Tyndall correspondence which referred to papers by Rennie. However, in both previous letters in this volume from Snow Harris (1984 and 1986), Thomson is discussed. The Rennie referred to is probably George Rennie (1791–1866), a British mechanical and civil engineer (*ODNB*).
5. *Should my life be spared*: Snow Harris survived until 1867.

From Thomas Sterry Hunt[1] 1 February 1864 2067

ON THE EARTH'S CLIMATE IN PALAEOZOIC TIMES. | *To Dr. John Tyndall, F.R.S., &c.* | Montreal, February 1, 1864.

My Dear Sir,

IN a Note which appeared in the Philosophical Magazine for October 1863 (page 323),[2] I endeavoured to explain the warmer climate of the earth in palaeozoic times by the aid of your remarkable researches on radiant heat, by which you have shown that small quantities of various gases, when diffused in the atmosphere, greatly retard cooling by radiation. I suggested that the large amount of carbon which was then diffused in the atmosphere as carbonic acid, and which has since been separated in the forms of carbonate of lime and carboniferous matters, must have had a powerful influence upon the temperature of the earth's surface during the palaeozoic period; I also indicated the influence of marsh-gas from the decay of the abundant vegetation of the period. I was not at the time aware that my suggestions had been anticipated by you in the Bakerian Lecture for 1861,[3] where you say that, from its influence on terrestrial radiation, all variation in the amount of aqueous vapour in the atmosphere must produce changes in climate; and you then add, "Similar remarks would apply to the carbonic acid diffused through the air, while an almost inappreciable admixture of any of the hydrocarbon vapours would produce great effects on the terrestrial rays, and corresponding changes of climate. It is not, therefore, necessary to assume alterations in the density and height of the atmosphere to account for the different amounts of heat being preserved to the earth at different times; a slight change in its variable constituents would account for this. Such changes, in fact, may have produced all the mutations of climate which the researches of geologists reveal." (Philosophical Magazine, Oct. 1861, p. 277)

In the American Journal of Science for November 1863 (p. 398), I have in this connexion referred to a paper by the late Major E. B. Hunt,[4] United States Engineers, "On Terrestrial Thermotics,"[5] published in the Proceedings of the American Association for the Advancement of Science for 1849. In this paper Major Hunt argues that the temperature of the earth's surface increases with the weight of the barometric column, and that, as the atmospheric mass must have been greater in the earlier geological periods by the amount of carbon and of carbonic acid since abstracted, the temperature of the earth must then have been higher. To this mass Prof. Dana[6] would add the abundant moisture of the insular climate of former times. The augmentation of barometric pressure dependent upon these additions to the atmosphere would, however, probably be very inadequate to explain the considerable differences in climate of which geology affords us evidence; and it is only in the light of your recent discoveries that the chemical constitution of the early atmosphere of our globe enables us to explain the warmth of ancient climates.

The object which I have in writing this note is thus not only to set forth major Hunt's early attempts to solve the problem, but to call public attention to the fact that what I conceive to be its true solution was first suggested by yourself.

I have honour to be, my dear Dr. Tyndall, | Very faithfully yours, | T. Sterry Hunt.

Phil. Mag., 27 (1864), pp. 236–7

1. *Thomas Sterry Hunt*: Thomas Sterry Hunt (1826–1892), an American geologist and chemist (*ADB*).
2. *Note which appeared . . . (page 323)*: T. S. Hunt, 'On the Earth's Climate in Palaeozoic Times', *Phil. Mag.*, 26 (1863), pp. 323–4.
3. *the Bakerian Lecture for 1861*: J. Tyndall, 'On the Absorption and Radiation of Heat by Gases and Vapours, and on the Physical Connexion of Radiation, Absorption, and Conduction', *Phil. Trans.*, 151 (1861), pp. 1–36 (read at the RS on 7 February 1861; Hunt was referring to pp. 28–9).
4. *Major E. B. Hunt*: Edward Bissell Hunt (1822–63), an American engineer who died of suffocation as the result of an accident, while engaged in the prosecution of investigations incident to his profession, and immediately connected with the American Civil War. Hunt worked primarily on rocket torpedoes (G. W. Cullum, *Biographical Register of the Officers and Graduates of the U.S. Military Academy at West Point, N.Y.*, 3rd edn, 3 vols (Boston and New York: Houghton, Mifflin and Company, 1891), vol. 2, p. 209).
5. *"On Terrestrial Thermotics"*: E. B. Hunt, 'Remarks on Terrestrial Thermotics', in *Proceedings of the American Academy for the Advancement of Science. Second Meeting, held at Cambridge, August, 1849* (Boston: Henry Flanders & Co., 1850), pp. 135–40.

6. *Prof. Dana*: James Dwight Dana (1813–95), an American geologist, mineralogist, volcanologist, and zoologist. He made pioneering studies of mountain-building, volcanic activity, and the origin and structure of continents and oceans around the world. Dana was the author of *System of Mineralogy* (1837), *Manual of Mineralogy* (1848), and *Manual of Geology* (1863) (*ANB*).

From François-Napoléon-Marie Moigno 1 February 1864 2068

Mon cher Monsieur

Je vous ai fait envoyer vendredi dernier trois nouvelles feuilles de ma traduction. Je désire instamment que vous y jetiez un coup d'œil pour me les renvoyer ensuite, pour attendre les bonnes feuilles imprimées.

Je vous remercie de la confiance que vous mettez en moi, et vous pouvez être sûr que je l'ai méritée. Il est impossible de mettre plus de soin, plus de patience, plus d'amour aussi à un travail quelconque que j'en ai mis à traduire votre admirable ouvrage. Plus je relis ma traduction, plus j'avance, et plus je suis content; je vous dirai même franchement que je me fais un titre de gloire de ma lutte avec votre texte anglais, lutte dans laquelle je crois être resté vainqueur, sauf quelques imperfections que je conquerrai par des *[1 word illeg]*.

Je terminerai mon travail cette semaine; et je vous prie instamment de m'indiquer courrier par courrier ce que je dois ajouter comme appendice à ce que contient déjà le volume anglais. Faut-il faire imprimer la *[1 word illeg]* de *[1 word illeg]*; les lettres à Mr. Thomson et Joule, etc, etc.

Répondez-moi bien vite, car nous n'avons plus que quelques instants pour achever.

Recommandez bien mes Mondes à mes amis, pour assurer de plus en plus leur succès. J'avais espéré que l'excellent docteur Carpenter continuerait son abonnement, mais il m'a averti qu'il ne le pouvait pas. Au reste le succès a dépassé mes espérances et je suis content. Grâce au Monde la première édition de <u>La Chaleur</u> considérée comme une forme de mouvement sera bientôt épuisée quoique tirée à deux mille exemplaires, et nous en ferons promptement une autre. Je compte sur un très grand succès.

Pourrait-on avoir le texte des deux leçons faites par vous *[à]* Osborn.

Aura-t-on pour les traduire au fur et à mesure les douze leçons d'optique que vous devez commencer bientôt.

A vous de cœur | l'abbé F. Moigno | 1er février 1854 | 2 rue d'Erfurt

My dear Sir,

Last Friday I sent you three new pages of my translation.[1] I urge you to glance them over straight away and then send them back to me, so we can proceed to the good printed copies.

I thank you for the trust that you are placing in me, and you can be certain that I have earned it. It would be impossible to put more care, more patience, more love into any kind of work than what I have done to translate your admirable book.[2] The more I read over my translation, the more I progress and the more I am satisfied. I will tell you frankly that I take pride in my struggle with your English text; a battle in which I believe I will remain the winner, except against a few flaws that I will [conquer] through *[1 word illeg]*.

I will finish my work this week, and I urge you to tell me, post by post, what I add as an appendix to the existing contents of the English volume. Should the note about *[1 word illeg]* of *[1 word illeg]* be printed; the letters to Mr. Thomson and Joule, etc, etc.

Answer me quickly because we only a short time to complete it.

Please recommend my Mondes[3] to my friends, to further ensure its success. I was hoping that the excellent Dr. Carpenter[4] would continue his subscription, but he warned me that he could not. As for the rest, our success has been unbelievable, and I am happy for it. Thanks to the Mondes, the first edition of Heat considered a mode of motion[5] will soon run out, despite printing two thousand copies, and we will soon be doing another print run. I am expecting a huge success.

Could we have the text of the two lessons that you did in Osborn?[6]

Will we have, to translate them one by one, the twelve lessons on optics[7] that you will surely be starting soon?

Yours truly | Father F. Moigno | 1 February 1854 | 2 Erfurt Street

RI MS JT/1/M/103

1. *Last Friday I sent you three new pages of my translation*: probably a reference to Moigno's translation of *Heat as a Mode of Motion*. See letters 1978, 1990, 1996, and 1999.
2. *translating your admirable book*: the French translation of Tyndall's *Heat as a Mode of Motion* appeared as J. Tyndall, *La chaleur considérée come un mode de mouvement*, trans. F. Moigno (Paris: Étienne Giraud, 1864).
3. *my Mondes*: see letter 1978, n. 1.
4. *Dr. Carpenter*: William Benjamin Carpenter (1813–85), a prominent physiologist and devout Unitarian who dedicated much of his intellectual life to advocating for the compatibility of science and religious belief. He had many friends among the scientific community. Although in his youth his views on natural law had been viewed as theologically suspect, he rose to become one of the most eminent men of science of his day. He was awarded the RS's Royal Medal in 1861, was elected FRS, succeeded Faraday as Lecturer at the RI, and was appointed Registrar of the University of London in 1856. From 1869 he was a prominent member of the Metaphysical Society. While he was on friendly terms with Huxley and Clifford, he and Tyndall did not enjoy a friendly relationship (*ODNB*).

5. *the first edition of* <u>*Heat*</u> *considered a mode of motion*: see n. 2.
6. *two lessons that you did in Osborn?*: probably a reference to the lectures that Tyndall gave at Obsorne House to the Royal family. See letters 2052, 2055, and 2056.
7. *the twelve lessons on optics*: Tyndall delivered 'Twelve Lectures On Experimental Optics' at the RI from 19 January to 25 February 1864 (*Roy. Inst. Proc.*, 4 (1862–6), p. 157). Moigno may not have known that Tyndall had already given this lecture series.

From George Biddell Airy 3 February 1864 2069

Royal Observatory, Greenwich, | London, S.E. | 1864 February 3

My dear Sir,

Would you do the favour to the Astronomical Society's[1] Club and myself to dine with us on Friday 5th at 5 ½ h, at the Freemason's Tavern?[2]

I have few opportunities of meeting you, and should be glad to use this.

I am, my dear Sir, | Yours very truly | <u>G B Airy</u>

I have but just returned from the country, or you should have had this before.[3]

Professor Tyndall

RI MS JT/1/A/38
RI MS JT/1/TYP/1/28

---

1. *Astronomical Society*: the Royal Astronomical Society formed in 1820 to support astronomical research.
2. *Freemason's Tavern*: 61 Queen's Street London, built to host Freemasons and other societies in the eighteenth century.
3. *I have but just returned . . . had this before*: this is written at the top of the letter.

From Gustav Kirchhoff 3 February 1864 2070

Heidelberg, den 3ten Febr. 1864.

Verehrter Freund!

gleich nachdem Sie mir die Vollmacht dazu gegeben hatten, habe ich bei Steinheil in München für Sie einen Spectralapparat, ähnlich dem meinigen, bestellt. Steinheil schreibt mir eben, dass er den Apparat für Sie bereits in Arbeit gegeben habe, dass er hoffe, im April ihn Ihnen liefern zu können, und dass er nicht vergessen werde, im Maerz Ihnen die Rechnung zukommen zu lassen. Für Ihre Photographie, über die ich mich sehr gefreut habe, danke ich Ihnen vielmal, und sende Ihnen die meinige.

Mit herzlichen Grüssen | Ihr | G Kirchhoff. Heidelberg, Febr. 3, 1864.

Dear friend!

As soon as you gave me your authorisation,[1] I ordered a spectral apparatus similar to mine from Steinheil in Munich.[2] Steinheil just wrote that work on your apparatus has already begun, that he hopes to deliver it in April, and that he will not forget to send you the invoice in March. Many thanks for your photograph. I was very pleased to receive it and send you mine.

With kind regards | Your | G Kirchhoff. Heidelberg, February 3, 1864.

RI MS JT/1/K/9

1. *As soon as you gave me your authorisation*: letter missing.
2. *Steinheil in Munich*: see letter 2065, n. 2.

From Friedrich Reusch 28 January 1864[1] 2071

XXXIII. *On Some Properties of Ice.* | *By* Professor REUSCH *of Tübingen.* | *To John Tyndall, Esq., F.R.S. &c.* | Tübingen, Jan. 28, 1864 | (Würtemberg)

Sir,

Your book[2] has awakened my interest for that wonderful body ice; and I have devoted the recent cold weather to a more accurate determination of its coefficient of refraction for ordinary and for extraordinary light, as well as to the observation and measurement of the coloured rings of polarized light. In particular your interesting observation upon the layer of snow which sank away from the roof of the greenhouse and curved itself below (page 202), led me to make a direct experiment upon the deformation of ice—an experiment which answered perfectly, and a description of which will perhaps be more interesting to you than to anyone else.

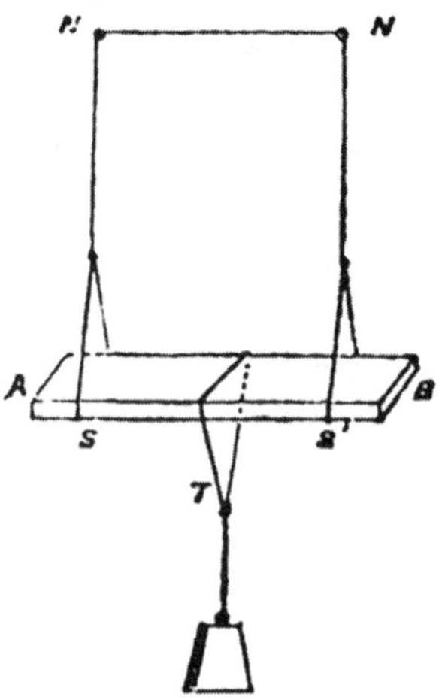

A plate of clean ice, A B, about 100 millims. in length, 12 to 15 millims. in breadth, 3 to 4 millims. in thickness, prepared by sawing and subsequent rubbing with a warm metal plate, is laid in the loops S S of a cotton or silken string which pass above over two pegs, N, N'; a third loop, T, carries a weight of about 200 grms. The temperature of the circumambient space should not much exceed 4° to 6° C. After the lapse of 20 to 30 minutes a bending may be plainly seen, which is at first regular, but afterwards greater in the middle, in consequence of the middle string becoming imbedded in the ice, and of the diameter of the plate being diminished. I have thus in different experiments, in which the distance S S' was 46 to 80 millims., observed bending of 6 to 8 millims. Once, indeed, I was able to bend a thin lamella of ice between the fingers of both hands.

In one experiment I allowed the ice to break, and measured the dimensions of the surface of fracture as well as I could; the greatest tension and pressure in the outermost layers I found to be = 690 grms. to the square millimeter. It would not be without interest to determine the constants of elasticity and rigidity (*Festigkeit*) for ice. Putting 2P for the greatest weight in the middle, $2l$ for the length S S', b for the (horizontal) breadth, h for the (vertical) thickness of the plate of ice, A for the greatest tension in the outermost layer, we have

$$P = \frac{1}{b} A \frac{bh^2}{l}$$

Our values are, for example, A = 690 grms.; b = 10 millims.; h = 3 millims.; 21 = 100 millims.; P = 107 grms.; 2P = 214 grms. If about 300 grms. be suspended from the middle, the plate would therefore soon bend, owing to the melting away consequent upon that degree of pressure. I believe that this experiment could be carried out in summer.

It has certainly happened to you, that in sawing through ice, the saw after a short time has ceased to act, the spaces between its teeth becoming filled up with freshly-formed ice, so that it passes along almost without friction. In truth the saw melts through the ice, the heat necessary for that being the equivalent of the work applied to the saw.

The finest plates of ice adapted for optical observations, and about 6 to 8 millims. in thickness, were found in my wash-hand bowl, a tolerably large tin vessel, in which the water (80 millims. in depth) froze during the very cold nights in my bedroom. (River- and spring-ice are always too little homogenous for optical purposes.)

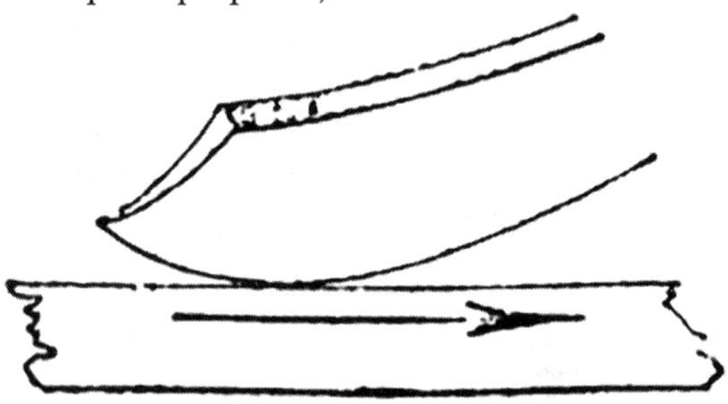

In dividing these precious plates of ice, I soon found that it was necessary to handle them like glass. If the convex blade of a knife be passed over a piece of ice with a certain pressure, a sharp crack will result, and the plate may be broken in the direction of this crack, provided that the temperature of the ice and of the air be below 0° C. Obviously the knife acts in this instance like a diamond, which depresses minute particles of glass, and through the wedge-action of which a progressive linear cracking is produced which renders fracture possible. A mere scratch suffices neither for glass nor for ice. It is worthy of notice that these clear sharp cracks are possible in ice which is in contact with water in a warm room—a proof that ice preserves its brittle character at temperatures lying close to its melting-point. The breaking off at the crack is, on the other hand, no longer possible on account of the regelation in upper part of the crack.

I remain, | Yours truly, | E. REUSCH, | *Professor of Natural Philosophy*, | *Tübingen.*

E. Reusch, 'On some Properties of Ice', *Phil. Mag.*, 27 (1864), pp. 192–4
RI MS JT/1/R/16

---

1. In this translation of Reusch's letter to Tyndall published in the *Phil. Mag.*, the following passage which began the handwritten letter was omitted: 'Above all I would like to express my sincere admiration for the chivalrous way that you fought for and continue to fight for the rights & achievements of my friend Dr. Mayer of Heilbronn, even against your own

esteemed countrymen. You have done a noble deed in the service of truth and it will bring you blessings. Further, I would like to thank you for the great pleasure it gave me to read your delightful book "Heat, considered . . ." I do not recall having ever read a book from beginning to end with such satisfaction' (RI MS JT/1/R/16). Here Reusch was referring to *Heat as a Mode of Motion*.

2. *Your book*: *Glaciers of the Alps*.

To James Croll 10 February 1864 2072

10th. Feb. 1864

Dear Sir

I have forwarded your paper[1] to Mr. Francis,[2] altering nothing therein. Mr. Gyll[3] is of course merely kicking against the pricks.[4] But it was thought desirable to allow a practical writer to state his own case.

Yours very truly | John Tyndall

James Croll Esq[ui]re

BL Add MS 41077, f. 9

1. *your paper*: J. Croll, 'On Supposed Objections to the Dynamical Theory of Heat', *Phil. Mag.*, 27 (1864), pp. 196–8.
2. *Mr. Francis*: William Francis.
3. *Mr. Gyll*: Joseph Gill; see letter 2062.
4. *kicking against the pricks*: Acts 9:4–6; meaning to fight against authority in a futile situation.

From Friederich Eduard Reusch 18 February 1864 2073

Tübingen | 18 Febr. 1864

Hochgehrter Herr!

Sie haben mir durch Ihr freundliches Schreiben vom 3 Febr. eine grosse Freude gemacht. Dass Sie sich die Mühe genommen haben, meine Notizen über das Eis in das philos. Magazine zu befördern, ist fast zu viel Ehre. Die classische Arbeit von Hopkins über die Bewegung der Gletscher habe ich unterdessen auch mit grosser Freude gelesen & es ist meine Sehnsucht, auch einmal einen Gletscher in der Nähe zu sehen, durch Ihr köstliches Buch „on heat" u. durch jenen Aufsatz sehr gesteigert werden. Eine sehr zahlreiche Familie & die Sorge für dieselbe hat mir bisher nicht erlaubt, grössere Reisen anders als mit Staatsunterstützung zu machen; diesen Herbst aber gehe ich sicher wenigstens in's Berner Oberland.

Was ich zur Ehre meines Freundes Dr Mayer gethan habe ist gering im Vergleich mit dem, was Sie gethan haben. Mayer ist auf meinen Antrag Ehrendoctor der hiesigem philos. Facultät geworden, ebenso ist er einer der ersten Ehrendoctoren der im verflossenen Jahre hier neugegründeten naturwissenschaftlichen Facultät.

Ihr Buch war für mich die Veranlassung die Versuche mit der „singenden Flammen" anzustellen, & ich glaube Einiges gefunden zu haben, was auch Ihres Interesses nicht ganz unwürdig sein dürfte.

Der Wheatstone'sche verticale, um eine vertikale Achse rotierende Spiegel, ist zur Beobachtung der schwingenden Flamme nicht ganz geeignet. Bei der Geschwindigkeit die man ihm geben muss, um einen continuirlichen Lichteindruck zu erhalten, wird das Bild der schwachen Flamme so sehr in die Breite gezogen, dass die lichtschwachen Partien verschwinden. Ich verbinde den Spiegel mit einer horizontalen Ache, mit welcher die Normale des Spiegels einen kleinen Winkel von 5–7° macht. Das nahe beim Spiegel befindliche Auge übersieht dann die ellipsenartige Bahn, welche das Bild der Flamme zu durchlaufen scheint. Die singende

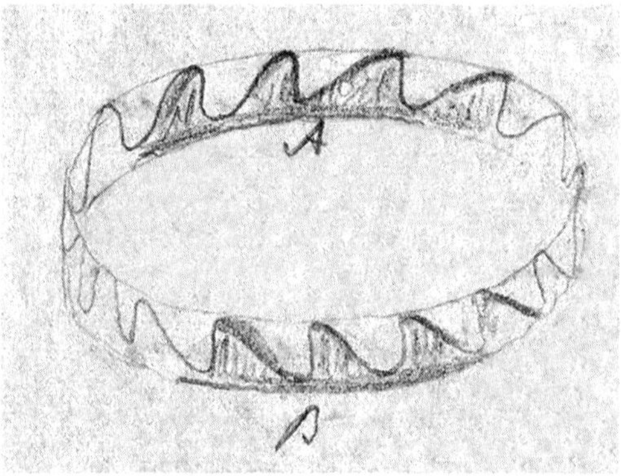

Flamme giebt eine gezakte Krone von grosser Schönheit. Die Zacken sind einseitig, auf einer Seite langsam ansteigend, auf der anderen steil abfallend, was wahrscheinlich dem Momente der Explosion entspricht. Natürlich sind nur die horiz. Partien A,B des Bilds zur Beobachtung geeigent.

Ein Erlöschen der Flamme ist weder bei Anwendung von metallenen noch gläsernen Brennern zu beobachten, bei den letzteren also, (mit denen ich am liebsten operiere, weil man sie in jedem Augenblick am Blastisch von jeder Länge & Weite machen kann) ist die blaue Minimumsordinate grösser als bei den metallenen Brennern.

<u>NB.</u> Die Krone lässt sich auch auf einen Schirm projicieren. —Die Geschwindigkeit des Spiegels braucht nur mässig gut zu sein. Räderwerk ist zu vermeiden, weil dessen Geschnarr die Flamme stört.

Seit einiger Zeit kommen Patentbrenner von Richard Meyn (verfertigt von Lorenzen & von Waldau in Altona) in den Handel. Diese Brenner haben oben ein eisernes Netz, und unten

eine Vorrichtung zur Regulierung des Luftzugs. Reducirt man den Gaszufluss immer mehr, so kommt ein Moment, wo ein im Innern der Flamme befindlicher blauer Saum immer mehr herabsteigt & endlich das Netz N berührt. Von diesen Augenblick schwingt die Flamme & ein grünliches Licht füllt den ringförmigen Raum GG. Durch Regulieren des Zuflusses von Gas & Luft lässt sich das Tempo der Vibrationen in mannigfaltigster Weise verändern.

Hält man einen resonierenden Körper in die Nähe der Flamme, so bemerkt man, wie sich dieselbe mit dem Tone des Körpers in Harmonie zu setzen sucht. Ein 6–8' langes Rohr über den Brenner geschoben giebt mit Leichtigkeit seinen tiefsten Ton. Dabei kommt aber häufig das Netz in's Glühen & die Flamme schlägt durch dasselbst, wenn die Maschen zu weit sind. —Ich habe zu diesen Versuchen einen besonderen Brenner construiert, dessen Anordnung die seitliche Skizze zeigt; bei O sind 6 Oeffunungen die durch Drehen des Rohrs R in beliebiger Weise reguliert werden können. Das Netz N wird sehr leicht in die gewölbte Form dadurch gebracht, dass man ein rundgeschnittenes ausgeglühtes Stück von ebenem Netz mit einer Holzform in ein Rohr drückt das oben etwas ausgeweitet ist. Das Netz hält doch Reibung & Federung im Rohr R.

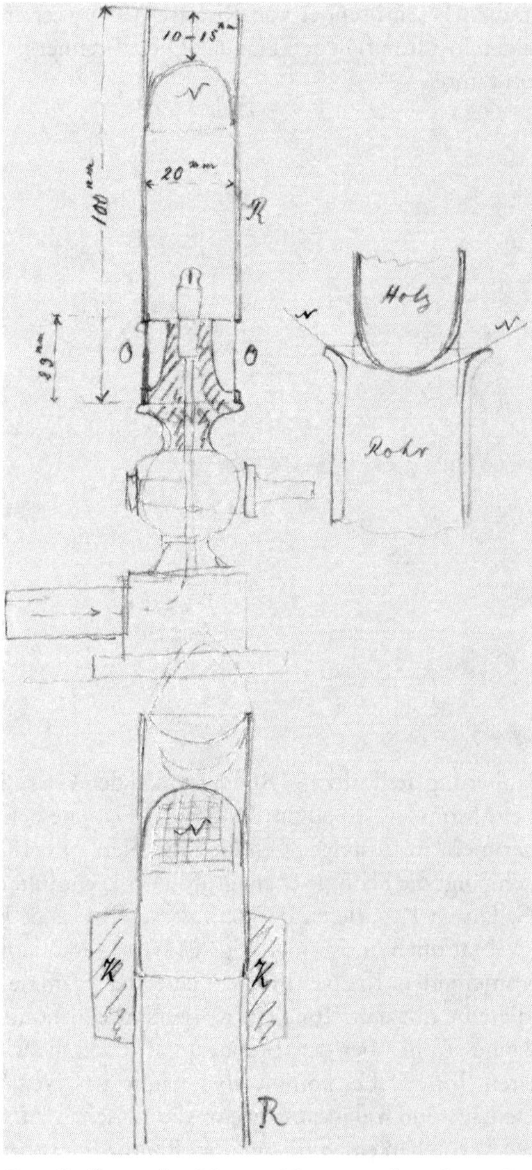

Die Stellung des Netzes N gegen die obere Oeffnung ist von grossem Einfluss auf das Tempo. Je tiefer N um so langsamer die Schwingungen; am zuweckmässigsten scheint eine Tiefe des Netzes von 10–15 ^{millim}.

Später habe ich auf das Rohr R mittels eines Korks KK ein Glasrohr gesetzt & das Netz in letzterem angebracht. Die schon oben beim Meyn'schen Brenner bezeichneten Erscheinungen, lassen sich hier noch viel schöner beobachten. Der ganze Raum über dem Netz erfüllt sich bei jeder Explosion mit dem schönsten grünen Licht, zugleich zeigen sich bei gehöriger Tiefe des Netzes & reducirtem Gaszufluss merkwürdige Intermittenzen, die schwer zu beschreiben sind, die Sie aber alle mit grosser Leichtigkeit bei Anstellung des Versuchs herausfinden werden. —Es habe diese Erscheinungen, in Dunkeln angestellt, etwas Zauberhaftes.

Sie müssen mir versprechen, diese Versuche zu wiederholen, da ich überzeugt bin, dass sie Ihnen viel Vergnügen machen werden. Ich selber werde so bald nicht dazu kommen, etwas darüber zu publiciren, da mich eine Arbeit über den Agat, wozu ich alle Präparate selber schliefen musste, vor Allem fertig bringen möchte. Das oben Mitgetheilte stelle ich daher vollkommen zu Ihrer Verfügung. Manches davon ist vielleicht von Anderen schon beobachtet worden & auch ich bin mehr durch Zufall als systemat. Agieren darauf geführt worden.

Darf ich Sie bitten, den Herren Story Maskelyne & Becker bei Elliot br. freundliche Grüsse zu sagen?

Mit aller Hochachtung | Ihr ergebenster | E. Reusch

Tübingen | 18 February 1864

Dear honoured Sir!

You have given me great pleasure with your letter of 3 February[1] It is almost too great an honour that you took the trouble to convey my notes on ice[2] to the philos. Magazine. In the meantime I have read the classic work of Hopkins[3] on the motion of glaciers with great pleasure, & my yearning to finally see a glacier up close, has been greatly increased by your delightful book "on heat"[4] and through that essay. A very sizable family & responsibility for it has prevented me from making longer journeys without state support; this autumn I will certainly at least get to the Bernese Oberland.[5]

What I have done to honour my friend Dr Mayer is little in comparison to what you have done. Mayer was made honorary doctor at the local philos. faculty at my proposal. He is in addition one of the first honorary doctors of the new faculty of sciences, which was founded in the past year.

Your book inspired me to conduct experiments on the "singing flame", & I believe I have made findings that could be not entirely unworthy of your interest.

The Wheatstonian vertical, a mirror that rotates around a vertical axis, is not perfectly suited for the observation of the singing flame. At the velocity that it must reach to achieve a continuous light effect, the image of the weak flame is distorted so much in width, that the parts with weak luminosity disappear. I connect the mirror to a horizontal axis, which makes a small angle of 5–7° with the perpendicular of the mirror. The eye that is close to the mirror does not see the ellipse-shaped orbit that seems to pass through the image of the flame. The singing flame exhibits a serrated crown of great beauty.

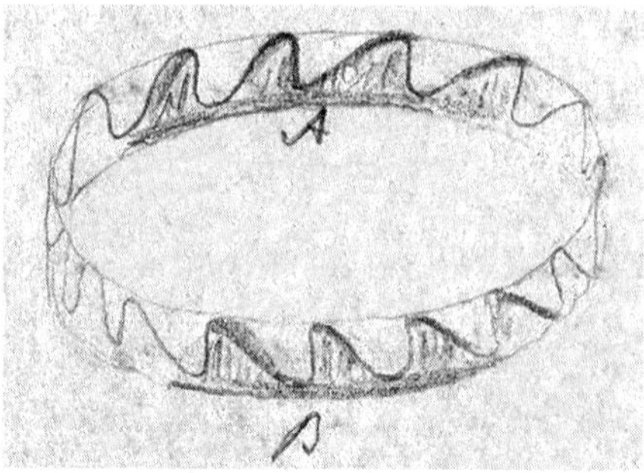

The spikes rise slowly on one side, fall sharply on the other, which probably corresponds to the moment of the explosion. Naturally only the horiz[ontal] parts A,B. of the image lend themselves to observation.

It has not been observed that the flame extinguishes using either metal or gas burners. With the latter though (which I prefer to use, because they can be made at any moment at the blowing table in any length or breadth), the blue of the minimum ordinate is greater than with the metal burners. ["Heat" p. 277][6]

NB The crown can also be projected onto a screen. —The speed of the mirror only needs be moderately good. Avoid wheelworks, because their vibrations disturb the flame.

Patent burners from Richard Meyn[7] have been for sale for some time (manufactured by Lorenzen & von Waldau in Altona[8]). These burners have an iron mesh on top and below a devise for regulating airflow.

If one reduces the flow of gas more and more, a moment arrives when the blue seam that is inside the flame descends & finally touches mesh N. From this moment onward, the flame swings & a green light fills the ring-shaped space GG. By regulating the flow of gas and air, the tempo of vibrations can be modified in the most various ways.

If you hold a resonating body near the flame, you will notice that it tries to set itself into harmony with the tones of this body. A 6–8' long tube shoved over the burner easily gives its deepest tone. In the process the mesh often begins to glow & the flame licks through it, when the weave is too wide. —I have constructed a special burner for these experiments, the arrangement of which is shown in the side sketch;

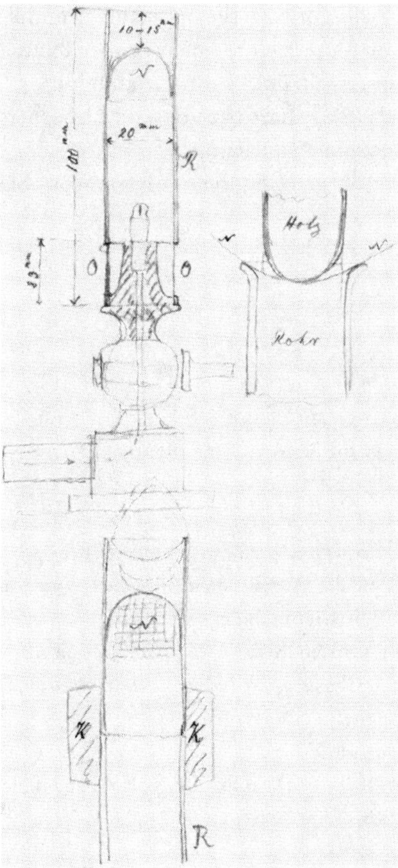

At O are 6 openings that can be regulated in any way by turning the tube R. Mesh N can easily be brought into the arched form, if you cut a round, tempered piece of the mesh and press it with a wood mould in a tube that is somewhat enlarged on top. The mesh holds through friction & suspension in tube R. The position of mesh N against the upper opening has a great influence on the tempo. The deeper N is, the slower the oscillations; most advantageous seems to be a mesh depth 10–15^{millim}.

Later I placed a glass tube on tube R by means of cork KK, and set the mesh into the former. The phenomena that I described above with the Meyn burner, can be observed even better here. With every explosion the entire space over the mesh is filled with the most beautiful green light, at the same time, with proper mesh depth & reduced gas flow peculiar intermittencies, which are difficult to describe, exhibit themselves, as you will easily find when you conduct the experiment. —Performed in darkness, these phenomena have something magical.

You must promise me to repeat these experiments, since I am convinced that you will find them very entertaining. I will not have the opportunity in the near future to publish on them, because I want above all to finish a paper about agate for which I had to grind all preparations myself. I place this communication completely at your disposal. Some of it may already have been observed by others & I was led to it more by accident than by systemat[ic] action.

May I request that you pass my friendly greetings on to Messrs Story Maskelyne[9] & Becker[10] by Elliot br.[11]?

With great respect | yours truly | E. Reusch

RI MS JT/1/R/17

---

1. *your letter of 3 February*: letter missing.
2. *my notes on ice*: see letter 2071.
3. *classic work of Hopkins*: W. Hopkins, 'On the Theory of the Motion of Glaciers', *Phil. Trans.*, 152 (1862), pp. 677–745.
4. *"on heat"*: *Heat as a Mode of Motion*, p. 277.
5. *Bernese Oberland*: the geographically higher region of the Swiss canton of Bern.
6. In margin: 'Heat p. 277', a reference to Tyndall's *Heat as a Mode of Motion.* A vertical line indicates that the note is a reference to this passage in the letter.
7. *Richard Meyn*: not identified; possibly Carsten Richard Meyn or Johann Christian Carsten Meyn, both who are listed in the *Chronological Index of Patents Applied for and Patents Granted* for 1862 and 1863, respectively.
8. *Lorenzen & von Waldau in Altona*: not identified.
9. *Story Maskelyne*: Nevil Story Maskelyne (1823–1911), an English geologist and mineralogist who was Keeper of Minerals at the British Museum from 1857–80 (*ODNB*).
10. *Becker*: Ernst Becker.
11. *Elliot br.*: Elliot Brothers was a British scientific instrument maker, founded by William Elliot (1780/1–1853) and run by his two sons from 1857 forward (*ODNB*). The company served many prominent academics such as Maxwell, Wheatstone, and Lord Rayleigh. Most of their products included galvanometers and batteries, though their later products were for avionics. See H. R. Bristow, 'Elliott, instrument makers of London: Products, customers and development in the 19th century', *Bulletin of the Scientific Instrument Society*, 36 (1993), pp. 8–11.

To Harriet Moore[1] [19 March 1864][2] 2074

Saturday

My dear M[rs]. Moore

There were many kind things said to me after the lecture last night,[3] but nothing that was said pleased me half so much as the greeting which you sent me by your kind daughter[4] & which I returned with all my heart.

Yours affectionately | John Tyndall

RI MS JT/1/T/1091
RI MS JT/1/TYP/3/868/2

1. *Harriet Moore*: Harriet Moore (née Henderson, 1779–1866), widow of the surgeon James Carrick Moore (1763–1834), and mother of five children, among whom were Harriet Jane Moore and Julia Moore. The wealthy Moores had close connections with the RI and frequently entertained Tyndall at their Wimbledon home.
2. *[19 March 1864]*: although LT gave Saturday, 9 March 1864 as a date for this letter, there was no such date (9 March was a Wednesday that year). A more likely date is Saturday, 19 March 1864, a day after Tyndall lectured on 'Contributions to Molecular Physics' at the RI (see n. 3), and perhaps LT mistakenly wrote '9' for '19'.
3. *the lecture last night*: on 18 March 1864, Tyndall delivered a Friday Evening Discourse at the RI, 'Contributions to Molecular Physics' (*Roy. Inst. Proc.*, 4 (1862–6), pp. 233–40). Tyndall had given this as the Bakerian Lecture the evening before at the RS (J. Tyndall, 'The Bakerian Lecture. —Contributions to Molecular Physics. —Being the Fifth Memoir of Researches on Radiant Heat', *Phil. Trans.*, 154 (1864), pp. 327–68; read 17 March 1864). Hirst noted in his journal entry for 20 March 1864 that in the RI lecture, Tyndall drew out heterodox conclusions 'on the danger of ascribing motives to Providence—apropos of the beneficial effects of the vapourous envelope of the earth—and of the groundlessness of the fears that the investigation of truth in science should be injurious to religion' (Brock and Macleod, *Hirst Journals*, p. 1666–7; also see *Ascent of John Tyndall*, p. 181). In the Chair at this RI lecture was Albert Edward (1841–1910), Prince of Wales, eldest son of Queen Victoria and Prince Albert (*ODNB*).
4. *your kind daughter*: a reference to either Harriet Jane Moore or Julia Moore.

From C. K. Akin [late March 1864][1] 2075

Cavendish Club | Regent Street | Monday

Dr. Akin presents his compliments to Professor Tyndall, and having seen in the last number of the Reader[2] an account of his late communications to the Royal Society and at the Royal Institution, in which he observes reference has been made to his own attempts on the subject of the converse phenomenon of Fluorescence, he begs leave to enclose to him the proof sheets of his introductory paper on the subject,[3] now printing in the Reports of the British Association, and of which hitherto extracts only have been published

Dr. Akin, being now in town for a short time, would gladly avail himself of this opportunity to call on Prof. Tyndall, and to have the pleasure of making his acquaintance, if it was convenient for him to see him.

RI MS JT/1/TYP/1/59
LT Typescript Only

1. *[March 1864]*: the month and year are given based on Tyndall's description of this letter in J. Tyndall, 'On the History of Calorescence', *Phil. Mag.*, 29 (1865), pp. 218–31.
2. *the Reader*: this was published in the 26 September 1863 issue of *The Reader* (vol. 2, no. 39). It was later published, with some additions in brackets, as C. K. Akin, 'On Calcescence', *Phil. Mag.*, 29 (1865), pp. 28–43.
3. *his introductory paper on the subject*: enclosure missing, but presumably C. K. Akin, 'Account of Preliminary Experiments on Calcescence', *Brit. Assoc. Rep. 1863*, p. 11.

To C. K. Akin [late March 1864][1] 2076

Clarendon Hotel, Chale, Isle of Wight | Wednesday

My Dear Sir,—I have read the proof which you have been kind enough to send me[2] with extreme interest. Being a truant from the Section at Newcastle,[3] I did not know that you had brought the subject forward there. Professor Stokes, however, informed me that you were engaged on the question,[4] and I had similar information from the 'Saturday Review.'[5]

There is certainly a most singular coincidence in the thoughts which have passed through both our minds on some of the points in question. As to the possibility of converting the Herschellic into the Newtonic rays I do not entertain a doubt; and indeed the piece of work which I had set before me for attack on my return from this place was a series of experiments on this very subject.

I have devised a very perfect means of sifting the Herschellic from the Newtonic rays, and with the latter alone I have strong hopes of being able to produce incandescence. But you have entered this field before me, and although the question lies directly in the way of my own inquiries, I should like to do in the matter what would be the most pleasing to you. I shall be in town on Saturday next, at the Royal Institution at noon; could you not look in upon me there?

Yours very sincerely | John Tyndall

J. Tyndall, 'On the History of Calorescence,' *Phil. Mag.*, 29 (1864), pp. 221–2

1. *[March 1864]*: the month and year are given based on relation to letter 2075.
2. *which you have been kind enough to send me*: see letter 2075, n. 3.
3. *the Section at Newcastle*: see letter 2024, n. 3.
4. *Professor Stokes, however, informed me that you were engaged on the question*: Tyndall later wrote to George Busk to check the dates that both he and Stokes had both been in attendance at the Athenaeum Club, presumably in order to verify the date at which Stokes had in fact first mentioned Akin's work to him. While Tyndall's presumed letter is missing, Busk's reply is letter 2173.
5. *Saturday Review*: probably 'Calcescence', *Saturday Review*, 17 (1864), pp. 105–6. This is a report of Akin's 4 December 1863 lecture before the Natural Science Society at the University of Cambridge.

To the Editors of *The Reader* 2 April 1864 2077

SCIENTIFIC CORRESPONDENCE. | RAY-TRANSMUTATION.
Athenaeum Club, 2nd April.

THROUGH the kindness of its author, I have this day become acquainted with Dr. Akin's communication to THE READER of the 26th of September, 1863.[1] From conversations with Professor Stokes I had obtained a general notion of the experiments in which Dr. Akin has been for some time engaged; but I now learn that, on one of the points touched upon in the report of my last investigation—the proposed explanation, namely of the glowing of a platinum wire in a hydrogen flame—he has anticipated me by several months.
JOHN TYNDALL.

The Reader, 461 (1864), p. 461

1. *THE READER of the 26th of September, 1863*: see letter 2075, n. 2. Also see letters 2076, 2079, 2165, 2167, and 2170, as well as the Introduction for further commentary.

From William Hopkins 2 April 1864 2078

Cambridge | April 2/64

My dear Tyndall,

My health has been sufficiently precarious during a great part of this winter to render considerable caution necessary so that I have been but very little away from home of late, and in London only on one or two occasions of official necessity. It is not likely that I shall be there for the election you mention,[1] but if so your recommendation of your friend Dr. Debus shall have its due weight. I shall probably resign before long the noble office of Senator of the London University though I should have liked very much to retain it, had I been resident in town, but living here I can't attend the meetings without giving up the whole day to each, which is expensive both in time and money. In fact I am getting tired of scientific work (except in my own study) without a farthing of scientific pay. After nearly 40 years of hard work I may say literally that I do not owe a farthing of my present income to science. I don't speak of this as a matter of complaint, but as a matter of ill luck, which seems likely in these my latter days, to circumscribe my intercourse with my London scientific friends, a prospect which I have been looking to for some time with deep regret—However, if I should give a vote when your friend is a candidate, though I make no absolute promise, I shall be very much disposed to give him my support.

Health is essential to a man who works as hard as you do. Take care of yourself.

Yours very truly, | W. Hopkins.

RI MS JT/1/TYP/2/646

---

1. *the election you mention*: not identified (it does not refer to the election of Debus to the Athenaeum, which was in 1873); possibly, however, this was related to Debus' position of Examiner in Chemistry at the University of London from 1864–9, which he held again from 1871–6 and 1878–82 (*Who's Who 1907: An Annual Biographical Dictionary* (London: Adam and Charles Black, 1907), p. 462).

From C. K. Akin [after 2 April 1864][1] 2079

Cavendish Club | Thursday

My dear Sir,

I hope to call on Saturday at the Royal Institution, but, as I suppose you may be interested to learn some particulars concerning what I may call the history of the matter in question, which I can state much more readily in writing, I herewith send you a short account of it.

So early as in the spring of /62 the main ideas contained in my paper[2] had occurred to me. Being at Cambridge, I then communicated with Prof. Stokes on the subject with a view to try some experiments, but there was no apparatus to do it with. Under these circumstances I was obliged to let the subject lie over for nearly a year, until, about the end of /62, having then gone to Oxford, I broached the matter to Mr Griffith[3] with whom subsequently I made some trials. I had hoped to have made some decisive experiments against the close of the meetings of the Royal Society in /63, and then against the meeting of the British Association, but circumstances beyond my control prevented my doing so. At last I communicated my plans as such, together with an account of the trials made and the experimental methods intended to be followed to the British Association—of the former you have the account, and of the latter you will find an outline in the Review[4] for the last 26th Sept., of which I hope to send you a copy

Believe me | my dear Sir | Yours very truly | C. K. Akin

Prof. Tyndall

RI MS JT/1/TYP/1/60
LT Typescript Only

1. *[late 1864]*: letter dated in relation to letter 2077.
2. *my paper*: see letter 2075, n. 3.
3. *Mr Griffith*: Akin completed these experiments with George Griffith (*c.* 1834–1902), an English science educator who taught at Oxford between 1857–67 (B. P. L., 'George Griffith', *Nature*, 66 (1902), p. 64).
4. *Review*: see letter 2075, n. 2. Akin presumably wrote 'Review' instead of 'Reader' by mistake.

To Thomas Archer Hirst [4 April 1864][1] 2080

The Athenaeum[2] | Monday

My dear Tom.

I have read your letter[3] with great sorrow. It was not <u>your</u> friend John Tyndall who thus passed through Kingstown[4] but some other of the Tyndall tribe. —I have lectured this morning[5] & am now preparing with a very bad head to lecture again tonight,[6] but as far as I can see at present it would be wise to give up your post at University College[7]—Your health must be attended to above all things.

Yours affectionately | John Tyndall

RI MS JT/1/T/952
RI MS JT/1/HTYP/526

1. *[4 April 1864]*: the date is given by relation to letters 2083 and 2084, both dated 5 April 1864 (a Tuesday).
2. *The Athenaeum*: see letter 1907, n. 7.

3. *your letter*: letter missing. Hirst had left London for Ireland on 22 March, returning home on 12 April (Brock and Macleod, *Hirst Journals*, p. 1667–8). Tyndall hoped to be able to join him (see letters 2083 and 2084).
4. *Kingstown*: a coastal town and major port near Dublin, Ireland (previously named Dunleary and now Dún Laoghaire). It was named 'Kingstown' in 1821 to commemorate the visit of King George IV.
5. *I have lectured this morning*: Tyndall delivered weekday lectures for students at the Royal School of Mines (see letter 1987, n. 4).
6. *to lecture again tonight*: also through the Royal School of Mines, Tyndall delivered a series of six lectures to working men in April and early May 1864.
7. *it would be wise to give up your post at University College*: Hirst taught at University College School from 1860–4, and did indeed give up his post at that point in order to focus upon both his mathematical research and health, as he shared in his journal entry for 17 April 1864 (Brock and Macleod, *Hirst Journals*, pp. 1667–8). The following year he was appointed Professor of Physics at University College London.

From Joseph Gill 4 April 1864 2081

ON THE DYNAMICAL THEORY OF HEAT. | *To John Tyndall, Esq., F.R.S., &c.*

SIR,

I AGREE with the Editors of the Philosophical Magazine, that it is not desirable that my remarks on the dynamical theory[1] should give rise to prolonged discussion. What is wanted is experimental proof of the direct conversion of heat into mechanical work; and after what has been so ably achieved by Dr. Joule for physical science, in proving beyond a doubt the conversion of work into heat, and determining the numerical equivalent of the transformation, surely it is reasonable to expect that every effort should be made to obtain positive physical evidence of the reality of the conversion of heat into work. I thank the Editors for kindly allowing me to state the difficulties which I have encountered in this pursuit, and the consequent doubts which have presented themselves of the completeness of the dynamical theory as it now stands; and I beg to avail myself further of their courtesy to clear up in a few lines some want of perspicuity in my last letter, which may in some degree have influenced the remarks of Dr. Rankine and of Mr. Croll in the Philosophical Magazine for March.[2]

Dr. Rankine is right in pointing out my inadvertence in overlooking the application of the second law of the action of heat in two instances. In the first, namely, my statement that the regenerator might take up and restore all the heat remaining in air after it has done work by expansion, my omission was but half unconscious, and arose from my having uppermost in my mind the ideas of thermodynamic action published by me some years ago; and had I entered more into detail on this point in my last letter it would have been seen that my oversight was more in appearance than in reality, as the difficulty of the hypothetical case proposed by me would not have been removed by a formal application of the second law. In the second case, the estimate I gave of the useful effect of steam-engines was the old statement of M. Regnault; and I should have stated that it was quoted by me from M. Verdet's Second Lecture on the Thermo-dynamical Theory of Heat (Paris, 15th February 1862, section 2). Of course, in this case the application of the second law made it requisite to suppose (in addition to M. Regnault's deductions from the mere calculation of the latent heat) the condensation of some steam as a result of the actual conversion of heat into work; but as this is the very point at issue, and a point on which the grand question of the practical improvement of our heat engines mainly depends, I may be allowed to state that, as far back as 1843, I published results of my experiments,[3] showing that the whole heat resulting from the condensation of steam in the cylinder of an engine working expansively should be found in the condenser, after allowing for known losses from other causes—a result which subsequent experience has steadily confirmed.

I allow the propriety of Dr. Rankine's disapprobation of my remark, that M. Hirn[4] was borne down by scientific authority in altering his opinions on the results of his own experiments, as, after all, the publication of this change of opinion was his own free act. Though I said "apparently borne down," the expression was inconsiderate, and I freely retract it. I cannot forget that I am indebted to the courteous kindness of Dr. Rankine for spontaneously sending me valuable scientific information, which in my comparatively isolated position here I could not otherwise have easily obtained; and though I reluctantly differ from him on some points of scientific theory, my high appreciation of his intellectual superiority will always cause me to respect and value his opinion, whether for censure or approbation.

While thanking Mr. Croll for his remarks on my uncertainty about the estimate of work done by heated air in the shape of raising the weight of the atmosphere, and on the specific heat of air, I would beg to refer back to my chief difficulty on this latter point, which perhaps I did not state with sufficient clearness, viz. the disappearance of energy in the supposed experiment. What I meant to express was this:—There must be more heat in the air after expansion than before, because the energy which might have been availed of by allowing the atmosphere to fall back into the vacuum, and thus produce work, has disappeared, and we can imagine no other equivalent for it. And the specific heat of the expanded air, or its capacity for heat, is greater because it does not show this increase of quantitative heat.

Not having been able to make the result of experiment agree with the dynamical theory as to the direct conversion of heat into work, I have been long pondering on some consistent explanation of the physical facts, which have continued to present themselves in my researches. On this subject I had a thing to say, but let it go for the present, and now I will return with renewed energy to the experimental investigation of this very interesting but still perplexing subject.

Believe me Sir, your most obedient Servant, | JOSEPH GILL. | Palermo, April 4, 1864.

J. Tyndall, 'On the Dynamical Theory of Heat', *Phil. Mag.*, 27 (1864), pp. 477–8

---

1. *my remarks on the dynamical theory*: J. Gill, 'On the Dynamical Theory of Heat', *Phil. Mag.*, 26 (1863), pp. 109–16 (this is letter 2062).
2. *remarks of Dr. Rankine and of Mr. Croll in the Philosophical Magazine for March*: W. J. M. Rankine, 'On the Dynamical Theory of Heat', *Phil. Mag.*, 27 (1864), pp. 194–6; and J. Croll, 'On supposed Objections to the Dynamical Theory of Heat', *Phil. Mag.*, 27 (1864), pp. 196–8.
3. *I published results of my experiments*: J. Gill, 'Hints on some Improvements of the Steam-Engines', in J. Weale (ed), *Quarterly Papers on Engineering* (London: John Weale, 1843), pp. 1–15 (papers in this volume are numbered individually).
4. *M. Hirn*: Gustave-Adolphe Hirn (see letter 2062, n. 7).

To William Francis 5 April 1864 2082

5th. April 1864

Dear Francis.

This cheque has been drawn for some time, and indeed placed in its envelope to be sent to you without any accompanying remarks. I delayed it thinking I might have something to say. All however I have to say consists in a request that you would have the kindness to cast your eye over the account and assure yourself that it is all right.[1]

Yours truly | John Tyndall

Dr. Wm. Francis.

St Bride Library, Authors' letters

---

1. *cast your eye over the account and assure yourself that it is all right*: see letter 2085 for Francis's reply.

To Thomas Archer Hirst 5 April 1864 2083

Tuesday | April 5 1864

My dear Tom

I wrote to you yesterday[1] in the midst of my preparations—Now the 1st working man's lecture[2] is a thing of the past and this morn[in]g I have leisure to write to you more fully.

My first impulse on reading your letter[3] was to start off at once to Ireland; but chained as I am here this is impossible, and your letter encourages me to hope that it is now unnecessary. Were I in your case I would take no decided step regarding University College until this fit of feebleness is past. We are different beings when ill and ought to distrust ourselves. I confess you do things that I should not dare to do, although I am not wanting in toughness. Mrs Tunstal[4] told me how ill you were previous to going to Ireland, and I reconciled myself with a sigh to what I conceived to be a mistaken act on your part.

If when your strength augments you still retain the idea that it would be desirable to give up your post in University College[5] then I would say in God's name relinquish it. Even if you have not enough to carry you on I could readily supplement it. Longman[6] for instance writes to me that the book[7] is going off <u>fast</u> and Dumas[8] has been introducing it to the French Academy.

Write to me again immediately. I do not understand the Kidney as I have never suffered in that region, and an evil that I do not understand is always formidable to me.

Good bye dear Tom | Yours affectionately | John Tyndall

Mrs. Lubbock[9] I am sure will be grieved to hear that you are so ill.

RI MS JT/1/T/659
RI MS JT/1/HTYP/525

---

1. *I wrote to you yesterday*: see letter 2080.
2. *lecture*: see letter 2080, n. 6.
3. *your letter*: letter missing.
4. *Mrs Tunstal*: see letter 2028, n. 7.
5. *give up your post in University College*: see letter 2080, n. 7.
6. *Longman*: letter missing; from either Thomas Longman (1804–79) or William Longman (1813–77), brothers who ran the publishing firm Longman, Green, Longman, and Roberts.
7. *the book*: *Heat as a Mode of Motion.*
8. *Dumas*: Jean-Baptiste Dumas (1800–84), a French chemist best known for his work on atomic masses (*CDSB*).
9. *Mrs. Lubbock*: Ellen Lubbock.

To Thomas Archer Hirst 5 April 1864 2084

Tuesday | April 5 1864.

My dear Tom

I have been to Dr. Bence Jones to talk to him about you: he blames you much for not returning[1] to London when you noticed things at Chester,[2] and he now advises you to return as soon as you can & he will visit you & put you right.

I wrote to you this morning,[3] but I suppose both letters will reach you at the same time. This then is No. 2.

Yours affectionately | John Tyndall

RI MS JT/1/T/658
RI MS JT/1/HTYP/525

1. *not returning*: Hirst was ill during his trip to Ireland (see letter 2080, n. 3, and letter 2083, in which Tyndall wrote that Hirst's illness was related to the kidneys).
2. *Chester*: a city in Cheshire, England, on the River Dee close to the border with Wales.
3. *I wrote to you this morning*: letter 2083.

From William Francis 7 April 1864 2085

Printing Office, | Red Lion Court, Fleet Street, E.C. | April 7th /64.

Dear Tyndall

I have been over the inclosed account[1] with our books and can find nothing wrong as regards the number of copies supplied to you,[2] or in the charges according to the scale which is attached to the proofs and on which the several orders for separate copies are written—But what I find very wrong is that Mr Askham[3] should have allowed the account to run on for so many years without ever sending it in to you—I can only account for this neglect from my having given him instructions not to charge you and some other friends according to the usual rate, but to carry a portion of half the expense of the separate copies to the Magazine, which he has failed to do—Mr Gyde,[4] knowing nothing of this arrangement, and finding these accounts open made out the bill in the usual manner and sent it in without mentioning the subject to me; and it was only subsequent to the receipt of your request to be supplied with a copy of our scale of charges that I became partly aware of what had happened—I now inclose the account as it ought to have been sent to you and if you find it correct, as I believe it to be, and will return it for receipt, I will forward you a cheque for the difference due on your cheque.

Yours truly | William Francis

Prof. Tyndall, F.R.S. &c &c

PS. I had given this letter to M[r] Gyde to send to you when he informed me that he had omitted to bring forward a previous unsettled account but for which bills had been sent in. This he has added on at the end

St Bride Library, Authors' letters

---

1. *inclosed account*: enclosure missing.
2. *I have been over . . . copies supplied to you*: this is in reply to letter 2082.
3. *M[r] Askham*: William Askham, a printing manager at the Taylor and Francis printing office (W. H. Brock and A. J. Meadows, *The Lamp of Learning: Two Centuries of Publishing at Taylor & Francis* (London: Taylor & Francis, 1998), p. 135).
4. *M[r] Gyde*: Charles Gyde, another printing manager at Taylor & Francis (Brock and Meadows, *The Lamp of Learning*, p. 135).

From Charles Gyde[1] 11 April 1864 2086

April 11/64.

Sir/ I beg to inform you that Dr Francis has gone out of town for a few days, but on his return he shall have your letter.[2]

Would you have the goodness when you are in this direction to look in & I could then show you the books with the receipts & payments made on your account at different times.

The only charges we have now to make are two items—200 copies of "Laplace's corrections for the velocity of Sound",[3] & the Subscription to "Poggendorff for 1863.[4]

I am, Sir, | Yours most respectfully | Charles Gyde | for Taylor & Francis[5]

Professor Tyndall, F.R.S. | &c

200 Laplace's corr[ections]. for the velocity of sound. | ¼ sh[eet]. Phil. [in verso] Mag—| Poggendorff 1863.

St Bride Library, Authors' letters

---

1. *Charles Gyde*: see letter 2085, n. 4.
2. *your letter*: letter 2082; see also letter 2085 from Francis to Tyndall.
3. *"Laplace's corrections for the velocity of Sound"*: J. Tyndall, 'Note on Laplace's Correction for the Theoretic Velocity of Sound', *Phil. Mag.*, 26 (1863), pp. 384–7.
4. *Poggendorff for 1863*: Johann Christian Poggendorff (1796–1877) edited the *Annalen der Physik* between 1824–76; the journal was often referred to by his name in this period.
5. *Taylor & Francis*: a publishing house founded in 1852 when William Francis joined his

father Richard Taylor (1781–1858) in his business. Taylor initially founded his company in 1798. Their subjects covered include agriculture, chemistry, education, engineering, geography, law, mathematics, medicine, and social sciences. It was not public knowledge that Francis was Taylor's son (see W. H. Brock and A. J. Meadows, *The Lamp of Learning: Two Centuries of Publishing at Taylor & Francis* (London: Taylor & Francis, 1998).

6. *200 Laplace's corr[ections]. . . . Poggendorff 1863.*: this is written on the reverse of the letter.

From Julius Haast 14 April 1864 2087

[1 word illeg] Christchurch NZ. | 14 April 1864

My dear Sir

Many thanks for your letter of July last[1] enclosing your photograph, the receipt of which gave me infinite pleasure and which I shall always value highly. I only returned a few days ago from a journey into the Alps, my principal object being to examine once more carefully the terraces, old moraines, roches moutoneés[2] etc. Not trusting to my eye or the Aneroid only, I used extensively the spirit level and thus obtained some very curious and interesting results, which I hope to lay before the scientific world towards the end of this year. There is not now in my mind the least doubt, that the former greater extension of glaciers in NZ. was due not to depression and the lowering of the atmosphere, but to upheaval through which the mean height of land came above the perpetual snow line and by which the extracted moisture from the prevailing westerly winds was converted into snow which filled for many miles any depression and plateau amongst the craggy peaks giving rise to enormous glaciers.

Many terraces and lines in our Alps which I mistook formerly for horizontal ones are only apparently so whilst the few really horizontal ones are derived from local lakes. I sent to our mutual friend Dr Hooker an account of a West coast glacier perhaps you will see it and you will agree with me that it *[lends]* to my strange conclusions. I followed with great interest your lectures on heat[3] but was unable to see them all, being so often absent from town and not being able to obtain afterwards the English papers with the report of them, and I may assure you, that not only did I take great interest in them, but should be delighted to receive them from your hands—Any book or pamphlets etc with which you may gratify me, will reach me safely if you want to send it to <u>Miss S. Dobson Collet.</u>[4] 13 Lodowick Terrace Regents park London NW.

Since I wrote last a new pass across the Alps at the head of the river Waimakariri[5] has been discovered by Mr Arthur Dobson,[6] formally my assistant surveyor and an energetic young man. This pass is about 3000 feet above the sea, leads towards the sources of the Arahaura[7] and is bounded on both sides by mountains covered with perpetual snow from which large glaciers descend. It is the same pass of which I spoke in several reports of mine as existing according to Maori information. So you see, that with every year we advance more and more with the knowledge of our interior and what formerly was only an unknown wilderness begins already now to offer us assistance in explaining many phenomena in physical geography & geology which within the last few years were only imperfectly understood.

I shall be delighted to hear again from you & with best wishes believe me my dear Sir ever yours | very sincerely | Julius Haast
Prof J. Tyndall FRS | etc | London.

RI MS JT/1/H/6

---

1. *your letter of July last*: letter 2021.
2. *roches moutoneés*: rock formations created by glacial erosion, resembling 'sheepbacks' (French).
3. *your lectures on heat*: see letter 1920, n. 4.
4. *S. Dobson Collet*: Sophia Dobson Collet (1822–94), an English feminist freethinker and writer (*ODNB*). She was an aunt of Haast's friend in New Zealand, Arthur Dudley Dobson (see n. 6). Haast must have had mail sent from London to New Zealand through these friends.
5. *river Waimakariri*: a large river on the South Island of New Zealand.
6. *Mr Arthur Dobson*: Arthur Dudley Dobson (1841–1934), a pioneer surveyor, engineer, and explorer. He is best known for taking the first party of Europeans to New Zealand's West Coast and assisting on provincial explorations (Suzanne Starky, 'Dobson, Arthur Dudley and Dobson, Edward', in *Dictionary of New Zealand Biography* (first published in 1990). Te Ara—the Encyclopedia of New Zealand, https://teara.govt.nz/en/biographies/1d14/dobson-arthur-dudley (accessed 20 November 2019)).
7. *Arahaura*: the Arahura river on the South Island of New Zealand.

To Thomas Archer Hirst 19 April 1864 2088

April 19th 1864.

My Dear Tom

I came up to consult you on a calamity which has befallen me.[1] I leave the letter as I must smother its effect—being pledged to a little party at the Athenaeum[2] today. Some other particularities have reached me which make the thing more sad.[3] I wish I could get away but unfortunately I cannot.[4] I am badly able to meet this stroke—it weighs with awful heaviness upon my poor brain.

Yours aff[ectionate]ly | John

RI MS JT/1/T/660
RI MS JT/1/HTYP/527

---

1. *a calamity which has befallen me*: Tyndall's mother had fallen ill.
2. *Athenaeum*: see 1907, n. 7.
3. *Some other particularities have reached me which make the thing more sad*: Tyndall's mother's mental health was deteriorating, and both she and Tyndall's sister Emma were living in one room, despite the generous allowance he provided for them. When Tyndall did get to Ireland in May he arranged for Annette Roche, a local friend, to manage his mother's affairs. See also letters 2094 and 2101.
4. *I wish I could get away but unfortunately cannot*: Tyndall had started a series of six lectures for working men at the Royal School of Mines earlier in April 1864. Hirst offered to go in his stead, but this proved unnecessary. Tyndall finished his lectures in early May, and was able to head to Ireland on 4 May (see letter 2091).

From James Clerk Maxwell 20 April 1864 2089

8 Palace Gardens Terrace | W | 1864 Ap 20

Dear Tyndall

It would give me much pleasure to belong to the Philosophical Club.[1] I got your letter[2] at King's College today.

Yours truly | J. Clerk Maxwell

P. M. Harman (ed), *The Scientific Letters and Papers of James Clerk Maxwell*, 3 vols (Cambridge: Cambridge University Press, 1995), vol. 2, p. 226

---

1. *Philosophical Club*: Maxwell was indeed elected a member of the Philosophical Club of the RS in April 1864 (T. G. Bonney, *Annals of the Philosophical Club of the Royal Society, written from its Minute Book* (London, Macmillan & Co., 1919), p. 57). For more about the club, see letter 2045, n. 3.
2. *your letter*: letter missing.

To Thomas Archer Hirst 27 April 1864 2090

London | 27th Ap. 1864

My dear Tom

I have been rather laid up since you went away. Last week during one of the warm days I laid off some of my under clothing. On Sunday I went down to Lubbock's—the day was cold, and towards evening particularly bitter. We had a long game of croquet, and was chilled. My back was a mere superficies of pains and aches on the following day. I tried a Turkish Bath, but it did not do much good and I had to lie by on Tuesday. Yesterday I had another Turkish bath, and today I am better.

Spottiswoode was at Lubbock's and they all enquired about you. Mrs. Lubbock[1] wished much to write to you. She also told me that she wished to come here and bid you goodbye. Why did you not let her?

I have had some letters from the physician at Leighlin.[2] The Dean[3] who first wrote to me[4] put the disease as a consequence. It is quite evident that he was in error; what he deemed a consequence was merely an accompaniment. The doctor never mentions it. Years ago I thought if she lived long enough she must finally give way. Her mind was so feeble, so rambling, so unoccupied, and so saturated with the enthusiasm of what people call religion. I shall try to get to London as soon as my lectures at Jermyn Street[5] are concluded, that is in about ten days from the present time.

Your note[6] reached me this morning and tomorrow (Friday) your letters shall be posted according to your directions. I feel great interest in the matter to which you refer. I confess it often crossed my mind—that is to say the possibility of such a thing crossed my mind, but I did not seriously dwell upon it.

During my illness I have been reading the life of Sir William Napier[7]—a fine ardent spirit, but utterly volcanic—it makes me feel better when I read the life of such a man; for I have volcanic elements within me, and I fear it is pure cowardice which represses them sometimes. I have been also reading a work entitled Aristotle by our friend Lewis[8] who used to write in the Leader.[9] He is a very clever fellow, but the tone of his book is decidedly unpleasant. He has no reverence and a vast deal of conceit.

I propose devoting two or three days to a ramble in Wicklow,[10] working my way on thus to Carlow.[11] I have never been through Wicklow, but I believe it is very beautiful. Glendalough[12] has always been a weird region to me, so I intend to visit Saint Kevin's bed.[13] I shall probably also look in upon the Waterford Tyndall's[14] before my return. My Friday evening[15] is a great bother to me. I have no matter of my own and I do not know where to find matter.

I trust you are getting strong. St. Malo[16] is a fine old place. I have a pleasant memory of it, and of that journey that I accomplished through Normandy with Debus.[17] Dear me! what a number of things are becoming memories, and how long the vistas of the past are beginning to appear! The Abbé[18] writes to me that the book 'se repond assez rapidement, and que tout le monde le fait le plus grand éloge'.[19] An "Atlantic Monthly" has been sent to me in which its merits are also trumpeted forth.[20] In fact except in the case of that unchivalrous Scotchman[21] I have been only too well treated. Goodbye my dear Tom. Walk about as much as possible, get strong, and come back to me tough & renewed in body & mind.

Yours affectionately | John

RI MS JT/1/T/661
RI MS JT/1/HTYP/528

1. *Mrs Lubbock*: Ellen Lubbock.
2. *the physician at Leighlin*: presumably a physician to Tyndall's mother, identified as 'Dr Phillips' in letter 2101.
3. *The Dean*: James Lyster (1810–91), Dean of Leighlin (1854–64), the incumbent of the Cathedral Church of St Laserian, Old Leighlin in the united Diocese of Cashel and Ossory within the Church of Ireland. Lyster emigrated to Canada in 1864 to take up the post of Dean of Ontario, serving in Kingston until 1885 ('An Exchange of Deaneries', *Public Opinion: A Weekly Review of Current Thought and Activity*, 20 August 1864, p. 222).
4. *wrote to me*: letter missing.
5. *my lectures at Jermyn Street*: see letter 2080, n. 6.
6. *Your note*: letter missing.
7. *Sir William Napier*: William Francis Patrick Napier (1785–1860), a British army officer and writer (*ODNB*). W. F. P. Napier, *Life of General Sir William Napier*, ed. H. A. Bruce, 2 vols (London: John Murray, 1864).
8. *Aristotle by our friend Lewis*: G. H. Lewes, *Aristotle. A Chapter from the History of Science including analyses of Aristotle's scientific writings* (London: Smith, Elder & Co., 1864). George Henry Lewes (1817–78), an English philosopher and critic of literature and theatre, from the 1850s he published popular science works, on seashore life, for *Blackwoods Magazine*, and also in John Chapman's *Westminster Review*. He also served as editor of the *Leader* for a short time. He was an intimate friend of Herbert Spencer and Marian Evans. Although he was unable to divorce his wife, from the mid-1850s he and Evans lived together as man and wife, which scandalized many of their friends. From 1854 Lewes wrote fewer popular pieces and turned to scientific investigations, and to physiology in particular, but also wrote extensively on the philosophy and science of mind (*ODNB*).
9. *the Leader*: Lewes co-founded *The Leader*, a weekly radical newspaper that was published in London between 1850 and 1860, with Thornton Leigh Hunt. Lewes primarily wrote theatre criticism in the *Leader*, but also ventured into popular science writing, and later into science as a practitioner (see n. 8).
10. *Wicklow*: see letter 2091, n. 9.
11. *Carlow*: County Carlow, Ireland, location of Leighlinbridge where Tyndall was born and his mother lived.
12. *Glendalough*: a glacial valley in County Wicklow, Ireland.
13. *Saint Kevin's bed*: originally a bronze-age tomb, Saint Kevin's bed is an artificial cave in Glendalough, Ireland named after Saint Kevin (498–618). Glendalough is one of the most important sites of monastic ruins in Ireland.
14. *Waterford Tyndall's*: Tyndall had been interested in his family tree and had learnt of a branch of the family in Waterford.
15. *Friday evening*: see letter 2091, n. 7.
16. *St. Malo*: Saint Malo is a historic French port in Brittany on the Channel coast.

17. *that journey that I accomplished through Normandy with Debus*: Tyndall and Debus were together in Normandy during the summer of 1855.
18. *The Abbé*: François-Napoléon-Marie Moigno.
19. *'se repond assez rapidement, and que tout le monde le fait le plus grand éloge'*: essentially, that the response Tyndall's *Heat as a Mode of Motion* (1863) or Moigno's French translation of it, *La chaleur considérée come un mode de mouvement* (1864), has been relatively quick and that everyone has given it the greatest praise. This letter from Moigno is missing.
20. *An 'Atlantic Monthly'*: 'Review of *Heat as a Mode of Motion*', *Atlantic Monthly*, 13 (1864), pp. 512–6.
21. *that unchivalrous Scotchman*: a reference to William Thomson (1824–1907). See letter 2002 and W. Thomson, 'Note on Professor Tyndall's "Remarks on the Dynamical Theory of Heat"', *Phil. Mag.*, 25 (1863), p. 429.

To Thomas Archer Hirst [3 May 1864][1] 2091

Tuesday

My dear Tom

I start for Ireland tomorrow morning early,[2] and have but a portion of today to put my things in order. Hence I cannot write much to you at present. I have written to Debus[3] telling him of your position[4] and desiring him to inform you without delay whether he can get the papers done.[5] You and he will therefore be able to arrange it between you. It would however be a cause of vexation to me to see you return from Paris[6] on *[their]* account. If Debus cannot do them I will employ somebody though whom to employ is not now known to me. Were it another portion of the year I might do them myself: but my absence, and my Friday evening[7] the subject of which I am still ignorant of, will disable me from doing any thing at *[them]*. I must see whether Frankland or somebody else will undertake them for me. This I should prefer to stopping you in your journey to Italy.

Yours ever | John Tyndall

Your letters & papers shall be all sent by this post.

I shall be at Leighlin Bridge[8] towards the end of next week say on Friday. I shall walk the Wicklow[9] previously—I wish you were with me. I am sure if together we should both get up strength; for I find that mere bodily motion is not sufficient—the mind must at the same be healthily exercised—You may come to this creed in the end.

I should not like to ask permission to send the papers to France. It would be too bold to put in writing & I have no opportunity of seeing Donnelly.[10]

RI MS JT/1/T/977
RI MS JT/1/HTYP/529

1. *[3 May 1864]*: the date is given by relation to letter 2095, dated 14 May, in which Tyndall wrote that he went to Ireland a week ago, so around 7 May, a Saturday. The Tuesday before that was 3 May, so Tyndall would have left for Ireland on 4 May.
2. *I start for Ireland tomorrow morning early*: following two failed attempts to head to Ireland in April, Tyndall was finally able to escape London on Wednesday, 4 May 1864. He had previously wished to join Hirst in Ireland to support his friend during ill health, and then to go to Ireland upon the news of his own mother's ill health, but his duties tied him to London. While in Ireland he made an effort to visit other members of the Tyndall family, and he returned to London on 14 May (see letters 2094 and 2095).
3. *I have written to Debus*: letter missing.
4. *your position*: see letter 2080, n. 7.
5. *get the papers done*: Tyndall hoped that Heinrich Debus could help to mark examination papers that he had Hirst lined up to do (see letter 1926, n. 3).
6. *return from Paris*: after his trip to Ireland and finishing his teaching duties, Hirst travelled to France, Italy, and Switzerland (where Tyndall joined him). He left London on 23 April, and returned on 22 August, having spent four months away (Brock and MacLeod, *Hirst Journals*, pp. 1669–86).
7. *my Friday evening*: probably a reference to his Friday Evening Discourse at the RI the following month, 'On a Magnetic Experiment' on 10 June 1864 (*Roy. Inst. Proc.*, 4 (1862–66), pp. 317–22).
8. *Leighlin Bridge*: Leighlinbridge, County Carlow, Ireland, where Tyndall was born and his mother lived.
9. *Wicklow*: a town of County Wicklow, Ireland; presumably this is a reference to Wicklow gap, one of the country's highest mountain passes.
10. *Donnelly*: most likely John Fretcheville Dykes Donnelly (1834–1902), an Army officer and promoter of scientific education (*ODNB*). Donnelly was involved in the marking of examination papers, hence Tyndall's comment about sending papers to France.

To unidentified[1] [6 May 1864][2] 2092

Dear Friend,

Between 6 and 7 yesterday[3] we entered Kingstown Harbour,[4] with the western sun flashing up from the water on our faces. I noticed for some time before reaching Kingstown, a finely formed purple mountain & on enquiry found it to be the Great Sugar Loaf.[5] I spent last night in Bray,[6] and this morning with a boy to carry my pack I sallied forth to Roundwood & the Seven Churches.[7] But still that finely formed purple mountain loomed in the air above me, and I attacked it and now sit alone upon its top; whence these lines are sent to you. Interpret them as being filled with my love to you & yours[8]
J. T.

A most extraordinary haze is covering all the mountain; the air throughout is gradually thickening to opacity, and still the sun shines upon the head of the Sugar Loaf. Far adown its slopes the lambs are bleating, and I must descend lest this haze should thicken round me & cut off my retreat. I have no compass and the land is strange to me. Lebwohl![9]

RI MS JT/3/32/2

1. *unidentified*: the reference to 'you & yours' and the German salutation 'Lebewohl!' imply an older woman, possibly Juliet Pollock.
2. *[6 May 1864]*: the date is given as Tyndall left London for Ireland on 4 May 1864, and returned to London on 14 May 1864. During this trip to Ireland, Tyndall visited Wicklow and then south to Gorey and on to Carlow. See letter 2091.
3. *yesterday*: presumably 5 May 1864.
4. *Kingstown Harbour*: now named Dún Laoghaire, it is effectively the harbour for Dublin.
5. *the Great Sugar Loaf*: also known as the Big Sugar Loaf and often as simply the Sugar Loaf, a mountain in east County Wicklow in Ireland.
6. *Bray*: a coastal town in north County Wicklow, Ireland. It is situated about 12 miles south of Dublin city centre on the east coast.
7. *Roundwood & the Seven Churches*: Roundwood is a village in County Wicklow about 12 miles southwest of Bray. The Seven Churches is a monastery complex located near the village of Eoghanacht on the Aran Islands. Tyndall appears to have made a diversion on the way to climb the Great Sugar Loaf. The Seven Churches in the Glendalough valley nearby are the remains of seven early medieval churches, making up one of the most important monastic sites in Ireland.
8. *you & yours*: if this letter is to Juliet Pollock (see n. 1), 'yours' would refer to her husband William Frederick Pollock and their sons Frederick, Walter, and Maurice Pollock.
9. *Lebwohl!*: goodbye! (German).

From Rudolf Clausius [11 May 1864] 2093

Zürich 11/5/64.

Lieber Tyndall,

Ich habe die bezüglichen Stellen der Séguin'schen Schrift durchgelesen, und daraus die Überzeugung gewonnen, dass Séguin nicht den geringsten Anspruch auf das Verdienst, das mech. Aequivalent der Wärme bestimmt zu haben, machen kann.

Er betrachtet auf S. 385 bis 389 eine Quantität Dampf welche bei 100° das Volumen von 1 Cubikmeter einnimmt, und denkt sich dass diese zusammengedrückt oder ausgedehnt und dadurch um je 20° erwärmt oder abgekühlt wird, während von Aussen keine Wärme zugeführt oder entzogen wird. Séguin bestimmt nun (freilich auf sehr unvollkommenen Weise) die zur Zusammendrückung nöthige mechanische Arbeit. Aber von einer Bestimmung der entsprechenden Wärmemenge ist gar nicht weiter die Rede, denn man in der Tabelle unter der Rubrik: "Températures correspondantes à l'effet produit" steht: 20°, so sind das nicht 20 Wärmeeinheiten, denn wir haben es ja nicht mit einem Kilogramm Wasser, sondern mit einem Cubikmeter Dampf zu thun. Erst in einer Note, welche sich in den Comptes rendus von 1847 t. XXV p. 420 befindet, spricht Séguin von Wärmeeinheiten, indem er die specifische Wärme des Dampfes in Betracht zieht, und findet dadurch Zahlen, welche ungefähr den Joule'schen entsprechen. Wie er hier die Rechnung gemacht hat, ist nicht näher angegeben (es scheint mir dabei noch ein besonderes Versehen abzuwalten) aber jedenfalls hat er als specifische Wärme des Dampfes einen solchen Werth angenommen, wie man ihn damals für richtig hielt (de la Roche u. Bérard hatten gefunden 0,847). Nachdem nun in neuerer Zeit Regnault gefunden hat, dass der früher für die spec. Wärme des Dampfes angenommene Werth viel zu gross ist, u. statt dessen, 0,48 zu setzen ist, würde dieselbe Rechnung welche Séguin angestellt hat, um das mech. Aequivalent der Wärme zu bestimmen, zu ganz anderen Zahlen führen.

In der That aber ist die ganze Art der Rechnung unwichtig. Sie setzt nämlich unter anderen voraus, dass Wasserdampf, wenn er, ohne Wärme zu empfangen oder abzugeben, zusammengedrückt wird oder sich ausdehnt, dabei gerade dampfförmig und im Maximum der Dichtigkeit bleibt. Dieses ist aber nicht der Fall. Ich selbst und gleichzeitig auch Rankine haben gezeigt, dass der Dampf unter diesen Umständen bei der Ausdehnung sich zum Theil niederschlägt, und bei der Zusammendrückung überhitzt wird. Hierdurch wird die ganze Betrachtung von Séguin unhaltbar.

In einem Artikel der North British Review "The Dynamical Theory of Heat", welchen Sie jedenfalls kennen, ist auf S. 52 von den von Séguin und Mayer ausgeführten Bestimmungen des mech. Aequivalent der Wärme so gesprochen, als ob sie ungefähr gleich gut wären. Von Séguin ist gesagt er habe gefunden: 363. Dieses ist die Zahl welche man erhält, wenn man in der Tabelle auf Seite 389 die Zahl 20° als 20 Wärmeeinheiten betrachtet, und demgemäss den Bruch $\frac{7870}{20}$ = 369,5 bildet. Diese Rechnung ist aber, selbst wenn man sonst Séguin's Betrachtungen als zulässig ansehen wollte, ganz falsch. Denn wenn eine Dampfmenge, welche bei 100° den Raum von 1 Cubikmetter nimmt, und welche nach Séguin (Siehe S. 382) dem Gewichte nach 0,588 kil beträgt, ihre Temperatur um 20° ändert, so entspricht das nicht 20 Wärmeeinheiten. Überhaupt ist es mir unbegreiflich, wie man Séguin's Art des mech. Aequivalent der Wärme zu bestimmen mit Mayer's auch nur vergleichen kann. Wenn auch in Mayer's Betrachtungen in dem Aufsatze von 1842 noch einigen Unklarheiten vorkommen, so ist doch sein Verfahren, das mech. Aequivalent der Wärme zu bestimmen, dem Principe nach richtig. Séguin's Verfahren dagegen ist geradezu falsch.

Ich würde Ihnen rathen, einfach die Partie aus Séguins Buch von Seite 385, "Supposons donc etc" bis inclusive der Tabelle auf Seite 389 abdrucken zu lassen, damit das wissenschaftliche Publicum dem die Schrift von Séguin meistens nicht zugänglich ist, selbst urtheilen kann, ob dieses eine wichtige Bestimmung des mechanischen Aequivalentes der Wärme ist.

Ich bin eben dabei beschäftigt, von meinen Abhandlungen über die mechanische Wärmetheorie eine Gesammtausgabe heraus zu geben, und zwar zunächst in einem Bande die Abhandlungen, welche die eigentliche Begründung der Theorie, u. ihre Anwendung auf die in die Wärmelehre gehörigen Eigenschaften der Körper und auf die Dampfmaschine enthalten. Dagegen die Anwendungen auf die Electricität und die Betrachtungen über die Molecularbewegungen will ich erst später folgen lassen. Ich lasse die Abhandlungen wörtlich abdrücken, aber unter Zufügung von erläuternden Anmerkungen und Zusätzen, welche vom Texte getrennt sind, um das ursprünglich gedruckte von dem jetzt hinzugefügten zu unterscheiden. Ich muss mir nämlich meine Priorität wahren, besonders gegenüber von Thomson und Rankine, welche unterstützt von Tait, sich immer gegenseitig loben und die Verdienste für sich in Anspruch nehmen, und meine Arbeiten möglichst unerwähnt lassen, oder, wenn das nicht geht, nur ganz obenhin berühren.

Wenn Sie und Hirst sich noch entschliessen zu unserer Naturforscherversammlung zu kommen, werden wir uns sehr freuen. Sobald Sie Sich darüber sicher entschlossen haben, sind Sie wohl so gut es mir zu schreiben. Ich wiederhole hiermit die Bitte, dass Sie für diesen Fall beide bei uns logiren.

Mit herzlichen Grüssen von meiner Frau und mir,

Ihr | Clausius

Die Partie aus Séguin's Buch schicke ich beifolgend wieder zurück. Für die Zusendung des Artikels von Akin danke ich Ihnen sehr. Den letzten Theil welcher sich auf mich bezieht, hatte er selbst mir schon zugesandt. Ebenso danke ich Ihnen sehr für die Zusendung Ihres interessanten Vortrages, der wieder viel dazu beiträgt, die wichtige Frage der strahlenden Wärme zu [*illeg*]

Zurich 11/5/64.

Dear Tyndall,

I have read the respective passages in Séguin's work[1] and am convinced that Séguin cannot make the least claim for the achievement of determining the mechanical equivalent of heat.

On p. 385 to 389 he considers a quantity of steam that takes up 1 cubic meter at 100° and imagines that it is compressed or expanded and in this way heated or cooled by 20°, during which no heat is externally introduced or withdrawn.

Séguin then determines (freely in a very incomplete manner) the amount of mechanical work necessary for the compression. But there is no mention of the determination of a corresponding amount of heat, since if one looks at the table under the heading "Températures correspondantes à l'effet produit" it says 20°, but that does not mean 20 units of heat, since we are not dealing with a kilogram of water, but with a cubic meter of steam. Séguin first speaks of units of heat in a note that is found in the Comptes rendus[2] from 1847 t. XXV p. 420, in which he considers the specific heat of the steam and comes to figures that are somewhat analogous to Joule's. How he did the calculations is not described more closely (this seems to me to be peculiar oversight), but in any case he assumed as the specific heat of the steam a value that was considered to be correct at the time (de la Roche and Bérard[3] had found 0.847). After Regnault discovered quite recently that the value that had previously been assumed for the spec[ific] heat of steam was too large and instead it should be set at 0.48, the same calculation that Séguin had employed to determine the mech[anical] equivalent of heat would lead to completely different figures.

In fact, the entire mode of calculation is unimportant. It assumes, namely, among other things that when water steam is compressed or expands without receiving or giving off heat, that it remains in the state of steam and the maximum of density. But this is not the case. I, myself, and Rankine simultaneously have shown that under these circumstances a part of the steam condenses during expansion, and is overheated during compression. In this way Séguin's entire observation becomes untenable.

In an article[4] in the North British Review "The Dynamical Theory of Heat", which you[5] surely know, the author speaks on p. 52 about Séguin's and Mayer's determination of the mech[anical] equivalent of heat, as if they were both more or less equally good. He says in regard to Séguin, he found: 363. This is the figure one gets, when the number 20° in the table on page 389 is taken as 20 units of heat and accordingly makes the fraction of $\frac{7870}{20} = 369.5$. This calculation, however, is incorrect, even if one was inclined to consider Séguin's observations otherwise as acceptable. If a quantity of steam that has a volume of 1 cubic meter at 100° and which, according to Séguin (see p. 382) has a weight of 0.588 kil changes its temperature by 20°, that does not correspond to 20 units of heat. On the whole, I cannot comprehend how one can compare Séguin's method of determining the mech[anical] equivalent of heat with Mayer's at all. Even if there is some obscurity in Mayer's observations in the papers from 1842, his method of determining the mech[anical] equivalent of heat is still correct in principle. Séguin's method in contrast is downright wrong.

I would advise you to simply have the passages out of Séguin's book[6] from page 385 "Supposons donc etc" up to and including the table on page 389 printed so that the scientific audience, most of whom do not have access to the writings of Séguin, can judge for itself whether this is a significant determination of the mechanical equivalent of heat.

I am occupied at the moment with editing a collected edition of my papers[7] on the mechanical theory of heat, that is, firstly, one volume that includes the papers that lay the actual foundations for the theory, and its application to the characteristics of bodies that belong to the theory of heat and to the steam engine. I intend to follow this later, however, with the applications to electricity and observations on molecular motion. I will have the papers printed verbatim, but with the inclusion of explanatory notes and addendum, which will be separate from the text, so that the original version can be differentiated from what was newly added. I must guard my priority, in particular from Thomson and Rankine, who are supported by Tait, and constantly praise one another and claim for themselves the achievement, and do their best not to mention my work, or, when that is not possible, only briefly touch upon it.

It would make us very happy, if you and Hirst would decide to come to our meeting of natural scientists.[8] Please be so good to let me know as soon as you have reached a decision. I repeat again my request, that in this case, you both reside with us.

With fond regards from my wife[9] and I,

Your | Clausius

I am sending the passages from Séguin's book back with this. I thank you very much for sending Akin's article.[10] He sent me the final part, which refers to me, himself already. I thank you as well for sending your interesting lecture,[11] that once again contributes much to supporting important questions on radiant heat.

RI MS JT/1/TYP/7/2264–6
LT Typescript Only

---

1. *Séguin's work*: see letter 1990, n. 2.
2. *note that is found in the Comptes rendus*: M. Séguin, 'Note a l'appui de l'opinion emise par M. Joule, sur l'identite du mouvement et du calorique', *Comptes Rendus*, 25 (1847), pp. 420–2.
3. *de la Roche and Bérard*: François-Étienne de la Roche (1743–1812), a Genevan physician, naturalist, and chemist (D. D. Damkaer, *The Copepodologist's Cabinet: A Biographical and Bibliographical History* (Philadelphia: American Philosophical Society, 2002), pp. 144–7). Jacques Etienne Bérard (1789–1869), a French naturalist, chemist, and physicist (*CDSB*). The reference is to F. Delaroche and J.-E. Bérard, 'Memoire sur la determination de la chaleur spécifique des différens gaz', *Annales de chimie et de physique*, 85 (1813), pp. 72–110, and 113–182.
4. *an article*: P. G. Tait, 'The Dynamical Theory of Heat', *North British Review*, 1 (1864), pp. 337–68.
5. *you*: Clausius here used the formal address form 'Sie'.
6. *Séguin's book*: see n. 2.

7. *collected edition of my papers*: in 1864, Clausius began collecting his papers on theories of heat, published as *Abhandlungen über die mechanische Wärmetheorie* (Braunschweig: Vieweg und Sohn, 1864). It was later published in an edited and translated edition by Tyndall and Hirst respectively, as R. Clausius, *The Mechanical Theory of Heat with its Application to the Steam Engine and to the Physical Properties of Bodies* (London: John van Voorst, 1867).
8. *our meeting of natural scientists*: see letter 2048, n. 2.
9. *my wife*: Adelheid Clausius.
10. *Akin's article*: C. K. Akin, 'Note on Ray-Transmutation', *Phil. Mag.*, 28 (1864), pp. 554–60. This letter and enclosure have not been found.
11. *your interesting lecture*: see letter 2074, n. 3.

To Annette Roche 14 May 1864 2094

London, 14th, May, 1864.

My dear Miss Roche

Yesterday at this hour I was under your roof[1]—to day I am in London, having travelled all night by the mail packet and train. I am anxious to express to you more fully than I did yesterday how kind I thought it on your part to write to me as you did,[2] and also to show you that I put a substantial estimate on your kindness by asking you to help me in a little matter where a lady alone can render proper aid.

It has been my strong desire that my mother[3] should spend the evening of her life, possessed of every comfort—that she should be neatly apparelled, carefully tended, and in want of nothing that could contribute to the pleasure of her existence. I confess that the appearance both of her and my sister[4] yesterday fell short of what I could have desired. Probably recent trials,—which I have a strong hope are merely evanescent—may account for this, but I want you to help me to alter it. I want you in your kindness to choose proper and becoming apparel both for my mother and sister.

Very many years ago I asked my mother to state the sum on which she could live respectably in Leighlin Bridge.[5]—She mentioned 40 pounds a year—I made it 60, and besides this I was ever ready to meet such extra demands as special circumstances might cause to arise. Within the last 12 months 80 pounds have been sent to her. How it is disposed of I do not know, but I was a little distressed yesterday to find one particular on which I have always laid some weight so imperfectly attended to.

My mother informed me yesterday that she had plenty of money. On the 1st of March I sent her her quarterly allowance of fifteen pounds, and about three weeks ago I sent her ten pounds without any application on her part for money. I paid the physician's[6] bill of five pounds yesterday myself. I moreover greatly fear that I shall be 60 pounds out of pocket purely through obliging a gentleman whom my mother reported to be very kind to her. In fact I have been most anxious to forward her wishes and promote her comfort in every way, and my visit to Leighlin 4 years ago induced me to believe that my desires were fulfilled. But yesterday I was not satisfied, and your note to me was all the more precious on that account.

I would ask you to allow me to send you the money necessary for such dresses for my mother and sister as your taste will suggest to you as fit and becoming—dark, sober and decent—but it is I fear worse than useless to leave money in my mother's hands, and she has or ought to have funds sufficient for this purpose. I write to her by to day's post[7] desiring her to place these funds at your disposal, and I should be very grateful to you if you would dispose of them so as to enable both her and my sister to appear in a manner not unworthy of them and me.

I was also cast down to find that both my mother and sister spent their time mainly in a bedroom. I knew nothing of this before, and determined to have it remedied. Miss Wall[8] undertook to have a vacant room downstairs papered and carpeted, the windows repaired and the room rendered quite comfortable. But Miss Wall mentioned to me that she was going to quit Leighlin for six weeks, and if this be the case I must put the matter in other hands, for the room will be most needed during the coming summer weather. I this day write to Miss Wall[9] enquiring whether she can set about the work at once and if not I must seek some other means of having the work speedily accomplished. The only thing however that I should venture to trouble you with is the appareling question, and this I trust to your kindness and goodness to solve for me.

If I am wrong in supposing that my mother has ample funds in her hands pray be good enough to draw upon me for the necessary amount.

Believe me dear Miss Roche | Most truly Yours | John Tyndall

Now I think I have given you the very strongest proof in my power to give of faith in your kindness—a faith founded as much upon that intuitive sympathy by which one human being divines the character of another as upon any thing else.

RI MS JT/1/TYP/11/3848–9
LT Typescript Only

---

1. *under your roof*: Tyndall had travelled to Ireland earlier in May in order to visit with members of his family.
2. *to write to me as you did*: letter missing, although it is clear that it was Roche that alerted Tyndall to his mother's ill health.
3. *my mother*: Sarah Tyndall.

4. *my sister*: Emma Tyndall.
5. *Leighlin Bridge*: see letter 2091, n. 8.
6. *physician's*: presumably a physician to Tyndall's mother, identified as 'Dr Phillips' in letter 2101.
7. *I write to her by to day's post*: letter missing.
8. *Miss Wall*: presumably the landlady.
9. *I this day write to Miss Wall*: letter missing.

To George Hector Tyndale 14 May 1864 2095

My dear Hector.

I pray that this may find you unharmed amid the perils of your present vocation.[1] I scan the papers eagerly in search of some intelligence regarding you, but thus far in vain.

Tom Hirst sent you some time ago a little biographical narrative written by himself[2]—I hope it reached you safely, I have been recently in correspondence with and paying visits to some people of our name and family[3] an account of which I will now give you: it may amuse you in camp.

The Tyndall de Veer who wrote to me from Holland has written to me frequently since.[4] You know he wished to learn whether we were related, and I replied that if he belonged or was related to the Irish Tyndall's there was very little doubt of it; but on this point he knew nothing but he wrote to his uncle in the West Indies, and he, after giving a genealogy of the Tyndalls stated that he himself had gone to Dublin in 1819 to visit a relation of his, one of the Irish Tyndalls.

There is a fine gentlemanly fellow named John Warre Tyndale resident here in London: I have dined at his house a few times: his wife and daughters are extremely intelligent and perfect ladies in demeanour. His walls are covered with portraits of old Tyndalls including one of the Martyr.[5] He and his wife believe firmly that we are related. He has a brother-in-law a barrister named Greenfield,[6] an extremely intelligent and accomplished man—and in 1843 he drew out a detailed genealogy of the family of Tyndall in which he carried them by rigid links up to the line of Plantaganet.[7] He is now at work in connexion with the martyr, and he has traced the emigration of a branch of this family into Ireland. He came to me not knowing that I was an Irish Tyndale, and pointed out the William Tyndale, residing in Dublin in 1676 as the origin of the Irish Tyndalls.

A week ago I went over to Ireland,[8] in part to relieve my confused head; and while there I ran down along the eastern coast, and met some of the representants of our family now resident there. One of them is a physician in Gorey:[9] a very respectable fellow, who with small means has managed to bring up a heavy family in a highly creditable manner. He has a nice house and a charming garden—altogether my visit to him pleased me much. There are some farmers of the name also in the neighbourhood, and some illegitimate children. I went on to New Ross,[10] and near that town I found the biggest of the Tyndalls that I had met with. His name is Robert Tyndale, the proprietor of a beautiful place named Oaklands: He has a large estate in the Co. Kilkenny and was some years ago High Sheriff of the County. His place is beautiful and I had a delightful ramble with his wife and a most charming daughter. —she is quite a beauty—I did not care very much for her mother; she was a large strong-minded woman, and I do not feel much interest in women of this class Tyndall himself seems an exceedingly nice fellow.

And now I have told you briefly what I have recently learned. Are we never to have done with that fearful war? I wish you would write to me,—for many months I have heard nothing about you. Your name has not appeared in the papers, and hence I draw the consolation that at all events you are still alive. My brave Swiss guide Bennen was smashed a few weeks ago by an avalanche:[11] he was climbing a mountain in company with two gentlemen and two local guides—when near the top the snow gave way and went down with a fearful rush carrying the party with it. Bennen and one of the gentlemen were killed.

Hirst's health has been very unsatisfactory during the past half-year: he is now on the continent, and I hope he will return invigorated. He has a great affection for you, and like myself is very anxious to hear something about you.

And still a life of battle would be preferable to me to a life of stagnation. Our brains were never made to be stagnant—action of some kind, taxing both brain and muscle is necessary to our health, and I would sooner be fighting at your side, or rather on the other side than living a life of inaction on the fairest portion of the earth's surface, I told the Count of Paris[12] some months ago that I had had a cousin at Antietam[13]—"Oh said he "he was there under a fine general—that was McClellan's battle"[14]—the Count evidently rates McClellan high.

Now sit down, if you are alone, and write to me—tell me how you spend your days—tell me what you think regarding the continuance of the war. The exponent of Federal news in the House of Commons[15] is a personal friend of mine. We shall probably make a tour together this year in the Caucasus—I mean Forster the member for Bradford—He is a very able fellow and perfectly straightforward. Give my kindest regards to Mrs Tyndall[16] and to your sister[17]—goodbye dear Hector and believe me always

Your affectionate cousin | John Tyndall

RI MS JT/1/TYP/5/1661–3
LT Typescript Only

1. *your present vocation*: Tyndale was a general in the American Civil War (1861–5).
2. *little biographical narrative written by himself*: Hirst contributed a brief life of Tyndall to L. Reeve (ed), *Portraits of Men of Eminence in Literature, Science, and Art, with Biographical Memoirs. The Photographs from Life, by Ernst Edwards, BA* (London: Lovell Reeve & Co., 1864), vol 2, pp. 25–32 (the portrait of Tyndall faced the first page of the memoir). Hirst noted in his journal entry of 22 March 1864 that 'On Saturday I commenced writing a sketch of Tyndall's life to be published in the series of lives with Portraits by Edwards. I had not the time to give it the finish I desired for I had but the two days to finish it. Nevertheless I warmed towards the work as I wrote' (Brock and Macleod, *Hirst Journals*, p. 1665). It is likely that Hirst sent Tyndale a copy of what he wrote rather than a copy of the published volume.
3. *paying visits to some people of our name and family*: Tyndall left London for Ireland on 4 May 1864 and returned during the night before he wrote this letter. See also letters 2091 and 2094.
4. *The Tyndall de Veer who wrote to me from Holland has written to me frequently since*: letters missing; see letter 2019.
5. *one of the Martyr*: William Tyndale.
6. *Greenfield*: B. W. Greenfield (dates unknown), who in 1843 privately printed 'Genealogy of the Family of Tyndale (of Hayling Island, Hants)' (see 'The Tyndalls of Hockwold', in R. E. C. Waters, *Genealogical Memoirs of the Extinct Family of Chester of Chicheley, their Ancestors and Descendants*, 2 vols (London: Robson and Sons, 1878), vol. 1, pp. 263–77, in footnotes of p. 267).
7. *line of Plantaganet*: a royal house in France from the twelfth to fifteenth centuries.
8. *A week ago I went over to Ireland*: see n. 3. See also letter 2094.
9. *a physician in Gorey*: John Tyndall (of Gorey).
10. *New Ross*: a town in southwest County Wexford, Ireland.
11. *was smashed a few weeks ago by an avalanche*: this is well described in C. Gos, *Alpine Tragedy* (London: George Allen and Unwin, 1948), pp. 18–23.
12. *the Count of Paris*: Prince Philippe of Orleans (1838–94), Count of Paris, grandson of King Louis Philippe I. He was Count of Paris from 1848–94.
13. *Antietam*: see letter 1946, n. 7.
14. *McClellan's battle*: shorthand for the Battle of Antietam (see note 13). General George Brinton McClellan (1826–85) stopped General Robert E. Lee's invasion of the north at Antietam. Although this has subsequently been recognised as a decisive battle in the war, McClellan's tactics and caution have subsequently been heavily criticized.
15. *The exponent of Federal news in the House of Commons*: William Edward Forster (1818–86), a British politician; M.P. for Bradford (*ODNB*).
16. *Mrs Tyndall*: Julia Tyndale.
17. *your sister*: Elizabeth Mitchell.

To Thomas Archer Hirst 17 May 1864[1] 2096

17th May 1864

My dear Tom

It pleases me exceedingly to hear such a good account of you.[2] It shews me that there is nothing whatever organically wrong with you. That like myself you simply suffer from a weak stomach which will not bear too much intrusion from the head. I hope you will continue to improve and that when I see you next, the pallor will have disappeared from your cheeks.

Debus is getting on manfully with the examination papers:[3] I have no doubt as to his ability to complete the task.

I called at Gorey[4] & saw Dr. Tyndall[5] and his family. They pleased me much. They are nice people, have a nice house and charming garden. I then went on to New Ross[6] and saw the big Tyndall.[7] Unhappily he and two of his daughters were going out to spend the day at the home of the young ladies' grandfather. But I met them on the way & conversed with them. I liked Tyndall very much. He was a fine looking gentlemanly fellow. His wife and one daughter were at home, and them I visited and had a stroll with them through their grounds. The place is a very fine one. The house large, the grounds well planted, and commanding a beautiful view of the valley of the Barrow.[8] The young lady was beautiful and clever:—dressed with great piquancy and altogether a charming creature. I did not like her mother quite so much. She was large and I thought a little coarse.

I went then to Leighlin[9] and found things better than I expected. In fact while I was there I noticed not the slightest difference in the deportment of the invalid[10] from what I had always observed. She looked a little more haggard. The doctor told me the Dean[11] was quite wrong in taking the matter I mentioned to you seriously. It was simply a portion of the delusion, and a portion which has now quite passed away. On the whole I am in hopes that by proper care the malady may be cured.

I have just had an extremely interesting letter from Clausius.[12] I had sent him[13] the work of Seguin[14] and asked his opinion on it. He is very decided. He has not the slightest claim, and how any man could think ever of comparing him with Mayer Clausius is at a loss to conceive. He complains much of the conduct of those northerners towards himself. "Sie loben sich gegenseitig"[15] and pay no attention to the just claims of others. And now good bye dear Tom.

Yours ever | John Tyndall

I quite enter into your feelings with regard to the book.[16] It was a matter that I could not alter;[17] still it was not to my taste.

It is here out of print.

Dear Tom. This was written long ago. I now send it. I will write again to you tomorrow or next day.[18] I send you everything that has come except mere book circulars & Railway Reports. Monday 30th. May

RI MS JT/1/T/663
RI MS JT/1/HTYP/528

---

1. This letter, although written on 17 May 1864 was, as Tyndall indicated at the end of the letter, sent on 30 May 1864.
2. *such a good account of you*: letter missing.
3. *Debus is getting on manfully with the examination papers*: Tyndall sought Debus's help in grading examination papers for him (that Hirst was originally lined up to do) while Hirst went to Ireland and then France and Italy. See letter 2091, n. 5 and n. 6.
4. *Gorey*: a market town in north County Wexford, Ireland.
5. *Dr. Tyndall*: John Tyndall (of Gorey).
6. *New Ross*: see letter 2095, n. 10.
7. *the big Tyndall*: Robert Tyndale (see letter 2095).
8. *the Barrow*: a river that flows through Leighlinbridge, one of the Three Sisters and the second-longest river in Ireland.
9. *Leighlin*: Leighlinbridge (see letter 2091, n. 8).
10. *invalid*: Tyndall's mother Sarah Tyndall, who had fallen ill.
11. *the Dean*: see letter 2090, n. 3.
12. *letter from Clausius*: letter 2093.
13. *I had sent him*: letter missing.
14. *work of Seguin*: see letter 1990, n. 2.
15. *"Sie loben sich gegenseitig"*: 'They praise each other' (German).
16. *the book*: probably a reference to *Heat as a Mode of Motion.*
17. *It was a matter that I could not alter*: reference unclear.
18. *write again to you tomorrow or next day*: letter 2098.

To George Gabriel Stokes [30 May 1864][1] 2097

Royal Institution of Great Britain | Monday

My dear Stokes,

All that you suggest[2] shall be done.

The lengths of the chambers <u>are</u> recorded in the tables from which they are extracted. If you think it would improve the paper[3] to restate them <u>that</u> shall also be done.

It is the only paper that I have ever had wholly done by my assistants. I have severely checked them, but still I had not the same mastery over it as if I had made the experiments myself.

Almost all the alterations are due to one simple cause—namely the calibration value of the total heat was somewhat incorrect, and this ran through all the reckoning.

Yours very truly | John Tyndall

RI MS JT/1/T/1430
RI MS JT/1/TYP/4/1479

1. *[30 May 1864]*: the date is given by relation to letter 2099, which is Stokes' reply of Tuesday, 31 May 1864. This letter, written on a Monday, would then be 30 May 1864.
2. *All that you suggest*: letter missing.
3. *the paper*: this is a reference to J. Tyndall, 'The Bakerian Lecture. —Contributions to Molecular Physics. —Being the Fifth Memoir of Researches on Radiant Heat', *Phil. Trans.*, 154 (1864), pp. 327–68; this had been received by Stokes on 17 March. See also letters 2099, 2112, 2114, 2115, 2117, 2119, and 2136.

To Thomas Archer Hirst 31st May 1864 2098

31st. May 1864.

My dear Tom

I wrote to you yesterday[1] and enclosed such documents as had come here for you. This morning some proofs have come from Longman[2] which I send you, and a long memoir from Signor Cremona,[3] which, as you are now probably near its author, I have detained.

My health has been very shaky[4] of late, but by care I am petting it and making it strong. It has been a mixed disturbance of body and mind; but the body as in most cases has been the chief cause of my suffering. My mental ills I have driven away by a kind of occupation which has interested me for the last week or so. In connexion with the new edition of my book I have had to look over the second Essay of Mayer,[5] which is as yet unpublished in England. And I was so much struck by its beauty and importance that I set to to make a kind of resume[6] of it. I have connected it with the work of others, and put Mayer in the position which he ought to occupy in the current of history. In doing this I have had occasion to refer to controverted points and at first feeling the power which the strength of my case conferred upon me I was disposed to use a strong hand; but I returned to my work again and again, discharging more and more personal feeling from my heart, and finally attaining such perfect philosophical tranquility that no one could detect a trance of temper in what I have written. And I have thus purged my writing of personal feeling without diminishing its strength in any degree. As it at present stands I think it will produce general conviction, and make my enemies somewhat ashamed of themselves.

This piece of work has thrown my preparations for my Friday evening[7] in the back ground; but I will commence them to day and God knows ten days are sufficient to expend upon a thing of the kind. I intend to give them an evening on magnetism,[8] in which I will gather a number of interesting experiments round a kernel of original matter which belongs to myself.

Lubbock and Mrs Lubbock[9] are going to Turin this week. Write to her Poste Restante[10] there. She speaks of leaving Lubbock in Turin to transact his banking business and of going to Bologna to pay you a visit! I have not yet decided any thing regarding my summer campaign, but today I must try to meet my friend Forester[11] and to arrange something.

I have been very much alone of late and have but rarely dined out. After my slice of mutton at dinner I have strolled up the Finchley Road and sought the groves of Hampstead. Hampstead is lovely just now, and I like a peaceful stroll there. I return between nine and ten, and find that exposure to the air helps me to sleep.

Sylvester has been enquiring about you many times since you went away.

I don't think I have any thing more to say to you worth hearing. The news from Leighlin Bridge[12] is better on the whole. I hope the cloud will pass away.

Believe me dear Tom | Ever yours affectionately | John Tyndall

RI MS JT/1/HTYP/531
LT Typescript Only

---

1. *I wrote to you yesterday*: letter 2096, written on 17 May but not sent until 30 May.
2. *Longman*: either Thomas Longman (1804–79) or William Longman (1813–77), brothers who ran the publishing firm Longman, Green, Longman, and Roberts.
3. *Signor Cremona*: Antonio Luigi Gaudenzio Giuseppe Cremona (1830–1903), an Italian mathematician and professor of geometry at the University of Bologna (*CDSB*).
4. *My health has been very shaky*: see letter 2090.
5. *second Essay of Mayer*: see letter 1904, n. 3.
6. *kind of resume*: J. Tyndall, 'Notes on Scientific History', *Phil. Mag.*, 28 (1864), pp. 25–51.
7. *preparations for my Friday evening*: see letter 2091, n. 7.
8. *an evening on magnetism*: see letter 2091, n. 7.
9. *Lubbock and Mrs Lubbock*: John Lubbock and Ellen Lubbock.
10. *Poste Restante*: general delivery (French); when a post office holds mail until the recipient requests it.
11. *Forester*: probably William Edward Forster (see letter 2095, n. 15), whom Tyndall had been planning to climb with. See also letter 2110.
12. *The news from Leighlin Bridge*: a reference to Tyndall's mother's health. See letters 2090 and 2095.

From George Gabriel Stokes 31 May 1864 2099

Lensfield Cottage Cambridge | 31st May 1864

Dear Tyndall,

I send you by book post your paper[1] for the insertion of the lengths of chamber in p. 24 and the following few pages.

As the change in the numbers is very much of the nature of a mere numerical correction, <Hereafter LT Typescript Only> I hardly think the foot note I proposed[2] is required.

Yours very truly | G. G. Stokes

RI MS JT/1/S/233
RI MS JT/1/TYP/4/1388

---

1. *your paper*: see letter 2097, n. 3. See also letters 2112, 2114, 2115, 2117, 2119, and 2136.
2. *the foot note I proposed*: letter missing.

From Roderick Murchison 1 June 1864 2100

16. Belgrave Square. | June 1 1864

Dear Tyndall

In my last Anniversary Address to the Geographers[1] now going through to press I have a long article versus the power of ice to excavate solid rocks.

It was with glacier subjects brought under my *[notice]* by descriptions of the glaciers of Tibet & New Zealand I felt compelled to make a declaration of my faith & to differ in toto from my valued friend Ramsay.

In speaking of what Ramsay suggests & believes[2] I have added: and Tyndall an eminent physicist, who has opened out new views on the regelation & other properties of ice, has even speculated on its power to have excavated the chief vallies of the Alps as well as the deeper depressions occupied by water"[3]

I have not read anything of yours to this effect,[4] but if you have so spoken or written pray let me know the fact & reference.

Or if not, pray convey to me in a few words as may be, what your opinion is.

Yours sincerely | Rod. I Murchison

In another place I allude to you as writing that "a glacier if not an elastic or viscous bodies, is at all events plastic through the regelation of its fragments" If Hopkins's fragmentary theory[5] was truth y[ou]r line shews how. Is this the fact & if not enlighten me.

My arguments are all purely geological.

RI MS JT/1/M/148

1. *Anniversary Address to the Geographers*: R. Murchison, 'Address to the Royal Geographical Society', *Proceedings of the Royal Geographical Society of London*, 8 (1863), pp. 170–254.
2. *In speaking of what Ramsay suggests & believes*: see letter 2146.
3. Murchison neglected to use an opening quotation mark for this sentence.
4. *I have not read anything of yours to this effect*: two years before, Tyndall had published a short note on the conformation of the Alps (see letter 1921, n. 6). His longer article on the subject would be published later in October 1864: J. Tyndall, 'On the Conformation of the Alps', *Phil. Mag.*, 28 (1864), pp. 255–71. In this Tyndall gave the following postscript: 'The foregoing paper was in the printer's hands before it was my privilege to read the last Anniversary Address to the Geographical Society by its President, Sir Roderick Murchison. I have since considered the arguments, and given, I trust, due weight to the authorities urged and cited in that excellent Address against the theory of erosion, as applied to the valleys of the Alps. But the effect on my mind is not such as to induce me to alter the opinions, based on observed facts, which I have ventured to express in these pages' (p. 271).
5. *Hopkins's fragmentary theory*: see letter 1908, n. 1.

To Annette Roche 2 June 1864 2101

Royal Institution | 2nd. June. 1864.

My dear Miss Roche

Thank you a thousand times for your kind note.[1] I send you two abominable photographs. Nature did not give me the power of sitting or standing still and when I attempt to do so I am a mere log of wood.

I have written to Dr Phillips[2] asking him to have the kindness to push forward the preparation of the room. Miss Wall[3] was very good to consent to superintend it—but her absence prevents me from availing myself of her kindness. When I mentioned the room I meant to have it done at once and not by any means to wait.

Yours most truly | John Tyndall

RI MS JT/1/TYP/11/3849
LT Typescript Only

1. *your kind note*: letter missing.
2. *Dr Phillips*: presumably a physician to Tyndall's mother.
3. *Miss Wall*: see letter 2094, n. 8.

From George Wynne 4 June 1864 2102

London | 4 June 1864

My dear Tyndall

I have just returned to town and have to thank you very much for your kind present to Frank.[1] Coming from you it will be an additional inducement to him to study the books you send him.

I should like very much to hear your lecture.[2] Whilst I am in town could you give me tickets for myself and a friend, provided the request is not unusual. I do not call upon you because I know your time is so much occupied, but I will see you before I go.

I shall be very much gratified by being mentioned in your autobiography,[3] I can claim the small merit of having had some discernment, I was first attracted by your Gentleman-like address and I then perceived that I had no ordinary man to deal with, but I never anticipated that the Civil ass[istan]t. was so completely to give his old master[4] the go by—farewell.

I am most truly yours | Geo Wynne

RI MS JT/1/W/96
RI MS JT/1/TYP/5/1861

1. *Frank*: Frank George Wynne.
2. *your lecture*: see letter 2091, n. 7.
3. *your autobiography*: this is perhaps the first indication of Tyndall's intent to write an autobiography. He tried to write it later with his wife Louisa, but their efforts were never completed.
4. *his old master*: a reference to Tyndall having worked under Wynne as a surveying assistant in the Irish Ordnance Survey.

To Lyon Playfair[1] 7 June [1864][2] 2103

Royal Institution of Great Britain | 7th June

My Dear Playfair

Thanks for your letter,[3] and the same ten times over for your kind offer. It would delight me to lecture at Edinburgh, but it is out of the question. I am here in a net which cannot be broken.

In a few days I hope to have the pleasure (& I hope she will grant it to me) of sending to Mrs Playfair[4] a copy of my book[5] from which she will learn all that I have to say upon glaciers—There is a type of romance in my relationship to Edinburgh. I little thought 10 years ago when I read my first paper there[6] that this interval would give me so many kind friends there.

I am very weary at present—giddy & confused—when I get my head in order I will ponder on your ice theory.[7] I am now aiming to get off to the lakes of Killarny[8] in a few days.

Believe me dear Playfair | most sinc[ere]ly yours | John Tyndall

Letter 671, John Tyndall, Lyon Playfair Letters, Imperial College Archives

---

1. *Lyon Playfair*: Lyon Playfair (1818–98), a Scottish chemist and politician. Playfair worked in Justus von Liebig's laboratory in Giessen (1839–41). From 1842, he was Professor of Chemistry at the Royal Manchester Institution, and from 1845, chemist to the Geological Survey of Great Britain and professor of chemistry at the School of Mines in London. In 1853 Playfair was appointed secretary for science in the Department of Science and Art. In this position, he attempted to establish a national system of science education. Playfair maintained a close relationship with the Royal family and especially Albert, the Prince Consort, and for a time educated the Queen's sons Albert and Alfred. In 1858, he became Professor of Chemistry at Edinburgh University (*ODNB*).
2. *[1864]*: the date on this letter is from Playfair's record-keeping system.
3. *your letter*: letter missing.
4. *Mrs. Playfair*: Jean Ann Playfair (née Millington, 1834–77), married in 1857. She was Playfair's second of three wives (L. G. Pine (ed), *The New Extinct Peerage 1884–1971: Containing Extinct, Abeyant, Dormant and Suspended Peerages with Genealogies and Arms* (London: Heraldry Today, 1972), p. 219).
5. *my book*: *Glaciers of the Alps.*
6. *my first paper there*: Tyndall read his first paper at the Edinburgh meeting of the BAAS in 1850: J. Tyndall, 'On the Magneto-Optical Properties of Crystals', *Brit. Assoc. Rep. 1850*, p. 23 (no abstract was provided).
7. *your ice theory*: Playfair's letter to Tyndall is missing, and thus anything he may have written about ice. In two previous letters in 1860, however, Playfair shared with Tyndall his view about the differences between ice and water as chemical bodies (see letters 1702 and 1719, *Tyndall Correspondence*, vol. 7).
8. *Killarny*: Killarney, a town in Country Kerry, in southwest Ireland.

To Capel Henry Berger[1] 11 June 1864 2104

11th June 1864

Dear Sir.

I thank you warmly both for your letter & for the accompanying circular.[2] As regards science, & indeed as regards human action in general, I always fall back upon the creed of Gamaliel—If it be of God—that is if it be true—I cannot overthrow it—if it be of men—that is if it be untrue—it will come to naught.[3]

As far as I am concerned therefore I desire that science should proceed as it has hitherto done, slowly & surely laying hold of the convictions of thinking men. I do not think the amount of opposition which it meets with in the world very detrimental to its health & progress; and I should prefer allowing it to bide in own natural time to making an organised declaration in its favour.

While therefore I sympathise with the spirit of liberality which characterises your present movement[4] I should not feel disposed to take any part in it myself.

Believe me dear Sir | very faithfully yours | John Tyndall

Cambridge University Library MS.Add.5989:f.24, in *Declaration of students of natural sciences*

---

1. *Capel Henry Berger*: Capel Henry Berger (1839–68), a British chemist and paint manufacturer (W. H. Brock and R. M. Macleod, 'The Scientists' Declaration: Reflexions on Science and Belief in the Wake of "Essays and Reviews," 1864–5', *British Journal for the History of Science*, 9 (1976), pp. 39–66). Brock and Macleod label Berger as the main person who organised and coordinated the 'declaration' referred to here (p. 43; see n. 4). He died in 1868 by an accidental poisoning of carbonic acid used to alleviate tooth pain ('Death of Mr. Capel Henry Berger, F.C.S.', *Chemical News and Journal of Industrial Science*, 26 June 1868, p. 307).
2. *your letter & for the accompanying circular*: letter and circular missing. The circular was very likely a copy of the declaration.
3. *the creed of Gamaliel . . . it will come to naught*: in the Christian tradition, Gamaliel the Elder (also spelt Gamliel), is described in Acts of the Apostles, as a Pharisee and Doctor of Mosaic Law, and a member of the council (Acts 5: 34–40); he notably spoke against the execution of the Apostles, saying, 'Refrain from these men, and let them alone: for if this council or work be of men, it will come to nought: But if it be of God, ye cannot overthrow it' (Acts 5: 38–9).
4. *your present movement*: on 16 April 1864, a declaration organised by Berger was signed by over two hundred 'Students of the Natural Sciences', expressing regret that scientific researches were being 'perverted by some . . . into occasion for casting doubt upon the Truth and Authenticity of the Holy Scriptures', and claiming that if scientific results appeared to contradict the Bible, investigators ought to 'leave the two side by side till it shall please God to allow us to see the manner in which they may be reconciled'. The document was widely circulated to members of learned societies in Britain, eventually receiving 717 signatures. It was published as *The Declaration of Students of the Natural and Physical Sciences* (London: Simpkin, Marshall & Co., 1865). See W. H. Brock and R. M. Macleod, 'The Scientists' Declaration: Reflexions on Science and Belief in the Wake of "Essays and Reviews," 1864–5', *British Journal for the History of Science*, 9 (1976), pp. 39–66.

From Juliet Pollock 11 June 1864 2105

59 Montagu Square W.[1] | June 11th. 1864.

My dear Mr. Tyndall

Not without the fear you spoke of last night—of boring a friend by dwelling on a subject of deep interest to oneself—I write to express myself a little upon the delight of the R.I. and upon your discourse[2] which ended them. I must really thank you for the quantity of interest you have added to our life in so many ways. I must tell you, now that the season is over, how much the hours spent at the R.I. have done—how much of the highest and most lasting kind of gratification they have given to us; and how I have found this kind of gratification stretch and grow with growing years. When I look back to the time of your first lectures I am sensible of improved powers in myself while listening to you & I am grateful for the higher appreciation which I bring to your themes and to the treatment of them.

While such feelings grow we are still young; for I suppose antiquity is the arrest of progress. You stimulate the thoughts, the imagination and all the intellectual powers of your pupils and I believe all who are worthy to be called so, must rejoice in improvement under your influence.

I felt very strongly the beauty of your closing passages and if you have undergone fatigue in our service it is something for you to know that the labour has been spent upon not ungrateful soil, and so I subscribe myself—

ever truly & gratefully yours | Juliet Pollock

RI MS JT/1/P/171

1. *59 Montagu Square W.*: see letter 1954, n. 2.
2. *upon your discourse*: see letter 2091, n. 7.

To Richard Dawes [before 12 June 1864][1] 2106

Royal Institution | Monday.

My dear Mr Dean,

I have just received your letter:[2]—At this season of the year it is all but impossible for me to get more than three successive days: these three in your case were to have been Tuesday, Wednesday, and Thursday. I must be in town on Thursday at 5.P.M., and I must be in town tonight at 10. P.M. Thus am I circumstanced. I may be more fortunate when I come from Switzerland at all events we will defer matters as you propose until then.

I wish most earnestly that you were nearer to London.[3] I have often wished this, for I scarcely know a roof under which it would give me greater pleasure to take shelter than yours.

Give my kindest regards to Mrs Dawes,[4] and tell her a man may have a heart to appreciate kindness and to love his friends though he may not be prepared to join in prayers for rain and for fair weather.[5]

Yours ever | John Tyndall.

I cannot say whether it will be possible for me to attend the British Association.[6]

I have given up Norway in favour of the Engadine and the Dolomite Alps.[7] I start as near the 1st of July as I can and hope to be back about the 10th or 12th of August.

RI MS JT/1/TYP/1/318
LT Typescript Only

---

1. *[before 12 June 1864]*: LT suggested June 1856 for this letter. However, this seems unlikely given Tyndall's expressed intentions for his climbing trips that year. Further, on 3 December 1863 he wrote to Clausius that he was thinking of going to Norway to climb (see letter 2048). Tyndall in fact went to Switzerland on 2 July 1864. It appears that Tyndall was able to visit Dawes later in June, since in letter 2107 to Juliet Pollock, dated 12 June 1864, Tyndall mentioned heading to visit Dawes for two or three days, saying 'I broke one engagement but I must keep this one'. In letter 2163 to Dawes, dated to October 1864, he wrote that he again was unable to visit Dawes and was not able to attend the BAAS meeting in Bath (see n. 6 below).
2. *your letter*: letter missing.
3. *nearer to London*: Dawes lived in Ledbury, Herefordshire in central-western England, near the border with Wales.
4. *Mrs Dawes*: Mary Ellen Dawes.
5. *to join in prayers for rain and for fair weather*: in 1860 there had been a national day of prayer for fine weather in hope of ending months of endless rain. The Anglican clergyman Charles Kingsley (1819–75) had refused to offer such a prayer, sermonising against interfering with God's plan. He wrote to the geologists Charles Lyell and Charles Bunbury (1809–86) about this, which became well known in scientific circles. See K. Anderson, *Predicting the Weather: Victorians and the Science of Meteorology* (Chicago: University of Chicago Press, 2005), pp. 161–7; and F. M. Turner, 'Rainfall, plagues, and the Prince of Wales', in *Contesting Cultural Authority: Essays in Victorian Intellectual Life* (Cambridge: Cambridge University Press, 1993), pp. 151–70.
6. *British Association*: although LT noted that the BAAS held its 1856 meeting in Cheltenham, it is more likely that this letter is from 1864, when the BAAS met in September in Bath. Tyndall did not attend the meeting. See letter 2163, also to Dawes.
7. *Engadine and the Dolomite Alps*: the Engadine Alps are in Switzerland; the Dolomites in Italy.

To Juliet Pollock 12 June 1864 2107

12th June 1864

My dear Mrs. Pollock.

I well remember the first night I ever saw you. It was at Barlow's, and I compared you on that occasion (in my own mind) to a moon, shining with a pure mild light. I had afterwards reason to conjecture that you could also be hot like the Sun if you were sufficiently aroused. Well sun, or moon, you have continued to shine on me and brighten me and gladden me ever since, and this last note of yours[1] is a bit of your brightness—very pleasant to me, and I thank you for it.

I am going to rush down to Ledbury[2] in accordance with a heavy engagement for two or three days. Long ago I promised my friend the Dean of Hereford[3] to go down to him—I broke one engagement but I must keep this one.[4] Good bye

Yours ever | John Tyndall

RI MS JT/1/T/1141
RI MS JT/1/TYP/6/2010

1. *last note of yours*: letter 2105.
2. *Ledbury*: a small town in Herefordshire, in central-western England, near the border with Wales.
3. *Dean of Hereford*: Richard Dawes.
4. *I broke one engagement but I must keep this one*: it appears that Tyndall was unable to visit Dawes earlier in June (see letter 2106), but made arrangements to do so mid-month.

From Rudolf Clausius 13 June 1864 2108

Züruch, 13.6.64

Lieber Tyndall,

Ich habe Ihren Aufsatz, welchen ich gleichzeitig mit diesem Briefe wieder absende, mit grossem Vergnügen gelesen, und finde ihn sehr gut. Ich halte es für das beste Mittel, die Leute, welche so leichtfertig reclamieren ohne sich die Mühe zu geben die betreffenden Sachen zu lesen,—zum Schweigen zu bringen, wenn man das wissenschaftliche Publikum mit den Sachen bekannt macht, und ihm so Gelegenheit giebt, selbst zu urtheilen. Die Zusammenstellung von Séguin's Rechnungsweise mit Mayer's wird ihre Wirkung gewiss nicht verfehlen, u. Thomson wird wohl einsehen, dass es nicht die rechte Art ist, sich, nachdem man einen Streit angefangen hat, in vornehmes Schweigen zu hüllen.

Die Versammlung der naturforschenden Gesellschaft findet am Montag, Dienstag und Mittwoch den 22, 23, und 24ten August statt, und wir rechnen darauf dass die Gäste spätestens am Sonntag den 21sten hier ankommen. Wenn Sie es möglich machen können, herzukommen, werden wir uns sehr freuen. Dass ich Präsident der Gesellschaft sei, ist aber nicht der Fall, sondern Professor Heer, der Botaniker, ist Präsident. Sagen Sie doch auch Hirst noch einmal dass wir uns sehr freuen werden, wenn er kommt und bei uns logiert. Ich hoffe er wird es mir nicht übel genommen haben, dass ich nicht noch an ihn besonders geschrieben habe. Da Sie beide zusammen wohnen, glaubte ich auch die Einladung an Sie beide gemeinsam schicken zu können. Ein Freund von Mayer, welcher hier Professor ist, will auch an Mayer schrieben und ihn auffordern herzukommen.

Mit herzlichen Gruessen an Sie und Hirst | Ihr | Clausius

Zurich, 13.6.64

Dear Tyndall,

I have read your paper,[1] which I am sending back to you simultaneously with this letter, with the greatest pleasure and I find it to be very good. I consider it to be the best means to silence the people who so frivolously protest without having taken the trouble to read the works in question, when one acquaints the scientific public with the material and gives it the opportunity to make its own judgement. Bringing together Séguin's and Mayer's methods of calculation will certainly not miss the mark, and Thomson will have to admit that it is not proper behaviour to veil himself in noble silence after having begun an argument.

The meeting of the Society of Natural Science[2] takes place on Monday, Tuesday and Wednesday, the 22, 23, and 24th of August, and we expect that the guests will arrive here on Sunday the 21st at the latest. We would be very pleased, if it would be possible for you to come[3] here. I am not president of the society, but Professor Heer,[4] the botanist, is. Tell Hirst again that we would be very pleased if he comes and stays with us. I hope it will not irritate him that I did not write to him personally. Since you both live together, I thought I could send the invitation to both of you. A friend of Mayer's,[5] who is professor here, will also write to Mayer and encourage him to come.

With fond regards to you and Hirst, | Your | Clausius

RI MS JT/1/TYP/7/2267
LT Typescript Only

1. *paper*: see letter 2098, n. 6.
2. *meeting of the Society of Natural Science*: the Naturforschende Gesellschaft, or Swiss Society of Natural Science, held its meeting in Zurich from 22–4 August 1864. See letter 1906, n. 6.
3. *if it would be possible for you to come*: he did (see letter 2130).
4. *Professor Heer*: Oswald Heer (1809–83), a Swiss geologist and naturalist (*HLS*).
5. *A friend of Mayer's*: not identified.

From Henry Hill[1] 13 June 1864 2109

Black Swan Passage | Halifax 13 June 1864

Mr John Tyndall

Dear Sir

I have just seen Mr Roby Pridie[2] who has informed me that my friend[3] is very poorly—in Paris.

Is it so? I should like to write to him, may I? You know what a kind friend he has been to me, and if he dies, there is no man living who can ever hold the place in my heart that he has.

If I was a single man I would most willingly die for him.

Forgive me Sir if I intrude for I do not wish to do so Your reply will much oblige me

I am yours sincerely | Henry Hill

RI MS JT/1/H/123

1. *Henry Hill*: a friend of Hirst's. See I. Hesketh and E. Sera-Shriar (eds), *The Correspondence of John Tyndall, Volume 4: The Correspondence, January 1853–December 1854* (Pittsburgh: University of Pittsburgh Press, 2018), p. 529.
2. *Roby Pridie*: William Roby Pridie (1825–74), a Yorkshireman whom Hirst and Tyndall befriended when they joined the Mutual Improvement Society in Halifax. He was the son of the Reverend James Pridie (1786–1873), an independent minister in Halifax.
3. *My friend*: Thomas Hirst.

To Thomas Archer Hirst 14 June 1864 2110

14th June 1864

My dear Tom,

It was a great relief to me to see your handwriting last night as I was beginning to be very anxious with you. I had pictured to myself a thousand ugly possibilities, nor could the consciousness that somebody would assuredly write to me in case any thing serious had happened to you quite dispel my uneasiness. Well it is a relief to hear from yourself though I could wish that you were able to send me better intelligence.

I am aiming to get away from this place as near the first of July as I can. I go straight to the Engadin,[1] and make Pontresina[2] my headquarters. Ball[3] joins me there on the 20th & from the 20th to the 25th we spend in the Lower Engadin. We are afterwards to be found by Forster[4] at Santa Caterina[5]—not far from Bormio,[6] where we all there hope to attack the Ortler Group,[7] and spend a week together among its fastnesses. Afterwards Ball purposes that we should visit the Dolomite Mountains,[8] but this is a part of the expedition to which I have not yet committed myself.

Now what do you think of joining me at Pontresina?—I hope to be there about the 6th of July, & we could then have a clear fortnight together—afterwards if you thought well of it you might accompany us on our excursion. I am disposed to think that this would be of great & mighty benefit to you.

Only one letter has come for you: lots of suspicious looking documents have arrived, but all those I opened were either advertisements or something equally worthless. A large quantity of proofs has come from Spottiswoode, but I forwarded you the only copy of the Educational Trans[9] that came to me. This morning I received a characteristic note from Henry Hill.[10] Pridie[11] has evidently been giving him a bad account of your health—'If I was a single man' says this poor fellow 'I would most willingly die for him.' I read the note for Faraday this morning—it brought tears to his eyes—there is something very genuine about it—so genuine that I wish to keep it.

My Friday evening I am told passed off extremely well. In fact though the subject was a slender one, the success of every thing appeared to please the audience mightily—There was a Queen's ball, and a meeting of the Astronomical Society—still the house was full.

Many people have been enquiring after you—among others Sylvester whom I see frequently. Busk, Carpenter,[12] Huxley, Ramsay & many others constantly ask about you: so that you are certainly not forgotten in London.

I wish I could have you with me for one clear month. The belief has forsaken me that that month could do more for you than any thing else. My poor sister[13] was sent to the sea, and in the little town where she resides lives a namesake—a Miss Margaret Tyndall—a 'veritable third cousin' she calls herself. It is funny how these people turn up. I am also asked to spend next Friday on the Thames between Marlowe and Windsor with my friends the Tyndale's of London.[14]

I hope you will soon write to me and that when you write you will be able to send me better news.

Yours affectionately | John

BL Add MS 63092, ff. 31–2
RI MS JT/1/HTYP/532

---

1. *Engaden*: the Engadine, a long Alpine valley in southeastern Switzerland.
2. *Pontresina*: a small town about 3 miles to the east of St. Moritz, in the Engadine.
3. *Ball*: John Ball.
4. *Forster*: see letter 2098, n. 11.
5. *Santa Caterina*: a small commune in northern Italy, twenty kilometers to the southeast of Bormio.
6. *Bormio*: a town in northern Italy.
7. *Ortler Group*: a mountain range of the Southern Limestone Alps mountain group in the Central Eastern Alps, in Italy and Switzerland.
8. *Dolomite Mountains*: a mountain range located in northeastern Italy. They form a part of the Southern Limestone Alps and extend from the River Adige in the west to the Piave Valley in the east.
9. *Educational Trans*: not identified.
10. *note from Henry Hill*: letter 2109; Hill was a friend of Hirst's.
11. *Pridie*: see letter 2109, n. 2.
12. *Carpenter*: William Carpenter (see letter 2068, n. 4).
13. *my poor sister*: Emma Tyndall.
14. *Tyndale's of London*: possibly John William Warre Tyndale and his family who lived in London. See letter 2095.

From John Frederick William Herschel [18 June 1864][1] 2111

Collingwood June 18 1864

Dear Sir

I beg to thank you for your papers on "Scientific History—on "Molecular Physics and on "a magnetic experiment."[2] They are extremely interesting and the experiments by which they are illustrated must on those who were so fortunate as to witness them have made a great impression.

The beautiful experiment of the change of length of a bar of iron, and the parallel arrangement of iron dust in a liquid bring strongly to my recollection an hypothetical explanation of muscular action by nervous influence (assumed to be Volta-electric) on the pendicules of a muscle which I put forth (vivâ̲ voce,[3] on a black board) at one of the meetings of the British Association—either that at Southampton in 1846 or at Leeds (I forget which)[4] It was to this effect. Conceive a muscular fibre to consist of a string of (very oblate) spheroidal particles enclosed in an inextensible close fitting sheath or envelope—insulated from each other by contractions in the sheath and from the sheath itself by a film of lubricating fluid. Let each of these spheroids be, or be capable of becoming, under electric influence, a magnet having its axis of polarity in the direction of the short axis of the spheroid and suffer the whole film to be surrounded with a galvanic spire or spherical thread of nerve conveying an electric current. Thus,

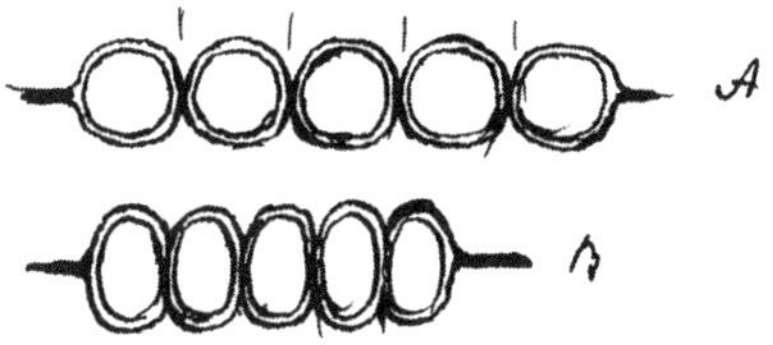

the arrangement A being that of the spheroid unexcited and B that when excited, and their axes arranged pole to pole

by the directive power of the spiral line, the envelope (being perfectly flexible and perfectly lubricated) would be swollen laterally & shortened longitudinally by the rotation through 90° of such enclosed spheroid. The thing is much clearer to conception if we consider the spheroids as mere plane elliptic discs surrounded with a continuous string, inextensible in length, where it is obvious that the rotation of the ellipses within their loops will shorten the interval between the extremities of the system.*

The magnetization of the spheroids might be conceived by regarding them each as a spiral coil, fed by a minute nervous filament long and loose enough (in the lubricating fluid) not to be broken by the rotation of the spheroids.

As this was explained incidentally in some discussion that arose on a paper read I do not think it found its way into any printed volume of proceedings—but I perfectly remember the fact of *having* so explained it though I cannot call to mind what was the particular subject of discussion.

It is very evident that this implies the converse principle that any cause which may temporarily dispose the *longer* axes of all the pendicules of a body in some one direction will lengthen the body in that direction just as any which so disposes their shorter, will shorten it. And this is if I apprehend you rightly (in p. 5.) precisely your principle of explanation of the lengthening of your bar.

Being on the subject of these papers I may notice an expression in your note on p. 12 in the "Notes on Scientific History" where you say "all this *and much more* was stated by Sir J. H. in 1833. &c but Mayer was the first &c &c. Among the "much more" is the broad statement that the source of the great deposit of dynamical efficiency laid up for human use in the coal strata, is the sun: which every body persists in ascribing to Stevenson[5] who if I mistake not puts it forth (of course independently) at a later date. —and there is, I am told a passage in some poem[6] published by Effingham Wilson[7] in 1831 (and which I had no knowledge of) which contains some similar suggestion. The term "vivifying action" is certainly vague but if light & heat be motion and nothing but motion it is difficult not to draw the conclusion that it is in some latent form of heat that the efficiency is stored up. And here I would refer you to a short paper[8] that I published in the London Ed. and Dublin Phil. Mag. for 1833 iii. p. 401 "On the absorption of light by coloured media" dated Oct 17 of 1833 in which occurs inter alia this expression (I cite from memory not having it by me) "Thus the question "what becomes of Light?" merges in the more general one 'what becomes of motion, and the answer is 'it continues for ever" The only further question then, as even yet, unanswered and so far as I see at present unanswerable being as to what portion of that motion in the form of rotary movement may remain confined to individual molecules, and what may continue to subsist in the form of propagated undulation reflected to and fro in the interior of bodies. Whether heat be vis viva seems to me a question whose answer depends very much on the notion people may form of vis viva. To me it conveys absolutely no other meaning than mass x square of velocity and in this sense its amount in a system of moving bodies or molecules is assuredly not a constant quantity. Take only the case of two bodies revolving round their common centre of gravity in elliptic orbit. Their masses are constant—their velocities variable, and variable in the same direction (i.e. both increase together or both decrease) so that the total vis viva $mv^2 + MV^2$ has a maximum & a minimum. If the Vis Viva's in different states of a system be not the same there must be a difference and this difference is, as is shewn in Mcc. Col. I. v. § 19. a function of the coordinates of the molecules (i.e. ultimately of their mutual distances)—To this difference if any one chuse to give the name "Potential vis viva" it then becomes a truism to say that the sum of the actual and the potential V. V. is invariable. If it be then said that this " Potential" expresses the attractive force expended in bringing the molecules from the one situation to the other—this again is mere assertion or equally reducible to a truism (if true) by a mere algebraic manipulation such as would insulate on one side of an equation the attractive force supposed to be "expended" and bring the masses, velocities, &c on the other: the fact being that refuge is here taken in the metaphysical conception of "Force" ie effort & its mode of supply & expenditure about which we know nothing and may assert anything.

I ought to apologize for running out into such a digression but the fact is that I cannot reconcile myself to the exceeding laxity of language into which I observe a tendency among physicists at present to fall—in confounding the terms force, motion, energy, meaning sometimes mechanical effort, sometimes vis viva, and sometimes actually mere momentum among whom heat, light, electricity, are often spoken of as forces &c &c. I think we owe much of this confusion to Groves title of his book "The Correlation of Physical Forces" and to the "Forces of nature" so often spoken of by Humboldt in his Kosmos.[9]

Your results about radiant heat and diathermancy are excessively interesting and open an immense field of more important results

There is a very curious consequence which would follow from the admission as a general & rigorous proposition that all expenditure of mechanical force which produced no other perceptible result, produces heat. —The precession of the equinoxes ought to heat the earth. For that movement which we call precession is the result of the Sun & Moon's mechanical force and consists in a conical motion of the Earth's axis of rotation. —Let the Earth stand with a given position of its axis. After 26000 years its axis will be again in the same position—the rotating motion is unaltered, and there is no sensible alteration in any particular cognisable to the senses. —What has become of all this "Energy" that has been expended in this interval?

Once more begging you to excuse these longueurs believe me

Dear Sir | Yours very truly | J. F. W. Herschel

*and that by a force the greater, the nearer the ellipses approach to circles

RS HS 393b
RI MS JT/1/H/101
RI MS JT/1/TYP/2/524–7

---

1. *[18 June 1864]*: this date is taken from the handwritten draft and LT's typescript version. A copy of the letter at the RS carries the date 18 July 1864.
2. *your papers on "Scientific History—on "Molecular Physics and on "a magnetic experiment."*: for these papers, see letter 2098, n. 6, and letter 2091, n. 7.
3. *vivâ voce*: in speech; orally (*OED*).
4. *one of the meetings*: this was at the 1846 annual meeting of the BAAS held in Southampton. Herschel's suggestion was in response to research presented by the Italian physicist and neurophysiologist Carlo Matteucci (1811–68), 'Summary of Researches in Electro-Physiology' (*Brit. Assoc. Rep. 1846*, pp. 28–9). On Herschel's suggestion, see 'British Association for the Advancement of Science', *Provincial Medical and Surgical Journal*, 28 October 1846, pp. 517–9, on p. 519; and 'Remarkable Electrical Agencies', *Chambers's Edinburgh Journal*, 13 March 1847, pp. 172–3, on p. 172.
5. *Stevenson*: George Stephenson (1781–1848), a British colliery and railway engineer (*ODNB*). Stephenson was known for making the suggestion that coal was essentially 'bottled sunlight' (see A. MacDuffie, *Victorian Literature, Energy, and the Ecological Imagination* (Cambridge: Cambridge University Press, 2014), pp. 34–6).

6. *some poem*: C. Whitehead, *The Solitary* (London: Effingham Wilson, Royal Exchange, 1831).
7. *Effingham Wilson*: Effingham Wilson (1785–1868), a nineteenth-century English radical publisher and bookseller (*ODNB*).
8. *a short paper*: J. Herschel, 'On the Absorption of Light by Coloured Media, Viewed in Connexion with the Undulatory Theory', *Phil. Mag.*, 3 (1833), pp. 401–12
9. *Kosmos*: A. Humboldt, *Kosmos: Entwurf einer physischen Weltbeschreibung*, 5 vols (Stuttgart: Cotta, 1845–62).

From George Gabriel Stokes 22 June 1864 2112

London 22nd June 1864

My dear Tyndall,

I am in fault for your paper.[1] I had it at Cambridge and looked for it there but not finding it and hearing from Mrs Airy[2] that I had left something behind me at the observatory[3] where I had been spending a couple of days (she did not say what it was) I thought most likely it was that and wrote to her accordingly. I recollected having had some papers (I do not recollect what) on the toilet table, nor was I aware that I left them behind. I heard from Mrs Airy today and though she does not actually name the mysterious article I gather from her description that it is what is vulgarly called a nightshirt, so my hopes in that quarter are disappointed. I am very sorry that I should have put you to this inconvenience and will have a good hunt when I get back—I miss also two <herafter LT Typescript only> proofs Phil. Trans. which I doubt not are with your paper wherever that is. During the Prince's visit[4] my house was a good deal put about to find room and it may have been thus that the paper was misplaced.

Yours very truly | G. G. Stokes

RI MS JT/1/S/234
RI MS JT/1/TYP/4/1389

1. *Your paper*: see letter 2097, n. 3. See also letters 2099, 2114, 2115, 2117, 2119, and 2136.
2. *Mrs Airy*: Richarda Airy (née Smith, 1804–75), wife of George Biddell Airy, the Astronomer Royal (*ODNB*).
3. *observatory*: the Royal Observatory, Greenwich.
4. *Prince's visit*: perhaps Albert Edward (1841–1910), Prince of Wales, eldest son of Queen Victoria and Prince Albert (*ODNB*).

To Annette Roche 24 June 1864 2113

24th, June, 1864.

My dear Miss Roche

The copy of the Alpine Journal[1] which you will have received before this note reaches your hands contains the account of the death of my brave guide.[2] I shall feel lonely without him this year in Switzerland. I can never hope to have another like him; we knew each other so thoroughly and trusted each other so completely.

I have been thinking at intervals during the last dozen years of placing a tomb over my father's[3] grave; but I put it off hoping to be able to choose one myself during a visit to Carlow.[4] Could you advise me as to how this matter might be arranged?—I should like something substantial,[5] an emblem of the man himself; and nobody knows but me how substantial he was.

If some man in Carlow would furnish me with a few sketches, with the cost annexed I could perhaps choose from them. I should like the tomb to be of some durable material.

I will write to you from the Alps.

Believe me | Yours most truly | John Tyndall

RI MS JT/1/TYP/11/3850
LT Typescript Only

---

1. *copy of the Alpine Journal*: the *Alpine Journal* is one of the oldest mountaineering journals in the world, published in London by Longman. It replaced *Peaks, Passes, and Glaciers*. The accident Tyndall referred to was published as Philip C. Gosset, 'Narrative of the Fatal Accident on the Haut-de-Cry, Canton Valais', *Alpine Journal*, 1 (1864), pp. 288–94.
2. *my brave guide*: Johann Joseph Bennen.
3. *my father's*: John Tyndall (1792–1847), who grew up in Leighlinbridge and married Sarah McAssey in 1813. They had two children that are definitely known: John and his older sister Emma (b. 1817). He probably served in the Royal Irish Constabulary at the police station in Nurney from 1828–33. The family then moved to Castlebellingham, County Louth for the next three years before returning to Leighlinbridge, where he worked as a boot and shoemaker. He may also have served as a land agent for William Steuart, one of the principal landlords in Leighlinbridge and also the main patron of the Tyndall family. His 1847 death was a blow to his son, who relied on his father for advice.
4. *Carlow*: a country town in County Carlow, Ireland.
5. *I should like something substantial*: see letter 2121.

From George Gabriel Stokes 24 June 1864 2114

Lensfield Cottage Cambridge | 24th June 1864

Dear Tyndall,

I sent you tonight by book post your original paper,[1] which I found at once, when I thought of looking for it, in a certain box in which the housemaid is in the habit of putting things for tidiness but which I do not myself use and had not thought of looking into. I had previously had a good hunt for it in several rooms. The copy of the MS has not yet been returned.

I was one of the referees. I did not keep the MS after my return to London till I should have completely looked over it, knowing that you wanted it, and therefore in my remarks I must trust a good deal to memory. There were three points on which I wished to make some remarks.

1. By operating on vapours, filling your tube, in tensions varying as you ensure a common height of column of the substance supposed liquefied. The reader would naturally like to know what that common height is. This is got by an easy numerical calculation from the data you have given combined with the ratio of the densities of air and water at a standard barometric height and the value of the height in inches. I make it about 0.00215 inch. I think this calculation should be made once for all by the author instead of being thrown on each individual reader. As it happens the order of opacity comes out the same for the vapours and liquids, because the latter were employed in a cell sufficiently thin to have its thickness comparable with 0.00215 in. so that there was no inversion in passing from the latter to the thickness employed. But to know whether any such inversion if it had occurred would or would not prove an inversion of order referable to the state (liquid or gas) of the substance, we should want to know whether the equivalent liquid thickness of the gas was equal to the actual liquid thickness in the experiments with liquids, or at least sufficiently nearly so to preclude the supposition of a change of order arising merely from that cause.

2. The statement as to the raising (shortening) of the periodic time of vibrations seems to me somewhat too absolute inasmuch as it is quite conceivable, be it even deemed improbable, that it is the vibrations of short period which we know to exist that are converted, without the wire into much longer ones, with the wire also into somewhat longer ones. But be that as it may, there is of course one broad distinction between this phenomenon and fluorescence, that in the latter case it is ethereal in the former molecular vibrations that are so converted.

By the bye fluorescence is not by any means confined to the more refrangible part of the spectrum. In the case of chlorophyll for instance, it begins near the line B and consequently very near the red extremity. Of course you are well aware of this, and I merely make the remark because (if I recollect rightly) your words would seem to imply that I had found it only in the more refrangible rays.

3. This is the most important of the points I have to mention. Your experiment on the quality of the heat radiated by the lamp-black film covering the thin metal face of a cube containing water at 2120 is sufficiently remarkable whatever be the explanation of it. But a difference of quality between the radiations from two different <u>opaque</u> bodies at the same temperature is what cannot be admitted without the clearest experimental evidence, seeing that it is in opposition to the extension B Stewart[2] and Kirchhoff[3] have made of Prevost's law of exchanges.[4] Prior to your experiment, had any one asked me I would have assumed with you that the temperature of the lamp-black was sensibly equal to 2120. But I now believe that so rapid is the loss of heat by radiation from the lamp-black and so bad the conducting power of the substance, thin as the film is its outer portion, from whence almost exclusively the radiation comes, is very sensibly lower in temperature than 2120. The point is well worth being made the subject of further experiments. If my explanation is right you ought to get different results from thin films of different thickness. Also it ought to make a difference if the face was if I may so speak <u>muffled</u> i.e. made to radiate in all directions but that wanted for the pile into a chamber kept to 2120, the chamber itself being screened off from the pile.

I hope you will find leisure to take up this subject, for unless my explanation, referring the result to known causes, be the true one (I myself however expect it will turn out so to be) a great deal may come out of the phenomenon. Meanwhile I think it would be well to insert some expression of doubt (I don't mean as to the accuracy of the experiment but as to the interpretation) in this part of your paper.

The other referee[5] after speaking very highly of the paper remarks. There is just one important point of theoretical connexion in which what is said in the paper appears to me altogether unsatisfactory.... It appears to me that the first quarter of § XIV is not at all conclusive; and further that it does not present any clear or probable view as to the real relation between radiation and conduction from which a reader would be led to see his way towards a conclusion. I do not think I say too much when I say that after reading it and considering it carefully I could not infer <u>from it</u> as probable that rock salt should be a better conductor than glass; or silver than platinum. That in these cases the worse radiator is the better conductor must, as far as it is proved, be looked on as a merely experimental conclusion and in my opinion the theoretical views in the particular passage referred to are less weighty than a mere statement of the fact, and a natural remark that the explanation is probably to be sought for in some relation between radiation and conduction. A full theory of this relation is conceivable as being within reach, but I can not think that the passage in the paper to which I object brings us nearer it, scarcely does as much as Huxley would call showing a finger post pointing to a way that might lead to it. I therefore think it would be a great improvement in the paper to remove or modify very much the first part of the concluding § XIV. But there is one more important matter noted in that §—the case of copper which of course must occur to any one as an exception when a table of the other metals is given in order of conductivities and reverse order of radiating power. It is very much to be desired that Tyndall should give a short description of the experiments he refers to in which he proved that silver is a better radiator than copper. A result so opposed to supposed results of previous experimenters is very striking, and particulars as to the evidence of it would be received with great interest. If not added to the present paper it is to be hoped Tyndall will take an early opportunity of publishing it.

With reference to the Akin experiment[6] he says. The platinum wire in the hydrogen flame may be said to be heated by conducted heat of the vapour and gas (whatever conducted heat may mean) and Tyndall is scarcely justified in his facts in saying that radiation from the flame could make the wire incandescent (unless by a concentration of the small amount of violet and ultra violet rays). It would be easy <Hereafter LT Typescript Only> to try by a good silver ellipsoidal reflector whether the whole radiation—violet rays and all included—could so be got to produce incandescence in a piece of platinum or other solid out of the flame, and I think this experiment would be worth trying. Mean while I think some slight change such as the insertion of a probably would improve Tyndall's statement, which as it stands is extremely interesting and suggestive, if to some extent open to objection.

Yours very truly | G. G. Stokes

RI MS JT/1/S/235
RI MS JT/1/TYP/4/1390–3

1. *your original paper*: see letter 2097, n. 3. See also letters 2099, 2112, 2115, 2117, 2119, and 2136.
2. *B Stewart*: Balfour Stewart (1828–87), a Scottish physicist and meteorologist known for his work on radiant heat and time as director of the Kew Observatory (1859–71) (*ODNB*).
3. *Kirchhoff*: Gustav Robert Kirchhoff.
4. *Prevost's law of exchanges*: in his experiments on heat, Pierre Prevost (1751–1839), a Genevan philosopher and physicist, explained that the *absolute equilibrium* of free heat is the state of this fluid in a portion of space which receives as much of it as it lets escape (P.

Prévost, 'Memoir on the Equilibrium of Heat', in D. B. Brace (ed), *The Laws of Radiation and Absorption: Memoirs by Prévost, Stewart, Kirchhoff, and Bunsen* (New York: American Book Company, 1901), pp. 3–13, on p. 7).

5. *The other referee*: not identified.
6. *Akin experiment*: C. K. Akin, 'Note on Ray Transmutation', *Phil. Mag.*, 28 (1864), pp. 554–60.

To George Gabriel Stokes 25 June [1864][1] 2115

Royal Institution of Great Britain | 25th June

My dear Stokes

All the points to which you refer me shall be attended to. I wish you would give me your opinion as to the last section of the paper.[2] I have such a clear physical image of the relation of radiation to conduction in my own mind that I think I must fail to express myself as I ought if what I have written could be read without exciting any intelligible idea in the mind of the reader.

However I will look over it again. With regard to the term "conducted heat" in the connexion quoted by your co-referee,[3] I am as much at a loss as he is to know what it could mean. The paper was copied by my assistant[4] and he possibly may have made some slips of the pen.

The manuscript[5] has arrived: but I want to take the copy to Switzerland with me. I suppose you can let me have it on Thursday next! I probably start on Saturday—certainly not later than Monday.

Yours very sincerely | John Tyndall

CUL SC—add 7656

1. *[1864]*: the year is given by relation to letters 2097, 2099, 2112, 2114, 2117, 2119, and 2136.
2. *your opinion as to the last section of the paper*: see letter 2099, n. 3. This relationship of radiation to conduction exercised Tyndall since his great 1861 paper ('On the Absorption and Radiation of Heat by Gases and Vapours, and on the Physical Connexion of Radiation, Absorption, and Conduction', *Phil. Trans.*, 151 (1861), pp. 1–36), in which he also had a whole final section on it (p. 33 onwards). Stokes was sceptical then as well, and there is a significant correspondence regarding this in *Tyndall Correspondence*, vol. 7.
3. *your co-referee*: not identified.
4. *my assistant*: not identified.
5. *The manuscript*: see n. 2.

From Elizabeth Dawson Steuart 28 June 1864 2116

June 28th 1864.

My dear John,

This morning a copy of "The Leader" Newspaper arrived here, for which I have to thank you: it is indeed most acceptable, for it tells me of your last lecture,[1] for the season, and I am glad they are over, that you may enjoy a little liberty and recreation. I trust the cessation from business, will be of great benefit to your health. I shall feel anxious to hear if you intend carrying out your ideas of visiting Norway[2] which will prove new and interesting. I regretted very much that I could not have the pleasure of seeing you alone, the day you so kindly called here: I had many things to say, for your own ear, solely, and it surprised me to hear that you had a companion[3] in all your visits to your old friends: whether such was your wish or not, I cannot tell, but it had a strange appearance: As if you required an introduction here, or had forgotten the way to this house! I have been staying with my Brother[4] at Kingstown[5] for the last 3 weeks, and only returned two days ago: before leaving home, I heard that Emma[6] had gone to Tramore[7] for the benefit of the sea air, and I truly hope it may restore her to health and strength of mind. I do not believe that medical attendance can assist her recovery in any way, at least not that of a general practitioner:

With every kind wish for your welfare, temporal and eternal, believe me, my dear John | most sincerely your friend | E. D. Steuart

A line from you is always most welcome

RI MS JT/1/TYP/10/3347
LT Typescript Only

1. *your last lecture*: letter 2091, n. 7.
2. *your ideas of visiting Norway*: in December 1863 Tyndall wrote to Rudolf Clausius that if 'I go to Switzerland next year I should make an effort to visit you, but I am thinking of going to Norway' (letter 2048).
3. *companion*: not identified, but it is clear that Steuart was quite put out that Tyndall had not visited her alone.
4. *my Brother*: Steuart had four brothers, William Duckett and three others. It is not clear here to which of them she is referring.
5. *Kingstown*: see letter 2080, n. 4.
6. *Emma*: Emma Tyndall.
7. *Tramore*: a seaside and resort town in County Waterford, Ireland.

From George Gabriel Stokes 30 June 1864 2117

Malahide[1] Ireland | 30th June 1864

My dear Tyndall,

I have read carefully over the few last pages of your paper[2] which you sent me, and must own that I am not and never was satisfied that the connexion between good radiation and bad conduction & v.v.[3] is established on theoretical grounds, or even rendered very probable. If a small portion of a heated mass were obliged to get rid of a given portion of heat in a given time, whatever the nature of the mass, and must do so either by conduction or radiation, then no doubt to show that the molecules could readily impart their motion to the ether would suffice to show the substance to be a bad conductor if by conduction be meant transference of motion (motion of the molecular kind which constitutes heat, not motion of the particles (or small parts) of the body as a whole, such as belongs to sound-waves) by pure molecular action, i.e. without going through the process of setting either in motion (internal radiation) and being set in motion by ether (internal absorption). Conduction however as usually understood includes the effects of molecular radiation and absorption (as for instance in the case of the metals which even in so small a thickness as the $\frac{1}{10000}^{th}$ part of an inch would I imagine be sensibly opaque to radiant heat as they are to light). Indeed Poisson[4] goes so far as to make conduction nothing but internal radiation and absorption. I think with you that it is more probable that a portion of the transfer of motion takes place simply by the action of molecular forces.

But suppose we could separate completely conduction from the effects of molecular radiation and absorption, so as to isolate the portion of the transfer due to purely molecular forces. Would your reasoning hold good then? I think not. I will use your own illustration a little modified

Let the O and ● represent molecules of the same mass but different size the O being more bulky, and let the | and || represent elastic connecting rods, the first the more slender. I should expect the big molecules to lose the more motion by imparting

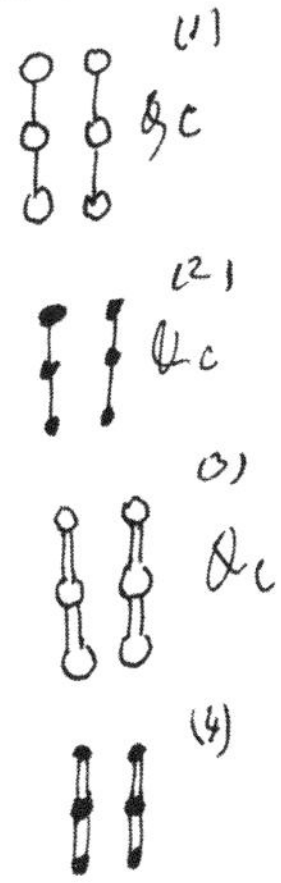

motion to the surrounding ether and the molecules connected by the stouter rods to lose more motion by communication to their neighbours. Then fig 1 would picture a body which was a good radiator and bad conductor, fig 2 a bad radiator and bad conductor, fig 3 a good radiator and good conductor, fig 4 a bad radiator and good conductor. Cases 2 and 3 seem to me a priori as likely to occur as 1 and 4. On the whole I <herafter LT Typescript only> must confess that what you have written seems to me hardly to show a finger post pointing to the true road.

Yours very truly | G. G. Stokes

I return the MS to Mr White.[5]

RI MS JT/1/S/236
RI MS JT/1/TYP/4/1394–5

1. *Malahide*: a coastal settlement in Fingal, County Dublin, Ireland.
2. *your paper*: see letter 2097, n. 3. See also letters 2099, 2112, 2114, 2115, 2119, and 2136.
3. *v.v.*: vivâ voce, meaning in speech; orally (*OED*).
4. *Poisson*: Siméon Denis Poisson (1781–1840), a French mathematician, engineer, and physicist, noted for his time in the Académie des Sciences and he was the final leading opponent of the wave theory of light and was proven wrong on that matter by Augustin-Jean Fresnel (*CDSB*).
5. *Mr White*: probably Walter White (1811–93), who had been appointed assistant secretary at the RS in 1861.

To Norman Macleod[1] 2 July 1864 2118

July 2, 1864. | Norman MacLeod, Esq.

SIR,

IN reply to your communication of the 29th ult.,[2] I beg to state that the results of the examination of candidates from the industrial classes, in Natural Philosophy, are, on the whole, highly satisfactory.

Judging from the answers, I should infer that the examinations cause a large amount of sound scientific knowledge to be diffused among the industrial classes. The importance of this result can hardly be over estimated.

The only point that I would strongly recommend to all teachers is, to aim at thoroughness of instruction rather than extent.

Your communication reached me as I was on the point of going to the Continent, my Report[3] is therefore more brief than it otherwise might have been. But I do not think I could add anything material to what I have here stated. The examinations, I would repeat, are most encouraging.

I am, &c. | (Signed) JOHN TYNDALL.

Twelfth Report of the Science and Art Department of the Committee of Council on Education. Presented to both Houses of Parliament by Command of Her Majesty (London: George E. Eyre and William Spottiswoode, 1865), Appendix A, pp. 41–2.

1. *Normal Macleod*: Norman Macleod (dates unknown), Assistant Secretary of the Department of Science and Art. He presumably acted for the Committee of Council on Education who oversaw the *Twelfth Report of the Science and Art Department of the Committee of Council on Education.*
2. *your communication of the 29th ult.*: letter missing.
3. *my Report*: enclosure missing.

To George Gabriel Stokes [5 July 1864][1] 2119

Royal Institution of Great Britain | Tuesday 64[2]

Dear Stokes.

I send you the concluding paragraph of my paper:[3] what I mean is this—suppose

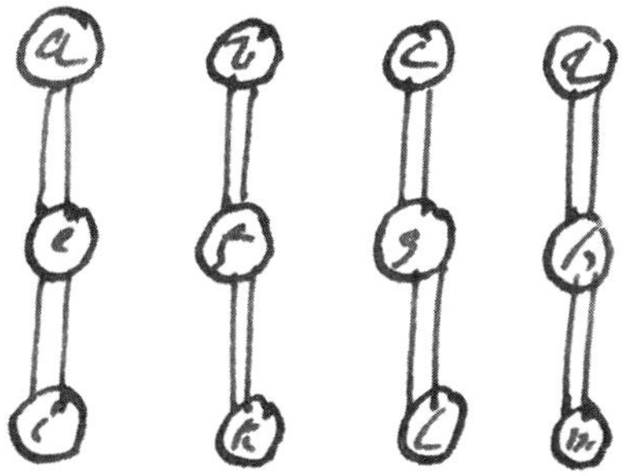

a b c d to be a series of atoms kept asunder by forces which we may visualise as longitudinally elastic rods which unite them suppose a vibratory motion imparted to a b c d: it will transmit itself thro' the rods to e f g h., and from these again to the next series & so on.

Now suppose these atoms plunged in a medium capable of accepting their vibrations & transmitting them in all directions. I reason that the existence of this medium must interfere with the direct transfer of motion from atom to atom, and that if anything in the physical character of one atom causes it to communicate motion more readily than another atom to the medium, then the direct transmission of motion it will be more impeded in an assemblage of atoms of the former kind than of the latter.

Yours very truly | John Tyndall

I shall [sadly] want a copy of my paper in Switzerland.

CUL SC—add 7656

1. *[5 July 1864]*: the date is given by relation to letter 2117 (dated 30 June 1864), in which Stokes asked Tyndall questions that he responds to in this letter regarding his argument in his fifth memoir (see letter 2097, n. 3). The following Tuesday was 5 July 1864. See also letters 2097, 2112, 2114, 2115, 2117, and 2136.
2. *64*: the year is handwritten onto the letter, probably by LT.
3. *my paper*: see letter 2097, n. 3.

From Julius Haast 13 July 1864 2120

Christ Church, July 13th 1864

My dear Sir!

Last mail brought me a copy of your admirable work, Heat considered as a mode of motion,[1] which has given me infinite pleasure and for which I am to offer you my best thanks. I have read it with great *[profit]*, *[&]* it is my intention to give a lecture on it next winter and to try some of your experiments in illustration—

I think that your work was a step in the right direction and that a great deal of practical advantage will also accrue from it in due course of time. —

Two mails ago I sent to Sir Rod. Murchison to be read to the Geol[ogical] Society a paper[2] on the causes which have led to the excavation of our Alpine Lakes & with this mail I have sent an appendix.

On the climate of the Pleistocene epoch of New Zealand,[3] I am exceedingly anxious to hear your opinion and if on mechanical grounds it can be objected to, I know Murchison & many of the older geologists will fight against it, but nevertheless I am certain, serious objections can not be raised, being based upon observations in the field.

Dr Falconer's theory[4] is ridiculous and I cannot understand how a man like Murchison could adopt it. Of course if it can be proved that the rocky bottom was below <u>freezing point</u> all the time & that no water could sink down through the ice, then I will give up my objection to it

A section across our glacier lakes shows clearly that the water below the glacier on the bottom of the present lake had when the glacier existed to rise nearly 200 feet before it could flow of & thus to rise to such an altitude it had to be forced up by the gravitation of the glacier itself. —

With next mail I shall have the pleasure to send you a paper of mine[5] on the formation of the Canterbury plains,[6] which will probably interest you as I had to treat of some glacier questions. I think that never before the real power of glaciers has properly been taken into accounts, because I am certain that principally during the latter part of the tertiary period they have played a much greater role than ever has been suspected and that the configuration of many a country can only be satisfactorily explained by admitting their enormous scooping power. —

I suppose you are taking some interest in alpine history, therefore it would perhaps interest you to have a look at my alpine collections in Kew, which our friend D[r] Hooker will gladly show you. —

I shall be delighted to hear soon again from you.

Believe me my dear Sir, | ever very sincerely yours | Julius Haast.

Prof. John Tyndall | FRS | London

RI MS JT/1/H/7

1. *Heat considered as a mode of motion*: *Heat as a Mode of Motion.*
2. *a paper*: J. Haast, 'Notes on the Causes which have led to the Excavation of Deep-Lake Basins in Hard Rocks in the Southern Alps of New Zealand', *Quarterly Journal of the Geological Society of London*, 21 (1865), pp. 130–2.
3. *On the climate of the Pleistocene epoch of New Zealand*: see J. Haast, 'Note on the Climate of the Pleistocene epoch of New Zealand', *Quarterly Journal of the Geological Society of London*, 21 (1865), pp. 135–7 (this was published as an appendix to yet another paper by Haast).
4. *D[r] Falconer's theory*: Hugh Falconer (1808–65), an Edinburgh-trained geologist, paleontologist, and naturalist, who had made significant contributions to botany and paleontology, and had been significant in establishing the antiquity of man, which Lyell later made use of in his *The Antiquity of Man* (1863). Further though, Falconer had considerable stratigraphic skill, and had argued that the Sewalik fossil material he had been working on in India was not of New Red Sandstone age (now called the Triassic) as had previously been supposed, but late Tertiary (it is now known to be predominantly Plio-Pleistocene), and that the deposit was of a similar sedimentary composition as the molasse of Switzerland (*ODNB*).
5. *paper of mine*: J. Haast, *Report on the Formation of the Canterbury Plains* (Christchurch: New Zealand Geological Survey, 1864).
6. *Canterbury plains*: an area south of the city of Christchurch, which has remained a highly debated subject of geological research because of its land and river formation left by glacial movement.

To Annette Roche 18 July 1864 2121

Pontresina, Canton Graubünden[1] 18th July, 1864.

My dear Miss Roche

I quitted London on the 2nd, and hence was far away from Albemarle Street[2] when your last letter[3] reached it. It followed me to this place which is a pretty village—one of the highest in the beautiful Ober Engadin.[4] I halted in Paris two days and was very cordially received by the savants there.[5] I especially liked the kindness of the old men, for they are proud of their position in the Academy[6] and in the world of science, and their recognition has therefore some significance. From Paris I proceeded to Bale,[7] thence to Coire[8] through Zurich. I had promised a friend of mine in Zurich[9] to attend a meeting of the Swiss Naturalists[10] to be held at Zurich next month, and this promise involved the bother of bringing a portmanteau instead of a simple knapsack which usually suffices for my summer expeditions. From Coire I sent on my luggage to Pontresina, and made a diversion to Tusis[11] myself to visit the celebrated gorge of the Via Mala.[12] It is very grand. The earth's crust appears to have been split to form this profound and high walled chasm, which approached from Tusis forms a grand doorway into the mountains. The general opinion I believe is that it is really a great crack, produced by some convulsion, but the human imagination is too liable to fall back upon such hypotheses, and to neglect the action of ordinary causes. I think good evidence could be adduced to prove that this famous chasm was cut out by the stream which now rushes through it, and that all ordinary actions required to enable them to assume the most stupendous forms is <u>sufficient time</u>.[13] From Tusis I crossed in the clear morning light along the beautiful valley of the Albula[14] to Tiefenkasten,[15] here I met the diligence, persuaded the conductor to let me occupy his seat for a consideration, and thus rode in the open air, and commanded from my perch the lovely scenery of the Ober Engadin. I halted a night at Samaden,[16] and next morning made my way to Pontresina where I now dwell.

The village is somewhat more than 6000 feet above the sea, the air is cool and this to me is important, for I am preparing a second edition of my last book "on Heat,"[17] and I brought a quantity of work with me which the coolness of the position will enable me to perform pleasantly. I am surrounded by noble mountains, and within reach of noble glaciers. From the window of my room I command a view of a group of magnificent snowpeaks, which feed the Roseg glacier.[18] I brought some instruments with me to make observations on the glaciers and commenced the week's operations energetically. Last Thursday however the weather was bad, and to shelter me from the rain I got beneath a roof of ice which I think must have chilled me in my damp clothes. So soon after my arrival I did not make sufficient allowance for my London constitution, and was consequently not able to do much for a day or two. The chill however is now past and I hope to be upon the glacier tomorrow executing my measurements.

In a week I purpose crossing into the Tyrol,[19] to meet my friend Ball,[20] who was I believe once member for Carlow.[21] We shall stay about the Ortler Group[22] for 10 or 12 days, after which he goes on to the Dolomite mountains,[23] and I am not sure that I shall be able to accompany him.

I am extremely obliged to you for the information you have obtained for me[24] regarding the tomb. I think a plain slab of granite on four pillars and bearing the words

John Tyndall[25] father of | John Tyndall F.R.S. | born October 1792; died ++ 1847

or something to this effect would be what I should require. However I should be very glad to receive the suggestions of the sculptor.[26] I should be willing to go as far as £25. —Perhaps however granite is not to be obtained on these terms.

I return to London at the end of next month and hope then to be able to arrange something definite regarding the tomb. This day's post brought me a letter from Dr. Phillips.[27] I fear we have little hope as to the recovery of my sister,[28] the energy of her constitution is hardly sufficient to carry her over such an attack.

Believe me with many thanks | Yours most truly | John Tyndall

Might I ask you to let Dr Phillips have the enclosed note?[29]

RI MS JT/1/TYP/11/3850
LT Typescript Only

---

1. *Pontresina, Canton Graubünden*: see letter 2028, n. 1.
2. *Albemarle Street*: the RI is located at 21 Albemarle Street, London.
3. *your last letter*: letter missing.
4. *Ober Engadin*: the Engadine, a long Alpine valley in southeastern Switzerland.
5. *very cordially received by the savants there*: Tyndall wrote in his journal: 'I stayed three nights, and two full days at Paris. I wished to see some of the Savants, and called upon a good many of them on Sunday. I missed nearly all the younger ones, who were out for the day: but I met two fine old sages, who were very kind to me. The freemasonry of science has something extremely pleasant about it: I needed no introduction, and still I did not know the men personally—I simply sent in my card, and they shewed me everything that was to be seen, and explained to me the character of the work on which they were engaged. The Academy of Sciences meets on the Monday, and I went there. My reception was exceedingly cordial. The younger men made their appearance, and most of them protested against my quitting Paris so soon. My little book I am told "has been crowned with an immense success". By the time of my return from Switzerland, it is probably that an edition of 2000 vols. will be sold' (RI MS JT/2/13c/1259A).
6. *Academy*: the Académie des Sciences, founded in 1666.
7. *Bale*: see letter 1909, n. 4.
8. *Coire*: the French spelling of the Swiss town of Chur.
9. *a friend of mine in Zurich*: Rudolf Clausius.
10. *meeting of the Swiss Naturalists*: see letter 2108, n. 2.

11. *Tusis*: the French spelling of the Swiss town of Thusis.
12. *Via Mala*: Latin for 'bad path', which refers to a notoriously treacherous path between Zillis-Reischen and Thusis. Tyndall wrote about this visit in 'On the Conformation of the Alps' upon his return (see letter 2100, n. 4).
13. *I think good evidence . . . is sufficient time*: in her own transcription of this letter LT noted 'This sentence is bad, but correctly copied'.
14. *Albula*: the Abula River and its valley is located in eastern Switzerland in the Central Eastern Alps.
15. *Tiefenkasten*: an early spelling of the Swiss town of Tiefencastel in eastern Switzerland.
16. *Samaden*: a small town in Swiss canton of Grisons.
17. *second edition of my last book "on Heat"*: the second edition of *Heat as a Mode of Motion* would be published in 1865.
18. *Roseg glacier*: a glacier in the Bernina range of mountains in Switzerland.
19. *Tyrol*: a region of the Alps in northern Italy and western Austria.
20. *my friend Ball*: John Ball.
21. *Carlow*: a country town in County Carlow, Ireland.
22. *Ortler Group*: see letter 2110, n. 7.
23. *Dolomite mountains*: see letter 2110, n. 8.
24. *for the information you have obtained for me*: see letter 2113 in which Tyndall had asked Roche to provide him with information regarding having a grave stone erected over his father's grave.
25. *John Tyndall*: see letter 2113, n. 3.
26. *sculptor*: presumably a 'Mr. Diegan', as indicated in letters 2142 and 2171.
27. *Dr. Phillips*: letter missing; for Phillips, see letter 2101, n. 2.
28. *my sister*: Emma Tyndall.
29. *enclosed note*: letter missing.

To Juliet Pollock 19 July 1864 2122

Pontresina[1] 19th July 1864.

My dear Mrs. Pollock.

Tennyson speaks somewhere of thunder peals "whenever thought has wedded fact,"[2] intimating that the processes by which the social and political condition of men have been ameliorated, have usually been processes of violent revolution. I might almost apply the maxim to my constitutions. In Switzerland I am always ameliorated; I pass from weakness to strength; but very often through a revolution. I very often break down for a few days at the commencement pass a stage of fever or something else and emerge from it a strong man. This year I have done so. Since I came here I have been engaged with measurements on the glaciers. Last Thursday I was on the ice and got slightly wet: to shelter myself I crept underneath a big stone till the drifting rain drove me away. I then retreated to a large hollow with a floor of granite and a roof of ice. Here I was protected from the rain, but my London body was chilled, and for four days I was worth nothing. Yesterday I felt that the molecules had shaken themselves into their proper positions—Today I feel still stronger, and tomorrow I hope to be a man of iron.

This is a most lovely country, close to the Tyrol[3] and resembling it more than any other part of Switzerland. The mountain forms are very fine, and the valleys crowded with pines very beautiful. At a distance noble snow peaks rise, and form the feeders of noble glaciers. There is a frankness about the people also which reminds one of the unspoiled Tyrolese. I quitted London very soon after I last saw you, and passed through Paris, Bale,[4] Zurich, & Coire[5] to my present position. From Coire I made a diversion to see the famous Via Mala[6], and reached this place last Saturday week (the 9th). Ever since I came I have been making observations upon the glaciers & getting instruction on other matters. The main problem which has occupied me is the formation of these mountains and valleys, and on this subject Nature has taught her pupil a great deal. In fact I do not think I should enjoy Switzerland were it not for this teaching. Mere beauty, mere grandeur, the enchantment of emotion—are not sufficient to fill a human life. Intellectual occupation must be added; work for the mind which the mind feels a pleasure in performing: the growth of knowledge, the acquisition of new truth. These are the things which keeps Switzerland perennially fresh to me. These are the things which shaken by the emotions make the music of the Alps to your friend.

A Mr. Bradley,[7] his wife and sister are staying here now. They know Tennyson, and have a house at Freshwater[8]—They know Easton[9] and thus they bring you to my mind. Mr. Bradley is principal of Marlborough College[10] and apparently a very capable man. He knows Carlyle & [Hereafter LT Typescript Only] sympathises with Essays and Reviews;[11] hence we get on very [Hereafter handwritten transcription resumes] well whenever we meet in conversation. The weather for several days has been dreary, these happily have been in part the days of my illness—but I have grown strong and the weather appears to have changed. Tomorrow I finish my glacier measurement and I may this be tempted to try the highest peak in the neighborhood—It is slightly dangerous, but danger is overcome by caution, and I should like in this way to assure myself that my strength is really restored.

Next week I pass the Bernina,[12] and go to Santa Catarina[13] to join my friend Ball[14]—We shall prowl together about the Ortler group of mountains for some days.

<u>Thursday</u>

The weather is heavenly this morning and the mountains magnificent. I am on the point of starting up the Piz Languard[15] from which point the eye can range over all the Alps—goodbye. Love to your husband[16] & the boys.[17]

Yours ever | John Tyndall

BL Add MS 63092, ff. 210–1
RI MS JT/1/TYP/6/2011–2

1. *Pontresina*: see letter 2028, n. 1.
2. *Tennyson speaks somewhere of thunder peals "whenever thought has wedded fact"*: this refers to lines 51–2 of Tennyson's poem 'Love Thou Thy Land' (1834): 'A bridal dawn of thunder-peals, | Wherever Thought hath wedded Fact' (published in A. Tennyson, *Poems* (London: Edward Moxon, 1842)).
3. *Tyrol*: see letter 2121, n. 19.
4. *Bale*: see letter 1909, n. 4.
5. *Coire*: see letter 2121, n. 8.
6. *Via Mala*: see letter 2121, n. 12.
7. *M^r^. Bradley*: George Granville Bradley (1821–1903), an Anglican clergyman and schoolmaster, who was headmaster of Marlborough College (1861–70), Master of University College, Oxford (1870–81), and Dean of Westminster (1881–1902). In the 1860s Bradley bought a house near Farringford on the Isle of Wight and soon developed a close friendship with Tennyson (*ODNB*).
8. *Freshwater*: see letter 2031, n. 28.
9. *Easton*: Easton Farm, Isle of Wight.
10. *Marlborough College*: a school in Marlborough, Wiltshire, founded in 1843.
11. *Essays and Reviews*: J. W. Parker (ed), *Essays and Reviews* (London: John W. Parker and Son, 1860). This was a controversial collection of seven broad church essays that considered the insights and implications of Higher Criticism for biblical literalism.
12. *Bernina*: the Bernina range of mountains in the Alps.
13. *Santa Catarina*: a village in the Eastern Alps.
14. *my friend Ball*: John Ball.
15. *Piz Languard*: a mountain overlooking Pontresina.
16. *your husband*: William Frederick Pollock.
17. *the boys*: Frederick, Walter, and Maurice Pollock.

From Friedrich Eduard Reusch 24 July 1864 2123

Hochgeehrter Herr!

Die beiliegende Platte ist Hydrophan von Czernowitza. Sie dient zur Ausstellung des folgenden Versuchs: in Alkohol getaucht & gewartet bis sie hell geworden ist, zeigt sie nach dem Herausnehmen & Abtrocknen sofort eine Trübung, bei welcher Spuren von dendritischen Gebildeten erscheinen. Wird aber die trüb gewordene Platte nun in Wasser getaucht, nach der Aufhellung herausgenommen & abgetrocknet, so bewegen sich grosse Dendriten rasch über die Fläche hin. Wieder in Wasser getaucht etc. kann man den Versuch sehr oft wiederholen. Die Erscheinung rührt her von der in den Poren des Steins vor sich gehenden Mengung von Wasser & Alkohol, welche bekanntlich von einer Contraction des Fluidum begleitet ist. Es erinnert dieser Versuch an Ihren schönen Versuch am Eis (Heat, pag 109).

Die obige Erscheinung habe ich zufällig beobachtet, als ich, durch Grahams Diffusionsversuch mit Graphitplatten veranlasst, Hydrophanplatten zu gleichem Zwecke bearbeitete.

Die Bilder der singenden Flamme im rotirenden Spiegel erscheinen nur dann als Continuum, wenn die Flamme während des Schwingens eine gelbe Spitze zeigt; ist sie aber ganz blau so sieht man die dunkeln Zwischenräume schon bei mässigen Geschwindigkeiten des Spiegels.

Der Versuch mit dem 2 Meter langen Blechrohr habe ich unterdessen feiner eingestellt; ich vermag den schönsten Orgelton, begleitet von einer Menge von Obertönen, durch die Flamme meines Brenners zu unterhalten. —Das Rohr wird zu diesem Zwecke an einem Stative fixirt & Flammenhöhe, Gaszufluss etc. auf den besten Effekt regulirt.

Ihre histor. Notizen im letzten Hefte des Phil. Mag. habe ich mit grösster Freude gelesen & das Heft sofort Freund Mayer mitgetheilt. Sie haben sich dadurch weitere Ansprüche auf unsern Dank & unsere Verehrung für Sie erworben.

Darf ich Sie bitten, Herrn Beker bei Elliot brthrs freundlich zu grüssen u. ihm zu sagen, dass jetzt der Apparat zur *[Verschaulichung]* des Gesetzes der Lichtbrechung im Reinen sei, dass man sogar die Farbenzerstreuung einigermassen damit demonstriren & durch eine besondere Einrichtung den Gang eines Lichtstrahls durch ein Prisma mechanisch verfolgen könne. Vielleicht komme ich dazu, Herrn Beker selber zu schreiben.

Mit aller Hochachtung grüsst Sie | Ihr ergebenster | Prof. Reusch

Tübingen | 24 Juli 1864

Dear Sir!

The accompanying sheet is hydrophane from Czernowitz.[1] It is used to conduct the following experiment: immersed in alcohol & waited until it lightened, after taking it out & drying it off, it showed a cloudiness, in which traces of dendritic structures appeared. If the now cloudy sheet is immersed in water, removed after brightening & dried off, large dendrites move quickly across the surface. Immersed again in water etc., the experiment can be repeated very often. The phenomenon comes from the mixture of water and alcohol taking place in the pores of the stone, which, as is known, is accompanied by a contraction of the fluid. This experiment is reminiscent of your beautiful experiments on ice (Heat,[2] page 109).

I observed the phenomena described above by chance, as I was preparing hydrophane sheets for the purpose of repeating Graham's[3] diffusion experiments on graphite sheets.

The images of the singing flame in the rotating mirror only appear as a continuum, if the flame exhibits a yellow tip during oscillations, but if it is entirely blue, you will see the dark interstices already at a moderate speed of the mirror.

In the meantime I have adjusted the experiment with the 2 meter long tin pipe more finely; I can obtain the most beautiful organ tones, accompanied by numerous overtones, through the flame of my burner. —For this purpose, the tube is affixed to a stand & the height of the flame, gas flow etc. regulated to achieve the best effects.

I read your histor[ical]. note[4] in the last issue of the Phil. Mag. with great pleasure & communicated the issue immediately to my friend Mayer. You have earned with it more claims to our gratitude and our respect.

May I request that you give my friendly greetings to Mr Beker[5] by Elliot br[o]th[e]rs[6] and tell him that the apparatus for illustrating the law of refraction is now straightened out, that it is possible to tolerably demonstrate the colour dispersion &, through a special arrangement, to mechanically follow the path of a light beam through a prism. Perhaps I will be able to write to Mr Beker myself.

With most respectful greetings | yours truly | Prof. Reusch

Tübingen | 24 July 1864

RI MS JT/1/R/18

---

1. *Czernowitz*: a city in western Ukraine, also spelled Chernivtsi.
2. *Heat*: *Heat as a Mode of Motion*.
3. *Graham*: Thomas Graham.
4. *your histor[ical]. note*: see letter 2098, n. 6.
5. *Mr Beker*: Ernst Becker.
6. *Elliot br[o]th[e]rs*: see letter 2073, n. 11.

To Michael Faraday 25 July 1864 2124

Pontresina.[1] | 25th, July, 1864.

My dear Mr Faraday

Thus far this year has been a bad year for Switzerland; the weather has been so capricious that even the few clear days which have been interspersed among the shaded and rainy ones could not, for want of faith in their continuance, be turned to proper account. I came here more than a fortnight ago. Immediately after my arrival I began a series of measurements on the two accessible glaciers—the Morteratsch[2] and the Rosegg[3] glaciers—which have yielded instructive results. I quitted London much worn down; I am now strong,—but in passing from weakness to strength I had to go through an illness of four days, which seemed to produce a revolution in my system. In fact, I got wet upon the glacier before I was able to bear a wetting, and suffered in consequence. This is a very beautiful portion of Switzerland, and were the weather favourable this village would be a capital headquarters for a month. There are many delightful excursions in the neighbourhood; the mountains are noble and finely formed; the distant snow fields beautiful, and the valleys, clothed with pines and enlivened by musical rivers, have an air of *comfort* about them which I have not seen elsewhere in Switzerland. The people also are frank and civil. This year I have been interested to a greater extent than ever in the question as to how the Alps have assumed their present forms. Two rival opinions now divide geologists:—one is that the valleys of the Alps and the lakes of Switzerland and Italy are open fissions, produced by the cracking of the earth's crust; the other is that the valleys have been produced in the main by erosion, and that even the lakes themselves have been scooped out by the long continued action of the ancient glaciers. There are difficulties in the way of each of these hypotheses; but I cannot help thinking that the theory of erosion stands upon the most solid basis.[4] It appeals to agencies which are known to exist or to have existed, and which, if granted sufficient time, are certainly competent to produce the effect ascribed to them. Whereas the crack and fissure hypothesis labours under the disadvantage of being a creature of the imagination; it invokes agencies which we have never seen in action, and, even granting the hypothesis, there are various phenomena in the Alps which cannot without the most artificial assumptions be reconciled with it. My own notion is that the swelling from which the Alps have been sculptured has been produced during those slow changes of form which a mass like the earth may be expected to undergo. That from the moment the first portion of this protuberance emerged from the sea, erosion agents have been acting upon it. Water courses cut their way into the mass, and formed the first ravines; then in process of time came the stupendous glaciers of the so called glacial epoch, ploughing out the valleys with irresistible might. The shapes of the mountains, and the forms of the valleys suggest immediately, in nine cases out of ten, an origin of this kind. You may take your stand upon a commanding peak and see before you valleys which the most thorough going plutonist[5] must ascribe to erosion. You may pass from these smaller valleys by insensible gradations to the largest valleys of the Alps, and the plutonist shall not be able to point out the place where erosion ceases to be the principal agent and where cracks and fissions begin. It is by thus acquiring the key among the smaller phenomena that we proceed with from steps to the larger ones. And what a hoary antiquity does this process of erosion indicate! This antiquity may appear trivial in comparison with the other epochs conceived in the consolidation of our planet; those other epochs, however, cease to impress us because they so much surpass the range of our conceptions. But the sculpturing of the Alps by ice and water comes in some measure within this range; the mind can rise to the level of the problem, and realize the might of *Time* without being crushed by the sense of its incompetence to seize upon the Infinite to which all those changes must ultimately be referred. Kindest regards to Mrs Faraday[6] and Miss Barnard.[7]

Yours ever sincerely | John Tyndall

I leave this place the day after tomorrow but hope to return to it in about 10 days. Would you kindly ask Smith[8] to direct my letters here as before?

Faraday Correspondence, 6:4484

1. *Pontresina*: see letter 2028, n. 1.
2. *Morteratsch*: the Morteratsch Glacier is the largest glacier by area in the Bernina Range of the Bünder Alps in Switzerland.
3. *Rosegg*: the Roseg Glacier is in the Bernina Range of the Bünder Alps in Switzerland.
4. *I cannot help thinking that the theory of erosion stands upon the most solid basis*: Tyndall would publish on this later in the year; see letter 2100, n. 4.
5. *plutonist*: Plutonic theory in geology ascribed most geological phenomena to the action of internal heat; this view was advocated by the geologists James Hutton (1726–97) and Charles Lyell.
6. *Mrs Faraday*: Sarah Faraday.
7. *Miss Barnard*: Jane Barnard.
8. *Smith*: not identified; possibly an assistant or porter at the RI.

From Michael Faraday 1 August 1864 2125

Hampton Court, | 1 Aug. 1864.

My dear Tyndall,

I do not know whether my letter will catch you, but I will risk it, though feeling very unfit to communicate with a man whose life is as vivid and active as yours; but the receipt of your kind letter[1] makes me to know that, though I forget, I am not forgotten, and though I am not able to remember at the end of a line what was at the beginning of it, the imperfect marks will convey some sense to you of what I long to say. We had heard of your illness through Miss Moore,[2] and I was therefore very glad to learn that you are now quite well. Do not run too many risks, or make your happiness depend too much upon dangers, or the hunting of them. Sometimes the very thinking of you and what you may be about wearies *me* with fears, and then the cogitations pause and change, but without giving me rest. I know that much of this depends upon my own worn out nature, and I do not know why I write it, save that when I write to you, I cannot help thinking it, and the thoughts stand in the way of other matters.

We are progressing in the cleaning at the R.I., and I hope you will find things there all right presently. Mr. Barrett's[3] letter[4] I have left for him, according to your instructions. Mr. Lubbock was enquiring for your address a few days ago, but I could not give him any, and now I conclude he knows it. I am so forgetful that I am afraid to write to him from here, for I have no other address for him, save that of the list of members. I think I shall send him your letter to me as the information.

See what a strange desultory epistle I am writing you, and yet I feel so weary that I long to leave my desk and go to the couch.

My dear wife[5] and Jane[6] desire their kindest remembrances. I hear them in the next room, but even they weary me—for I forget—but not you, my dear Tyndall, for I am

Ever yours, | M. Faraday

Faraday Correspondence, 6:4487

1. *your kind letter*: letter 2124.
2. *Miss Moore*: either Harriet Jane Moore or Julia Moore.
3. *Mr. Barrett's*: William Barrett.
4. *letter*: letter missing.
5. *My dear wife*: Sarah Faraday.
6. *Jane*: Jane Barnard.

From Juliet Pollock — 2 August 1864 — 2126

59 Montagu Square W[1] | August 2. 1864

My dear Mr. Tyndall

It added to all our happiness to hear of your well being, but I am sorry that your molecules should have been chilled: they must not go into any more ice caverns. Our molecules, on the contrary, have run the risk of dissolution by heat since you left us, and have hardly known how to find the firmness of purpose to hold together. But we had our pleasant garden to shelter them in and under the shade of boughs, but not melancholy ones, I have lost and neglected some of the creeping hours of time.

Those that were not neglected were passed in reading sonnets and in playing croquet. The sonnets are by Charles Turner,[2] Tennyson's Brother (you have seen his early ones) and are very lately published in one thin volume, which I intend to present to you as soon as you return home, for I feel very sure of your being charmed with their sweet music and with their delicate appreciation and fine observation of nature's beauties, and with their power of expressing thoughts strong as well as tender, in the best possible words. There is one that brought you to my mind as being 'like' what you would feel and think and tell. It is called the Boatman of the Moselle.[3]

We have made little excursions from London on several occasions, to St. Julians[4] and last Sunday, to the Chief Baron's[5] place at Hatton.[6] We had a delightful day there, indeed it is always delightful to be with him. He is the fulfilment of Sir H. Wottons character of a happy man;[7] his mind is all light and love and truth, and he has no physical infirmity to interfere with it. He walked with me in the sunshine of his garden till I was obliged to ask for shelter & rest and then when we came in doors he read out to me some of his favourite passages in Burke[8]—and talked to me a good deal about the Queen's English on which subject he has been lately in correspondence with the Dean of Canterbury.[9] At dinner he said grace—"God be praised for all his mercies upon the table, and round the table"—surveying the large gathering of children & grandchildren there present with happy tears in his eyes. He was as happy at Walter's[10] having come in first, in the last Eton examination[11] and having been sent up for good and having got two prizes as if he were his only child instead of one among an immensity of grandchildren. He appreciates every little drop of the rolling ocean with all its reflection of light and colour.

The Miss Speddings[12] are at Spa[13] till the 15th. Aug[us]t. and afterwards go to Switzerland you may possibly come across them. Our croquet meetings lately have been under the influence of Miss Caruac[14] who is very charming and lively & plays as skilfully and gracefully as any player I ever saw with such a pleasant temper that it seems a favour to be beaten by her. We have become friends and her mother has called upon me but their way of life is rather more in fashion and dissipation than ours & for that reason we shall probably never be very much together.

We go to St Julians on the 15th. August staying there probably till the 1st Sept[embe]r On the 3d. we go to Freshwater[15] to Easton & I hope you will be able to come to us there.

The Moores[16] are at Esher[17]—the Barlows[18] at Sydenham[19]. <T>he poor dear man daily worse—the Erskines[20] (in Town at present)—go to Fir Grove[21] next Saturday.

My husband's[22] love—he has borne the heat this year better than usual. Walter and Moss[23] join in affec[tiona][te] greetings and Moss asks you how many days it would take you to eat 24 biscuits if you eat one half per day.

Yours my dear M[r]. Tyndall | Ever most truly | Juliet Pollock

RI MS JT/1/P/172
RI MS JT/1/TYP/6/2013–4

---

1. *59 Montagu Square W*: see letter 1954, n. 2.
2. *Charles Turner*: Charles Tennyson Turner (1808–79), an Anglican clergyman and poet, and elder brother of the poet Alfred Tennyson. Pollock was referring to C. Turner, *Sonnets* (London: Macmillan and Co., 1864).
3. *Boatman of the Moselle*: the full title was 'The Moselle Boatman and his Daughter' (Turner, *Sonnets*, pp. 9–10).
4. *S[t]. Julians*: see letter 1954, n. 14.
5. *Chief Baron*: Jonathan Frederick Pollock (1783–1870), William Frederick Pollock's father and the Chief Baron of the Exchequer.
6. *Hatton*: Hatton was Pollock's family estate in Hatton, Hounslow in the county of Middlesex.
7. *Sir H. Wottons character of a happy man*: a reference to British diplomat and author Henry Wotton's (1568–1639) poem 'Character of a Happy Life' (J. Hannah (ed), *Poems by Sir Henry Wotton, Sir Walter Raleigh, and Others* (London: William Pickering, 1845), pp. 28–31).
8. *Burke*: Edmund Burke (1730–97), Member of Parliament and philosopher who was widely regarded for his opposition to the French Revolution, which he saw as undermining the natural institutions of society. Burke's *Reflections on the Revolution in France* (1790) prompted a number of responses, among them Tom Paine's *Rights of Man* (1791) and Mary Wollstonecraft's *A Vindication of the Rights of Men* (1790) and *Vindication of the Rights of Woman* (1792).
9. *Dean of Canterbury*: Henry Alford (1810–71), an English theologian and author, Dean of Canterbury from 1857–71 (*ODNB*).
10. *Walter*: Walter Pollock.
11. *Eton examination*: Walter Pollock was a student at Eton College from 1862–5.
12. *Miss Speddings*: the Speddings were a large family; it is thus not clear exactly which Miss Speddings are referred to here.
13. *Spa*: likely Cheltenham Spa, a fashionable spa town at this time.
14. *Miss Caruac*: or possibly 'Carnac'; not identified.
15. *Freshwater*: see letter 2031, n. 28.
16. *The Moores*: this is likely a reference to Harriet Jane Moore and Julia Moore, although they had other siblings: John Carrick Moore and Graham Francis Moore (1806–83), a lawyer; another sister was deceased, Louisa Moore (1802–53). Their mother, Harriet Moore (née Henderson, 1779–1866) might also be implied here. Her husband, James Carrick-Moore

(1762–1860), a surgeon and promoter of vaccination, and brother of the feted war hero John Moore of Corunna, had died four years earlier. The wealthy Moores had close connections with the RI and frequently entertained Tyndall at their Wimbledon home.

17. *Esher*: a small, then fashionable, town in Surrey.
18. *the Barlows*: the family of John Barlow.
19. *Sydenham*: a district in the southeastern boroughs of London.
20. *the Erskines*: probably the legal theorist Thomas Erskine May (1815–86) and his wife Johanna Laughton (*ODNB*).
21. *Fir Grove*: Fir Grove House, Eversley, built *c.* 1736; home of the judge Thomas Erskine (1788–1864) and his wife Henrietta Elizabeth Erskine (née Trail) (*c.* 1757–*c.* 1815) (*ODNB*).
22. *My husband's*: William Frederick Pollock.
23. *Moss*: Maurice Pollock.

From William Hüber[1] 3 August 1864 2127

Paris 3 Août 1864

Monsieur,

La lecture de votre intéressant ouvrage "The Glaciers of the Alps" J. Murray 1860, m'a captivé au point que je regrette vivement de ne pas le voir traduit en français, comme vos leçons sur la chaleur l'ont été par l'Abbé Moigno. La glaciologie, cette brochure nouvelle de la science qui peut dévoiler tant de mystères, n'est en France qu'à état de lacune. Aucun livre ne peut mieux contribuer à la combler que le vôtre.

Ayant personnellement habité pendant plusieurs années les hautes régions des Alpes et spécialement le glacier du Rhône, lorsque j'étais chargé par le gouvernement de mon pays de l'étude des routes militaires de la Furka, du Grimsel, de l'Oberalp et du réseau Grison, j'ai rempli mes rares moments de loisir par l'étude de ces grands phénomènes glaciaires que vous décrivez si bien et que vous expliquez avec tout de clarté et de talent.

Je viens donc vous demander, Monsieur, l'autorisation de traduire ce livre en français. Si cette proposition était acceptée, quelles seraient les conditions pécuniaires que je devrais remplir vis-à-vis de vous et de votre éditeur.

Monsieur le Docteur Kolb (rue Miroménil 7 Paris) avec lequel vous avez voyagé en Suisse me connaît personnellement et pourra, si vous le désirez, vous donner, avec M. Maunoir, secrétaire de la Société de Géographie de Paris, les renseignements que vous êtes en droit d'exiger sur mes faibles talents. Permettez-moi seulement d'ajouter que, fils d'Anglaise, je crois posséder suffisamment votre langue pour pouvoir vous garantir une traduction telle que vous devez la désirer.

Agréez, Monsieur, l'expression de ma parfaite considération | William Huber | 67 Boulevard Malesherbes Paris

Ingénieur Civil. Ancien élève de l'École Centrale des Arts & Manufactures—Capitaine à l'État. Major du Génie de la Confédération Suisse—Ex-Ingénieur en chef des toutes militaries des Hautes-Alpes—Membre des Sociétés: de Géographie à Paris, d'Utilité publique Suisse, de la classe d'Industrie de Genève et autres.

Chevalier de la Légion d'Honneur. —

Paris 3 August 1864

Sir,

Reading your interesting book "The Glaciers of the Alps"[2] J. Murray 1860, captivated me so much that I deeply regret not seeing it translated into French, as your lessons on heat were by Father Moigno. Glaciologie, that new scientific brochure[3] in France that could unveil so many mysteries, is lacking. No book can contribute to filling it more than yours.

Having personally lived for many years in the high regions of the Alps and especially the Rhône Glacier,[4] when I was asked by my country's government to study the military routes of the Furka, Grimsel, Oberalp and the Grison network,[5] I occupied my rare moments of leisure with the study of those great glacial phenomena that you describe so well and explain with all clarity and talent.

So I would like to ask, Sir, for your authorization to translate this book into French.[6] If you accept this proposition, what would be the monetary conditions that I would have to fulfil in regards to you and your publisher?

Doctor Kolb[7] (Rue Miroménil 7, Paris), with whom you travelled in Switzerland, knows me personally and he could give you, if you so desire, with Mr Maunoir,[8] secretary of the Geographical Society of Paris, the information that you are entitled to demand about my humble talents. Only allow me to add that as a son of an Englishwoman I believe that I possess your language sufficiently well to be able to guarantee you the translation that you desire.

Please accept, Sir, my best regards. | William Huber | 67 Boulevard Malesherbes, Paris.

Civil Engineer. Alumnus of the Central School of the Arts & Manufacturing—Captain of the State. Staff Engineer of the Swiss Confederation—Former Chief Engineer of all the militaries of the High Alps—Member of the Societies: of Geography in Paris, Public Utility in Switzerland, of the Industrial Class in Geneva, and others.

Knight of the Legion of Honour. —

RI MS JT/1/H/509

1. *William Hüber*: William Hüber (dates unknown), a French civil engineer and author of *Les Glaciers* (Paris: Challamel Ainé, 1867).
2. *"The Glaciers of the Alps"*: *Glaciers of the Alps.*
3. *Glaciology, that new scientific brochure*: not identified, but presumably a French work on glaciology that Huber felt did not compare to Tyndall's own work *Glaciers of the Alps*, which had not been translated into French.
4. *Rhône Glacier*: see letter 1915, n. 2.
5. *the Furka, Grimsel, Oberalp and the Grison network*: the Furka Pass is a high mountain pass in the Swiss Alps which connects the villages of Gletsch (in the Valais) and Realp. The Grimsel Pass is a mountain pass in southern Switzerland that crosses the Bernese Alps. The Oberalp Pass is a mountain pass in the Swiss Alps in central Switzerland that connects the cantons of Graubünden and Uri.
6. *to translate this book into French*: this did not happen, but Huber did share information about Tyndall's glacier research in *Les Glaciers* (Paris: Challamel Ainé, 1867).
7. *Doctor Kolb*: possibly Karl Kolb (dates unknown), a German doctor.
8. *Mr Maunoir*: Charles Maunoir (1830–1901), a Swiss geographer known for his surveys of France and Africa. He served as the Secretary of the Société de Géographie between 1867–96 ('Obituary. M. Charles Maunoir', *The Geographical Journal*, 19 (1902), p. 513–4).

From Jean François Léonard Alexandre de Colnet d'Huart[1] 9 August 1864 2128

Luxembourg le 9 Août 1864

Monsieur,

Je viens de lire vos douze leçons sur la théorie de la chaleur, traduites par l'Abbé Moigno.

Le sujet que vous traitez me préoccupe depuis sept ans. Je parviens de mon côté aux mêmes conclusions que vous; mais en suivant une marche toute différente, celle du raisonnement rigoureux de l'analyse mathématique. Je ne puis vous exprimer, Monsieur, combien la lecture de vos leçons m'a enthousiasmé.

Je vous adresse par la poste la première partie de mon travail. Je démontre rigoureusement, si la théorie de Young et de Fresnel est vraie, que la chaleur est la force vive résultant d'une rotation moléculaire; que cette vitesse de rotation est un très grand nombre de fois plus considérable que la vitesse de vibrations; que la molécule doit se déformer comme la terre s'est déformée par la rotation autour de son axe; que la surface de la molécule augmentait ainsi, elle repousse les molécules avoisinantes et par suite que le corps se dilate. Mais, chose singulière, en ayant égard à cette force répulsive, les formules de Young et de Fresnel (perfectionnées par l'illustre géomètre françois Lamé), qui régissent le mouvement vibratoire (calorifique), me conduisent aux formules de

Fourier, qui régissent le mouvement de la chaleur de conductibilité dans les corps [athermones]. C'est là la plus précicuse vérification de mes calculs. La seconde parti de mon travail est [ma] [preuve], et si ce genre de traites <de> la partie de la physique dans laquelle vous vous êtes illustré peut vous intéresser je m'empresserai de vous l'adresser par la poste. Mon équation différentielle des mouvements calorifiques me conduit à des théorèmes qui jettent une vive lumière sur la coloration des corps [disphanes], sur les résultats obtenus par Melloni, Kirchhoff et Plücker; sur la force appelée affinité chimique; sur les limites du spectre complet (calorifique et lumineux); enfin sur la transformation de l'électricité en chaleur.

Mais, Monsieur, je n'ai pas comme vous le talent de donner de l'attrait à la science; je ne suis qu'un pauvre analyste qui chérit la nature. Je vous dois beaucoup de reconnaissance, d'abord pour le plaisir que m'a procuré la lecture de votre livre; ensuite parce que j'y trouve [indiquer] tant de problèmes à résoudre par la méthode que j'ai adoptée. S'il est vrai que tout le problème bien posé est à moitié résolu, vous m'avez facilité ma tâche.

Il me reste à vous demander pardon, Monsieur, pour le temps précieux que vous enlève la lecture de ma lettre.

Je vous prie, de vouloir bien agréer l'expression de la considération la plus distinguée de votre très humble et très reconnaissant serviteur | De Colnet-d'Huart | Membre de quelques académies et sécretaire de la Société royale des Sciences du Grand Duché de et à Luxembourg

à Monsieur John Tyndall, Membre de la Société royale à Londres.

Luxembourg 9 August 1864

Sir,

I have just read your twelve lessons on the theory of heat, translated by Father Moigno.[2]

I have been busy for seven years with this very subject. For my part, I have reached the same conclusions as you, but using a very different method: the rigorous reasoning of mathematical analysis. I cannot say, Sir, how much I enjoyed reading your lessons.

I am sending you the first part of my work[3] by post. I rigorously demonstrate, if the theory of Young[4] and Fresnel[5] is true, that heat is the strong force resulting from a molecular rotation; that this speed of rotation is many times greater than the speed of vibrations; that the molecule is distorted like how the earth is distorted by the rotation around its axis; that the surface of the molecule grew in this way. The molecule pushes the neighbouring molecules and then the body expands. But, remarkably, regarding the repulsive force, the formulas of Young and Fresnel (perfected by the famous French geometer Lamé[6]) governing the vibrating (calorific) movement, lead me to the formulas of Fourier,[7] governing the movement of heat conductivity in athermic bodies. This is the most invaluable verification of my calculations. The second part of my work[8] is *[my proof]*, and if you are interested in texts on the part of physics in which you became famous, I will send it to you by post. My differential equation for calorific movement lead me to theorems that shed much light on the coloration of *[disphanous]* bodies, on the results obtained by Melloni, Kirchhoff and Plücker;[9] on the force called chemical affinity; on the limits of the complete spectrum (calorific and luminous); and finally on the transformation of electricity into heat.

However, Sir, I do not have your gift for making science appealing; I am only a poor analyst who loves nature. I am very grateful to you, firstly for the pleasure of reading your book; next because it provided so many problems to solve by the method I chose. If it is true that any problem clearly explained is a problem half-solved , then you made my task easier.

Nothing remains but for me to beg your pardon, Sir, for the precious time you are taking to read my letter.

Please accept your most humble and grateful servant's deepest regards | De Colnet-d'Huart | Member of several academies and secretary of the Royal Society of Sciences of the Grand Duchy of Luxembourg

To Sir John Tyndall, Member of the Royal Society in London.

RI MS JT/1/D/136

1. *Jean François Léonard Alexandre de Colnet-d'Huart*: Jean François Léonard Alexandre de Colnet d'Huart (1821–1905), Professor of mathematics and physics at la Athénée de Luxembourg and author of *Mémoire sur La Théorie Mathématique de la Chaleur et de la Lumière* (Luxembourg: Pierre Bruck, 1870). See 'Alexander de Colnet d'Huart', *Luxemburger Wort*, 12 June 1905, p. 2.
2. *your twelve lessons . . . by Father Moigno*: see letter 1978, n. 3.
3. *first part of my work*: probably J. F. L. A. Colnet d'Huart, *Determination de la relation qui existe entre la Chaleur rayonnante, la Chaleur de conductibilité et la Chaleur latent* (Luxembourg: V. Buck, 1863).
4. *Young*: Thomas Young (1773–1829), an English physician and natural philosopher whose work helped to establish the wave theory of light. Young had held the RI professorship of natural philosophy (1801–3) (*ODNB*). .
5. *Fresnel*: Augustin-Jean Fresnel.
6. *Lamé*: Gabriel Lamé (1795–1870), a French mathematician and the chair in mathematical physics at the Sorbonne, was most notable for the mathematical theory of elasticity (*CDSB*).
7. *Fourier*: Joseph Fourier (1768–1830), a French mathematician and physicist known for his mathematical investigations into heat transfer and vibrations, and author of *Théorie Analytique de la Chaleur* (Paris: Firmin Didot, 1822) (*CDSB*).

8. *second part of my work*: probably J. F. L. A. Colnet d'Huart, *Nouvelle théorie mathématique de la chaleur et de l'électricité*, 2 vols (Luxembourg: V. Buck, 1864–5).
9. *Plücker*: Julius Plücker (1801–68), a German mathematician, experimental physicist, and professor of physics at the University of Bonn (1847–68). He first established a reputation for his work on analytical geometry, but under the influence of Faraday's research he began experimental work in electricity and magnetism in the 1840s. His ideas about diamagnetism and the behaviour of crystals in magnetic fields were challenged by Tyndall, which led to ill feeling and controversy (*CDSB*).

From Johann Bohn[1] 14 August 1864 2129

To Professor Tyndall, Esq., F.R.S.

I READ your "Notes on Scientific History"*[2] with great pleasure and satisfaction. I agree perfectly with all you say respecting Mayer's researches as compared with those of others; in some respects, indeed, I am inclined to go further than you do. Seven years ago I studied the history of the principle of the Conservation of Energy, and arrived at the same conclusion, with respect to the *modern* development of this theorem, as the one for which you so ably and warmly contend. Your recent disinterested advocacy of Mayer's claims, and my own conviction that historic truth is the sole object of your research, inspire the hope that the following remarks will be found worthy of attentive perusal.

Descartes,[3] so far as I know, was the first to give expression to the thought that whatever is not material must necessarily be indestructible. This non-material something he called "Force," a word which subsequently had, for a long period, divers and consequently vague meanings.

In Descartes's *Principia Philosophiae* (Pars II., § xxxvi.) we find the following:[4]—

"Deum esse primariam motus causam et eandem semper motus quantitatem in universo conservare.

"Motûs naturâ sic animadversâ, considerare oportet ejus causam, camque duplicem: Primò scilicet universalem et primariam, quae est causa generalis omnium motuum qui sunt in mundo; ac deinde particularem, à quâ fit, ut singulae materiae partes motus, quos priùs non habuerunt, acquirant. Et generalem quod attinet, manifestum mihi videtur illam non aliam esse, quàm Deum Ipsum, qui materiam simul cum motu et quiete in principio creavit†,[5] jamque per solum suum concursum ordinarium, tantundem motûs et quietis in eâ totâ quantum tunc posuit conservat. Nam Quamvis ille motus nihil aliud sit in materiâ motâ quàm ejus modus; certam tamen et determinatam habet quantitatem, quam facilè intelligimus eandem semper in totâ rerum universitate esse posse, quàmvis in singulis ejus partibus mutetur. Ita seilicet ut putemus, cùm una pars material duplò celeriùs movetur, quàm altera, et haec altera duplò major est quàm prior, tantundem motûs esse in minore quàm in majore, ae quantò motus unius partis lentior fit, tantò motum alicujus alterius ipsi aequalis fieri celeriorem. Intelligimus etiam perfectionem esse in Deo, non solùm quòd in se ipso sit immutabilis, sed etiam quòd modo quàm maximè constant et immutabili operetur : Adeò ut iis mutationibus exceptis, quas evidens experientia, vel divina revelatio certas reddit, quasque sine ullâ in creatore mutatione fieri percipimus, aut credimus, nullas alias in ejus operibus supponere debeamus, ne qua inde inconstantia in ipso arguatur. Unde sequitur quàm maximè rationi esse consentaneum, ut putemus ex hoc solo, quòd Deus diversimodè moverit partes materiae, cùm primùm illas creavit, jamque totam istam materiam conservet, codem planè modo, eâdemque ratione quâ priùs creavit, eum etiam tantundem motûs in ipsâ semper conservare."

Descartes therefore, precisely like Colding,[6] bases the proof of his theorem on a divine attribute. The unsatisfactory nature of such a proof is manifest. Are we not equally justified in asserting that the assumption of a constant quantity of motion involves a limitation of divine power? The almightiness of God must manifest itself by actual achievement, new motion must incessantly be created; therefore, assuming with Descartes the indestructibility of that which exists, the quantity of motion must increase.

Every attempt to deduce a natural law from an à priori conceived attribute of God must inevitably be utterly fruitless.

Leibnitz[7] was the first to publish, in its proper form, the general theorem of the conservation of *vis viva*,[8] and to demonstrate the same by empirically ascertained and rationally established theorems. He at once opposes Descartes's views, and introduces the important conception of *vis viva.* All this will be found in his article in the *Acta Eruditorum*, Lips. 1686, entitled:

> "Brevis demonstratio erroris memorabilis Cartesii et aliorum circa legem naturae, secundum quam volunt a Deo eandem semper quantitatem motus conservari, qua et in re mechanica abutuntur."

In the warm discussion which arose, Leibnitz argued that the assumption of the incorrectness of his views involved the necessity, or at least the possibility of a perpetual motion; which later he urged is manifestly absurd. Colding employs the same argument; and Helmholtz, in 1847, in his well-known work 'On the Conservation of Force,'[9] attributes great importance to his theorem concerning the absurdity of perpetual motion.

John Bernoulli,[10] however, gave the clearest exposition of the principle of the conservation of *vis viva*, as will be admitted after a perusal of his correspondence with Leibnitz (*Vivorum cel. G. C. Leibnilii et Joh. Bernoulli commercium phil. et math.*), of his "Discours sur les lois de la communication du movement" (*Opera omnia*, tom. iii. p. 1), and especially of his memoir entitled "De vera notione virium vivarum earumque usu in dynamicis" (*Opera Omnia*, tom. iii. p. 239).

Allow me to draw your particular attention to the following two passages of the last-named memoir:—

"§ III. Hine patet *vim vivam* [quae aptius vocaretur *facultas agendi*, Gallice *le pouvoir*] esse aliquid reale et substantiale, quod per se subsistit, et quantum in se est, non dependet ab alio. Unde concludimus, quamlibet vim vivam habere suam determinatam quantitatem, de qua nihil perire potest, quod non in effectu edito reperiatur. Hine sponte fluit, vim vivam semper conservari; adeo ut quae ante actionem residebat in uno pluribusve corporibus, nisi quid in prioribus remanserit. Atque hoe est, quod vocamus *conservationem virium vivarum.*"

One fact of peculiar interest is John Bernoulli's assertion that the *vis viva* which apparently disappears—that is to say, the *vis viva* which is not employed in external work such as the raising of a weight—may be consumed in molecular work. The following extract from § 9 of the above memoir will establish this point.

"Si corpora non sunt perfecte elastica, aliqua pars virium vivarum, quae periisse videtur, consumitur in compressione corporum, quando perfecte se non restituunt; a quo autem nune abstrahimus, concipientes, compressionem illam esse similem compressioni elastri, quod post tensionem factum impediretur ab aliquo retinaculo, quo minus se rursus dilatare posset, et sic non redderet, sed in se retineret vim vivam, quam a corpore incurrente accepisset; unde nihil virium periret, etsi periisse videretur."

The conversion of *vis viva* into heat, or at least the possibility of such a conversion, was first asserted by Augustin Fresnel.[11] The French translation of Thomson's 'Chemistry'[12] contains an appendix "On Light" from Fresnel's pen, wherein he says:—

"C'est un principe général du movement des fluides élastiques, que, de quelque façon que l'ébranlement s'étende ou se subdivise, la somme totale des forces vives reste constante. Et voilà principalement pourquoi la force vive doit être considérée comme la mesure de la lumière, dont la quantité totale reste toujours à trés peu près la meme, tant qu'elle ne traverse due moins que des millieux très transparens. Les corps noirs et même les surfaces métalliques les plus brillantes ne réflechissent pas à beaucoup près la totalité de la lumière qui tombe sur leur surface; les corps imparfaitement transparens, et même les plus diaphanes, quand ils sont assez épais, absorbent aussi (pour me servir de l'expression usitée) une quantité notable de la lumière incidente; mais il n'en faut pas conclure que le principe de las conservation des forces vives, n'est plus applicable à ces phénomènes; il résulte au contraire de l'idée la plus probable qu'on puisse se faire sur la constitution mécanique des corps, que la somme des forces vives doit toujours rester la même (tant que les forces accélératrices qui tendent à ramener les molécules à leurs positions d'équilibre n'ont pas change d'intensité), et que *la quantité de forces vives qui disparaît comme lumière est reproduite en chaleur*."

Should you consider the contents of this letter suitable for the pages of the Philosophical Magazine, I should feel honoured by its publication in that journal.

Giessen, August 14, 1864.

J. C. Bohn, 'Historic Notes on the Conservation of Energy', *Phil. Mag.*, 28 (1864), pp. 311–4.

---

1. *Johann Bohn*: Johann Conrad Bohn (1831–97), a German physicist. From 1858 Bohn held a private lectureship at the Physics Institute of the University of Munich having received his doctorate for his historical treatise on the conservation of force. In 1860 he moved to the University of Giessen as associate professor of mathematics and physics. He translated Tyndall's *Heat as a Mode of Motion* into German but letters in 1866 show that Tyndall refused him permission to publish. In 1866 he became a professor in Aschaffenburg at the Centralforstlehranstalt (School of Forestry) for the Kingdom of Bavaria, where he taught mathematics and physics and later geodesy until his death in 1897 (*CDSB*).
2. The following footnote appeared here in the original: '* Phil. Mag. S. 4. Vol. xxviii. P. 25' (see letter 2098, n. 6).
3. *Descartes*: René Descartes (1596–1650), a French philosopher and mathematician, known primarily for his publication of *Discourse on Method* (1637) and his work on analytical geometry. He is widely held as the one of the founders of seventeenth-century continental rationalism (*CDSB*).
4. *the following*: Bohn here assumed that Tyndall could read Latin.
5. The following footnote appeared here in the original: '† Phil. Mag. S. 3. Vol. xxiii. P. 442 (1843): Mr. Joule, "That the grand agents of nature are by the Creator's fiat indestructible"'.
6. *Colding*: Ludwig August Colding (1815–88), a Danish civil engineer and physicist, who articulated the principle of conservation of energy contemporaneously with, and independently of, James Joule and Julius Mayer (*CDSB*).
7. *Leibnitz*: Gottfried Wilhelm Leibniz (1646–1716), a German polymath and philosopher most noted for his independent development of integral and differential calculus (*CDSB*).
8. *vis viva*: see letter 1981, n. 2.
9. *On the Conservation of Force*: H. Helmholtz, 'The Conservation of Force: A Physical Memoir' [1847], in R. Kahl (ed), *Selected Writings of Hermann von Helmholtz* (Middletown, CT: Wesleyan University Press, 1871), pp. 3–55. This work was translated into English in 1863 by Edmund Atkinson.
10. *John Bernoulli*: Johann Bernoulli (1667–1748), a Swiss mathematician known for his work on infinitesimal calculus (*HLS*).
11. *Fresnel*: Augustin-Jean Fresnel.
12. *French translation of Thomson's 'Chemistry'*: T. Thomson, *Systèm de Chimie: Supplément* (Paris, Chez Méquignon-Marvis, 1822), pp. 536–9.

To Ellen Lubbock 18 August 1864 2130

Pontresina[1] Friday 18th Aug. 1864

My dear Mrs. Lubbock

Your note to Hirst caused me rather to blame myself for letting you know anything about that avalanche adventure;[2] for I did not want to add an iota to your anxiety. I will now tell you the termination of the adventure, and with it my notes will end, for tomorrow I shall be on my way to Zürich to join the meeting of the Naturforscher,[3] who assemble there on the 22nd. I shall remain at Zürich three days, then start for Paris, where I remain a day—thence to London which I hope to reach by the 28th or 29th. Soon afterwards I hope to have the happiness of seeing you all in the enjoyment of health and strength.

Well, regarding the avalanche, I told you I had lost my watch. It might have been jerked into a crevasse, but it might also be covered up in the snow, and if in the snow the action of the sun might by melting the snow bring it to the surface. It would also be possible to root in the snow in search of it. Well I went to Santa Catarina,[4] leaving my poor little ticker in the cold snow. The day before yesterday an expedition was organized to seek the watch. Some very agreeable people were staying here, among the number Mr. North,[5] an exceedingly pleasant old gentleman of 64, member I believe for Hastings. He was one of the party. There were four other gentleman & two guides. We got into difficulty upon the Morteratsch galcier;[6] were often hemmed in by apparently impracticable crevasses & had to cross places which must have rather terrified the unpractised members of our party. Still all went happily and we reached the side of the glacier in safety. From this spot we had to ascend to a higher and smaller glacier which was the scene of the accident. Leaving three of our party who were tired upon the mountain side, three others accompanied by the two guides continued the ascent. It was very rough, but we finally reached the snow, roped together and while doing so were shown what might become of us had we not been able to arrest our motion downwards. A huge stone, many tons weight detached itself from its resting place on the upper snows and rolled swiftly down the slope along which we had hidden. Its velocity accelerated until it began to take huge bounds through the air, and when it reached the hole near which we pulled up it flew with violence down to the lower glacier and was finally swallowed in a chasm.

Well we crossed the slope, and found a bit of rope which informed us that we had hit the exact line of our glissade: one guide went up the slope, a second one went down, examining the surface. I did not wish to quit my two companions so remained with them and left the hunt to the guides. I had but poor hope of success, still I thought we ought to make the efforts. We had not been 20 minutes there when a choir & a wave of the hat from the higher guide announced that the watch was found. The sun had done its duty & melted it out. I thought it would have been spoiled as it has lain there for nearly three weeks. It pointed to half past 5. I remembered that this was about the hour at which it stopped where I forgot to wind it up at home. I wound it up & found that I had to turn the key the full number of time that indicated running down. The little creature had continued to beat underneath the snow, and now when wound up to my great delight commenced ticking as if nothing had happened to it.[7] It is now ticking in my pocket. I have put it in an inner pocket, near my heart so that its present warmth may make amends for the tremendous cold it must have endured. Love to John[8] & love to the children.[9]

Yours ever | [John Tyndall]

BL Add MS 63092, ff. 212–3

---

1. *Pontresina*: see letter 2028, n. 1.
2. *avalanche adventure*: on 30 July 1864, Tyndall and his party were victim to an avalanche while descending the Piz Morteratsch. See letter 2145.
3. *meeting of the Naturforscher*: see letter 2108, n. 3.
4. *Santa Catarnia*: Santa Caterina di Valfurva is in the northern Italian province of Sondrio.
5. *Mr. North*: Frederick North. See letter 2145, n. 15, and letter 2157.
6. *Morteratsch glacier*: the largest glacier by area in the Bernina Range of the Bünder Alps in Switzerland.
7. *commenced ticking as if nothing had happened to it*: Tyndall recounted this story in a letter to the *Times* (letter 2145).
8. *John*: John Lubbock.
9. *the children*: at this time the Lubbocks had five children: Amy Harriet Lubbock (1857–1929), John Birkbeck Lubbock (1858–1929), Constance Mary Lubbock (1860–92), Norman Lubbock (1861–1926), and Gertrude Lubbock (1863–1934). A sixth child (third son) would come the following year, Rolfe Arthur Lubbock (1865–1909) (R. Birkbeck, *Family of Lubbock*, pp. 36–7).

To Theodore Maxwell[1] — 31 August 1864 — 2131

31st Aug. 1864

Dear Sir

I think you misinterpret the third law of motion. Mathematically used to assure special qualities on the part of matter: in order to deal with mechanical questions they supposed bodies to be perfectly hard or perfectly elastic, but such bodies never occur in nature—Your target will <u>not</u> take up all the moving force of the ball—the quantity of vis viva[2] possessed by the target after collision is less than the quantity possessed by the ball before collision, and the difference is immediately converted into heat. The whole of the vis viva is subsequently converted into heat. You set the target swinging, and this swinging is finally abolished by the resistance of the supports & of the air. When the target finally comes to rest; all the moving force, or vis viva possessed by the ball before collision has been converted into heat.

Yours very truly | John Tyndall

American Philosophical Society, Letter 7

---

1. *Theodore Maxwell*: not identified.
2. vis viva: see letter 1981, n. 2.

From William Hüber[1] 1e September 1864 2132

Paris 10 Sept. /64

Mon cher Monsieur,

J'ai l'honneur de vous accuser réception du livre que vous avez bien voulu m'envoyer. Je vous remercie de cette très aimable attention. Permettez moi seulement de le lire attentivement, avant de répondre à la demande que vous m'adressez, relative à sa traduction.

La science, en France, est loin de ce qu'elle devrait être. A part quelques hommes d'élite, la majeure partie des gens se contentent d'à peu près et d'une [teinture] générale de chaque question. C'est vous dire que votre "Mountaineering in 1861" aurait plus de chance par son volume seul de se vendre dans ce pays que votre intéressant et savant travail des "Glaciers of the Alps". Permettez moi cependant de le lire avant de formuler une réponse positive.

Je vois dans ce travail que vous étiez au Grimsel en 1861. Je regrette vivement de ne pas avoir eu l'honneur de vous connaître à cette époque, car j'ai passé moi-même tout l'été de 1860 et 61 à Gletsch, et les observations restreintes que j'ai fait sur les glaciers auraient été à votre entière disposition, n'ayant rien d'assez complet à publier moi-même.

J'ai passé une grande partie de l'hiver dans ces hautes régions, au 9 Décembre 1860 j'étais à étudier les neiges au Grimsel même, puis à la Furka dans le but d'établir la relation entre les quantités de neige tombée ou accumulée par le vent et l'orientation des différents passages des Alpes de la grande chaîne—question intéressante que je n'ai encore vu traitée nulle part.

Vous m'obligeriez si, dans les nombreux ouvrages que votre expérience a pu consulter, vous vous rappeliez avoir lu quelque chose relatif à cette question.

Pardonnez moi, Monsieur, mon indiscrétion, mais votre aimable envoi m'enhardit au point que je vous écris comme si je vous connaissais depuis longtemps.

Veuillez croire, mon cher Monsieur, à mes sentiments bien dévoués. | W. Hüber

Paris 1 Sept. /64

My dear Sir,

I have the honour of acknowledging that I received the book[2] that you kindly sent me. I thank you for your most welcome attention. I only need to read it carefully before answering the question that you asked me[3] about its translation.

Science, in France, is far from what it should be. Apart from a few elite men, the majority of people are content with approximation and with a general *[sense]* of each question. This means that your "Mountaineering in 1861", being a single volume, will have a better chance of selling in our country, rather than your interesting and scholarly work, "Glaciers of the Alps".[4] However, allow me to read it before formulating a positive response.[5]

I see in this work that you were at Grimsel[6] in 1861. I deeply regret that I did not have the honour of knowing you then, since I also spent the summer of 1860 and 61 at Gletsch,[7] and the limited observations that I made on the glaciers would have been at your disposal, as I have nothing complete enough to publish myself.

I spent much of the winter in the higher regions. On 9 December 1860 I was studying the snow at Grimsel, then at Furka,[8] in order to establish a connection between the quantities of snow that fell or accumulated due to wind and the orientation of the different passages of the main chain of the Alps—an interesting question that I have not yet seen being addressed anywhere.

You will oblige me if, among the many books that you may have consulted in your vast experience, you remember reading something in relation to that question.

Forgive me, Sir, for my boldness, but your kind parcel[9] emboldened me to write to you as if I had known you for a long time.

Please accept, my dear Sir, my most devoted sentiments. | W. Hüber

RI MS JT/1/H/510

---

1. *William Hüber*: see letter 2127, n. 1.
2. *receipt of the book*: *Mountaineering in 1861.*
3. *that you asked me*: letter missing.

4. *"Glaciers of the Alps"*: *Glaciers of the Alps.*
5. *a positive response*: letter missing.
6. *Grimsel*: see letter 1913, n. 20.
7. *Gletsch*: a hamlet in the German-speaking half of the canton of Valais, located in the upper Rhone valley.
8. *Furka*: see letter 2127, n. 5.
9. *your kind parcel*: Tyndall sent Hüber a copy of *Mountaineering in 1861.*

From Juliet Pollock 8 September 1864 2133

Easton Farm | Freshwater[1] | Sep. 8th 1864.

Where is our friend?[2] Is he in England, in London, in Albemarle St.,[3] among sweet airs or un-sweet gases—where is he? We want him here. We have been here since the third Sep., and we want him to come to us before Walter[4] is gone. Walter goes on Wednesday week. You might undertake some adventures with him, climb some cliffs, and face the worst winds upon our pastoral slopes.

Our abode has all its old charms, with the addition of Jane[5] to our comfort. Do you know Jane?—it is the one who is not Polly.[6] The Tennysons are perhaps at Faringford[7] now, for they were expected home yesterday. He was unwell abroad, and so they cut short their intended tour.

Our last evening in London was spent with Carlyle who was at his best, and made grand perorations—he thinks if half mankind had its tongue cut out it would take to doing not being able for talking—and we might get some profit from that. And he thinks nobody ought to write any more verses about love—we have heard all about that long ago.

In passing through London, I had a glimpse of poor Macready[8] who is parting (parted now) with a favourite daughter (Cecilia) ordered to Ceylon on account of incipient disease of the lungs poor man.

Yours ever | my dear Mr Tyndall | most truly | Juliet Pollock.

RI MS JT/1/TYP/6/2018
LT Typescript Only

1. *Freshwater*: see letter 2031, n. 28.
2. *our friend*: Tyndall himself.
3. *Albemarle St.*: the RI is located at 21 Albemarle Street, London.
4. *Walter*: Walter Pollock.
5. *Jane*: presumably a governess for the Pollocks's sons.
6. *Polly*: see letter 2063, n. 7.

7. *Faringford*: Farringford, the Tennyson family's home on the Isle of Wight from 1853–72.
8. *Macready*: William Charles Macready (1793–1873), a famous English actor who performed in many Shakespearean plays. He was retired and unwell at this point (*ODNB*). Juliet Pollock was a close friend who later wrote a memoir about him, *Macready as I Knew Him* (London, 1884). Her son Frederick Pollock would later edit Macready's memoir: F. Pollock (ed), *Macready's Reminiscences, and Selections from His Diaries and Letters*, 2 vols (London: Macmillan & Co., 1875).

To Juliet Pollock [10 September 1864][1] 2134

Saturday

Your friend[2] is here, and has been here for the last ten days working as hard as his brain and fingers will let him. The day after my arrival I called at Montagu Square[3] and found the shutters up and the house dead. I knew of course what had happened, and gleaned further from the Moores,[4] on Thursday that you were safely housed at Freshwater.[5]

I must be grim with myself this year. I prolonged my stay in Switzerland till my conscience became a terror to me. I waited to join the Swiss Savants at their annual meeting[6] and this made my holiday nearly a fortnight longer than it ought to have been. I am now working up my lost time, and must resist all seductions that would take me from my work. I must shut my ears even to you, though when I picture the Cliffs and Downs I am sorely tempted. But I find total abstinence as regards roaming necessary. I am correcting the proofs of a paper[7] that I wrote since my return; I have also just finished the proofs[8] of a long one which I had written before I went, and my book[9] is wholly out of print and the publishers demanding a second edition.[10] I am now writing for it. —endeavouring to incorporate in it all that I have done during the last two years. Then my laboratory experiments are calling for me. So that on the whole if I don't overcome my work it will overcome me.

You got my little note[11] from the Stelvio,[12] I hope, and you posted the note[13] for me to Sir John Herschel. After I wrote to you I was within a hairsbreadth of annihilation—carried away by an avalanche and deposited on the brink of a Chasm which would have made a horrible grave.[14] I lost my watch—The little creature lay for three weeks under the snow. I went to seek it & found it, wound it up there & the and the creature came to life & has ticked merrily ever since.[15] It is close to post time—give my love to the boys[16] & to your husband,[17]

Goodbye | Yours ever | John Tyndall

I wonder would Carlyle growl at me if I went to see him.

RI MS JT/1/T/1534

1. *[10 September 1864]*: the date is given by relation to letter 2133 (dated 8 September 1864, a Thursday) and letter 2135 (dated 11 September 1864, a Sunday). This letter, written on a Saturday, would then be 10 September.
2. *Your friend*: Tyndall is referring to himself in reply to Pollock's enquiry in letter 2133.
3. *Montagu Square*: see letter 1954, n. 2.
4. *the Moores*: see letter 2126, n. 16.
5. *housed at Freshwater:* a town on the Isle of Wight, near where the Pollocks leased Easton Farm.
6. *Swiss Savants at their annual meeting*: see letter 2108, n. 2.
7. *a paper*: probably J. Tyndall, 'On the Dynamical Theory of Heat', *Phil. Mag.*, 27 (1864), pp. 84–91.
8. *the proofs*: see letter 2097, n. 3.
9. *my book*: *Heat as a Mode of Motion.*
10. *second edition*: see letter 2121, n. 17.
11. *my little note*: letter missing. In letter 2135, Pollock wrote that 'I never got your letter from the Stelvio'.
12. *the Stelvio*: a village near the mountain pass of the same name in the Ortler Alps on the border of Switzerland and Italy.
13. *the note*: in letter 2135, after noting that she did not receive Tyndall's letter from the Stelvio, she wrote that 'consequently I have forwarded nothing to Sir J. Herschel'.
14. *carried away by an avalanche . . . a horrible grave*: in letter 2130, Tyndall recounted this 'avalanche adventure' to Ellen Lubbock.
15. *has ticked merrily ever since*: this story is told in more detail in Tyndall's letter to the *Times* (letter 2145).
16. *the boys*: Frederick, Walter, and Maurice Pollock.
17. *your husband*: William Frederick Pollock.

From Juliet Pollock — 11 September 1864 — 2135

Easton Farm[1] | Sunday | Sep. 11th 1864.

My dear Mr Tyndall,

Only a few lines before Church to say it is a grief that you cannot come, and to tell you that I never got your letter from the Stelvio and that consequently I have forwarded nothing to Sir J. Herschel.[2]

I hope you got mine answering yours[3] from Pontresina.[4]

Why do you tell me anything so dreadful as the risk you run? it makes me shudder—true you are well now—but such perils repeated, as I fear they may be, are subjects for just alarm.

What! not a Saturday and Sunday to spare?—not two little days? Bear this in mind—the place is lovely, and Tennyson[5] is come back, and Mrs Cameron[6] is in and Lucy Frankland[7] is here, and will not be here again—neither perhaps shall we—and I should like a moonlight walk with you if only to show off my courage.

Tell me what you think of Enoch Arden and Ayliners field.[8] You will receive in a day or two the promised token from me of Charles Turner's little volume of tender beautiful Sonnets[9]—And now good bye; God bless you, but if I were to say that I would remember you in my prayers you would smile as at a weak saying for you don't esteem prayers—and there are many arguments to be used against them—and I understand them but still I do pray and must pray and could not be happy otherwise though I don't consider churches as essential to the act.

Yours most truly | Juliet Pollock.

RI MS JT/1/TYP/6/2020
LT Typescript Only

1. *Easton Farm*: the Pollock's rented Easton Farm near the town of Freshwater on the western end of the Isle of Wight.
2. *Stelvio*: in letter 2134, Tyndall has asked Pollock if she received his letter he had written to her from Stelvio, a village near the mountain pass of the same name in the Ortler Alps on the border of Switzerland and Italy. However, as Pollock wrote here, she did not receive it and thus forwarded nothing to Herschel.
3. *mine answering yours*: Pollock's letter is 2126; Tyndall's is missing.
4. *Pontresina*: see letter 2028, n. 2.
5. *Tennyson*: Alfred Tennyson.
6. *Mrs Cameron*: Julia Margaret Cameron (1815–79), a British photographer who was well known in the nineteenth century for her style of portraits. Notable subjects were Alfred Tennyson, Thomas Carlyle, Charles Darwin, John Herschel, Henry Taylor, and Ellen Terry. Born in India, she and her husband bought a house on the Isle of Wight in 1860, which they named Dimbola Lodge (*ODNB*).
7. *Lucy Frankland*: not identified.
8. *Enoch Arden and Ayliners field*: Alfred Tennyson's poems 'Enoch Arden' (1864) and 'Ayliners Field' were published in *Enoch Arden and Other Poems* (London: Edward Moxon & Co., 1864).
9. *Charles Turner's little volume of tender beautiful Sonnets*: see letter 2126, n. 2.

From George Gabriel Stokes 15 September 1864 2136

Malahide[1] Ireland | 15th Sept[embe]r 1864

My dear Tyndall,

Reading your paper (5th Memoir)[2] for press recalls to my mind the importance, as it seems to me, of instituting experiments to determine whether we can fairly deem the radiating surface of the lamp-black coating of a Leslie's cube[3] containing water at 2120 to be really at 2120. Were it so, it seems to me that the relative deportment of chloroform and iodide of methyl with lampblack[4] and heated platinum is just the reverse of what we should have expected. For as lampblack transmits the low rays well, it ought to emit them badly, and therefore it seems to me that the radiation from lamp-black at 2120 ought to be related to that from platinum at 2120, as to quality, as the radiation from a higher to that from a lower source.

I suspect that from the excellent radiating and bad conducting power of lampblack, the really radiating layer is considerably below 2120 in spite of the small thickness of the whole stratum.

By the bye how do you get over this difficulty as to a necessary physical connexion between conduction and radiation? We know of plenty of bodies that are highly transparent with regard to light of one kind A, and highly opaque with regard to another kind B, differing, it may be, not greatly in refrangibility; and the absolute identity of the laws of radiant heat and light lead us to suppose that there are probably bodies similarly related to two kinds of heat A', B', both it may be invisible. Such a body must be a very bad radiator for heat of the kind A', and a very good one for heat of the kind B'. What sort of a conductor is it?

Whether such a body exists or not seems to me of small moment. Unless we reject the identity of laws of light and radiant heat, we may imagine such a body to exist without any violation of existing laws.

Yours very truly | G. G. Stokes

P.S. At p. 361 I have inserted the words "subsequently to the presentation of this Memoir" after the words.

The change of quality produced. . . . by the introduction of a platinum spiral. . . . My assistant, Mr Barrett[5]

Please write to the printers to say if you approve as I desired them to insert subject to your approval. If my words [Hereafter LT Typescript Only] won't do please alter them to what will be right.

RI MS JT/1/S/237
RI MS JT/1/TYP/4/1396

1. *Malahide*: see letter 22117, n. 1.
2. *your paper (5th Memoir)*: see letter 2097, n. 3. See also letters 2099, 2112, 2114, 2115, 2117, and 2119.
3. *Leslie's cube*: an instrument invented by John Leslie (see letter 2012, n. 2) for measuring variations in thermal radiation emitted from different surfaces at the same temperature (see J. Leslie, *An Experimental Inquiry Into the Nature and Propagation of Heat* (London: J. Mawman, 1804), pp. 6–8).

4. *lampblack*: a pigment consisting of almost pure carbon in a state of fine division; made by collecting the soot produced by burning oil or gas (*OED*).
5. *Mr Barrett*: William Barrett.

From Heinrich Debus 16 September 1864 2137

Queenwood Coll[1] | Sept[em]b[e]r 16th 64

My dear John,

I was very much pleased to see your handwriting again;[2] during the last two or three months your existence has been manifest to me only through indirect channels.

Did you see Joule's letter in the July number of the Phil Mag.[3] It is remarkable how distinguished men will bear resemblance to the weaker vessels of creation. It is no uncommon occurrence that correct ideas are uttered by men without the proper understanding of the same,, or in other words, without that the consequences and working of said ideas is in the least appreciated by their authors. The utterance of a principle is undoubtedly something, but the working of the principle, its application in various cases, the statement of the legitimate deductions from it, is quite another, a much greater, in fact the thing. If Mr Seguin[4] said something that sounds like the principle involved in the conservation of forces then it remains to be shown that he understood himself what he said.

The conversation (you refer to in your letter[5]) about the examinations of South Kensington[6] have partly been forgotten by me; however, I think you proposed 3/6 per hour. This sum I shall only accept on the understanding that you have had your fair share in the transaction.

As soon as I can manage some time I shall translate a little of the book[7] and let you see the specimen.

Poor Wright[8] is getting better; He is going to Ryde[9] for a short time and we will hope that it may do him good. This news was communicated to me by Mr Conwill[10]

Ever yours | H. Debus

[11]Mr. McLeod,[12] the Assistant in Hofmann's Laboratory,[13] has sent also to me a copy of the declaration about the matter of science and religion.[14] There have been orthodox, bigoted, pig headed, religions, bad and hypocritical, good and sincere men and women, to whom any of the above objections, either single, or combined with any other, might be given, and will be as long so the world stands. To find narrowminded bigotry amongst parsons, lawyer's and tradesmen is a common phenomenon and not to be wondered at. But to find a set of would be Philosophers putting their heads together for the

sake of stifling free inquiry (for this it must be or their circular is absurd) is indeed quite a phenomenon. It would have been a phenomenon in the middle ages, how much more then in the 19th Century. Do the fools believe that they are about to stop the inquiring tendencies and the reasoning powers of the human mind with their declaration? Do they not know that it is a sin to circumscribe the greatest gift of God to man by their narrow notions; Is it not one of the noblest attempts of man to solve the riddles around him, say, life itself? The ink freezes in my pen when I behold the man with a piece of paper, crying so far and not further. How is one to explain it? Of many explanations one of all events holds good. The signer's are not Philosophers, whatever else they may be. They give themselves a testimonium paupertatis[15] that will last a good time. Now enough of my prattle

Ever yours | H. Debus

RI MS JT/1/D/31

---

1. *Queenwood Coll*: see letter 2051, n. 4.
2. *your handwriting again*: letter missing.
3. *Joule's letter in the July number of the Phil Mag.*: J. P. Joule, 'On the Dynamical Theory of Heat', *Phil. Mag.*, 26 (1863), pp. 145–7.
4. *Mr Seguin*: Marc Seguin.
5. *your letter*: letter missing.
6. *the examinations of South Kensington*: see letter 1926, n. 3.
7. *a little of the book*: not identified.
8. *Wright*: see letter 2051, n. 5.
9. *Ryde*: an English seaside town on the Isle of Wight.
10. *Mr Conwill*: John Conwill (1802–80) was Tyndall's schoolmaster at the National School at Ballinabranna from *c.* 1836. He was central in Tyndall's early education particularly with regard to mathematics (see G. Cantor and G. Dawson (eds), *The Correspondence of John Tyndall, Volume 1: The Correspondence, May 1840–August 1843* (Pittsburgh: University of Pittsburgh Press, 2016)).
11. It is unclear if the following text, which is written on a separate sheet, is a postscript, enclosure, or a separate letter.
12. *Mr. McLeod*: Herbert McLeod (1841–1923), a British chemist who served as lecture assistant to August Wilhelm Hofmann at the Royal College of Chemistry (1860–5), and later assistant to Edward Frankland (*ODNB*).
13. *Hofmann's*: August Wilhelm Hofmann (1818–92), a German chemist who was the first director of the Royal College of Chemistry, London (1845–65) (*CDSB*).
14. *declaration about the matter of science and religion*: see letter 2104, n. 4.
15. *testimonium paupertatis*: proof of poverty, or involuntary admission of ignorance (Latin).

From Thomas Archer Hirst 16 September 1864 2138

21 Daniel Street | Bath | Sep[tembe]r 16th 1864

My dear John,

So far the Bath meeting[1] has been a very pleasant one; to me it is so chiefly in consequence of the presence of our dear Mrs Lubbock[2] and the Spottiswoodes.[3] Daily they make one dine and breakfast with them and daily Mrs Lubbock asks if I cannot induce you to come. Although on the committee of section A I am not able to provide for the reading of a paper of sufficient interest to demand your presence should such a one be presented I will of course let you know in time to be present. I have this day taken an unusually active part, for me, in public proceedings. I have had a discussion with Cayley, the president,[4] on Chasles'[5] late work.[6] It was of course an amicable one and what I had to say was well received both by the President and the audience. I found (entre nous[7]) that I was rather more conversant with Chasles method than anyone else in the room; and that I was enabled to inform Cayley himself, that his own communication had been anticipated and stated even in a more general form by Chasles himself.

I then made a communication on my own present researches,[8] which was equally well received and drew from the President some excellent remarks, geometrical as well as complimentary. Ellis[9] followed me and was by no means so successful. He is a man full of crotchets, too crotchetty to do good work with present methods and ambitious to invent methods of his own; he lacks the faculty of seeing that the world perfectly indifferent about his crotchets and thus makes himself a bore. Colenso[10] is here we put him on the committee of Section A (in consequence of his good arithmetical work I suppose). He is to dine with us at Lubbocks in half an hour so I must say Good Bye. You will come if you can.

Yours ever | T. A. Hirst

RI MS JT/1/H/250

---

1. *Bath meeting*: the BAAS held its thirty-fourth annual meeting in Bath from September 15–20 1864 (*Brit. Assoc. Rep. 1864*).
2. *Mrs Lubbock*: Ellen Lubbock.
3. *Spottiswoodes*: William Spottiswoode and his wife Eliza Taylor Spottiswoode (née Arbuthnot, 1837–94).
4. *Cayley, the president*: Arthur Cayley (1821–95), a British mathematician who helped found the modern British school of pure mathematics (*ODNB*). During this meeting he was President of Section A, mathematics and physics.

5. *Chasles*: see letter 1994, n. 7.
6. *late work*: Chasles published a number of mathematical papers in 1864, so we cannot be certain to which this refers (see *Catalogue of Scientific Papers, Compiled and Published by the Royal Society of London (1864–1873)*, 6 vols (London: Eyre and Spottiswoode, 1877), vol. 7, pp. 375–6).
7. entre nous: between us (French).
8. *communication on my own present researches*: T. A. Hirst, 'On a Generalization of the method of Geometrical Inversion', *Brit. Assoc. Rep. 1864*, p. 3–4).
9. *Ellis*: Alexander John Ellis (see letter 1991, n. 1).
10. *Colenso*: John William Colenso (1814–83), a British mathematician, theologian, Biblical scholar, and social activist, who was the first Church of England Bishop of Natal. He was influenced by the liberal theologian Frederick Denison Maurice, and sparked controversy over his interpretation of scripture and outspoken stance on such issues as polygamy, the doctrine of eternal damnation, and the route to salvation. He was a polygenist, arguing that the different races of humankind were in fact different species, a view held by a number of notable men of science at this time, although this too was very much a point of contention (*ODNB*).

To Juliet Pollock 16 September 1864 2139

Royal Institution of Great Britain | 16th Sept. 1864

My dear Mrs. Pollock

Had I trusted the impulse of my soul on reading your last letter[1] I should at once have replied to you promising to go to the Island.[2] But as I get older I learn more and more the necessity of waiting to see whether the impulse is realizable. Now I do not see my way clearly as yet. You speak of two little days. It cannot be done in two days. Saturday is used up in going down: Sunday I spend there, and Monday is occupied in coming up. Then I quit the Island with you and your husband and all the occurrences that take place during my visit in my mind. to expel these costs more than a day, so that really your two little days is only a fraction of the time needed. I am now working incessantly at my book.[3] It would be wrong in me to quit it till it is complete. If afterwards I can take a run down to you I will do so. Probably I shall work myself into a condition which will compel me to run away and if so compelled I will run off to you.

I got your letter in Bormio[4]—It was sent after me from Pontresina,[5] and it cheered me so much, not that I needed cheering for I was very jolly at the time, but it was so pleasant to me that I then and there sat down and replied to it. My letter contained a note to Sir John Herschel,[6] and I posted the letter myself. But the post office people in the Italian valleys of the Alps are ineffably stupid and careless. They were very near causing my friend Ball[7] to quarrel with me through the retention of a letter which he ought to have received a fortnight before it reached him.

Your letter received at Bormio caused me to long to see Turner's poems:[8] since I came back I have learned from Reviews that he is "theological".[9] I am sorry for this—His theology is the more "rotten pole" as your husband[10] once helped me to express it round which the fresh tendrils of his heart cling—I can hardly excuse the intellectual mistake of a man who confounds such a rotten pole with the permanent and eternal.

Good bye | Yours ever | John Tyndall

I have made myself an exile from Bath,[11] notwithstanding all its seductions.

RI MS JT/1/T/1142
RI MS JT/1/TYP/6/2021

1. *your last letter*: letter 2135.
2. *the Island*: the Isle of Wight.
3. *my book*: Tyndall had recently begun work on the second edition of *Heat as a Mode of Motion* (in letter 2134, he had written to Pollock that the publishers were 'demanding a second edition').
4. *your letter in Bormio*: letter 2126. Bormio is a town and commune located in the Lombardy region of the Alps in northern Italy.
5. *Pontresina*: see letter 2028, n. 1.
6. *a note to Sir John Herschel*: Tyndall noted in his next letter to Herschel (2140) that he was now aware that he would not have received this earlier letter. See also letters 2134 and 2135.
7. *Ball*: John Ball.
8. *Turner's poems*: see letter 2126, n. 2.
9. *Reviews that he is "theological."*: possibly 'Reviews: Sonnets by the Rev. Charles Turner', *Saturday Review*, 6 August 1864, pp. 184–5.
10. *your husband*: William Frederick Pollock.
11. *Bath*: see letter 2138, n. 1.

To John Frederick William Herschel 17 September 1864 2140

Royal Institution of Great Britain | 17th Sep. 1864.

Dear Sir John,

Your excellent letter[1] in the Athenaeum, which I trust will have the effect of stifling that contemptible 'Declaration'[2] reminds me that I received in a remote corner of the Italian Alps a most interesting letter[3] from your pen; that I on my arrival in Bormio[4] posted a note to you in reply to that letter; that I enclosed the note in a letter to my friend Mrs Frederick Pollock,[5] but that I now find that the letter never reached M[rs] Pollock,[6] and that consequently the enclosure never reached you.

I told you in that note that I purposed reprinting that fine passage from the Outlines of 1833[7] regarding the operations of the sun, in the 2nd edition of my book[8] now in preparation, I have already copied it, and it will duly reach the printer's hands.

I also referred to the extremely philosophical views of muscular contraction contained in your letter,[9] and suggested the desirability of publishing that portion of the letter which referred to this subject.

If delay therefore has occurred in replying to a letter which I valued very much it has been caused by the ineffable stupidity and carelessness of the post office people in the Italian valleys of the Alps.

Yours most truly | John Tyndall

RS HS 394a
RI MS JT/1/TYP/2/528

---

1. *Your excellent letter*: Herschel wrote to the *Athenaeum* to protest against being pressed to sign a 'religious manifesto', on the grounds that it was both an infringement of his freedom of religious opinion, and an action likely to sow discord in the Christian world. He also insisted that his refusal to sign should not be seen as a profession of atheism or infidelity (see 'Science and Scripture', *Athenaeum*, 17 September 1864, p. 375).
2. *'Declaration'*: see letter 2104, n. 4.
3. *most interesting letter*: letter 2111.
4. *Bormio*: see letter 2139, n. 4.
5. *M[rs] Frederick Pollock*: Juliet Pollock.
6. *I now find that the letter never reached M[rs] Pollock*: see letter 2135.
7. *Outlines of 1833*: see letter 2111, n. 8.
8. *the 2[nd] edition of my book*: see letter 2139, n. 3.
9. *your letter*: letter 2111.

From Heinrich Debus 18 September [1864][1] 2141

Sunday[2] Sept[em]b[er] 18

I a chemist of great renown, a physicist of considerable repute, and a mathematician of great attainments have the misfortune to be charged by John Tyndall, Esq[ui][r][e], a Doctor of philosophy (not of artificial make but by the make of the higher powers), a sweet orator, an eminent professor in the Royal Institution of Great Britain, a [magnificent][2] [3]physicist, a member of X learned societies, a teacher of Royalty, a lover of nature, and of beauty, and finally of many other titles and attainments, of having returned an insufficient answer to a question. Have I, Mr Professor; with all due respect your memory seems to have gone somewhat out of gear. Here is the letter with the question. Did I not answer with mathematical accuracy your question? Did I not say exactly what you wanted to know according to your question? You with your great and accurate mind ought to know your own questions. However not merely has the Professor forgotten his own question but also the conversation we had some weeks ago. Nothing remains then for me, the Professors small and humble friend, but to tell him, that there were 100 hours of 3/6 a piece, according to his own proposition.[4] Herewith closes the epistle about the question and now to something else.

RI MS JT/1/D/32

---

1. *[1864]*: this letter is archived at the RI with letter 2144, also from Heinrich Debus, and Debus may have originally enclosed it with that letter. In 1864, Tyndall had asked Debus to grade papers for him.
2. This letter was an enclosure with that above.
3. *[magnificent]^2*: Debus here placed brackets around 'magnificent' and added a 'squared' symbol.
4. *100 hours of 3/6 a piece, according to his own proposition*: this is the sum that Tyndall and Debus had discussed as payment for marking the South Kensington examinations (see letter 2137).

To Annette Roche 20 September 1864 2142

Royal Institution. 20th, Sep, 1864.

My dear Miss Roche

Thank you a hundred times over for your great kindness. Of the two designs[1] which you have been kind enough to send me I think I like No 1 best. I should like it as solid as possible, and with as little unnecessary ornament as possible.

I shall write to Mr Diegan[2] and ask him to undertake the work for me. I shall also write to my mother[3] as you suggest, though hitherto I have been accustomed to have my own way.

I grew in strength during my stay in Switzerland and did very little in the way of adventurous mountaineering. Nevertheless I was in greater danger this year than ever I have been in before. Descending a mountain I found myself for a time riding on an avalanche to apparent destruction.[4] There were five of us roped together and the slipping of one of the party on a steep ice slope produced all the mischief. Nevertheless we pulled up just in time after having shot over a space of about 1000 feet. We were somewhat knocked about, and my watch was jerked out of my pocket; I found the broken chain dangling about my neck, and a fragment of the chain in my pocket. The watch was gone.

I went to Italy and stayed there nearly a fortnight—returned to Pontresina,[5] and made an expedition in search of the watch. I thought that if it had not fallen into a crevasse (we were shot over two of them) the sun might melt the snow and thus bring it to the surface. The watch was found after having lain three weeks embedded! It had simply ran down, was in no way damp within—I wound it up and the creature came to life once more and has continued ticking ever since. Perhaps no watch in the world has had such a romantic adventure.[6]

Believe me | Yours most truly | John Tyndall

RI MS JT/1/TYP/11/3852
LT Typescript Only

1. *Of the two designs*: Tyndall had asked Roche if she could oversee the erecting of a tombstone on his father's grave. See letters 2113 and 2121.
2. *Mr Diegan*: letter missing. He was the sculptor hired by Tyndall to create the tombstone.
3. *I shall also write to my mother*: this letter to Sarah Tyndall is missing.
4. *an avalanche to apparent destruction*: in letter 2130, Tyndall recounted this 'avalanche adventure' to Ellen Lubbock.
5. *Pontresina*: see letter 2028, n. 1.
6. *such a romantic adventure*: this story is told in more detail in Tyndall's letter to the *Times* (letter 2145).

From Felix Foucou[1] 21 September 1864 2143

Paris—21 septembre 1864

Mon cher Monsieur Tyndall,

J'ai reçu votre excellente lettre du 16 sept, et j'y aurais répondu tout de suite pour vous en remercier, si vous n'aviez été absent de Londres pour le congrès de British Association à Bath. Je voulais aussi attendre d'avoir quelque chose de décidé avec notre éditeur, M. Baudry. Ce gentleman est, comme tous les éditeurs, exceedingly selfish. Cependant il est très désireux de publier votre livre et si vous acceptez la proposition que je vais vous faire, nous commencerons, je pense, la mise en œuvre dès le mois prochain, afin de pouvoir faire paraître le livre le 1er Janvier 1865.

D'abord je dois vous dire que, tout en appréciant la délicatesse qui vous a fait agir en me laissant entièrement libre et en voulant que je sois rémunéré avant vous, il m'est impossible d'y consentir, si le livre doit avoir du succès, c'est à votre nom qu'il le devra et non au mien. Au contraire, c'est déjà un grand avantage pour moi de pouvoir placer mon nom à côté du votre. Ainsi, en toute justice, c'est vous qui devez passer avant moi.

Maintenant, M. Baudry veut conclure avec moi un traité d'après lequel je n'aurai droit à une rémunération qu'après que les frais auront été couverts. Il ne serait, dans ce traité, nullement question de l'auteur, et M. Baudry qui vient de publier une traduction de la Métallurgie du Dr. Percy, prétend que le docteur ne touche pas un centime de droit d'auteur en France. C'est possible, mais c'est injuste. Enfin, M. Baudry paierait à M. Murray votre éditeur, une somme pour les gravures. Lorsque les frais seront couverts je partagerai avec M. Baudry les bénéfices.

En conséquence, je vous proposerai de partager avec vous ce qui me reviendra sur ces mêmes bénéfices; c'est-a-dire que M. Baudry vous remettra le même relevé de compte qu'a moi-même et vous adressera la même somme qui pourra me revenir. Si vous acceptez, je ferai rédiger le traité entre M. Baudry et moi, de façon à ce que vous y soyez mentionné à ce titre: vous aurez une copie de ce traité, de manière à pouvoir faire valoir vos droits vous-même, au cas où je viendrais à mourir.

En ce qui concerne M. Murray, avant de recevoir votre lettre qui me dit qu'il se contentera de 10 ou 12 Livres Sterling pour les bois, ce qui est très modéré, j'avais dit à M. Baudry qu'il faudrait probablement lui payer 20£ ou 500 francs. Ce prix a été accepté. Je ferai donc, si vous n'y voyez pas d'objection, insérer au traité que cette somme vous sera remise contre l'envoi des bois: vous donnerez alors à M. Murray ce qui lui revient, et le reste, quoique minime représentera la part de prélèvement préalable qui doit vous appartenir en toute justice. C'est moins pour la somme, que pour le maintien d'un principe qui, à mon avis, devrait présider à toutes les transactions vis-à-vis des auteurs. Comme je vous l'ai déjà dit, ce sont les auteurs qui font la fortune du livre, et les traducteurs ne peuvent réclamer qu'en seconde ligne, puisque déjà ils tirent avantage de se produire à côté d'autres déjà connus.

Seriez-vous assez bon pour me faire parvenir les sept ou huit numéros d'un Journal de Bath rendant compte du meeting, cela me servira pour notre préface, que je prépare en ce moment.

Je recevrai avec bien du plaisir votre travail sur l'érosion dans le "Philosophical Magazine" et quelque communication récente sur les Alpes, si vous en avez. Ce sera pour former un *[appendix]* au livre. D'ailleurs, une fois que le traité sera signé avec M. Baudry, nous combinerons la confection du volume au point de vue matériel et d'arrangement.

Mille amitiés | Felix Foucou

Professor Tyndall | London

Paris—21 September 1864

My dear Mr. Tyndall,

I received your kind letter of 16 September.[2] and I would have responded immediately to thank you, had you not been away from London for the British Association conference in Bath.[3] I also wanted to wait until something was decided with our publisher, Mr Baudry.[4] This gentleman is, like all publishers, exceedingly selfish.[5] However, he is very eager to publish your book[6] and if you accept my proposition, I think we can start the process next month, so the book could debut on January 1st 1865.

Above all, I have to tell you that while I appreciate your delicacy in leaving me entirely free and in wishing for me to be paid before you, I cannot possibly to agree to it. If the book is successful, it will be thanks to your name and not mine. Indeed, it is already very advantageous for me to be able to place my name next to yours. As such, in all fairness, it is you that ought to come before me.

Now, Mr. Baudry wants to sign a contract with me, according to which I will not be entitled to a payment until the costs have been covered. In this contract there will be no mention of the author, and Mr. Baudry, who has just published the translation of Dr. Percy's[7] Metallurgy,[8] claims that the doctor did not touch a penny of authorial rights in France. That may be so, but it is unfair. Finally, Mr. Baudry will pay Mr. Murray, your publisher, a sum for the prints. When the costs have been covered, I will share the profits with Mr. Baudry.

Consequently, I propose to split my share of those profits with you. In other words, Mr. Baudry will give you the same receipt as me, and will send you the same amount that will come to me. If you accept, I will rewrite the contract between Mr. Baudry and me, so that you would be mentioned in this manner. You will have a copy of the contract, so that you can claim your rights in the case of my death.

Regarding Mr. Murray, before receiving your letter that told me that he would be content with 10 or 12 Pounds Sterling for the woodblocks,[9] which is very reasonable, I told Mr. Baudry that he would probably have to pay him 20£ or 500 francs. This price was accepted. If you have no objection, I will have it added to the contract that this sum will be paid to you for the shipment of the woodblocks. You will have to ask Mr. Murray what is due to him, and the rest, though minimal, will represent the share of the advance that in all fairness should be yours. It is less for the sum than for maintaining a principle, which in my opinion should govern all transactions towards authors. As I already told you, it is authors who make the book, and the translators can only claim second in line, since they already have the benefit of publishing alongside known authors.

Would you be kind enough to send me the seven or eight issues of the Bath newspapers that report on the meeting. It will help me for our preface, which I am in the process of preparing.

I would be very much pleased to receive your work on erosion in the "Philosophical Magazine"[10] and some recent writings about the Alps, if you have any. They would be used to form an appendix for the book. Moreover, once the contract is signed with Mr. Baudry, we will combine the making of the volume in both a material and organizational sense.

With deepest friendship | Felix Foucou

Professor Tyndall | London

RI MS JT/1/F/35

---

1. *Felix Foucou*: Félix Foucou (1831–70), a French civil engineer, inventor, and geologist, with a particular interest in petroleum deposits (which led him to the United States in 1870, where he died). Foucou also wrote for and edited scientific periodicals, and was the author of *Histoire du travail: La nature et l'homme* (Paris: J. Hetzel, 1868) (P. Larousse, *Grand Dictionnaire universel du XIXe siècle*, 17 vols (Paris: Larousse, 1866–90), vol. 8 (1872), p. 652).
2. *your kind letter of 16 September*: letter missing.
3. *the British Association conference in Bath*: see letter 2138, n. 1.
4. *Mr. Baudry*: Jules Baudry (1815–98), of the Paris-based publisher Gide et Jules Baudry ('In Memorium. M. Jules Baudry', *The Publisher's Circular*, 28 May 1898, p. 629).
5. *exceedingly selfish*: this is written in English in the original.
6. *your book*: this is probably a reference to *Glaciers of the Alps*, which William Hüber had been reluctant to publish in France. However, it may also be a reference to Tyndall's *Mountaineering in 1861*, which would suggest that Tyndall was 'shopping around' to see if he could get a better deal with Foucou than with Hüber. See letter 2132 and Foucou's subsequent letter to Tyndall, 2156.
7. *Dr. Percy's*: John Percy (1817–89), an English metallurgist (*ODNB*).
8. *Metallurgy*: J. Percy, *Metallurgy: The Art of Extracting Metals from their Ores and Adapting them to Various Purposes of Manufacture* (London: John Murray, 1864).
9. *woodblocks*: woodcut blocks for illustrations in the book.
10. *your work on the erosion in the "Philosophical Magazine"*: J. Tyndall, 'On the Conformation of the Alps', *Phil. Mag.*, 28 (1864), pp. 255–71.

From Heinrich Debus 22 September 1864 2144

Sept[em]b[e]r 22[d] 1864

My dear John,

Thanks for your liberality.[1] I sincerely hope you have not given me too much and deprived yourself of a portion of your own. If you have time would you have the kindness and improve the English of some examination questions I have prepared? I should feel very thankful if you would. I have read the accounts about the Br. Association[2] published from day to day by the papers, and have been struck by the superiority of interest manifested by the public in the labours of the geographical section over that in the proceedings of the other departments. Livingstone[3] is no doubt a very courageous and enterprising man but it appears to me as if there was a bit of the Humbug about him.

Ever yours | H Debus

RI MS JT/1/D/32

---

1. *Thanks for your liberality*: Tyndall had arranged to pay Debus for his work marking examinations for South Kensington (see letter 1926, n. 3 and letter 2137). Tyndall had clearly responded to Debus in a subsequent letter which is missing.
2. *Br. Association*: see letter 2138, n. 1.
3. *Livingstone*: David Livingstone (1813–73), a Scottish physician, Christian missionary, and explorer in Africa (*ODNB*). Livingstone had just returned to England from the Second Zambesi Expedition (1858–64), which had been organised by the Royal Geographical Society to explore South East Africa for mineral and other resources and to open up the Zambesi River to navigation. Although Livingstone was lauded by many as a national hero (which only increased after his death), a number of expedition members recorded that Livingstone was an inept leader incapable of managing a large-scale project.

To the *Times* 30 September 1864 2145

TO THE EDITOR OF THE TIMES

Sir,—Various accounts of the accident here described being afloat, I have thought it worth while to give the correct one. If it should have the effect of checking the tendency to underrate the dangers of the Alps, and, without damping courage, augment caution, my chief object in writing the account will be accomplished.

I am, Sir, your obedient servant, | JOHN TYNDALL | Royal Institution, Albemarle-street, Sept. 30.

Towards the end of last July, while staying at Pontresina,[1] in Ober Engadin,[2] I was invited by two friends to join in an expedition up the Pia Morteratsch.[3] This I willingly did, for I wished to look at the configuration of the Alps from some commanding point in the Bernina Mountains[4] and also to learn something of the capabilities of the Pontresina guides. We took two of them with us—Jenni,[5] who is the man of greatest repute among them, and Walter,[6] who is the head of the bureau of guides. We proposed to ascend by the Rosegg,[7] and to return by the Morteratsch glacier, thus making a circuit, instead of retracing our steps. About eight hours of pleasant, healthful exertion placed us on the Morteratsch Spitze,[8] where we remained for an hour, and where the conviction forced on my mind on many another summit was renewed—namely, that these mountains and valleys are not, as supposed by the renowned President of the Geographical Society,[9] ridges and heaps tossed up by the earth's central fires, with great fissures between them, but that ice and water, acting through long ages, have been the real sculptors of the Alps.[10]

Jenni is a heavy man, and marches rather slowly up a mountain, but he is a thoroughly competent mountaineer. We were particularly pleased with his performance in descending. He swept down the slopes and cleared the 'schrunds,'[11] which cut the upper snows, with great courage and skill. We at length reached the point at which it was necessary to quit our morning's track, and immediately afterwards got upon some steep rocks, which were rendered slippery here and there by the water which trickled over them. To our right was a broad couloir,[12] which was once filled with snow, but this had been melted and refrozen, so as to expose a sloping wall of ice. We were all tied together at this time in the following order:—Jenni led, I came next, then my friend H.,[13] an intrepid mountaineer, then his friend L.,[14] and last of all the guide Walter. L. had had but little experience of the higher Alps, and was placed in front of Walter, so that any false step on his part might be instantly checked. After descending the rocks for a time Jenni turned and asked me whether I thought it better to adhere to them or try the ice-slope to our right. I pronounced in favour of the rocks but he seemed to misunderstand me, and turned towards that couloir. I stopped him before he reached it, and said, 'Jenni, you know where you are going; the slope is pure ice.' He replied, 'I know it; but the ice is quite bare for a few yards only. Across the exposed portion I will cut steps, and then the snow which covers the ice will give us a footing.' He cut the steps, reached the snow, and descended carefully along it—all following him, apparently in good order. After a little time he stopped, turned, and looked upwards at the last three men. He said something about keeping carefully in the tracks, adding that a false step might detach an avalanche. The word was scarcely uttered when I heard the sound of a fall behind me, then a rush, and in the twinkling of an eye my two friends and their guide, all apparently entangled together, whirred past me. I suddenly planted myself to resist their shock, but in an instant I was in their wake, for their impetus was irresistible. A moment afterwards Jenni was whirled away, and thus all five of us found ourselves riding downwards with uncontrollable speed on the back of an avalanche which a single slip had originated.

When thrown down by the jerk of the rope, I turned promptly on my face and drove my baton through the moving snow, seeking to anchor it in the ice underneath, had held it firmly there for a few seconds, when I came into collision with some obstacle and was rudely tossed through the air, Jenni at the same time being shot down upon me. Both of us here lost our batons. We had, in fact, been carried over a crevasse, had hit its lower edge, our great velocity causing us to be pitched beyond it. I was quite bewildered for a moment, but immediately righted myself, and could see those in front of me half buried in the snow, and jolted from side to side by the ruts among which they were passing. Suddenly, I saw them tumbled over by a lurch of the avalanche, and immediately afterwards found myself imitating their motion. This was caused by a second crevasse. Jenni knew of its existence and plunged right into it—a brave and manful act, but for the time unavailing. He is over 13 stone in weight, and he thought that by jumping into the chasm a strain might be put upon the rope sufficient to check the motion. He was, however, violently jerked out of the fissure and almost squeezed to death by the pressure of the rope.

A long slope was below us which led directly downwards to a brow where the glacier suddenly fell in a declivity of ice. At the base of this declivity the glacier was cut by a series of profound chasms, and towards these we were now rapidly borne. The three foremost men rode upon the forehead of the avalanche, and were at times almost wholly immersed in the snow; but the moving layer was thinner behind, and Jenni rose incessantly and with desperate energy drove his feet into the firmer substance underneath. His voice shouting 'Halt! Herr Jesus, halt!' was the only one heard during the descent. A kind of condensed memory, such as that described by people who have narrowly escaped drowning, took possession of me; and I thought and reasoned with a preternatural clearness as I rushed along. Our start, moreover, was too sudden and the excitement too great to permit of the development of terror. The slope at one place became less steep, the speed visibly slackened, and we thought we were coming to rest; the avalanche, however, crossed the brow which terminated this gentler slope and regained its motion. Here H. threw his arm round his friend, all hope for the time being extinguished, while I grasped my belt and struggled for an instant to detach myself. Finding this difficult I resumed the pull upon the rope. My share in the work was, I fear, infinitesimal, but Jenni's powerful strain made itself felt at last. Aided probably by a slight change of inclination, he brought the whole to rest within a distance of the chasms over which, had we preserved our speed, a few seconds would have carried us. None of us suffered serious damage. H. emerged from the snow with his forehead bleeding, but the wound was superficial. Jenni had a bit of flesh removed from his head by collision against a stone; the pressure of the rope had left black welts on my arms, and we all experienced a tingling sensation over the hands, like that produced by incipient frostbite, which continued for several days. I found a portion of my watch-chain hanging round my neck, another portion in my pocket—the watch itself was gone.

This happened on the 30th of July. Two days afterwards I went to Italy, and remained there for ten or twelve days. On the 10th of August I was again at Pontresina and on that day made an expedition in search of the lost watch.

Both the guides and myself thought the sun's heat might melt the snow above it, and I inferred that if its back should happen to be uppermost the slight absorbent power of gold for the solar rays would prevent the watch from sinking as a stone sinks under like circumstances. The watch would thus be brought quite to the surface; and, although a small object, it might possibly be seen from some distance. I was accompanied up the Morteratsch glacier by five friends, of whose conduct I cannot speak too highly. One of them in particular, a member of the British Legislature, 64 years of age,[15] exhibited a courage and collectedness in places of real difficulty which were perfectly admirable.

Two only of the party, both competent mountaineers, accompanied me to the scene of the accident, and none of us ventured on the ice where it originated. Just before stepping upon the remains of the avalanche, a stone some tons in weight, detached by the sun from the snow slope above us, came rushing down the line of our glissade. Its leaps became more and more impetuous, and on reaching the brow near which we had been brought to rest it bounded through the air, and with a single spring reached the lower glacier, raising a cloud of ice-dust in the air. Some fragments of rope found upon the snow assured us that we were upon the exact track of the avalanche, and then the search commenced. It had not continued for 20 minutes when a cheer from one of the guides—Christian Michel of Grindelwald[16]—announced the discovery of the watch. It had been brought to the surface in the manner surmised, and on examination seemed to be dry and uninjured. I noticed, moreover, that the position of the hands indicated that it had only run down beneath the snow. I wound it up, hardly hoping, however, to find it capable of responding. But the little creature showed instant signs of animation. It had remained 13 days in the avalanche, but the application of its key at once restored it to life, and it has gone with unvarying regularity ever since.

JOHN TYNDALL

Wellcome, MS. 7777/108
Times, 1 October 1864, p. 10[17]

---

1. *Pontresina*: see letter 2028, n. 1.
2. *Ober Engadin*: the Engadine, a long Alpine valley in southeastern Switzerland.
3. *Pia Morteratsch*: Piz Morteratsch, a mountain in the Bernina range, bordered on the east by Morteratsch Glacier and the southwest by the Tschierva Glacier.
4. *Bernina Mountains*: the Bernina range is part of the Bünder Alps in Switzerland.
5. *Jenni*: one of Tyndall's guides on his Alpine expedition.
6. *Walter*: one of Tyndall's guides on his Alpine expedition.
7. *Rosegg*: by the Rosegg glacier.
8. *Morteratsch Spitze*: the peak of the Piz Morteratsch.

9. *President of the Geographical Society*: Roderick Murchison.
10. *the real sculptors of the Alps*: see letter 2100.
11. *"schrunds"*: *bergschrund* in full, a crevasse or series of crevasses often found near the head of a mountain glacier (*OED*).
12. *couloir*: a steep gorge or gully on a mountain side (*OED*).
13. *my friend H*: Thomas Neville Hutchinson (1826–99), a clergyman and botanist who served as Vice Principal of Chester Diocesan Training College (1854–60), and Second Master at King Edwards School in Birmingham (1860–5). In February 1865 he was appointed assistant master in natural science at Rugby School, a post he held until 1883 ('Canon Thomas Neville Hutchinson', *Wiltshire Archaeological and Natural History Magazine*, 30 (1899), p. 275). At Rugby, Hutchinson participated in botany with the school's Natural History Society (R. Desmond, *Dictionary Of British & Irish Botanists And Horticulturists including Plant Collectors, Flower Painters and Garden Designers* (London: Taylor & Francis/Natural History Museum, 1994), p. 369).
14. *his friend L.*: Henry Lee-Warner (1842–65), a former student at Rugby School, graduate and later Fellow of St. John's College (1865–9), and from 1864 an assistant master at Rugby School and Fellow of St. John's from 1865. Notably, while an undergraduate at Cambridge he was one of eleven students selected to sit with the future King, Edward VII (Albert Edward (1841–1910), Prince of Wales, eldest son of Queen Victoria and Prince Albert) in classes with the Anglican clergyman Charles Kingsley (1819–75).
15. *a member of the British Legislature, 64 years of age*: Frederick North. See letters 2130 and 2157.
16. *Christian Michel of Grindelwald*: Christian Michel, along with his brother Peter, were two well-respected alpine guides who lived in Grindelwald (J. Ball, *The Central Alps, including the Bernese Oberland, and All Switzerland excepting the Neighbourhood of Monte Rosa and the Great St. Bernard: with Lombardy, and the Adjoining Portion of Tyrol* (London: Longmans, Green, and Co., 1873), p. xxx).
17. This letter was also published as J. Tyndall, 'Accident on the Piz Morterasch', *Alpine Journal*, 1 (1863–4), pp. 437–40); it included a short letter from Hutchinson speaking to the accuracy of Tyndall's account. Tyndall later published this story with further details as chapter 19 in *Hours of Exercise in the Alps* (pp. 206–18).

From Andrew Ramsay 5 October 1864 2146

Corsewall, Stranraer | 5. Oct 1864.

My dear Tyndall,

I only received your note[1] of 1st last night on coming in from a long days work. I am exceedingly pleased that you like my reply[2] so well, and I am bound to confess that it gives me a very decided pleasure when anything I may chance to do is appreciated by one whose approval is worth having. Marry come up![3] as the novelists say, when one gets a compliment of the order of "my distinguished friend" my minds toe is very apt to go to work kicking the complimenter.

I hope Sir Roderick[4] will not think it incumbent on his dignity to reply to my reply. I do not want to have anything more to do with it, but if he does rejoin, I have left plenty of points untouched, and further if I chose to be savage I could show an utter want of consistency & coherence in the whole affair. But I will not allow myself under any circumstances to be tempted into doing that to my venerable Superior Officer. What a curious piece of vanity it is his going into the matter at all!

I am doing a bit of mapping here to hasten the completion of an area. It will only take a few days, & you will think me very jollily placed when I tell you I am staying with our friend John Carrick Moore. It is a charming house & they are delightful people.[5]

Ever sincerely | Andrew Ramsay

I am having 200 extra copies[6] printed & shall send you one. I have just seen your letter in the Times[7] & have it yet to read. I saw it just as we were starting out to day, & now I have barely time to dress for dinner.

RI MS JT/1/R/4
RI MS JT/1/TYP3/1006

1. *your note*: letter missing.
2. *my reply*: Ramsay's reply to Roderick Murchison's 1863 address to the Royal Geographical Society (see letter 2100, n. 1) was published as A. C. Ramsay, 'On the Erosion of Valleys and Lakes: a Reply to Sir Roderick Murchison's Anniversary Address to the Geographical Society', *Phil. Mag.*, 28 (1864), pp. 293–311.
3. *Marry come up!*: an expression of surprise or indignation.
4. *Sir Roderick*: Roderick Murchison.
5. *delightful people*: on the Moore family, see letter 2126, n. 16.
6. *200 extra copies*: presumably of his reply to Murchison.
7. *your letter in the Times*: presumably a reference to letter 2145.

From Rudolf Clausius 9 October 1864 2147

Zürich, 9. Oct. 64

Lieber Tyndall,

Ich habe diesmal eine eigenthümliche Bitte an Dich welche von dein, was sonst gewöhnlich Gegenstand unserer Correspondenz ist, sehr verschieden ist. Es steht in 14 Tagen, (um 23sten October) die Taufe unseres Söhnchens bevor. Es ist nun hier Sitte zwei Pathen dazu zu nehmen, einen Herrn und eine Dame. Die Pathen brauchen nicht nothwendig selbst zugegen zu sein, sondern sie können durch andere Personen die gegenwärtig sind, vertreten werden, und sie gelten dann doch als die rechten Taufpathen. Ich richte nun im Namen meiner Frau und in meinem Namen an Dich die Bitte, die eine der beiden Pathenstellen zu übernehmen. Der Knabe soll dann nach Dir "John" heissen. Da natürlich bei der weiten Entfernung zwischen hier und London nicht daran zu denken ist, dass Du selbst bei der Taufe anwesend sein kannst, so werde ich, wenn du uns auf unsere Bitte mit einer zusagenden Antwort erfreust, Deine Stellvertretung übernehmen. Die andere Pathenstelle gedanken wir Frau Professor Kenngott die Du kennst anzutragen.

Wenn Dein junger Freund mich aufsucht, werde ich ihm gern bei der Auswahl eines passenden Logie's behülflich sein, und jedenfalls wird er uns in unserem Hause immer sehr willkommen sein.

Mein Buch ist jetzt fertig, und es werden von der Buchhandlung zwei Exemplare an dich geschickt werden, eins für Dich, und ein zweites, welches ich Dich bitte, in meinem Namen der Royal Society zu übergeben. An Francis wird auch ein Exemplar gesandt, und ich werde wenn es angekommen ist, an ihn schreiben, um ihn zu fragen, ob er eine englische Uebersetzung unter Benutzung der schon in dem Phil. Mag. vorhandenen Uebersetzungen der einzelnen Abhandlungen Herausgeben will.

Mit freundlichsten Grüssen an Hirst, und herzlichen Grüssen von meiner Frau und mir an Dich,

Dein | Clausius

Zurich, 9. Oct. 64

Dear Tyndall,

This time I would like to make a singular request of you,[1] which is very different from the usual topics of our correspondence. In 14 days (23rd of October), our little son[2] will be baptized. It is the custom here to name two godparents, a gentleman and a lady. The godparents do not necessarily have to be present themselves, but can be represented by other people who in attendance, and they are still considered to be the rightful godparents. On behalf of my wife[3] and myself, I would like to ask you to become one of the godparents. The lad will be named John after you. Since, of course, due to the great distance between here and London, it is beyond consideration that you can be at the baptism yourself, I would take over your representation, should you delight us by granting our request. We intend to invite Frau Professor Kenngott[4] who you know to be the other godparent.

When your young friend[5] comes to me, I will gladly assist him in finding an appropriate accommodation, and in any case he will always be very welcome in our house.

My book[6] is now completed, and the bookseller will send two copies to you, one for you and a second, which I request that you donate to the Royal Society in my name.[7] A copy will also be sent to Francis,[8] and I will write to him when it arrives to ask him whether he would like to publish an English translation using the translations of the individual papers that already exist in the Phil Mag.

With the warmest greeting to Hirst and fond regards from my wife and I to you,

Your | Clausius

RI MS JT/1/TYP/7/2268–9
LT Typescript Only

1. Clausius here used for the first time the informal address form 'dein', a departure from using the formal form 'Sie' in previous letters to Tyndall. This suggests a move from a formal professional relationship between the two men, to one of a deeper personal relationship. That Tyndall recognised the significance of this change is evident from his reply to this letter (2153).
2. *our little son*: Rudolf John ('Johnny') Clausius.
3. *my wife*: Adelheid Clausius.
4. *Frau Professor Kenngott*: the wife of Gustav Adolph Kenngott (1818–97), a German mineralogist (*The Encyclopaedia Britannica: A Dictionary of Arts, Sciences, Literature and General Information,* 11th edn, 29 vols (New York: Encyclopaedia Britannica Company, 1911), vol. 15, p. 732). The Kenngotts were the godparents to Clausius' other children, Mathilde (1861–1907) and Helene (1863–1919). Future children were Hedwig (1866–1946), Alfred (1868–1950), and Else (1872–1954).
5. *your young friend*: Frank George Wynne.
6. *My book*: see letter 2093, n. 6.
7. *donate to the Royal Society in my name*: this book remains in the RS's collection.
8. *Francis*: William Francis.

From Jane Hogan[1] 10 October 1864 2148

37 Stanley Street | Macclesfield | Cheshire | 10th Oct. 64

Sir,

I have no doubt but you will wonder at receiving a letter from one who must appear a stranger to you, but if you will just look at the enclosed,[2] it will perhaps remind you of me—we have met before, but it is 18 years ago. I must first tell you that I am a sister of Mrs. Ginty's of Rio de Janeiro.[3] If you look back to the time when she was first married and living in Wakefield,[4] it was while I was on a visit with her there that I had the pleasure of seeing you.

You will think it strange that I write to you now, when so many years have passed away, but when you know my reason for doing so, you will I hope kindly excuse me. You have been a friend of Mr. Ginty's and my sister I hope you are still. I now write to ask you to be my friend—you will not I hope think I take too great a liberty in asking you, on account of the exalted position you are in, for the sake of the friendship that exists between Mrs. Ginty and you, you will I hope pardon me. I will now tell you how I thought of writing to you and my object in doing so.

A few weeks ago I was looking over some books and one was an album Mr. Ginty[5] had bought me and which I had in Wakefield, when you knew me as sister of John Roberts[6] and Mrs. Ginty. I was happy then and wanted for nothing—but years have brought sad changes to me. Shortly after I left my sister I got married. My husband[7] died four years ago leaving me with six children, quite unprovided for, were it not for Mrs. and Mr. Ginty helping me, I am sure I do not know what I would have done. I have not had any assistance now since last winter and consequently I have had a great deal of struggling, ever since my husband died I have had much to contend with—but more this year. I wrote to tell Mrs. Ginty of my trouble but I do not expect to have any help from her until Xmas. I owed my landlord £3.7.6 for rent and could not pay him consequently I was obliged to leave the house, all my things were taken, I am now in lodgings with nothing to live on but what I get for a little plain sewing and I am not able to do much for I have been an invalid for a long while and obliged to remain in bed almost constantly. I sit up now writing this with difficulty on reading the enclosed lines which you kindly wrote for me a thought suggested itself to me that if I wrote to you, you might perhaps, for the sake of the friendship that once existed between you and some of my family, that perhaps you would kindly be a friend to me—now in my great need. May I ask you will you give me a little help, I will indeed be most grateful to you if you would kindly lend me two pounds until I hear from Mrs. Ginty, I shall then repay you with many thanks. I am now in much distress and trouble, it will be a very great act of kindness indeed if you will, I shall ever feel grateful to you. I enclose the verses merely to remind you of that time and also to convince what I say is truth. I hope you will kindly pardon me for the liberty and that you will do what you can for me.

I am Sir | Yours most humbly | Jane Hogan

When I had the pleasure of seeing you, I was Jane Roberts—my address now is

Mrs. Hogan | 37 Stanley Street | Macclesfield | Cheshire

RI MS JT/1/TYP/11/3644–44a
LT Typescript Only

1. *Jane Hogan*: Jane Hogan (née Roberts), sister of Margaret Ginty.
2. *enclosed*: as she referred to further on in the letter, presumably some lines of verse Tyndall had written for her when they first met *c.* 1846.
3. *Mrs. Ginty's of Rio de Janeiro*: Margaret Ginty.
4. *Wakefield*: a city in the north west of England in the county of Yorkshire.
5. *Mr. Ginty*: William Ginty.
6. *John Roberts*: brother of Jane Hogen and Margaret Ginty.
7. *My husband*: not identified.

To Juliet Pollock [11 October 1864][1] 2149

Royal Institution of Great Britain | Tuesday

My dear M^rs^. Pollock

I wrote to you this morning[2] before quitting my perch in St. John's Wood[3]—I came to Albemarle Street[4] & there find your letter.[5] Saturday week I will try to get down to Newport & thence to Freshwater.[6] Sunday we could have a stroll, and Monday I could rush to town again. I have been living such a life of dissipation that my work has got awfully in arrear; and hence three days—Saturday, Sunday, and Monday—is all that my tyrant conscience will hear of.

Your interview with Mr Taylor[7] raises a question in my mind as to the right of forming an opinion of a man from a single interview. a crumb of bad cheese, or a glass of bad wine taken overnight; or an overdose of lobster salad—any thing in short that keeps one awake distorts the finest nature, makes the most sensible man stupid, the most eloquent taciturn. Do you not feel after a good night's rest exactly like a bird to which it is not trouble to sing—from which song rushes by a gravity as natural as that by which water falls down hill! But do you not also feel that after a bad night's rest that thought must be pumped up to its elevation before it can fall into expression! I do. To get the law of a man's nature, the range of his mind, the tendency of his feelings you must see him more than once. In fact I fear & deprecate that women's tendency to jump to conclusion regarding a man without any sufficient knowledge of the state of his digestion!

Yours ever | John Tyndall

RI MS JT/1/T/1185
RI MS JT/1/TYP/6/2027

1. *[11 October 1864]*: the date is given based on letter content. In letter 2154, dated 12 October 1864, William Frederick Pollock noted that 'Henry Taylor was here during my absence', having arrived on the day that he left. The letter suggests that the meeting with

Juliet Pollock occurred between 8 and 10 October. Pollock also noted that Tyndall was busy and could not visit, indicating that he had knowledge of the content of this letter. Thus, this letter is best dated as 11 October 1864, which indeed fell on a Tuesday.

2. *wrote to you this morning*: letter missing.
3. *St. John's Wood*: a district of northwest London, where since June 1861 Tyndall lived with Hirst at 14 Waverly Place. He would later move out on 25 January 1868 to occupy Faraday's rooms at the RI following his death.
4. *Albemarle Street*: the RI is located at 21 Albemarle Street, London.
5. *your letter*: letter missing.
6. *Newport & thence to Freshwater*: towns on the Isle of Wight.
7. *M[r] Taylor*: Henry Taylor (1800–86), a British poet and public servant (*ODNB*). A report of this meeting is recorded in William Frederick Pollock's letter (see n. 1), which suggests that the meeting between Juliet Pollock and Taylor likely occurred between 8 and 10 October 1864.

From Richard Owen — 11 October 1864 — 2150

British Museum, | October 11th 1864.

My dear Tyndall,

I derive so much pleasure & profit from a study of your writings that I must thank you for kindly adding your 4th. Memoir on Heat,[1] and the Paper on the Conformation of the Alps,[2] to my Series.

Your brain, physiologically speaking, is now at its best, & that it may long continue there is the hope & wish of.

Yours sincerely | R. Owen

RI MS JT/1/O/3

1. *4th. Memoir on Heat*: see letter 2014, n. 2.
2. *Paper on the Conformation of the Alps*: see letter 2100, n. 4.

From William Frederick Pollock — 12 October 1864 — 2151

Easton Farm | Freshwater[1] | Wednesday | 12th. Oct. 1864

My dear Tyndall

If the state of your work & other engagements really would allow the thing to be done, I wish you would come to us here on Saturday next. The place is still in undiminished beauty, & the weather promises to be fine—It will be our last Saturday & Sunday here—& it w[oul][d] be an enormous pleasure to have you here in your old quarters again. I may remind you that you w[oul][d] get here in time for a walk on Saturday by leaving the Waterloo station (via Lymington & Yarmouth) at 11 a.m. & by leaving this at 8 a. m. on Monday you w[oul][d] reach London early in the afternoon—The return ticket on Saturday is little more than a single fare.

Walter[2] left us for Eton[3] three weeks ago—& Fred,[4] for Cambridge[5] on Saturday, & I have been away for ten days on a little jaunt into your country—but without time for Killarney[6] w[hic][h] is reserved for another occasion.

Come if you can, we should be so glad to see you.

Yours ever truly | W. F. Pollock

RI MS JT/1/P/233
RI MS JT/1/TYP/6/2026

1. *Easton Farm* | *Freshwater*: see letter 2135, n. 1.
2. *Walter*: Walter Pollock.
3. *Eton*: Walter Pollock was a student at Eton College from 1862–5.
4. *Fred*: Frederick Pollock.
5. *Cambridge*: after his time at Eton College (1858–63), Frederick Pollock entered Trinity College, Cambridge, where he obtained his BA (1867) and MA (1870).
6. *Killarney*: a town in County Kerry in the south west of Ireland.

From Andrew Ramsay 12 October 1864 2152

Glasgow 12 Oct 1864

My dear Tyndall

When you wrote me[1] that you had been touched by the same wire I did not know what you meant. But now I do, for this morning in my brothers house I have found the Phil. Mag. and I am this moment engaged in reading your paper[2] which is most excellently reasoned, & comes besides in the very nick of time. It is curious that we should both have come out in the same number,[3] & it will be sure to have a good effect. But to me the curious part of the business is that it should at this time of day be necessary to write both papers at all. For as regards watery erosion there is not a word that we have written almost, that has not been expressed 50 & 70 years ago by Playfair[4] & Hutton.[5] They knew everything except the great extension of the glaciers, and considering how little of geological detail was known in his day, Hutton's book[6] reads like a piece of special inspiration.

I am glad you have given some calculations with regard to the possible width of cracks. The Weald[7] has always been considered a cracked area & Hopkins has a paper[8] on it which I have always thought a very poor affair, though many people think it a very great paper. For 12 years in my lectures I have constantly shown that any crack or cracks due to the bending of the Wealden Strata & chalk could only have been a few yards wide. I forget the number, but it is calculated. Hopkins too has another curiously erroneous paper on Cumberland,[9] which he treats as if it were a <u>dome</u> thrust up from below. The valleys & the lakes according to him are the result!! You will find in Mallets paper[10] on Earthquakes in the transactions of the Irish Academy some precise statements of the extension which different rocks can bear (if bent) without fracture, and if you look up Jukes paper[11] on river valleys in Ireland (Geol[ogical]: Journal 2 years ago or so) you will find that he gives a beautiful explanation of the bend of the Rhone at Martigny, which I fancy he never in reality saw. It is in the postscript.

Between ourselves I had Ball[12] in my eye when I perused that paragraph about men without exact knowledge having no right to have any opinion in the subject.

I am glad you have written that passage about the front of a glacier, for people have always assumed when one spoke of a glacier ploughing out <u>drift</u>, that the glacier was said as it were to drive all the material on in front of it.

Did I tell you about Loch Doune.[13] Such a lovely rock basin, some 7 miles long. There is no need for any inference there, for you <u>see</u> the rock all moutonnée[14] and striated right across its mouth, and forming very decent hills. The river runs out through a glorious gorge, at the end of which is another lake, dammed up by moraine rubbish, I saw half turned into peat & marsh. The Galloway hills[15] formed a grand glacier centre, even for the smaller glaciers that followed the emergence of the land after the great glacial epoch was in the decline.

I left Carrick Moore[16] yesterday. It is a jolly place to stay at. He has a Kaim[17] in his grounds, & the country near Stranraer is full of them. Kaim is Scotch for an Irish Eskir or a Swedish Ösar, & they have a great deal to do with the end of the glacial period, but what none of us yet know. They often enclose lakes. They are not Moraines, that is certain, & I am almost sure that they are of marine origin. We are mapping them everywhere.

Ever sinc[ere]^{ly} | Andrew Ramsay

RI MS JT/1/R/5
RI MS JT/1/TYP3/1007–8

---

1. *When you wrote me*: letter missing; this would probably have been Tyndall's reply to Ramsay's previous letter (2146).
2. *your paper*: see letter 2100, n. 4.
3. *both have come out in the same number*: for Ramsay's paper, see letter 2146, n. 2.

4. *Playfair*: John Playfair (1748–1819), a Scottish clergyman, geologist, and mathematician. He was professor of natural philosophy at the University of Edinburgh and popularized Hutton's work with his *Illustrations of the Huttonian Theory of the Earth* (1802) (*ODNB*).
5. *Hutton*: James Hutton (1726–97), a highly influential Scottish geologist known primarily for his work on uniformitarianism (*ODNB*). The author of *Theory of the Earth; or an Investigation of the Laws observable in the Composition, Dissolution, and Restoration of Land upon the Globe* (1788), Hutton is often considered as the father of modern geology (*ODNB*).
6. *Hutton's book*: Ramsay could be referring to the work mentioned in n. 5, or J. Hutton, *An Investigation of the Principles of Knowledge and of the Progress of Reason, from Sense to Science and Philosophy,* 3 vols (Edinburgh: A Strahan, 1794).
7. *The Weald*: an area of south east England between the parallel chalk escarpments of the North and the South Downs.
8. *Hopkins has a paper*: W. Hopkins, 'On the Geological Structure of the Wealden District and of the Bas Boulonnais', *Transactions of the Geological Society*, 7 (1845), p. 1–51.
9. *paper on Cumberland*: W. Hopkins, 'On the Elevation and Denudation of the District of the Lakes of Cumberland and Westmoreland', *Phil. Mag.*, 3 (1842), pp. 468–77.
10. *Mallets paper*: Robert Mallet (1810–81), an Irish geophysicist known for founding the science of seismology. R. Mallet, 'On the Dynamics of Earthquakes: Being an Attempt to Reduce their Observed Phenomena to the Known Laws of Wave Motion in Solids and Fluids', *Transactions of the Royal Irish Academy*, 21 (1846), pp. 51–106.
11. *Jukes paper*: Joseph Jukes (1811–69), a British geologist known primarily for his expeditions aboard the HMS *Fly*. J. Jukes, 'On the Mode of Formation of Some of the River Valleys in the South of Ireland', *Quarterly Journal of the Geological Society*, 18 (1862), pp. 378–403.
12. *Ball*: John Ball.
13. *Loch Doune*: Doune is a small town in the district of Stirling, Scotland. The lake is at the end of the Teith River, which flows near Doune.
14. *moutonnée*: designating or resembling a bare rock outcrop or rocky hillock that has the surface or form of a roche moutonnée as a result of glacial action (*OED*).
15. *Galloway hills*: part of the Southern Uplands of Scotland north of Galloway.
16. *Carrick Moore*: John Carrick Moore.
17. *Kaim*: one of the elongated mounds of post-glacial gravel, found at the lower end of the great valleys in Scotland and elsewhere throughout the world; an esker or osar (*OED*).

To Rudolf Clausius 13 October 1864 2153

Royal Institution of Great Britain | 13th Oct. 1864[1]

Lieber[2] Clausius

Du bist ein Verführer![3] I have been asked many times by my friends in England to stand sponsor for their children but I have always declined—stating that I did not believe one word of what our holy church says regarding the rites of Baptism. But I cannot refuse you, and it will give me great pleasure to stand Godfather for your little son.[4] I hope he may turn out twenty times as good a man as his sponsor.

If I could get away from London I would go over to you, but that is impossible. I expect to be a hale old man 15 or 20 years hence, and if that be the case I shall give little Johnny his first lesson in mountain climbing. His father may give him his first lesson in mathematics, and thus the little man will be well taught, both mentally and bodily.

Give my kindest regards to your wife,[5] and my respects to Madame Kenngott.[6]

Believe me dear Clausius | Yours ever truly | John Tyndall

RI MS JT/1/T/189

---

1. *13th Oct. 1864*: the date was later added by Tyndall.
2. *Lieber*: Dear (German).
3. *Du bist ein Verführer!*: You are a seducer! (German, using the informal address 'Du'). In his previous letter (2147), Clausius had for the first time addressed Tyndall with the informal address 'Du' rather than the formal 'Sie'. This suggests a move from a formal professional relationship between the two men, to one of a deeper personal relationship.
4. *to stand Godfather for your little son*: in letter 2147, Clausius had asked Tyndall to be godfather to his son Rudolf John ('Johnny') Clausius.
5. *wife*: Adelheid Clausius.
6. *Madame Kenngott*: see letter 2147, n. 4.

To William Frederick Pollock 13 October 1864 2154

13th Oct. 1864.

My dear Pollock,

When Mrs. Pollock[1] wrote to me last[2] I had to put a kind of strain upon my heart to hold myself back. I feel a similar strain upon it now, but I am pledged to the Editor of the Philosophical Magazine to have a paper[3] in his hands on the 18th and the experimental portion of this paper is not yet complete. I was up at 6 o'clock this morning working at it and shall be up at the same hour to-morrow morning—In fact without this devotion to it I cannot get it done.

I would sooner go down for a day after the 18th—a week day—than loosen my hold upon this paper on Saturday.

How long do you intend to remain upon the Island?[4]

Thank Mrs. Pollock for me in your most handsome phraseology for that beautiful little book[5] that she was kind enough to send me. The man's soul is all music. Some of the sonnets sadden me, for they cause me to ask "what is to become of the world when the old and tottering faith round which all this beauty is entwined has passed away." Is not man's moral and religious nature self-supporting, or shall we always need those outward fixings to prop it up? How are we to get the tendrils from the pole without injury I do not know, and still the pole is visibly giving way. I see nothing for it but to change the tendrils into trunks which shall be able to stand alone.

Yours ever | John Tyndall.

RI MS JT/1/TYP/6/2022
LT Typescript Only

1. *Mrs. Pollock*: Juliet Pollock.
2. *wrote to me last*: letter 2135.
3. *a paper*: probably J. Tyndall, 'On Luminous and Obscure Radiation,' *Phil. Mag.*, 28 (1864), pp. 329–41.
4. *the Island*: the Pollocks had been staying at Easton Farm near the town of Freshwater on the western side of the Isle of Wight.
5. *that beautiful little book*: Juliet Pollock sent Tyndall a copy of Charles Tennyson Turner's *Sonnets* (see letter 2126, n. 2).

To William Frederick Pollock 17 October 1864 2155

London | 17th Oct. 1864.

My dear Pollock,

I shall start by the 10 o'clock train on Wednesday morning and hope to reach Freshwater[1] sometime during the afternoon.

You will I trust give me an early breakfast on Friday so as to enable me to reach London by 2.35.

Let me tell you that you have generated a severe struggle between my head and heart;[2] and though heart has carried it head will be grim until I get to my work again.

Yours ever | John Tyndall.

RI MS JT/1/TYP/6/2028
LT Typescript Only

1. *Freshwater*: a town at the western end of the Isle of Wight where the Pollock's were renting Easton Farm.

2. *you have generated a severe struggle between my head and heart*: a letter from Pollock, replying to Tyndall's of 13 October and presumably convincing Tyndall to come visit them, is missing.

From Felix Foucou[1] 17 October 1864 2156

Paris—17 octobre 1864.

Mon cher Monsieur Tyndall,

J'ai reçu et je vous en remercie bien, 10 les trois numéros du "Reader", dont un envoyé par votre ami M. Hirst, auquel je vous prie de présenter mes bons souvenirs; 20 votre mémoire de "Royal Institution"; 30 l'extrait du "Philosophical Magazine" sur la conformation des Alpes. Tous ces documents me seront de la plus grande utilité et je ne saurais assez vous exprimer ma reconnaissance pour la peine que vous prenez ainsi à mon sujet. J'espère bien que si vous aviez, à Paris, besoin de quoi que ce soit, vous me feriez le plaisir de me le demander avant de vous adresser à personne.

Ma femme et moi vous félicitons bien d'avoir échappé au danger de l'avalanche.

Vous devez avoir reçu un numéro du Journal "Le Temps" contenant la traduction de votre lettre à l'éditeur du "Times". Si vous avez quelque erreur de traduction à me signaler, faites-le et vous me rendrez service. D'ailleurs pour votre livre je vous enverrai les épreuves avant le tirage.

La lettre ci-jointe vous dira où nous en sommes avec M. Hachette. J'ai causé avec cet éditeur, qui croit que l'ouvrage réussira, mais qui s'est interdit de publier pour le moment des ouvrages étrangers. Il m'a dit que si je le publiais à mes frais, il pourrait mettre le nom de sa maison, et qu'il ferait tout son possible pour nous assurer une large vente. C'est donc ce que je ferai, si la nouvelle tentative que je vais faire auprès de M. Charpentier ne réussit pas. Vous vous rappelez que c'est à lui d'abord que je m'étais adressé, et, sans positivement me refuser, il m'avait ajourné à une autre époque. Aujourd'hui le moment me parait plus favorable. Je vais commencer dans la "Revue Nationale", qui appartient à M. Charpentier, une série de lettres sur la Géographie Physique, lesquelles doivent être plus tard réunies en volume par le même éditeur. J'espère, dès la publication de la 1ère lettre, réussir à lui faire publier votre livre. Si vous n'y voyez pas d'objection, nous le publierions en deux parties séparées, l'une pour les gens du monde aimant les récits d'excursions et les descriptions de la nature; l'autre pour les personnes plus éclairées—Le format de la Bibliothèque Charpentier s'accommoderait bien à cet arrangement, et à l'aide d'une préface spéciale pour chaque partie, on arriverait à faire deux volumes de bonne dimension.

Puisque vous voulez bien m'offrir de m'adresser les Woodcuts je l'accepte volontiers, mais dites moi le prix que je devrai, avant tout, faire prélever pour ces gravures. Pour le reste, et comme transaction entre nos deux manières de voir, nous partagerons les bénéfices de la publication, puisque vous ne pensez pas devoir accepter une prime d'auteur préalable.

Serait-il possible d'avoir l'adresse inaugurale de Sir Roderick Murchison sur les montagnes? Votre théorie mérite un développement spécial et une discussion approfondie dans l'ouvrage.

Mille amitiés | Felix Foucou

Paris—17 October 1864.

My dear Mr Tyndall,

I have received, and I thank you very much for it: 1. the three issues of the "Reader",[2] one of which was sent by your friend Mr. Hirst, to whom I ask you to give my best wishes; 2. your article from the "Royal Institution";[3] 3. the excerpt from the "Philosophical Magazine" on the conformation of the Alps.[4] All of these documents will be very useful for me and I do not know how to thank you enough for all the trouble that you went through for me. I certainly hope that if you need anything in Paris, you would do me the pleasure of requesting it from me before contacting anyone else.[5]

My wife and I congratulate you on having escaped the danger of the avalanche.[6]

You should have received an issue of the journal "Le Temps"[7] which includes the translation of your letter to the editor of the "Times".[8] If you have any translation errors to report, you would do me a great service by doing so. Moreover, for your book,[9] I will send you the proofs before the printing.

The attached letter[10] shows you where we are with Mr. Hachette.[11] I chatted with this editor, who believes that the book will be successful, but is refraining from publishing foreign books at the moment. He told me that if I publish it at my own expense, he could attach the name of his publishing house and that he would do everything possible to ensure sales. This is what I will do, if the new attempt with Mr. Charpentier[12] does not succeed. You will remember that he was the first one I approached, and without absolutely refusing me, he postponed it to another day and age. Now conditions seem to be more favourable. In the "Revue Nationale", which belongs to Mr. Charpentier, I will put out a series of letters on Physical Geography, which must be collected later in a volume by the same publisher. I hope that, from the publication of the 1st letter, I succeed in making him publish your book. If you have no objections, we will publish it in two separate parts, one will be for the general public who love stories of expeditions and descriptions of nature, the other for more enlightened people. The format of the Charpentier Library[13] would suit this arrangement very well, and with the help of a special preface for each part, we could make two good-sized volumes.

Since you are willing to offer sending me the Woodcuts, I gladly accept them, but above all, tell me the price that I will have to, debit for these prints. As for the rest, and as a transaction between our two ways of seeing, we will share the profits of the publication, since you do not think you should accept a bonus authorial advance.

Would it be possible to have the inaugural speech by Sir Roderick Murchison[14] on mountains? Your theory deserves special development and a thorough discussion in the book.

With deepest friendship | Felix Foucou

RI MS JT/1/F/36

1. *Felix Foucou*: see letter 2143, n. 1.
2. *the three issues of the "Reader"*: it is not clear which issues are referred to here.
3. *your report of "Royal Institution"*: probably Tyndall's Friday Evening Discourse at the RI on 10 June 1864, 'On a Magnetic Experiment' (*Roy. Inst. Proc.*, 4 (1862–6), p. 317–22).
4. *the excerpt from the "Philosophical Magazine" on the conformation of the Alps*: see letter 1921, n. 6, and letter 2100, n. 4.
5. *I certainly hope . . . before contacting anyone else*: gift and service exchange has been highlighted as an important element of building working relationships; but it is likely that the specific reference here to 'anyone else' is to Foucou's rival in publishing Tyndall's works, William Hüber. See letter 2143, n. 6.
6. *on having escaped the danger of the avalanche*: on 30 July 1864, Tyndall and his party were victim to an avalanche while descending the Piz Morteratsch. See letters 2130 and 2145.
7. *an issue of the journal "Le Temps"*: *Le Temps*, 13 October 1864, p. 2.
8. *your letter to the editor of the "Times"*: Tyndall's letter of 30 September 1864 (*Times*, 1 October 1864, p. 10). This is letter 2145.
9. *your book*: a reference to *Glaciers of the Alps*, which Foucou was hoping to translate into French. See letter 2145 (especially n. 4), as well as n. 5 above.
10. *The attached letter*: letter missing.
11. *Mr. Hachette*: Louis Christophe François Hachette (1800–64), a Paris-based publisher known for publishing medical texts (*The Encyclopedia Americana*, 30 vols (New York: Encyclopedia American Corporation, 1923), vol. 13, pp. 606–7).
12. *Mr. Charpentier*: Gervais-Hélène Charpentier (1805–71), a French bookseller and publisher, often referred to as the father of the pocket-sized book ('Nécrologie', *Revue bibliographique universelle*, 6 (1871), p. 181).
13. *the Charpentier Library*: as early as 1841, Charpentier devoted himself entirely to his collection, de la Bibliothèque Charpentier, and, from 1860, to the publication of a magazine, *Revue nationale et étrangère*, in which Henri Brisson, Edouard Laboulaye, Pierre Lanfrey, and Eugene Despois collaborated. With this collection and work he focused on the creation of a new format of books, recognized as convenient, elegant, and portable by all.
14. *the inaugural speech by Sir Roderick Murchison*: on 9 November 1863, Murchison delivered

his inaugural address to the Royal Geographical Society of London ('Inaugural Address', *Proceedings of the Royal Geographical Society*, 8 (1864), pp. 1–5). Murchison more openly stated his views on the formation of the Alps and the role (or not) of ice in carving out valleys, in his anniversary address to the same society (see letter 2100, n. 1 and n. 4). See also letters 2146 and 2152.

From Frederick North 17 October [1864][1] 2157

Hastings Lodge—Hastings—Oct 17

My dear Sir

Thank you very much for letting me see your publication,[2] on the "Conformation of the Alps"—My knowledge is very slight & my opinion therefore worth little, on this subject, but it is impossible not to be struck with the lucidity of your treatment of it—& it is difficult not to fancy one knows something of geology oneself, from this remarkably clear way in which you bring out your facts, & the inferences from them, for the mind of your reader.

You do not, I suppose, intend to convey, that there has been no Plutonic action[3] on our crust, or that glacial & water influence worked wholly the inequalities on the Earthy surface!

However, it is idle for me to press on this.

My main object, in writing, was to suggest that, when you want some free time, or have a spare day or two, you would come & visit us here. We should be delighted to welcome you—& renew our Pontresina acquaintance[4]—

That "Instructive Charm" which you mention, produced a disruption in my family—my youngest daughter[5] then accepted young John Symonds[6]—& they are to be married on the 10th Nov[e]m[ber][7]—Could you not come to us, *[1 word illeg]* the end of it, then—It would be great fun, & you should have 8 Bridesmaids to prance about with.

Yours ever truly | Fredrick North

By the way, you were hard on me, publishing my Age in the Times,[8] the other day. Mrs Sabine[9] wrote my daughter,[10] that it was very delicate of you not mentioning names—but all my Constituents here recognize the old man.

RI MS JT/1/N/16

1. *[1864]*: the year is given based on references to Tyndall's paper on the conformation of the Alps, and Tyndall giving North's age in a letter to the *Times*.
2. *your publication*: see letter 2100, n. 4.
3. *Plutonic action*: see letter 2124, n. 5.
4. *Pontresina acquaintance*: Pontresina is a small mountain village in the Maloja Region

in the canton of Graubünden in Switzerland. North was here referring to their climb together (see letter 2130). Tyndall referred to North in his letter to the *Times* recounting their avalanche accident (see letter 2145).

5. *my youngest daughter*: Janet Catherine North (1837–1913), sister of the traveler and botanical illustrator Marianne North (1830–90) (*ODNB*). Janet would later edit her sister's autobiography, *Recollections of a Happy Life: being the Autobiography of Marianne North*, 2nd edn, 2 vols (London: Macmillan and Co., 1892).
6. *John Symonds*: John Addington Symonds (1840–93), an English poet, literary critic, and advocate of sexual reform (*ODNB*).
7. *married on the 10th Nov[e]m[ber]*: they were indeed married on 10 November 1864, in Hastings North, *Recollections of a Happy Life*, vol. 1, p. 34.
8. *publishing my Age in the Times*: in his letter to the *Times* (letter 2145), Tyndall revealed that North was 64 years of age.
9. *Mrs Sabine*: Elizabeth Juliana Leeves Sabine (1807–79), the British translator of Alexander von Humboldt's *Kosmos*. She also assisted her husband, Edward Sabine, in his scientific work (M. Brück, *Women in Early British and Irish Astronomy: Stars and Satellites* (London: Springer, 2009), pp. 101–4).
10. *my daughter*: this could be either Janet Catherine North or Marianne North.

From Jane Hogan[1] 18 October 1864 2158

37 Stanley Street | Macclesfield | 18th Oct 64

Sir,

I received your note[2] this morning, to say you will kindly lend me the two pounds. I shall indeed be most grateful to you for I am very much in need of it now. I shall if I am spared repay you soon. I am now in lodgings and am very anxious to remove into a small house for I find it impossible to get on, the longer I am here, the greater my difficulties become, it will be so much cheaper and better for me to go into a little house. If you can kindly lend me the money by return I shall be deeply grateful, it will enable me to get a few articles of furniture and remove at the end of this week. I do not think there is any occasion to send my note to Mr. Ginty,[3] he knows by this time of all the trouble I have had lately and my furniture gone, for I wrote telling them all. I expect some assistance from my sister before Xmas, perhaps by next Mail, I may get it, but that is uncertain, she promised to send me some near Xmas. I shall certainly return to you the money you have promised to lend me as soon as I receive help from them, at present, I am very much in want of it and shall be most grateful if you will kindly lend it to me. I must thank you much for your kindness.

I am Sir | Yours respectfully | Jane Hogan

I will feel most thankful indeed if you will kindly write by return for I have been in so much trouble for some time, I shall be so glad to get relief.

RI MS JT/1/TYP/11/3645
LT Typescript Only

1. *Jane Hogan*: see letter 2148, n. 1.
2. *your note*: letter missing. Hogan had written to Tyndall on 10 October asking for a loan (letter 2148).
3. *Mr. Ginty*: William Ginty.

From Hendrik Willem Schroeder van der Kolk[1] 20 October 1864 2159

Zutphen (Holland) 20 Oct. 64.

Hochgeehrter Herr!

Da ich Ihnen ganz unbekannt bin, hoffe ich, dass Sie mich erlauben werden mich an Sie zu richten, und hauptsächlich nach Anleitung Ihrer trefflichen Abhandlung on the heat as a mode of motion. Ich habe diese, so wie die Auseinandersetzung Ihrer eigenen höchst interessanten Untersuchungen mit der grössten Freude gelesen, und kam gerne mit Ihnen in näherer Bekanntschaft, da ich mit der nämlichen Theorie mich auch am meisten beschäftige. Deswegen füge ich hier eine kleine Abhandlung hinzu über die Mechanische Energie der chemischen Wirkungen. Wiewohl diese specielle Anwendung auf die Chemie bis jetzt, so viel ich weiss, nicht auseinander gesetzt worden war, glaube ich doch, dass die nämlichen Betrachtungen, in so fern sie richtig sind, Ihnen wohl nicht unbekannt sein werden. Wenn andere die Theorie so deutlich auseinander setzen, ist die Anwendung auf die verschiedenen Naturereichnissen keine so schwere Arbeit. Ich muss noch hinzufügen, dass in der nächstfolgenden Lieferung von Pogg. Annalen eine kleine Note folgt, wo *[mein]* letzter Satz, i.e., dass die Verbindungswärme nicht als Maass der chem. Affinität betrachtet werden kann, deutlicher auseinander gesetzt ist.—

Ich bin ganz mit Ihnen einverstanden, dass die Erscheinungen bei den Gasen vom höchsten Interesse sind, da diese uns den sichersten Aufschluss für die moleculären Eigenschaften geben können. Der nämliche Gedanke brachte mich dazu in einer im vorigen Jahre bei Pogg. erschienen Abhandlung, zu versuchen in den Regnaultschen Bestimmungen der Abweichungen der Gasen vom Mar. und G. L. Gesetz einigen Zusammenhang zu finden, wo ich in der Formel pv=kt, k als eine von Temperatur und Druck abhängige Constante betrachtete. Es ergab sich unter anderem, dass diese Constante beim Wasserstoff vom Druck unabhängig ist, und ein mit diesem Gase gefüllter Thermometer, wo die Ausdehnung bei constantem Drucke gemessen wird, würde also einem mit ideellem Gase gefüllten ganz gleich gehen, da in beiden Fällen k constant ist. Man findet dann für den absoluten Nullpunkt -273.15, welche Bestimmung mir genauer scheint, als die welche aus dem Ausdehnungscoefficienten der Atmosphäre abgeleitet ist.

Ich erlaube mich noch ein Exemplar meiner Dissertation hinzuzufügen, wovon in Pogg. t. 110 ein Auszug erschienen ist.

Da die Mechanische Wärmetheorie in England so viele Vertreter findet, hatte ich gewünscht meine Abhandlung der Redaction der Phil. Mag. zur Einschaltung anzubieten. Für den Fall, dass es Ihnen von hinreichendem Interesse zu scheinen möchte, füge ich ein zweites Exemplar meiner erstgenannten Abhandlung hinzu, adressirt an die Redaction der Phil. Mag. Ich würde es selbst wohl im Englischen übersetzt haben; das Schreiben im Englischen ist mir aber zu wenig geläufig, um nicht fürchten zu müssen, dass die Correction für die Redaction gleich schwierig sein würde als die etwaige Uebersetzung. Vielleicht hat aber die weitere Publication keinen Nutzen, und in diesem Falle bitte ich Sie das Exemplar zurück zu halten.

Innerhalb kurzer Zeit hoffe ich eine neue Berechnung der Versuche der Schallgeschwindigkeit von Moll u van Beek zu publiziren. Diese Versuche sind noch immer die genaueste, die Berechnung ist aber in vielen Hinsichten fehlerhaft. Unser Landsmann Dr. Simons, hat schon früher eine verbesserte Berechnung publizirt, welche in den Phil. Trans. damals abgedruckt worden ist; einige Fehler sind hierbei jedoch übersehen worden.

Da von den fremden Sprache die Deutsche mir beim Schreiben am geläufigsten ist, erlaube ich mir diese ins Deutsch zu schreiben, da diese Sprache Ihnen, wie ich aus Ihrer Abhandlung bemerkte, bekannt ist.

Mit der grössten Hochachtung habe ich die Ehre zu sein

[Ein Dienstfertigster] | Dr. H. W. Schroeder van der Kolk

Zutphen (Holland) 20 Oct. 64.

Dear Sir!

As I am completely unknown to you, I hope you will be so kind as to allow me to address myself to you, in particular for guidance on your splendid work on the heat as a mode of motion.[2] I read this, as well as the discussion of your highly interesting investigations, with the greatest pleasure and would be happy to make closer acquaintance with you, given that I mainly work on said theory myself. I therefore enclose a short paper[3] on the mechanical energy of chemical action. Although this special application to chemistry had not, as far as I know, been studied until now, I believe that the views alluded to, as far as they are correct, will not be unknown to you. When others present the theory in such a clear way, their application to different natural events is not a very complicated task. I must add that the upcoming issue of Pogg. Annalen contains a short report[4] where my last proposition, i.e. that the heat of combination cannot be taken as a measure for chem affinity, is explained more clearly.—

I agree with you that phenomena in gases are of the highest interest, for they can give us the most certain insights into molecular properties. This consideration motivated me to attempt to find, in an article[5] published last year in Pogg., some relations in Regnault's studies of the deviation of gases from the law of Mar. and G. L.,[6] where in pv=kt I took k as a constant dependent on temperature and pressure. I found that among other relations, this constant is independent of pressure for the case of hydrogen and that a thermometer filled with this gas, where one measures the expansion at constant pressure, would therefore behave exactly like one filled with an ideal gas, as in both cases <u>k</u> is constant. One then finds -273.15 as the absolute zero, a determination that appears to me more accurate than the one derived from the expansion coefficient of the atmosphere.

I take the liberty of including a copy of my dissertation, an excerpt[7] of which appeared in Pogg., v. 110.

As the mechanical theory of heat has so many supporters in England, I wished to offer my paper[8] to the editors of the Phil. Mag. for inclusion. For the case that you might find it of sufficient interest, I am enclosing <u>a second copy of my paper previously mentioned</u>, addressed to the editors of the Phil. Mag. I would have translated it into English myself, but I am not familiar enough with writing in English without having to fear that the editors would find it as difficult to correct my text as translating it. But it may be that further publication is of no use and in that case I would ask you to withhold the copy.

I hope to publish shortly a new calculation[9] of the experiments of Moll and van Beek[10] for the speed of sound. These experiments are still the most accurate ones, but the calculation is flawed in many aspects. Our compatriot, Dr Simons,[11] had previously published a better calculation that appeared in the Phil. Trans.[12] However, some mistakes were overlooked on that occasion.

Among the foreign languages, German is the one with which I am most familiar in writing and for that reason I allow myself to write you in German, as I noticed from your article that you are familiar with this language.

With the greatest respect I have the honour to be

[A servant] | <u>Dr. H. W. Schroeder van der Kolk</u>

RI MS JT/1/S/57

---

1. *Hendrik Willem Schroeder van der Kolk*: Hendrik Willem Schroeder van der Kolk (1836–67), Professor of Natural Philosophy at Zutphen who specialised in physical chemistry and thermochemistry (H. A. M. Snelders, 'Hendrik Willem Schroeder van der Kolk (1836–1867) en de fysische chemie', *Scientiarum Historia*, 13 (1971), pp. 185–97).

2. *heat as a mode of motion*: *Heat as a Mode of Motion.*
3. *a short paper*: H. W. Schroeder van der Kolk, 'Ueber die mechanische Energie der chemischen Wirkungen', *Poggend. Annal.*, 122 (1864), pp. 439–54, 658–9. This was also published as H. W. Schröder van der Kolk, 'On the Mechanical Energy of Chemical Actions', *Phil. Mag.*, 29 (1865), pp. 269–82.
4. *a short report*: H. W. Schroeder van der Kolk, 'Zusatz zu der Abhandlung über die mechanische Energie der chemischen Wirkungen', *Poggend. Annal.*, 122 (1864): 658.
5. *an article*: probably L. C. Stuart, 'Uber das gegenseitige Verhältniss des Gay-Lussac' schen Gesetzes zu dem Mariotte' schen und dem Mayer' schen Gesetze', *Poggend. Annal.*, 119 (1864), pp. 327–31.
6. *Mar. and G. L.*: Mar. refers to Edme Mariotte (1620–84), a French physicist. Mariotte's Law is, in English-speaking countries, more usually known as Boyle's Law, or the Boyle-Mariotte Law; G. L. refers to Joseph Louis Gay-Lussac (1778–1850), a French chemist and physicist.
7. *an excerpt*: possibly H. W. Schroeder van der Kolk, 'Ueber die Bestimmung des galvanischen Leitungswiderstandes', *Poggend. Annal.*, 110 (1860), pp. 452–76.
8. *my paper*: see n. 3.
9. *publish shortly a new calculation*: H. W. Schroeder van der Kolk, 'Ueber die Geschwindigkeit des Schalles', *Poggend. Annal.*, 124 (1865), pp. 453–70. This was also published as H. W. Schroeder van der Kolk, 'On the Velocity of Sound', *Phil. Mag.*, 30 (1865), pp. 34–49 (note and errata on 391–2).
10. *experiments of Moll and van Beek*: G. Moll and A. Van Beek, 'An Account of Experiments on the Velocity of Sound, Made in Holland', *Phil. Trans.*, 114 (1824), pp. 424–56. Gerard Moll (1785–1838), was a Dutch polymath and professor of natural philosophy at the University of Utrecht. Albert van Beek was his associate.
11. *Dr. Simons*: Gerard Simons, Assistant at the Observatory of the University of Utrecht.
12. *better calculation that appeared in the Phil. Trans.*: G. Simons, 'On the Theoretical Investigation of the Velocity of Sound, as Corrected from M. Dulong's Recent Experiments, Compared with the Results of the Observations of Dr. Moll and Dr. Van Beek', *Phil. Trans.*, 120 (1830), pp. 209–14.

From John Amory Lowell[1] 25 October 1864 2160

Boston, United States of America.

Dear Sir,

Your book on 'Heat in motion'[2] has attracted much attention in the United States, & my present motive in writing is to invite you to give a course of twelve lectures[3] before the Lowell Institute[4] in Boston in the autumn of 1865 or the winter of 1866.

For such a course I pay 1200 dollars, & to obviate all questions of currency,

I would pay it in gold. I am well aware that this would not be a sufficient inducement, unless you were disposed to visit our country, in which case I would suppose that you should come early & devote the summer to travelling. You might then lecture in October & November & be at home before the Xmas holidays.

Sir Charles Lyell who has twice visited the United States on my invitation[5] can give you all requisite information.

I am | yours resp[ectfull]y | John Amory Lowell | Boston | 25 Oct[obe]r 1864

RI MS JT/2/10/300

---

1. *John Amory Lowell*: John Amory Lowell (1798–1881), an American businessman and philanthropist. He became the sole trustee of the Lowell Institute once his cousin John Lowell Jr. died in 1836 (D. R. Lowell (ed), *The Historic Genealogy of the Lowells of America from 1639 to 1899* (Rutland, VT: Tuttle Company, 1899), p. 117).
2. *'Heat in motion'*: *Heat as a Mode of Motion.*
3. *invite you to give a course of twelve lectures*: Tyndall declined this offer in letter 2188. He would eventually visit the United States for a lecture tour (on light) from October 1872 to February 1873 (see M. D. Barton, 'The "efficient defender of a fellow-scientific man": John Tyndall, Darwin, and Preaching Pure Science in Nineteenth-Century America', MA thesis, Montana State University, 2010; *Proceedings at the Farewell Banquet to Professor Tyndall. Given at Delmonico's, New York, February 4, 1873* (New York: D. Appleton and Company, 1873); K. Sopka, 'An Apostle of Science Visits America: John Tyndall's Journey of 1872–1873', *The Physics Teacher*, 10 (1972), pp. 369–75; K. Sopka, 'John Tyndall: International Populariser of Science', in W. H. Brock, N. D. McMillan, and R. C. Mollan (eds), *John Tyndall: Essays on a Natural Philosopher* (Dublin: Royal Dublin Society, 1981), pp. 193–203; *Ascent of John Tyndall*, pp. 301–15; and J. Tyndall, *Six Lectures on Light delivered in America in 1872–1873* (London: Longmans, Green, and Co., 1873). The forthcoming thirteenth volume of *The Correspondence of John Tyndall* will include his correspondence during this trip.
4. *Lowell Institute*: an educational foundation in Boston, Massachusetts providing public and academic lectures, founded in 1840 by the American businessman and philanthropist John Lowell Jr. (1799–1836) (*ANB*).
5. *twice visited the United States on my invitation*: Lyell lectured for the Lowell Institute in 1841, 1845, and 1852, and in total visited the United States four times between 1841–53. On Lyell in America, see R. H. Dott, 'Charles Lyell's debt to North America: his lectures and travels from 1841 to 1853', in D. J. Blundell, & A. C. Scott (eds), *Lyell: the Past is the Key to the Present*, Special Publications (London: Geological Society, 1998), pp. 53–69; R. H. Dott, 'Lyell in America—His Lectures, Field Work, and Mutual Influences, 1841–1853',

Earth Sciences History, 15 (1996), pp. 101–40; R. H. Silliman, 'The Hamlet Affair: Charles Lyell and the North Americans', *Isis*, 86 (1995), pp. 541–61; and L. G. Wilson, *Lyell in America: Transatlantic Geology, 1841–1853* (Baltimore: Johns Hopkins University Press, 1998).

From Emile de Laveleye[1] 28 October 1864 2161

Liège 28 oct. 1864

Monsieur,

Je vois que vous venez de publier dans le Philosophical Magazine un article: On the Formation of the Alps.

J'ai l'intention de faire paraître dans le Revue des Deux Mondes, pour faire suite à un autre article sur la Suisse, un article sur le Mont-Rose. Votre travail pourrait m'être très utile. Pourrais-je vous demander de me le faire parvenir? Le meilleur moyen de me le faire venir serait de me l'envoyer sous bande affranchie. Je vous en serais bien reconnaissant.

Si vous aviez q.q. récit de vos intéressantes excursions dans les hautes alpes je serais bien heureux aussi de le posséder et d'en faire mention.

Emile de Laveleye | Professeur d'Économie politique à l'université de Liège (Belgique)

Liège, Oct. 28th 1864

Sir,

I see that you just published the article: On the Formation of the Alps in the Philosophical Magazine.[2]

I intend to publish, in the Revue des Deux Mondes, in response to another article on Switzerland, an article on the Monte Rosa.[3] Your work could be very useful to me. Could I ask you to send it to me? The best way to get it to me is to send it by prepaid post. I would be very grateful to you.

If you have some stories about your interesting excursions in the High Alps, I would be happy to have them also, and to mention them as well.

Emile de Laveleye | Professor of Political Economy at the University of Liège (Belgium)

RI MS JT/1/D/114

1. *Emile de Laveleye*: Émile Louis Victor de Laveleye (1822–92), a Belgian economist known for his works on political economy and for co-founding the Institut de Droit International

(1873) (*The Encyclopaedia Britannica: A Dictionary of Arts, Sciences, Literature and General Information*, 11th edn, 29 vols (New York: Encyclopaedia Britannica Company, 1911), vol. 16, pp. 291–2).

2. *On the Formation of the Alps in the* Philosophical Magazine: see letter 1921, n. 6, and letter 2100, n. 4.
3. *Monte Rosa*: the highest mountain in the Swiss Alps, located between Switzerland and Italy near the town of Zermatt. Tyndall climbed it twice in August 1858, the second time achieving the summit solo (see *Tyndall Correspondence*, vol. 6).

To George Gabriel Stokes 29 October [1864][1] 2162

| Saturday 29th Oct

My dear Stokes

I think you will be interested to learn that I have succeeded in raising platinum to vivid redness by black heat.

The same black heat sets paper instantly ablaze.

Dr. Akin[2] has seen some of the experiments.[3]

Yours very truly | John Tyndall

I hope in one of the meetings of the R.S. to be able to show you platinum raised to whiteness by black heat.

RI MS JT/1/T/1427
RI MS JT/1/TYP/4/1479

1. *[1864]*: the year is given based on mention of C. K. Akin and the experiments Tyndall was working on. Also, in 1864, 29 October indeed fell on a Saturday.
2. *Dr. Akin*: C. K. Akin. Previously Tyndall and Akin had become embroiled in a priority dispute; see letters 1925, 2075, 2076, 2077, 2079, and 2114. Also see letters 2165, 2167, 2168, 2170, 2173, 2182, 2183, 2190, 2197, 2198, 2201, 2204, 2208, and 2214, as well as the Introduction in this volume and in the forthcoming ninth volume of *The Correspondence of John Tyndall*.
3. *the experiments*: on this line of experiments, see: J. Tyndall, 'On Calorescence', *Phil. Trans.*, 156 (1866), pp. 1–24 (received 20 October 1865; read 23 November 1865); and J. Tyndall, 'Sixth Memoir on Radiation and Absorption. —Influence of Colour and Mechanical Condition on Radiant Heat', *Phil. Trans.*, 155 (1866), pp. 83–96 (received 21 December 1865; read 18 January 1866).

To Richard Dawes [October 1864][1] 2163

Royal Institution, | Saturday.

Dear good Friend,

I wish I could go to you, but I cannot.[2] I am in the midst of work[3] which it would be sinful to quit. I wished so much to go to Bath,[4] but was unable to do so with a clear conscience. Give my kindest regards to Mrs. Dawes.[5]

Goodbye. | Yours affectionately | John Tyndall.

RI MS JT/1/TYP/1/326
LT Typescript Only

1. *[October 1864]*: this date was suggested by LT, and seems reasonable given the mention of the BAAS meeting in Bath which had occurred in September 1864. Also, the previous letter in this volume, dated 29 October [1864], indicates an important series of experiments Tyndall was invested in.
2. *I cannot*: previous letters in this volume to Dawes and his wife show that Tyndall found difficulty in finding time to step away from his duties to visit them. In letter 1961, he wrote to Mary Ellen Dawes that 'I wish most earnestly you were nearer to London'.
3. *work*: see letter 2162, n. 3.
4. *Bath*: see letter 2138, n. 1.
5. *Mrs. Dawes*: Mary Ellen Dawes.

To Andrew Ramsay [October 1864][1] 2164

My dear Ramsay

I was delighted with your letter[2] & delighted with your paper.[3] I think we have given the enemy[4] something to think of. No doubt they will come to the front once more but their case is hopeless, & we may await their onset with the calmness of philosophers—God is sure to defend the right!

Yours ever | John Tyndall

I meet Logan[5] tomorrow at Sir Edmund Head's[6]

Bonhams, Knightsbridge, auction, 15 Nov 2017, lot 93

1. *[October 1864]*: the month and year are given by relation to letter 2152. This letter refers to the dispute about glacial/river versus fracture theory of formation of alpine valleys.
2. *your letter*: letter 2152.
3. *your paper*: see letter 2146, n. 2

4. *the enemy*: this refers to an ongoing debate between Ramsay and Roderick Murchison on the action of glaciers. See letters 2100, 2146, and 2152.
5. *Logan*: William Edmond Logan (1798–1875), a Canadian geologist most noted as the founder and director of the Geological Survey of Canada (C. G. Winder, 'Logan, Sir William Edmond', in F. G. Halpenny (ed), *Dictionary of Canadian Biography*, 22 vols (Toronto and Quebec: University of Toronto and Laval University Press, 1972), vol. 10 (1871–80), p. 444–9).
6. *Sir Edmund Head*: Edmund Head (1805–68), a British politician, diplomat, and governor-in-chief of British North America (1854–61) (*ODNB*). Head attended lectures and events at the Athenaeum Club with Tyndall and was an advocate for increased public awareness of science. Tyndall said that it was Head who had inspired him to write his popular science books (see *Ascent of John Tyndall*, p. 277).

From C. K. Akin [1 November 1864][1] 2165

DEAR SIR,—I have just received your letter of today,[2] and intend to shortly reply to such portions as seem to me to require it.

As regards the way in which your attention was turned to the subject which is in question, of course you are the best as well as the only authority, and I have no intention of challenging any statement on your part, deliberately made, and where there are no facts within my knowledge proving that your recollection misleads you. Such, however, I believe to be the case in reference to what occurred at the Royal Institution in April, in the course of our conversation. You then spoke of a 'triple alliance'[a] (those were the words you used). I replied, 'Leave the sun to us at Oxford,'[3] but added that, as regarded experiments on artificial sources of rays (the expressions, I believe, were 'the other two experiments,' meaning those described in my paper), I should be happy to cooperate with you separately. If there was 'no closing'[4] with any offer, it must have been on your part, and as regards this last-mentioned offer, which I made in answer to your own.[5]

[a] That was after I had said that I was bound to Mr. Griffith.[6]

J. Tyndall, 'On the History of Calorescence', *Phil. Mag.*, 29 (1865), pp. 218–31, on p. 224

1. *[1 November 1864]*: date provided by Tyndall in his recollections of his dispute with Akin recorded in J. Tyndall, 'On the History of Calorescence,' *Phil. Mag.*, 29 (1865), pp. 218–31, on p. 224. Tyndall noted that this was the second letter he received from Akin in this dispute out of five; the fifth, which has not been found, was dated 9 November 1864. See letter 2162, n. 2.

2. *your letter of today*: letter missing.
3. *'Leave the sun to us at Oxford'*: Tyndall could only work by electric light in the basement of the RI, while Akin and Griffith could work with natural light.
4. *'no closing'*: Tyndall indicated that the word 'closing' was placed between inverted commas, because in his reply to Akin's first letter he had reminded him that 'he had not "closed" with my proposal' (Tyndall, 'On the History of Calorescence', p. 224).
5. *your own*: Tyndall did not include the manner in which Akin closed this letter, although he stated in the article that the 'extract' appears '*in extenso*' (Tyndall, 'On the History of Calorescence', p. 224).
6. *Mr. Griffith*: George Griffith (*c.* 1834–1902), an English science educator. He taught at Oxford from 1857–67 (where he worked with Akin), but left after the professorship of Experimental Philosophy was given to Robert Clifton in 1865. He was afterward the first natural science master at Harrow (1867–93). Griffith also took an active role in the BAAS, serving as its assistant general secretary for thirty years (1862–78, 1881, and 1890–1903) (B. P. L., 'George Griffith', *Nature*, 66 (1902), p. 64).

To [Robert Fitzroy][1] 3 November 1864 2166

Royal Institution of Great Britain | 3rd. Nov. 1864

My dear Admiral

Long may you enjoy the power of wielding the pen in the varied fields of human culture that you have made your own.

Your beautiful volume[2] has duly reached me.

Yours ever | John Tyndall

Wellcome, MS. 7777/07

1. *[Robert Fitzroy]*: the recipient's identity suggested by the reference to 'your beautiful volume'. Robert Fitzroy's (1805–65) book (see n. 2) was published 1863, the year he was promoted to Vice-Admiral.
2. *Your beautiful volume*: probably R. Fitzroy, *The Weather Book: A Manual of Practical Meteorology* (London: Longman, Green, Longman, Roberts & Green, 1863).

To C. K. Akin [3 November 1864][1] 2167

And now for a few concluding words:—I cannot help expressing my deep regret that you did not write to me in a different manner.[2] I should have gone almost any length to gratify you, had you adopted another tone; and I shall prove my sincerity in writing thus, by doing that which still remains in my power to do. I have to say, then, that from this 3rd of November 1864 to the 3rd of November 1865 I shall not make known, publicly or privately, any experiments on 'ray-transmutation'. Nearly seven months ago I was, as you know, ready to take this subject up. Out of deference to you I did not do so; so that at the termination of the period referred to I shall have held back for nearly nineteen months. That you will be able to release me long before, by the actual performance of this experiment, is almost beyond a doubt; and in this event I hope to muster sufficient greatness of heart not to envy you.

J. Tyndall, 'On the History of Calorescence,' *Phil. Mag.*, 29 (1865), pp. 218–31, on p. 228

---

1. *[3 November 1864]*: the date is given by Tyndall within the letter. Akin's response to this letter is missing. See letter 2162, n. 2.
2. *in a different manner*: letter missing; although Akin had clearly written suggesting that Tyndall had acted improperly in pursuing his experiments on ray transmission.

To George Gabriel Stokes 3 November [1864][1] 2168

Thursday 3rd Nov

My dear Stokes,

Some days ago I wrote[2] to inform you that I had succeeded in making platinum red hot with obscure rays. Circumstances which have since occurred[3] cause me to ask you to consider that note unwritten.

Yours truly | John Tyndall

RI MS JT/1/T/1428
RI MS JT/1/TYP/4/1479

---

1. *[1864]*: the year is given by reference to letter 2162.
2. *I wrote*: letter 2162.
3. *Circumstances which have since occurred*: namely, the arrival of letter 2165 from Akin. Akin had suggested that Tyndall had acted outside the bounds of gentlemanly propriety. See letter 2162, n. 2.

From James Joseph Sylvester 4 November [1864][1] 2169

Athenaeum[2] | 4th Nov.

My dear Tyndall,

Allow me to add my congratulations to those of your numerous friends and well wishers at the result of this day's ballot at the Royal Society which led to the so justly merited award to you of the Rumford medal.[3] The decision was all but unanimous in your favour and I sincerely hope that you may live long to enjoy this distinction and to earn new claims to the estimation of your fellow laborers in science by the same course of strenuous exertion which has gained you your present most flattering success.

You will be pleased also to hear that Hirst has been elected on the Council and was one of the few who had his name on every one of the balloting papers.

Believe me with renewed congratulations | Yours ever truly | J. J. Sylvester

I think we may exchange congratulations at the award of the Copley Medal to Darwin[4]—It is long since I have felt such a thrill of satisfaction as I have experienced from the upshot of this day's proceedings. It is a day that will long be memorable in the annals of progress. The Royal Society has done itself honor and conferred honor on Old England by this public recognition of transcendent merit.

RI MS JT/1/TYP/4/1509
LT Typescript Only

---

1. *[1864]*: the year is given by reference to Tyndall receiving the RS's Rumford Medal and Darwin the Copley Medal.
2. *Athenaeum*: see letter 1907, n. 7.
3. *award to you of the Rumford medal*: the Rumford Medal is bestowed by the RS every alternating year for 'an outstandingly important recent discovery in the field of thermal or optical properties of matter made by a scientist working in Europe'. Tyndall received the medal in 1864 'for his researches on the Absorption and Radiation of Heat by Gases and Vapours' ('November 30, 1864. Anniversary Meeting', *Roy. Soc. Proc.*, 13 (1864), pp. 494–520, on pp. 515–7).
4. *Copley Medal to Darwin*: the Copley Medal, the RS's oldest and most prestigious award, is bestowed to its recipient for outstanding achievements in any branch of science. Darwin received the Copley Medal in 1864 'for his important researches in Geology, Zoology, and Botanical Physiology' ('November 30, 1864. Anniversary Meeting', *Roy. Soc. Proc.*, 13 (1864), pp. 494–520, on pp. 505–10).

To C. K. Akin [7 November 1864][1] 2170

I have now only to add, that when you have so far overcome the agitation of mind to which you so frequently refer, as to be able to look at this question with the eyes and mind of a philosopher, I shall be happy to communicate with you. But no possible good can arise from a continuance of this correspondence. Hence, as far as I am concerned, it is for the present at an end.

J. Tyndall, 'On the History of Calorescence', *Phil. Mag.*, 29 (1865), pp. 218–31, on p. 229

---

1. *[7 November 1864]*: this letter is incomplete. In his essay 'On the History of Calorescence', Tyndall noted that this letter preceded a reply from Akin, 'dated from Harley Street, November 9, 1864', which was, Tyndall noted, Akin's fifth letter to him (p. 229). Tyndall noted that it was three weeks before he heard from Akin again, receiving two letters in close succession, one twenty-four hours prior to, and the other on the same day that Akin's 'Note on Ray-transmutation' appeared in the December number of the *Phil. Mag.* All Tyndall said of these letters, which have not been found, is 'he wrote to me commencing his letter with "Dear Sir," and ending it with "Believe me yours very faithfully."' Tyndall quoted this extract in his recounting of his dispute with Akin in 'On the History of Calorescence,' *Phil. Mag.*, 4, 29 (1865), pp. 218–31, on p. 229. See letter 2162, n. 2.

To Annette Roche 7 November 1864 2171

The Athenaeum.[1] | 7th. Nov. 1864.

My dear Miss Roche

The thought of writing to you has occurred to me many times during the last six weeks; but my work during this period has so absorbed my energies, that my correspondence generally has suffered greatly.

I have been able to put things in train with Mr. Diegan.[2] We have made a definite agreement and you ought to be informed that it is out of respect to you that Mr. Diegan agrees to include in his estimate the transfer of the tomb to old Leighlin[3] and its proper erection there.

I ought to be in good spirits this week, for only a few days ago the Council of the Royal Society voted me a big medal[4] for my recent scientific researches. It consists of a lump of gold worth sixty pounds; and some separate and additional lumps of the same total value—So that if I should fall into poverty I can pledge the medal, and thus keep my head above water for a time.

A fortnight ago I flew down for a day to the Isle of Wight.[5] I spent an evening with Tennyson, and on the following day he accompanied me in a climbing excursion up the chalk cliffs. He leads a very retired life, as he does not like the gaze of the populace; but I have always found him very genial.

I wish you would help me in this little matter. There is such an appetite at present prevalent for news regarding every man of the least note in the world, that every such person is sure to be talked about and written about, and is often doomed to be very stupidly talked about. I wish you would get from my mother[6] some account of her grandfather and grandmother.[7] The latter I know was a quaker and the daughter of a wealthy man, but she was read out of the meeting for marrying beyond the pale of the Society I want to get some authentic account of both her and her husband. His name I think was Malone. I wish I could learn where any documentary evidence regarding this family is to be found.

I have succeeded in fixing the approximate time when the first Tyndall's went to Ireland: if one could only be allowed to live and die in peace, without raking up such matters, it would be a blessing. But the curious world will not let one work in peace; and it will have nonsense if it cannot have truth.

Believe me | Yours very sincerely | John Tyndall.

RI MS JT/1/TYP/11/3851
LT Typescript Only

---

1. *The Athenaeum*: see letter 1907, n. 7.
2. *Mr. Diegan*: the sculptor Tyndall hired to create the tombstone for his father (see letters 2121 and 2142).
3. *Leighlin*: Leighlinbridge, County Carlow, Ireland, where Tyndall was born and his mother lived.
4. *a big medal*: Tyndall received the RS's Rumford Medal in 1864 (see letter 2169, n. 3).
5. *Isle of Wight*: see letters 2154 and 2155 for Tyndall's plans to visit the Pollocks and Tennysons on the Isle of Wight.
6. *my mother*: Sarah Tyndall.
7. *her grandfather and grandmother*: Sarah Tyndall descended from a prosperous farming family with lands in Ballybrommell, County Carlow, that included in earlier generations Quakers by the name Malone (see G. Cantor and G. Dawson (eds), *The Correspondence of John Tyndall, Volume 1: The Correspondence, May 1840–August 1843* (Pittsburgh: University of Pittsburgh Press, 2016)). She wrote to Tyndall in October 1858 (in a missing letter but from an extract in Tyndall's journal for 1871: 'My grandmother's family were people of wealth & property. their name was Malone, but my grandmother marrying out of meeting she was denied by them. Her father left three townlands one to each of her brothers—Ballybromwell to William, Ballintrain to Thomas, and Killkay to Pim. I believe that is Nehemiah, and to my grandmother one angry shilling This however never altered

the affections of her husband towards her, and they had abundance while they lived. there were no such farms in the country as he kept. They had one son & that was my father and she reared him as she was reared herself as a Quaker' (see letter 1569, n. 8, *Tyndall Correspondence*, vol. 6).

From Heinrich Debus 9 November 1864 2172

Queenwood Coll[1] | Nov[em]b[e]r 9. 1864 | Tuesday.

My dear John,

I should be delighted to be present at the anniversary meeting of the Royal Soc and see you receive your well merited medal.[2] It is a very graceful act on the part of the Society to bestow this medal on you, because I know no other scientific man who has a greater claim for this distinction than you possess. Count Rumford[3] himself if he was still alive would have, I firmly believe, selected you with delight as the recipient of his mark of honour. I think there is no investigator at the present moment in existence who has laboured so patiently, so successfully, in Rumford's own scientific field as you have done. Even if I was not your friend I should like to be present at the bestowale[4] of a mark of honour on a man who has long ago deserved it. For it is well and becoming that such acts should be surrounded by all the solemnity at the command of earnest men and in the presence of all who are linked together by the great cause of science. My duties at Queenwood require my presence on Thursday morning at 9 o'clock, require me in fact the whole day as well as the following Friday. It is true I might get dispensation from this duty and accordingly I have been wavering about the course I ought to follow. After due consideration, remembering my late frequent visits to London, I have come to the conclusion that I ought to remain here and do my duties just as I have accepted to perform them. Under these circumstances it only remains for me to give you my hearty <u>congratulations</u> on this memorable occasion by letter and to express the hope that I may soon have the pleasure to meet you.

With my love to Tom—| Your ever H Debus

RI MS JT/1/D/35

---

1. *Queenwood Coll*: see letter 2051, n. 4.
2. *your well merited medal*: Tyndall received the RS's Rumford Medal in 1864 (see letter 2169, n. 3).
3. *Count Rumford*: Benjamin Thompson (see letter 1920, n. 6).
4. *bestowale*: bestowal.

From George Busk 10 November 1864 2173

Burlington House[1] | Nov 10th 1864

My dear Tyndall,

Being as you properly observe both civil & obliging I have looked over the attendance at the Club[2] since Oct 1863 & find that the two luminaries Tyndall & Stokes were in conjunction at the subjoined dates. As to any phenomena attending this conjunction I find no record, which is very strange—

Nov 26th

Feb 25th

The Comet Tyndall was also present on the

29th Oct 1863

26th May 1864 & the Planet Stokes on March 17th 1864

Hoping this may be satisfactory I am—

Yours devotedly | G. Busk

RI MS JT/1/B/162
RI MS JT/1/TYP/1/178

1. *Burlington House*: location of the RS.
2. *the Club*: the Philosophical Club of the RS (see letter 2045, n. 3). The significance of this letter is that it appears to be a reply to a question from Tyndall regarding when he and Stokes had both attended the Philosophical Club on the same evening (Busk had been Treasurer of the club since 1863). In the context of his ongoing debate with Akin, it can only be that he was trying to date the sequence of events regarding when he had first heard of Akin's interest in ray-transmission and his discussion of the same with Stokes. Tyndall wrote a history of his interactions with Akin in his article 'On the History of Calorescence', *Phil. Mag.*, 29 (1865), pp. 218–31. See letter 2162, n. 2.

From Edward Sabine 11 November 1864 2174

The Royal Society | Burlington House | W. | Nov. 11 1864

Dear Tyndall

Mr. John Ball is desirous of joining the Royal Society;[1] and I am sure that he would be greatly pleased if you would draw up his certificate & propose him. He stands I suppose in the highest place in Alpine Botany: he is generally extremely well learned in Alpine Geology; and has been a great promoter of attention of physical research amongst the Alpine Travellers. —Amongst those who know him well & would be delighted to sign his certificate are

Lord Dunraven,[2] and Sir Thomas Larcom;[3] to whom I could send the certificate if you were too busy.

Always truly yours | Edward Sabine.

RI MS JT/2/10/297

---

1. *desirous of joining the Royal Society*: Ball would be elected FRS on 4 June 1868 ('Annual Meeting', *Roy. Soc. Proc.*, 16 (1868), p. 419). Tyndall did sign the certificate. See letter 2175.
2. *Lord Dunraven*: Edwin Wyndham-Quin (see letter 2018, n. 1).
3. *Thomas Larcom*: Thomas Aiskew Larcom (1801–79), 1st Baronet and leading official in the early Irish Ordnance Survey (*ODNB*). He was elected FRS in 1844.

To Edward Sabine[1] 13 November 1864 2175

13th November 1864

My dear General

It is some years since I thought of Ball's[2] joining the Royal Society, but his nearest friends then expressed doubt about the matter. I then privately hoped that long before the year 1864 he would have accomplished something to render his election an act of necessary justice, but he has not done so. He has splendid talents and abundant leisure, and still he will not converge himself on any definite piece of work and complete it. You say he is about the highest Alpine botanist; I believe he is, but there his knowledge lies, in his own brain.

He can make this knowledge public & he ought to be urged to do it. A man of his popularity and attainments, backed by a small amount of original work, bearing the mark of thought and labour, will be elected by acclamation.

I will sign his certificate if he be brought forward,[3] but I am sure you would not counsel me to be the leader in a movement to which I could only give half my heart.

Believe me dear friend | Yours ever truly | John Tyndall

RI MS JT/2/10/298

---

1. This letter is a draft; the sent version is missing.
2. *Ball's*: John Ball.
3. *if he be brought forward*: see letter 2174, n. 1.

From John Frederick William Herschel 13 November 1864 2176

Collingwood Nov. 13 1864.

My dear Sir

Many thanks for your highly interesting paper on Luminous and obscure radiation.[1] I am very glad to see that you propose dividing some share of your attention to the extra spectral calorific rays which lie out very far beyond the extreme visible red as indicated in the note to my paper of 1840[2] to which you refer. And there are some further facts in that of 1842 §.205.[3] in connexion with this which seem to indicate a continuous extension of the spectrum in this direction not due to merely calorific rays. It might be worth while to try & trace the visibility of these rays. That of the chemical rays to an almost indefinite extent beyond the violet is clearly proved by the exp[erimen]ᵗ. recorded on Art. 56 of my 1ˢᵗ. paper.[4]

The dynamical difficulties is in the way of a change of period of vibrations of a ray (such as Prof. Stokes has adopted for his explanation of fluorescence and such as Dʳ Akin has used to explain the incandescence of platinum wire in hydrogen flame) seem to me very great in the face of a theorem which as I fancy I have pretty rigorously demonstrated in my treatise on sound[5] § 393 et seq. on the law of forced vibrations which seems quite to preclude such a change at least except through the intermedium of some indirect mode of action to me not easily imaginable. —Stokes, to whom I once mentioned this objection tried to see over it but I confess, as it appeared to me very unsatisfactorily. I once thought I could see my way to such a change by regarding the new vibration as some sort of secondary or tertiary harmonies (harmonies of harmonies) of the original—but nothing came of it.

As regards the incandescence of platinum[6] any metal or other body very violently heated discharges white light—or at least rays ranging over all the spectrum more or less interrupted—for so far as common observation goes does it seem to matter at all how the heat is communicated to it—If a platinum wire be held in the flame of any burning body not containing hydrogen (suppose carbonated of sulphur) urged by an oxygen fan blowpipe—I suppose it would become hot, 1ˢᵗ observing then red—then white hot, just as in the flame of hydrogen.

Your transcalescence of Iodine is very extraordinary—and as you have now the means of insulating the calorific rays would it not be possible to impress on moistened black paper an extension to make visible their rays of interference, so as actually to measure the lengths of their vibrations?

NB. In the propositions communicated in §. 26, 27 of your paper[7] ought not the word “calorific” or “thermal” to be introduced into the enunciation before the word "radiation?"—The pile used to measure the deflection is sensible only to such radiations.

I have written a very rambling note but I hope you will attribute it to the great interest excited by your beautiful experiments & believe me

Yours v[er]y t[rul]y | J F W Herschel

RS HS 394b
RI MS JT/1/TYP/2/529–30

---

1. *paper on Luminous and obscure radiation*: see letter 2154, n. 3.
2. *my paper of 1840*: J. Herschel, 'On the Distribution of the Calorific Rays in the Solar Spectrum', *Phil. Trans.*, 130 (1840): pp. 1–59.
3. *that of 1842 §.205*: see letter 1958, n. 3.
4. *my 1st. paper*: J. Herschel, 'On Light by Coloured Media, and on the Colours of the Prismatic Spectrum Exhibited by Certain Flames; With an Account of a Ready Mode of Determining the Absolute Dispersive Power of Any Medium by Direct Experiment', *Transactions of the Royal Society of Edinburgh*, 9 (1823), pp. 445–60.
5. *my treatise on sound*: J. Herschel, 'Treatises on Physical Astronomy, Light, and Sound,' in *Encyclopaedia metropolitana* (London and Glasgow: Richard Griffin and Company, 1834). The treatise on sound is on pp. 747–840, with the discussion on 'forced vibrations' on pp. 812–20.
6. *incandescence of platinum*: see letter 2162, n. 3.
7. *your paper*: see letter 2014, n. 2.

From John Frederick William Herschel — 14 November 1864 — 2177

Royal Institution | 14th Nov. 1864

My dear Sir John,

I wish I was at liberty to write freely on some of the points raised in your last letter[1] to me. But to my great regret this is not the case. Owing to one of those incidents[2] which ruffle the calm flow of science, I have voluntary renounced the idea of making any thing known, for some considerable time, regarding the experiments which I have recently executed on the change of refrangibility of the extra red rays.

There are other points however on which I am at the liberty to speak. I have devoted some time to the examination of the extra-red region with a view to your own results. I have operated with rock salt prisms, with flint glass prisms (from Munich) and with hollow prisms possessing rocksalt sides and filled with bisulphide of carbon. As yet I have failed to obtain the maxima & minima which revealed themselves on your thermo-graph

From this I inferred that the maxima & minima were of atmospheric origin, and I thought they might be caused to appear by sending the beam, before analysis, through a thin layer of water. The experiment was very successful.

I have tried several other liquids in the same way, but hitherto with negative results.

I have also interposed a number of gases in the path of the beam, but without success. The absorption by both liquids & gases appears to extend uniformly over the extra-red space. I have not, however, yet done with the subject, and should anything positive arise I will at once let you know.

With regards to § 26[3] I would remark that <u>all</u> the rays from the platinum are absorbed by the lampblack on the face of the pile; and all of them are therefore converted into heat. It is the <u>heat value</u> (if I may use the term) of the rays which is given by the deflection.

The other points to which you refer shall receive due attention.

Yours most truly | <u>John Tyndall</u>

RS HS 395
RI MS JT/1/TYP/2/531

1. *your last letter*: letter 2186.
2. *one of those incidents*: this is in reference to a dispute between C. K. Akin and Tyndall regarding a paper on ray transmutation Akin wanted published in the *Phil. Mag.* Between November 1864 and 1865, Tyndall did not write or publish on the subject. The ordeal is discussed in detail in J. Tyndall, 'On the History of Calorescence', *Phil. Mag.*, 29 (1865), pp. 218–31. See letter 2162, n. 2.
3. *§ 26*: see letter 2014, n. 2, and letter 2183.

From John Barlow 14 November 1864 2178

5 Berkeley St. | Nov[embe][r] 14th. 1864

My dear Tyndall.

I cannot express how rejoiced I am that the Council of the Royal Society have, by an almost unanimous award, conferred upon you the second distinction[1] which they have to bestow.

I earnestly trust that you will obtain the highest[2] as soon as you deserve it: you will not have long to wait.

M[rs]. Barlow[3] adds her cordial congratulations to mine—

Ever truly yours—John Barlow

RI MS JT/1/B/18
RI MS JT/1/TYP/1/143

---

1. *the second distinction*: the Rumford Medal is an award bestowed by the RS every alternating year for 'an outstandingly important recent discovery in the field of thermal or optical properties of matter made by a scientist working in Europe', second only to the Copley Medal. Tyndall received the Rumford Medal in 1864 'for his work on the Absorption and Radiation of Heat by Gases and Vapours' ('November 30, 1864. Anniversary Meeting', *Roy. Soc. Proc.*, 13 (1864), pp. 494–520, on pp. 515–7). By 'second distinction', Barlow was referring to Tyndall having previously been awarded, in 1853, the RS's Royal Medal 'for his paper on diamagnetism and magne-crystallic action, published in the Philosophical Magazine in 1851'. However, Tyndall declined the award, on account of some council members doubting his priority in the original research and insisting that Julius Plücker (1801–68) deserved the award instead (see letter 0828 in I. Hesketh and E. Sera-Shriar (eds), *The Correspondence of John Tyndall, Volume 4: The Correspondence, January 1853–December 1854* (Pittsburgh: University of Pittsburgh Press, 2018); and R. Jackson, 'John Tyndall and the Royal Medal that was never struck', *Notes and Records: The Royal Society Journal of the History of Science*, 68 (2014), pp. 151–64).
2. *obtain the highest*: Tyndall never did receive the RS's Copley Medal, which was bestowed every year for 'outstanding achievements in research in any branch of science'.
3. *Mrs Barlow*: Cecilia Anne Barlow (née Law, d. 1868), the daughter of a wealthy member of Parliament. They were married in 1824.

From John Frederick William Herschel 16 November 1864 2179

Collingwood | Nov. 16/64.

My dear Sir

I am much surprised at your failure to get my heat spots[1] α β χ δ ε. I have no idea that they can be of atmospheric origin—at least, if so, not arising from any variable or accidental states, as I have never failed to produce them whenever there was good sunshine.

I always imagined that they were due to some peculiarity in the scale of transcalescence of some one or more of the glasses used—and for that reason I had projected to get both lenses and prisms of Rock Salt but I never could procure a sufficiently large & clear piece to work a lens of sufficient diameter.

The prism I used was a Fraunhofer[2] one of Munich glass the lens an achromatic of 3 inches 1/8 clear aperture and about 24 inches focus by Cauche[3] and I never failed to get the spot nor did I ever perceive any difference in their relative magnitude other than might be fairly attributed to greater or less intensity of sunshine—or in their places in relation to the visible rays of the spectrum.

I have the glasses all still by me and if you would like to try them I will send them to you.

I have tried to photograph the spots permanently by using coloured alcoholic solutions and applying repeated washes so as to concentrate the coloring matter more on the dried or partially dried places than on the general ground—but could not succeed in getting any good results worth preserving—though there were faint images so imperfect–but not sufficiently disengaged from the general colour.

I remain | Dear Sir | Yours very truly | J. H. W. Herschel

RI MS JT/1/H/102
RS HS/396
RI MS JT/1/TYP/2/532

---

1. *your failure to get my heat spots*: see letter 2177.
2. *Fraunhofer*: a scientific instrument maker in Munich founded by Joseph von Fraunhofer (1787–1826), specializing in optics (M. W. Jackson, *Spectrum of Belief: Joseph von Fraunhofer and the Craft of Precision Optics* (Cambridge, MA: MIT Press, 2000)).
3. *Cauche*: Cauche and Bardou, French optical instrument makers.

From John Frederick William Herschel — 17 November 1864 — 2180

Royal Institution of Great Britain | 17th Nov. 1864

My dear Sir John

It was an oversight on my part not to mention to you[1] that sunshine is a luxury beyond my means—all the experiments which I have hitherto made have been executed with the electric light.[2]

Yours very sincerely | John Tyndall

I shall test the point more surely before I relinquish it.

Many thanks for your kind offer:[3] If all else should fail I shall gladly avail myself of it.

RS HS/397

---

1. *not to mention to you*: see letter 2177.
2. *sunshine is a luxury . . . the electric light*: see letter 2165, n. 3.
3. *for your kind offer*: Herschel had offered to lend Tyndall his lenses. See letter 2179.

To Henry Bence Jones 18 November 1864 2181

18th. Nov. 1864.

My dear friend

We are going to establish a sound healthy scientific organ and have purchased the Reader.[1]

I have taken a single hundred pound share.

There are only thirty-five such shares to be issued. The proprietary will therefore be small and we shall have the matter in our own hands—

Will you become a shareholder—say for a single share?

Yours ever | John Tyndall

RI MS JT/1/TYP/3/793
LT Typescript Only

1. *the Reader*: *The Reader: A Journal of Literature, Science, and Art* was a short-lived and London-based academic magazine. Tyndall, Huxley, Pollock, Lubbock, Spottiswoode, Herbert Spencer, Francis Galton, John Stuart Mill, and others purchased the paper in order to use it to advance modern science. The office for *The Reader* was located at 24 Tavistock St. in Westminster, London. William Frederick Pollock later recalled that Tyndall and Huxley were to be 'scientific editors', but that although significant work was involved from the start, he could not recall ever seeing Tyndall at the office of the paper (*Remembrances*, vol. 2, p. 129). *The Reader* would not last a year in their hands, having difficulties fincancially, with finding an audience, and the commitments of its editors. See the Introduction to this volume for more details, and letters 2184, 2186, 2187, 2189, 2193, 2196, 2200, 2202, and 2207. See also *Ascent of John Tyndall*, pp. 189–90; and A. Desmond, *Huxley: From Devil's Disciple to Evolution's High Priest* (Reading, MA: Perseus Books, 1994), pp. 330–1, 343.

From Heinrich Debus 18 November 1864 2182

Queenwood,[1] Stockbridge, Hants.[2] | Nov, 18th, 1864.

My dear Tyndall

On the 13th of December[3] 1863 I had a conversation with you on the refrangibility of heat emitted from different sources. You then described to me the beautiful experiments which you had made some time previously on the nature of heat radiated from hydrogen, carbonic oxide and other flames, having only one product of combustion, and the absorption of these kinds of heat by various gases and vapours. We conversed on the experiments of Prof Stokes, and on the possible conversion of long undulations into short ones.

You then proceeded to explain to me that the glowing of a platinum spiral in a hydrogen flame was an instance of such a change of period[4]

Yours very Truly | H. Debus.

RI MS JT/1/TYP/7/2399
LT Typescript Only

---

1. *Queenwood*: see letter 2051, n. 4.
2. *Hants.*: abbreviation for Hampshire.
3. *On the 13th of December*: in this letter Debus was clearly responding to a request from Tyndall for his recollection of the sequence of events regarding those of his experiments that were the subject of his dispute with the Hungarian physicist C. K. Akin. Tyndall sent it back to him as an enclosure with letter 2183, suggesting changes that he would like to make for the version he intended to publish. Note: the original handwritten letter is missing, and the text which appears here is the amended text which Tyndall enclosed back to Debus with his letter, 2183. See letter 2162, n. 2.
4. *You then proceeded . . . of such a change of period*: this sentence is clearly that which Tyndall substituted for a longer explanation than Debus had originally supplied. As noted above, the original handwritten version of this letter is missing.

To Heinrich Debus 18 November 1864 2183

Queenwood,[1] Stockbridge, Hants.[2] | Nov. 18th. 1864.

My dear Heinrich

I have struck out a few unnecessary sentences in your note,[3] and thus shortened it: You see no objection to my publishing it in its present condition?

Yours ever | John Tyndall

[enclosure]

My dear Tyndall

On the 13th of December 1863[4] I had a conversation with you on the refrangibility of heat emitted from different sources. You then described to me the beautiful experiments which you had made some time previously on the nature of heat radiated from hydrogen, carbonic oxide and other flames, having only one product of combustion, and the absorption of these kinds of heat by various gases and vapours. We conversed on the experiments of Prof Stokes, and on the possible conversion of long undulations into short ones.

You then proceeded to explain to me that the glowing of a platinum spiral in a hydrogen flame was an instance of such a change of period.[5]

Yours very Truly | H. Debus

RI MS JT/1/TYP/7/2399
LT Typescript Only

1. *Queenwood*: see letter 2051, n. 4.
2. *Hants.*: abbreviation for Hampshire.
3. *your note*: letter 2182. See also letter 2162, n. 2.
4. *On the 13th of December 1863*: see letter 2182, n. 3.
5. *You then proceeded . . . of such a change of period*: see letter 2182, n. 4

From John Frederick William Herschel 18 November 1864 2184

Royal Institution of Great Britain | 18th Nov. 1864

My dear Sir John

We sadly need a good weekly scientific organ, and the 'Reader'[1] which has just become purchased from its present proprietors promises to satisfy this want.

It is exceedingly well supported.

Would you object to adding your name to the list of shareholders. You may do so without any risk simply by capitalizing your contributions.

I myself have taken a single £100 share, but others (Huxley, for example) take no shares but are shareholders in virtue of their contributions.

There never was a better opportunity for establishing a healthy weekly scientific organ.

Believe me | Yours very truly | John Tyndall

RS HS/398a

1. *the 'Reader'*: see letter 2181, n. 1.

To George Gabriel Stokes 21 November 1864 2185

21st. Nov. 1864

My dear Sir

A large amount of patient labour is invested in the paper of Sir William Snow Harris[1] referred to me by the Council. Considerable skill in the mention and use of apparatus is moreover manifested. On account of these and the superior qualities evidenced in the paper, it would give me a sincere pleasure to be able to recommend it for publication in the Philosophical Transactions.

It is a source of regret to me that I am not able to do so. I am, however, not satisfied with the reasoning and the experiments of the author. It appears to me that he is mistaken in his interpretation of some of the fundamental laws of electricity, and that the mistake extends to the experimental arrangements, by which he proposes to demonstrate these laws

I should however be extremely sorry if the Council attached too much weight to my opinion on this matter: and, I trust, if the second referee should differ from me in opinion and recommend the publication of the paper, that the council will lean to his decision rather than to mine. The results, considered empirically, may be useful, and the apparatus is capable of interesting applications.

Yours very truly | John Tyndall

Prof Stokes | Sec. R.S. &. &.

Report for Sir W. S. Harris's paper On the laws & operation of Electrical Force.[2]

[in verso] 1864 | Harris | by Tyndall

RS RR/5/99 Tyndall 21 November 1864

---

1. *the paper of Sir William Snow Harris*: Harris's paper was not published in the *Phil. Trans.*, but an abstract had been published earlier in 1864 in the RS's proceedings: W. S. Harris, 'Further Inquiries concerning the Laws and Operation of Electrical Force', *Roy. Soc. Proc.*, 13 (1864), pp. 364–75. This was also published as: W. S. Harris, 'Further Inquiries concerning the Laws and Operation of Electrical Force', *Phil. Mag.*, 29 (1865), pp. 65–75.
2. *Force*: this line is written at the top left of the first page.

From John Frederick William Herschel 21 November 1864 2186[1]

Collingwood | Nov. 21. 1864

My dear Sir

I wish you & your coadjutors every success in your undertaking[2] and I wish it could be accomplished to unite the best *[matter]* which crops up, now being in the form of Memoirs presented to the great Scientific bodies of the country into one good monthly or bimonthly Journal of Science—and that I fear is hopeless. —As regards my cooperation in the way of taking shares in it whether paid up or as you suggest capitalised by contributions I am sorry to say I must decline it. I have no funds available for the one trade and have no inclination to pledge myself either expressly or implicitly to the permanent support of a periodical in the other.

Wishing I could give you a more satisfactory reply I remain | my dear Sir | Yours very truly | John Herschel

RS HS/398b

---

1. 'Draft copy' appears at the top of the letter in Herschel's hand; the original sent version is missing.
2. *your undertaking*: see letter 2184, in which Tyndall asked Herschel if he wished to take part in the ownership of *The Reader*. On *The Reader*, see letter 2181, n. 1.

From Henry Bence Jones 23 November 1864 2187

Brook Street, London

My dear Friend

I have not had a moment in which I could write to you until now.

I will share your venture[1] to the extent of £100

Yours most truly | H. Bence Jones

31 Brook St | Nov 23/64

RI MS JT/1/J/55
RI MS JT/1/TYP/2/686

---

1. *your venture*: in letter 2181, Tyndall had asked whether Henry Bence Jones would be interested in supporting *The Reader: A Journal of Literature, Science, and Art,* with a view to making it a vital organ for scientific articles. Tyndall had invested £100.00 and asked whether Jones would make a similar commitment. On *The Reader*, see letter 2181, n. 1.

To John Amory Lowell[1] [November 1864][2] 2188

Replied as follows:—[3]

Dear Sir

Your name and Institute[4] have long been household words with me, and the thoughts which I associate with both enhance the gratification which your invitation would in any case have given me.

It would be both an honor and a pleasure to me to lecture before the Lowell Institute, and it is a source of regret to me that my studies at present prevent me from closing with your offer.[5] I am however committed to a line of research which claims all my attention, and I could not, without seriously invading my duty towards this investigation undertake to lecture in America next year.

With much esteem | I am dear Sir | Yours very sincerely | John Tyndall
John Amory Lowell Esq[ui]^{r}[e] | &c &c &c

RI MS JT/2/10/301

1. *John Amory Lowell*: see letter 2160, n. 1.
2. *[November 1864]*: the month and year are given by relation to letter 2160 from America (dated 25 October 1864), of which this is a reply.
3. *Replied as follows*: this transcription taken from Tyndall's record of this correspondence.
4. *Institute*: see letter 2160, n. 4.
5. *prevent me from closing with your offer*: see letter 2160, n. 3.

From William Garland Barrett[1] 7 December 1864 2189

17 Cobden Road. | Brighton. | December 7. 1864.

My dear Sir.

Through my son[2] I have heard of the kindly interest you have taken in my present position, and your mention of my name in connexion with some vacancy in the literary department of the 'Reader.'[3]

If I understand my son aright the department of travel and of literature in the Reader is at present unoccupied. This I could undertake with full consciousness of my ability to fulfil its requirements.

Myself a rather extensive traveller, having spent 14 years in various parts of the West Indies and South America, journeyed through Holland, Belgium and part of France, and having very recently returned from Australia where I went into the far interior. These antecedents would enable me to speak from a practical standpoint on matters connected with physical and political geography.

For many years I have been connected with the press, and I beg to enclose a few testimonials received by me when I was a candidate for another position of a literary character. The shortness of the time since my son spoke to me on this subject, prevents my getting fresh ones, but in their place I send the Eclectic Review,[4] and St. James's Magazine,[5] both for this month in each of which are articles by myself. In the first on "New South Wales,"[6] and in the second on "Old and odd Inn signs."[7]

Should I obtain the appointment you may rely upon my throwing all my energies into it, and on my unremitting attention to the required duties.

I remain, | my dear Sir, | Yours respectfully | W[illiam]: G[arland]: Barrett.
D^{r}. Tyndall F.R.S. —

UCL-03—UCL-04b

1. *William Garland Barrett*: William Garland Barrett (1812–65), an English amateur naturalist and Congregationalist minister. As a member of the London Missionary Society he had formerly run a station to aid emancipated slaves in Kingston, Jamaica, before returning to live in Herefordshire in 1848 (*ODNB*).
2. *my son*: William Fletcher Barrett.
3. *the 'Reader'*: see letter 2181, n. 1.
4. *Eclectic Review*: a journal founded in 1805 in order to raise awareness of the role which literature plays 'in obstructing or in accelerating the progress of religious truth and human happiness' (Wellesley Index to Victorian Periodicals, 1824–1900).
5. *St. James's Magazine*: a long-running literary magazine.
6. *"New South Wales"*: possibly 'New South Wales', *Eclectic Review*, 7 (1864), pp. 626–36.
7. *"Old and odd Inn signs"*: probably 'Old Inn Signs and Old Saws', *St. James's Magazine*, 12 (1864), pp. 103–6 (the article is signed only with a 'B').

From William Frederick Pollock — 9 December 1864 — 2190

9th. Dec. / 64

Dear T.

I have looked at the Phil Mag Suppl.

From what A.[1] writes at p. 560 I hold him to be an unmitigated s—nd—l.[2]

When he wrote it he must have known from yourself that you had actually made the discovery in question.

Y[ou]rs &c. | W. F. P.

RI MS JT/1/P/234
RI MS JT/1/TYP/6/2034

1. *A.*: C. K. Akin. The reference is to C. K. Akin, 'Note on Ray-Transmutation', *Phil. Mag.*, 28 (1864): pp. 554–60. On p. 560, Akin discussed the closure of his experiments on spectral rays and states that any claims which Tyndall might make on the subject in the future are illegitimate. See letter 2162, n. 2.
2. *s—nd—l*: scoundrel.

From Rudolf Clausius 13 December 1864 2191

Zürich, 13/12/64

Lieber Tyndall,

Schon lange drängt es mich, Dir wieder zu schreiben, und Dir meinen herzlichen Dank zu sagen für die Freundlichkeit, mit der Du die Pathenschaft bei unserm Söhnchen übernommen hast. Die Taufe hat, wie es festgesetzt war, am 23sten October stattgefunden, und Du (vertreten durch mich) und Frau Kenngott habt als Pathen fungirt. Wenn der Knabe erst grösser ist und ein Verständniss für wissenschaftlichen und Mannes Werth hat, so wird er stolz auf seinen Pathen sein. Meine Frau wird Dir ebenfalls selbst Ihren Dank aussprechen.

Ganz vor Kurzem hatten wir einen Brief von Frau Poggendorff. Sie schreibt, dass ihnen ihre Reise sehr gut bekommen ist. Von hier waren sie auf etwa 3 Wochen an den Genfer See gegangen, dann haben sie die Naturforscher-Versammlung in Giessen besucht, und endlich haben sie noch mit Weber zusammen eine kleine Rheinreise gemacht. Ihr Aufenthalt in Zürich scheint ihnen viel Vergnügen gemacht zu haben. Besonders hebt Frau Poggendorff hervor, wie sie sich gefreut haben, einmal wieder mit dem "Gletscher-Kletterer" zusammenzutreffen.

Der junge Wynne ist hier angekommen und ist, wie ich glaube, gut untergebracht. Ich wusste nämlich, dass ein mir befreundeter Professor, Osenbrüggen, ein Jurist, eine Wohnung gemiethet hatte, die ihm zu gross war, und von der er zwei Zimmer wiederzuvermiethen wünschte, aber nur an solche junge Leute, welche ihm gut empfohlen waren. Von diesen beiden Zimmern hat nun der junge Wynne das eine für den mässigen Preis von 25 fr. für den Monat gemiethet, während das andere von einem jungen Holländer genommen ist. Er ist dadurch in einer anständigen Familie, in welcher ein sehr gutes Deutsch gesprochen wird, (was in schweizerischen Familien nicht der Fall ist) und ist doch in seiner Freiheit nicht beschränkt. Er besucht fleissig Vorlesungen, arbeitet im chemischen Laboratorium von Wislicenus und hat bei einem Bruder von Wislicenus deutsche Stunden. Wir haben ihn schon einige Male bei uns gehabt, z.B. am Tauftage, und er gefällt uns und allen die mit ihm zu thun haben sehr gut. Ich habe ihn auch aufgefordert, bei einigen Freunden von uns, die ihn sehr gern aufnehmen würden, Besuche zu machen. Er hat es auch zugesagt, hat aber bis jetzt, wie es scheint aus Schüchternheit, noch damit gezögert. Ich will es ihm nächstens wieder sagen.

Mein Buch hast du hoffentlich richtig erhalten. Ich schicke heute auch einen Brief an Francis ab.

Mit herzlichen Grüssen | Dein | Clausius.

Zurich, 13/12/64

Dear Tyndall,

I have long felt the urge to write to you[1] again and express my most sincere thanks for the kindness with which you accepted the godparenthood of our little son.[2] The christening took place, as had been scheduled, on the 23rd of October and you (represented by me) and Mrs Kenngott[3] acted as godparents. When the boy is older and knows to value science and a man, he will be proud of his godfather. My wife[4] will also communicate her thanks to you separately.[5]

We quite recently received a letter from Mrs Poggendorff.[6] She writes that their journey did them much good. From here, they went to Lake Geneva for some 3 weeks; then they visited the meeting of Natural Scientists[7] in Giessen,[8] and finally also made a short trip on the Rhine with Weber.[9] Their stay in Zurich seems to have pleased them very much. Mrs Poggendorff emphasizes in particular how much they enjoyed meeting the "glacier-climber" once again.

Young Wynne[10] has arrived here and is, I believe, well housed. Namely, I knew that a professor of my acquaintance, Osenbrüggen,[11] a professor of jurisprudence, had rented an apartment that was too large for him and he hoped to rent out two rooms, but only to such young people as were highly recommended to him. Young Wynne rented one of these two rooms for the reasonable rate of 25 fr. a month, while a young Dutchman took the other. In this way, Wynne is in a respectable family where very good German is spoken (which is not the case among Swiss families) and where his independence is nevertheless not restricted. He attends lectures assiduously, works in the chemical laboratory of Wislicenus[12] and has German lessons with a brother of Wislicenus's.[13] We have had him here a few times, on the day of the christening for instance, and we like him very much as does everyone who has to do with him. I have also encouraged him to visit some friends of ours who would be very glad to receive him. This he has also agreed to but hesitated until now—due to shyness it seems. I will tell him once again next time.

I hope that you have received my book[14] in hand. I will also send a letter to Francis today.

With kind greetings | Your | Clausius

RI MS JT/1/TYP/7/2271–2
LT Typescript Only

---

1. *you*: Clausius here used the informal address form ('du'). See letter 2147, n. 1.
2. *godparenthood of our little son*: in letter 2147, Clausius had asked Tyndall to be godfather to his son, Rudolf John ('Johnny') Clausius. See also letters 2153 and 2192.
3. *Mrs Kenngott*: see letter 2147, n. 4.

4. *My wife*: Adelheid Clausius.
5. *her thanks to you separately*: letter 2192.
6. *Mrs Poggendorff*: Charlotte Eleonore Poggendorff.
7. *meeting of Natural Scientists in Giessen*: the Versammlung Deutscher Naturforscher und Ärtze, or Society of German Natural Scientists and Physicians, met in Giessen in September 1864.
8. *Giessen*: Giessen, a town in central Germany north of Frankfurt, well-known as a centre for the study of mathematics.
9. *Weber*: Wilhelm Eduard Weber (1804–91), a German physicist known for his work on the electromagnetic telegraph (*ADB*).
10. *Wynne*: Frank George Wynne.
11. *Osenbrüggen*: Eduard Osenbrüggen (1809–79), a German-Swiss legal scholar. He was a professor at both the University of Dorpat and the University of Zurich (*ADB*).
12. *Wislicenus*: Johannes Wislicenus (1835–1902), a German chemist most famous for his work in early stereochemistry. He ran a laboratory at the University of Zurich (*CDSB*).
13. *brother of Wislicenus's*: not identified.
14. *my book*: see letter 2093, n. 7.

From Adelheid Clausius 13 December 1864 2192

Zürich, 13. Dec.[1] 64

Hochgeehrte Herr Tyndall!

Auch ich kann es nicht unterlassen Ihnen meinen herzlichen Dank auszusprechen, dass Sie die Pathenstelle bei unserem kleinen Jungen angenommen haben; Sie haben uns dadurch eine recht grosse Freude gemacht, und wenn unser kleiner Johnny, den wir Rudolph John haben taufen lassen, den beiden Namen einmal Ehre machen sollte, und die guten Eigenschaften von Ihnen und von seinem Papa vereinigt, dann kann ich gewiss dem lieben Gott nicht dankbar genug dafür sein. Unser kleiner Johnny gedeiht, Gott sei Dank! bis jetzt allerliebst, und soll seinen Papa recht ähnlich sein, von dem er sogar schon die gelehrten Falten über der Nase hat. Ja, es ist gewiss wahr, solch ein Kind ist eine Gottesgabe, und giebt dem Leben einen wunderbaren Reiz. Heute sind es gerade fünf Jahre, dass unsere Hochzeit gewesen ist, und an solch einem Tage empfindet man das Glück, das einem gegeben ward, noch mehr wie sonst, und wünscht es allen denen, für die man sich interessiert, wie innig würden wir uns freuen, wenn auch Sie sich einmal bezähmen liessen.

Ihr junger Freund, Mr Wynne gefällt hier allgemein durch sein offenes, natürliches Wesen. Ich hoffe, wir werden ihn recht oft bei uns sehen, er hat mir auch erzählt dass Sie so fürchterlich immer arbeiten, was mein Mann mir auch schon oft gesagt hat. Sie werden sich gewiss wieder nächsten Sommer hier in der Schweiz stärken und erholen müssen, und wenn Sie denn auch dabei wieder besuchen, und ihr Pathchen kennen lernen, so wäre das allerliebst; bis jetzt ist es noch ungefährlich, da können Sie unsern Johnny noch nicht auf die Berge nehmen, ob Sie ihn also in 15 oder 20 Jahren mit auf Ihre halsabrechenden Gletscherfahrten nehmen, das wollen wir uns denn doch später noch überlegen.

Mit herzlichen Grüssenn | Adelheid Clausius

Zurich 13. Dec. 64

Honoured Herr Tyndall!

I cannot refrain from thanking you[2] from the bottom of my heart that you have agreed to be godfather to our little boy.[3] You have made us very happy, and if our little Johnny, who we have baptized Rudolph John, is one day an honor to both names, and combines your and his father's good qualities, I will not be able to thank dear God enough. Our little Johnny is thriving, thank God!, until now wonderfully, and is said to quite resemble his father, from whom he even already has the scholarly wrinkles above his nose. Yes, it is true, such a child is a gift from God, and makes life so wonderfully appealing. Today, exactly five years have passed since our marriage, and on such days one feels the happiness that has been granted even more than usual and wishes it for all whom one cares about, how deeply we would rejoice, if you let yourself be tamed eventually.

Your young friend Mr. Wynne[4] is generally liked here for his open, natural character. I hope that we will see him often at our home, he also told me that you always work so terribly much, which my husband as well has told me often enough. You will certainly need to regenerate and gather strength again here in Switzerland next summer, and it would be wonderful if you came to visit again and got to know your godson; so far it is not at all dangerous, since you cannot take our Johnny into the mountains yet. Whether you take him in 15 or 20 years on your breakneck glacier tours, that is something that we will consider later.

With fond regards | Adelheid Clausius

RI MS JT/1/TYP/7/2270
LT Typescript Only

1. *Dec.*: the typescript of this letter (the original is missing) showed 'Nov.' for the month, but as this letter surely follows letter 2191, dated 13 December 1864, we believe the original transcription of 'Nov.' was in error.
2. *you*: here Adelheid Clausius used the formal address form 'Sie'.
3. *godfather to our little boy*: Rudolf John ('Johnny') Clausius; see letters 2147, 2153, and 2191.
4. *Mr. Wynne*: Frank George Wynne.

From William Frederick Pollock — 16 December 1864 — 2193

59, Montagu Square, | London, W.[1] | 16/Dec/64

Mon Cher Philosophe,

It is capital[2]—clear, strong & temperate—& the Reader[3] should be proud of it. It is not, I observe, quite finished—but only let it end, as it begins, & continues,—& nothing c[oul][d] be better—Y[ou][rs] | W. F. P.

RI MS JT/1/P/235
RI MS JT/1/TYP/6/2035

1. *59, Montagu Square, | London, W.*: see letter 1954, n. 2.
2. *It is capital*: probably a reference to J. Tyndall, 'Science and Spirits', *The Reader*, 4 (1864), pp. 725–6.
3. *the Reader*: see letter 2181, n. 1.

From George Busk — 17 December 1864 — 2194

15 Harley St. | Dec. 17th 1864

Divine Phoebus,[1]

I am desolated in being compelled to deprive myself of the pleasure and profit of accompanying you in your intended circuit[2] seeing that I am at present pretty completely tied up. But I do hope when you pervade the Pass of Llanberis[3] you will remember me with gratitude as its discoverer. I don't doubt that is & must be the direction to which your feet will not to say nose will be directed—but if not it ought to be, especially as it will be snowed up.

With the utmost reverence | Your worshipper | G. B.

P.S. It is a shame however to leave the Phil. Club[4] in the lurch—what shall I do for my precious 14,[5] if you all go away to Wales.

P.S. I hope there will be somebody in my place to take care of you.

P.S. When at Beddgelert[6] count the suspicious ones. Six I should say.

P.S. Your best route if Welsh will be the rail to Machynlleth[7] and round by Tan y Bulch[8] and Beddgellert to Carnarvon[9] only tho you will miss the Pass.

RI MS JT/1/B/163
RI MS JT/1/TYP/1/179

1. *Divine Phoebus*: Phoebus (Apollo), God of light, philosophy, poetry, and music.
2. *I am desolated . . . your intended circuit*: presumably there is a letter missing, in which Tyndall shared with Busk his plans of going to Wales.

3. *Pass of Llanberis*: a mountain pass in Snowdonia, a mountainous region in northwestern Wales. In the event, Tyndall and Huxley went to Derbyshire, not Wales.
4. *Phil. Club*: the Philosophical Club of the RS (see letter 2045, n. 3). The next meeting of the club was held just five days after Busk wrote this letter, on 22 December. He presented on the animal and human fossil discoveries from Windmill Cave in Gibraltar by Joseph Frederick Brome (1814–70), a British archeologist and governor of the Military Prison in Gibraltar (T. G. Bonney, *Annals of the Philosophical Club of the Royal Society, written from its Minute Book* (London, Macmillan & Co., 1919), p. 169; A. Menez, 'Custodian of the Gibraltar Skill: The History of the Gibraltar Scientific Society', *Earth Sciences History*, 37 (2018), pp. 34–62; G. Busk and H. Falconer, 'On the Fossil Contents of the Genista Cave, Gibraltar', *Quarterly Journal of the Geological Society*, 21 (1865), pp. 364–70).
5. *my precious 14*: reference unclear.
6. *Beddgelert*: a village in Snowdonia, Gwynedd, Wales.
7. *Machynlleth*: a market town in northwestern Wales.
8. *Tan y Bulch*: in Welsh, Tan-y-Bwlch, is a railway station that serviced the Snowdonia region in Gwynedd, Wales.
9. *Carnarvon*: in Welsh, Caernarfon, is a small town in northwestern Wales.

From George Gabriel Stokes 17 December 1864 2195

Lensfield Cottage Cambridge | 17th Dec[embe]r 1864

My dear Tyndall,

I feel confident I never expressed an opinion as the impossibility of a raising of refrangibility as a general physical law. I expressed a strong opinion as to the generality of the law as applied to the phenomenon of internal dispersion (fluorescence) (See my paper[1] in the Phil. Trans. for 1862 Part II arts. 80, 229, 240) from the uniform result of my experience.

As to whether the rendering wholly luminous by a concentration of rays of non-luminous heat should or should not be included under the term fluorescence, is of course mainly a question of definition of terms. It seems to me better not to include it. For though there is this feature in common—that luminous rays are emitted in consequence of the action of incident non-luminous rays, of course of a different refrangibility—there is this striking difference, that in the one case is body is raised, by non-luminous heat, to a temperature at which bodies <Hereafter LT Typescript Only> in general become luminous, in the other body not so raised in temperature, a cold liquid for example, gives out it may be blue rays, an effect to be obtained by temperature only by going much beyond a r<adiant> heat.

Yours very truly | G. G. Stokes

RI MS JT/1/S/238
RI MS JT/1/TYP/4/1397

1. *my paper*: G. G. Stokes, 'On the Long Spectrum of Electric Light', *Phil. Trans.*, 152 (1862), pp. 599–619.

To Juliet Pollock [19 December 1864][1] 2196

Monday

My dear Mrs Pollock,

I have heard of you almost daily—but I am no less glad to hear from yourself.[2] Give my love to Walter;[3] and my best regards to Mrs Cameron[4]—I do so like her for taking such good care of you.

I joined your husband[5] yesterday at 2. You would be proud to know how dismal he is without you. We drove to Carlyle's and found the Seer[6] at the top of the house with a sky-light above him—I do so like and reverence the old fellow, and we get on so well together. He did not know me at first, for it is three or four years since I have seen him. But he warmed up and harangued us cheerfully.

We have been just at the Annual Meeting of the Reader[7]—and all went off happily.

Yours ever | John Tyndall

De Mussy[8] is a gentleman and has a warm heart

RI MS JT/1/T/1180
RI MS JT/1/TYP/6/2077

1. *[19 December 1864]*: the date is given by the content of the letter. Pollock recalled being approached by Tyndall and Huxley about *The Reader* (see letter 2181, n. 1) in late October 1864, and noted in his *Remembrances* that the first number under their direction appeared 10 December 1864, the paper being wound up after several months (*Remembrances*, vol. 2, pp. 128–31). In his journal entry for 16 December 1864, Tyndall noted that there were several *Reader* meetings across the previous two weeks, although there's no clear indication that any of these was the general meeting (RI MS JT/2/13c/1270). However, given that Hirst noted in his journal entry of 25 December 1864 that Tyndall was in London between 14 and 20 December, leaving with Lubbock and Huxley on that day to hike in Derbyshire, it seems likely that this letter was written between these dates, with 19 December falling on a Monday (Brock and Macleod, *Hirst Journals*, p. 1711).
2. *to hear from yourself*: letter missing.
3. *Walter*: Walter Pollock.

4. *Mrs Cameron*: see letter 2135, n. 6.
5. *your husband*: William Frederick Pollock.
6. *Seer*: a reference to Carlyle's novel *Sartor Resartus: The Life and Opinions of Herr Teufelsdrockh* (London: Chapman and Hall, 1831).
7. *Reader*: see letter 2181, n. 1.
8. *De Mussy*: Noël-Francois-Odon Guéneau de Mussy (1813–85), a French physician known for his contributions to internal medicine and his publication of *Clinique médical* (1874–85). Huxley recommended him to Spencer, who found that his recommendation of hot bath treatments helped with his insomnia (H. Spencer, *An Autobiography*, 2 vols (London: Williams and Norgate, 1904), vol. 2, p. 92).

From Edward Sabine [before 20 December 1864][1] 2197

Dear Tyndall,—Viewing your inquiry simply as an administrative question, Dr. Akin's Report on the Transmutation of Spectral Rays[2] in the British Association Reports for 1863 indicated a very wide field of experimental researches. In aid of these the General Committee made a grant of £45 to Mr. Griffith[3] and himself. A part of this grant was expended in apparatus to be employed by those gentlemen in the spring or early summer of 1864, the results of which were designed by Dr. Akin to be presented by himself to the Meeting of the British Association at Bath in the September of that year. For this purpose Dr. Akin went to Oxford to join Mr. Griffith, but from some unexplained cause their purpose was frustrated, and 1864 passed without the experiments being made. The grant, however, did not lapse, as part of it had been expended in apparatus, which it appears to be still be Dr. Akin's intention to employ (though not at Oxford where there is sunshine, or in London where there is no sunshine suitable for the purpose). Such, so far as I can collect them is a statement of the facts.

The object of the British Association in making such grants is to aid the progress of science, by enabling those, who may be so disposed, to make experiments, but by no means to retard its progress by discouraging others from approaching the same ground.[4] One cannot read, I think, Dr. Akin's Report in 1863 without the impression that there is work enough in the subject for a dozen at least of active experiments. He may have had an advantage (and properly used it is a great one) in mediating longer than others on the laurels that may be gathered, but he who desires to be foremost in the field must be 'prompt in action' where there are so many competitors.

Always truly yours, | Edward Sabine

J. Tyndall, 'On the History of Calorescence', *Phil. Mag.*, 29 (1865), pp. 218–31, on p. 222

---

1. *[before 20 December 1864]*: Tyndall dated this letter to late December, and it most likely was written before Tyndall's letter to William Francis (2198). This letter refers to the ongoing dispute between Tyndall and C. K. Akin. See letter 2162, n. 2.
2. *Report on the Transmutation of Spectral Rays*: C. K. Akin, 'On the Transmutation of Spectral Rays', *Brit. Assoc. Rep. 1863*, pp. 93–105.
3. *Mr. Griffith*: see letter 2165, n. 6.
4. *but by no means to retard its progress . . . the same ground*: Akin had apparently suggested to Tyndall that the grant from the BAAS gave him exclusive privileges in researching the subject of ray transmutation. See also letter 2208 from William Armstrong.

To William Francis [20 December 1864][1] 2198

Royal Institution of Great Britain | Tues<u>day</u> Ev[enin]g.

My dear Francis.

Would you think the following alterations of a word worth making?

I have said in one of my paragraphs: 'Hitherto Dr. Akin[2] has done nothing but read & write &c' for the word 'nothing' use the word <u>little</u>. I intended making this change this morning but it escaped my memory—Could it appear as an erratum?[3]

If not for this number perhaps you would kindly take a note of it for the next.

Yours very sincerely | <u>John Tyndall</u>

St Bride Library, Authors' letters

---

1. *[December 20th 1864]*: in his journal entry for 16 December 1864 Tyndall noted that he showed his intended reply to Akin to Huxley, Pollock, and Hirst. On Saturday, 17 December, he noted that he had sent it to the printer. The following Tuesday was 20 December 1864.
2. *Dr. Akin*: this letter is a part of the ongoing dispute between Tyndall and the Hungarian physicist C.K. Akin (see letter 2162, n. 2).
3. *I have said in one of my paragraphs . . . an erratum?*: in fact this change was made before the article went to press, see J. Tyndall, 'On the History of Negative Fluorescence', *Phil. Mag.*, 29 (1865), pp. 44–55 (this line appears on p. 49). This article, an important one in the dispute between Tyndall and Akin, in which Tyndall recounted his account of the history of their respective works, directly followed C. K. Akin's 'On Calcesence', *Phil. Mag.*, 29 (1865), pp. 28–43.

To Thomas Archer Hirst [21 December 1864][1] 2199

Baslow Derbyshire | Wednesday

My dear Tom

I have sent this day the final proof[2] to Mr Mattus.[3] Would you, like a good fellow, take a last look at it before your departure.[4] The proof contains the corrections which have already been made and the only new places are marked with a cross thus X. The printer is to compare this proof, now sent, at these points with the revise, and he will then see what alteration is needed. I have written to Mr Mattus, and given him the needful instructions, but I should feel more satisfied if the final revise had passed under your eyes.

Yours affectionately | John Tyndall

RI MS JT/5/12 (loose enclosed in a grey folder)
LT Typescript Only

1. *[21 December 1864]*: the date is given by relation to Tyndall's visit to Baslow. According to Hirst in his journal entry of 25 December 1864, Tyndall had left London on 20 December (Brock and Macleod, *Hirst Journals*, p. 1711).
2. *final proof*: probably J. Tyndall, 'On the History of Negative Fluorescence', *Phil. Mag.*, 29 (1865), pp. 44–55. See letter 2198.
3. *Mr Mattus*: not identified, but presumably an assistant to William Francis who published the *Phil. Mag.*
4. *before your departure*: Hirst noted in his journal for 25 December 1864 that 'On Sunday the 25th which was Christmas day I walked to Tyers Hill and arrived in time to join Thomas Burkitt and his pretty and graceful little daughter Ada Mary at dinner' (Brock and Macleod, *Hirst Journals*, p. 1711).

To William Frederick Pollock [24 December 1864][1] 2200

My dear Pollock

Huxley would have written you had it not been agreed upon between us that I should call upon you and make some arrangement regarding the publication of his article.[2] I did call last night, but learned that you were at St. Julian's[3]

Your letter to him, which I have read, makes me feel more that ever how necessary it will be for us to cultivate a habit of mutual accommodation. I read his article before it was printed, and I thought it excellent. It seemed somewhat hard on Disraeli;[4] but I thought the hardness, such as it was, merited. The gentlemen of whose works D. had spoken so flippantly were at least his equals in ability, and far more than his equals in philosophic insight and devotion to truth. I cannot help thinking that a man who goes out of his own line to treat grave questions and great reputations with ridicule merits rebuke. However, this is a point which but slightly affects the real merits of the article and I should be quite willing to go a good way to meet your wishes regarding it.

But the article in the main is, I think, excellent, and calculated to do our paper good. What the Reader has hitherto lacked is the evidence of any power in those who write for it, and I think we ought to be careful to preserve this great element among us; to secure the strength, as well as the grace, of humanity. So long as a contributor writes as a gentleman, I think the strong expression of strong conviction ought to be rather encouraged than repressed. Nothing breaks a man's spirit more than finding its utterance hampered by too many conditions. Of course friendly suggestion, counsel, and admonition ought always to be current amongst us; but we must not exact too much of each other, but let each express himself as much as possible in that particular way in which he feels his strength to lie.

Tom Taylor's article[5] is extremely good; but its excellence to my mind mainly arises from its perfect straightforwardness. Plain speaking backed by knowledge; this is our greatest need. Your own article[6] I thought full of interest. The treatment was unique, and the point chosen exactly calculated to bring out the merits of your author with reference to his compeers. I read it with extreme pleasure. But this power <*1–2 words missing*> need not interfere with my appreciation of Huxley's noble truthfulness and strength, and I think his article, with such modification of detail as we might agree upon, ought to appear next Saturday. Doubtless through a mistake on his part, he thought it was needed for the last number, and I know he put himself to inconvenience to write it

Yours ever | J. T.

RI MS JT/2/13c/1272–3

1. *[December 24th 1864]*: the date is suggested by LT's typescript of Tyndall's journal, and this fits with letter 2202, Pollock's reply of 27 December 1864.
2. *his article*: T. H. Huxley, 'Science and "Church Policy"', *The Reader*, 4 (1864), p. 821.
3. *St. Julian's*: see letter 1954, n. 14.
4. *Disraeli*: Benjamin Disraeli (1804–81), a British statesman of the Conservative Party who served twice as Prime Minister (in 1868 and from 1874–80) (*ODNB*).
5. *Tom Taylor's article*: not identified; articles in *The Reader* were unsigned so we are unable to determine which article this refers to.
6. *Your own article*: not identified; articles in *The Reader* were unsigned so we are unable to determine which article this refers to.

To William Francis 26 December 1864 2201

26th Dec | The Athenaeum[1]

My dear Francis

In order to be early at Red Lion Court[2] this morning I quitted my house before the post arrived. The consequence is that I did not get your queries until after 2. P.M. I immediately drove down to Red Lion Court; but found all locked up.

Your queries are I think judicious,[3] and I have so modified the parts referred to, that I think they will now pass muster.

I will call down tomorrow morning to see whether there is any thing else that strikes you as desirable to modify.

When I think of the splendid work which I might be now getting done, it makes my heart ache to have to waste my time in replying to Dr. Akin.[4] It might shorten matters if proofs were demanded for statements of a personal bearing. I had to stop my correspondence with Akin because it simply descended to an undignified series of assertions on his part and of denials upon mine.

Yours ever | John Tyndall

St Bride Library, Authors' letters

---

1. *The Athenaeum*: see letter 1907, n. 7.
2. *Red Lion Court*: historic building in London that served as a long-time home to printers; at this time it was the offices of the Taylor & Francis publishing house.
3. *Your queries are I think judicious*: Tyndall had written asking if it was possible to alter the text of an article he had submitted for press. As Francis here suggests, these alterations were made. See letter 2198, especially n. 3.
4. *Dr. Akin*: The reference is to the protracted dispute that Tyndall had with C.K. Akin. See letter 2162, n. 2.

From William Frederick Pollock 27 December 1864 2202

St Julians | Sevenoaks[1] | Tuesday 27th Dec. 1864

My dear Tyndall

Thanks for your letter[2]—I shall be in town tomorrow & Thursday—& will try to see you at the R.I. tomorrow about 2—from 11 to 1—I shall be at the Equitable[3]—& at after 3—at the Reader Office[4]—probably dining at the

Athenaeum[5]—& going to Lady Mounteagle's[6] in the evening—on Thursday I shall of course be assisting at our achievement of the paper at 24 Tavistock St[7]—as I shall see you tomorrow I will now say nothing about Huxley's article,[8] except that I hope we may have the benefit of it. He can easily retain all its power & remove any ground of objection—

I thought we had a good number last week—the theatrical notices of the plays[9] at Westminster[10] & Eton[11] were by "our own correspondent,"[12] a friend of yours, & scholar of Trin. Coll. Cam.[13] He was naturally at Eton—& I especially wrote to Stanley[14] to get him a ticket for the Westminster performance[15]—The Homeric review[16] (w[hic]h I am glad you like) is not mine—but written by another friend of yours[17]—The first notice was a joint affair—but this was all her's. I have good things coming on for this week & the next—for the present I foresee no difficulty in keeping the larder well filled. We are all well here—but want sun. With all kind regards from Juliet & the boys.[18]—
Yours ever | W. F. Pollock

RI MS JT/2/10/322

1. *St Julians | Sevenoaks*: see letter 1954, n. 14.
2. *Thanks for your letter*: letter 2200.
3. *the Equitable*: the English Court of the Chancery.
4. *the Reader Office*: Pollock had been approached by Tyndall and Huxley to join in the purchase of the *Reader* and for his literary help (see letter 2181, n. 1). He wrote in his journal: 'Their object was to provide a better record than existed of the progress of science, and of the labours and opinions of scientific men, with reports of the proceedings of scientific institutions, and space for correspondence' (*Remembrances*, vol. 2, pp. 128–9). The office for *The Reader* was located at 24 Tavistock St. in London.
5. *Athenaeum*: see letter 1907, n. 7.
6. *Lady Monteagle's*: Marianne Spring Rice (dates unknown), Lady Monteagle, wife of Thomas Spring Rice (1790–1866), 1st Baron Monteagle; daughter of the Leeds industrialist John Marshall. She was Spring Rice's second wife, married in 1841. She brought much needed money into the family, which allowed them to keep a house in London as well as Mount Trenchard House, in County Limerick, Ireland.
7. *24 Tavistock St*: the office of *The Reader* (see n. 3).
8. *Huxley's article*: T. H. Huxley, 'Science and "Church Policy"', *The Reader*, 4 (1864), p. 821. In it, Huxley had been critical of Disraeli, and Pollock had obviously objected. Tyndall thought Huxley's criticism merited, and was wary of too much editorial control over Huxley's combative style for the paper. Tyndall noted in his journal entry for 28 December that 'Pollock and Huxley have arranged matters pleasantly' (RI MS JT/2/13c/1273).
9. *theatrical notices of the plays*: 'Miscellanea', *The Reader*, 4 (1864), p. 801.
10. *Westminster*: Westminster School in London.

11. *Eton*: Eton College in Eton, Berkshire.
12. *"our own correspondent"*: possibly William Hepworth Thomson (1810–86), Professor of Greek at Trinity College. Thomson succeeded William Whewell as Master of Trinity in 1866, following Whewell's untimely death (*ODNB*). Tyndall would dine as his guest at Trinity in May 1865, the day after dining with young Frederick Pollock, who was studying there (RI MS JT/2/13c/1289).
13. *Trin. Coll. Cam*: Trinity College, Cambridge.
14. *Stanley*: Arthur Penrhyn Stanley (1815–81), an Anglican clergyman, educated at Rugby School and Oxford University (1837) where he obtained a fellowship. He accepted a canonry at Canterbury (1851) before becoming professor of ecclesiastical history at Oxford (1856). In 1863 he married Lady Augusta, one of Queen Victoria's women of the bedchamber. They were subsequently frequently guests at Windsor Castle. From 1864–81 he was Dean of Westminster, during which time he continued to publish lectures and sermons (*ODNB*).
15. *Westminster performance*: *Phormic*, a comedic play by the Roman playwright Terence (*c.* 195/185–*c.* 159? BC).
16. *Homeric review*: 'Lord Derby's Iliad', *The Reader*, 4 (1864), p. 793–4.
17. *another friend of yours*: Juliet Pollock (as indicated in letter 2203).
18. *boys*: Frederick, Walter, and Maurice Pollock.

From Juliet Pollock[1] 27 December 1864 2203

My dear Mr. Tyndall

you must have my *[truly]* good wishes for your happiness and all the best blessings that belong to such a nature as yours in my own hand writing to day I rejoyce in your approbation of my Homeric performance[2] for I took pains with it. I am very happy here in the house of my youth with the dear relations and aspirations of those days & the addition of my Husband and boys . . . and I think with warmth and gratitude of the friends beyond whom I value that of none more my dear Mr. Tyndall than yourself. Yours ever most truly | Juliet Pollock

RI MS JT/2/10/322

1. Juliet Pollock's letter is cross-written over her husband's letter to Tyndall (letter 2002).
2. *my Homeric performance*: see letter 2202, n. 16.

To William Francis 28 December 1864 2204

28th. Dec. 1864

My dear Francis

I am going to ask a last favour of you, in reference to our conversation of yesterday.[1]

You then drew my attention to a passage commencing thus:—'My first interview with Dr. Akin has not ended'—and terminating thus: 'Let me here illustrate the correctness of my conclusion.' I have since pondered your remarks and have come to the conclusion that they are of weightier moment than I first supposed them to be.[2]

Is it too late to omit that passage?[3]

I would gladly wear any expense that this alteration would render necessary. If the sheets are already marked off I would cheerfully pay for the reworking of them & every other charge incidental to the alteration.

Yours very sincerely | John Tyndall

I should regard it as an extremely friendly act on your part if you would arrange this matter for me.

I will call in at 11 o'C. just to hear your fiat.

St Bride Library, Authors' letters

---

1. *our conversation of yesterday*: Tyndall and Francis had met to discuss the editing of the text of Tyndall's 'On the History of Negative Fluorescence', *Phil. Mag.*, 29 (1865), pp. 28–43; see Tyndall's letter to Francis, 2201.
2. *You then drew my attention . . . supposed them to be*: Tyndall had been in dispute with C. K. Akin, and had clearly discussed the best phrasing with his publisher, Francis, at the meeting referred to. See letter 2162, n. 2.
3. *to omit that passage*: this passage was indeed omitted.

To Adelheid Clausius 31 December 1864 2205

31st. December 1864

My dear Mrs. Clausius

I have reproached myself over and over again for not replying to your letter.[1] But I was first tempted to delay my reply in order that I might write to you in German: afterwards I was taken away from London for a time,[2] and was otherwise so entangled in my work as to be unable to write to you as I wished to do. Let me now say that though in England I have always declined becoming a godfather, when the letter of your husband[3] came to me, asking me to stand in this relation to your first boy,[4] it gave me the deepest gratification. It proved to me that he cared something about me, and among my many friends there is not one whose friendship I rate higher than his.

I hope Johnny is strong and hearty; I hope he is able to cry for his bread and milk; and to make his mother attend properly to his wants. If the silversmith has done his duty the little man will receive, in company with this letter, a goblet, which you are to fill with the proper beverage, and present to the little man in order that he may drink his godfather's health.

This is the last day of the old year and before it speeds away let me wish you & your husband & Johnny and the little girls,[5] a happy new one.

Let me also offer you my very best thanks for the great kindness you have shown to my friend Frank Wynne.[6] You have been kind to the son of a noble father,[7] and of one of the most excellent mothers[8] that I have ever known: She is gone,[9] but the father, believe me, is grateful to you for all that you have done for his boy.

With best love to Johnny. | Believe me | Yours ever truly | John Tyndall

RI MS JT/1/T/168

1. *to your letter*: letter 2192.
2. *taken away from London for a time*: Hirst noted in his journal entry for 25 December 1864 that Tyndall left London with Huxley and Lubbock for a hiking tour in Derbyshire on 20 December (Brock and Macleod, *Hirst Journals*, p. 1711).
3. *letter of your husband*: Rudolf Clausius; letter 2147.
4. *first boy*: Rudolf John ('Johnny') Clausius.
5. *little girls*: Mathilde (1861–1907) and Helene (1863–1919). Future children were Hedwig (1866–1946), Alfred (1868–1950), and Else (1872–1954).
6. *Frank Wynne*: Frank George Wynne. Tyndall had recommended Wynne to Clausius who helped him get settled in Germany.
7. *noble father*: George Wynne.
8. *most excellent mothers*: Anne Wynne (née Osborne, 1808–64), the daughter of Sir Daniel Toler Osborne of Beechwood, Tipperary. She married George Wynne in 1834. The Wynnes had three sons (Henry, Edward, and Frank), to whom Tyndall sometimes sent letters, and one daughter, Lucy.
9. *gone*: Anne Wynne died on 29 May 1864.

To Thomas Archer Hirst [late December 1864–early January 1865][1] 2206

My dear Tom

Francis[2] & myself have had a long chat about you, and the final result was not favourable to the idea of your going to Ireland. However it would be well not to make any definite resolve until we meet & have a further consultation on the matter

Yours aff[ectionate]ly | John Tyndall

RI MS JT/1/T/973

---

1. *[late December 1864–early January 1865]*: Hirst noted in his journal in December that he had intended to apply for the Mathematics chair at Queen's College, Cork in Ireland (Brock and Macleod, *Hirst Journals*, p. 1711). However, in his entry for Sunday, 8 January 1865, he noted that as a result of meeting with Tyndall, he was persuaded not to apply after all: 'Tyndall was at Lamas and goes there again to day. He has decided for me with respect to the Cork Professorship and my mental disquietude is at an end. I shall take no further steps in the matter and remain in London' (p. 1713). Hirst also mentioned in his journal entry of 1 January 1865 that 'I have had two letters from him unfavourable to my leaving London, but leaving the final decision until we meet' (p. 1712). This letter is one of those letters, and the other is missing.
2. *Francis*: probably William Francis, a close friend of both Tyndall and Hirst.

To Herbert Spencer [late December 1864][1] 2207

Royal Institution of Great Britain | Wednesday

My dear Spencer

Every channel in my brain is at the present moment discharging its maximum of current force—the conductors are heated the flow is so strong, and I am determined to avoid contact with anything and everything which could divert the currents from their present direction. I would not kiss the prettiest girl in London, through fear of the subsequent perturbation, and if I decline to join you at Westminster it is because I cannot afford to swerve in the slightest degree from the work to which I am present pledged & wedded.

A month hence I shall be a free man But I think you must count on me as an irregular I cannot undertake any responsibility of organisation: in 99 cases out of 100 I strike the blow & fire the shot of yourself and M^{r}. Mill.[2] The residual one percent I reserve with the view of having a shot at both of you.

Yours ever | John Tyndall

University of London—145.791/334

1. *[late December 1864]*: the month and year are given based on the assumption that Tyndall was replying to Spencer about his lack of involvement in *The Reader* (see letter 2181, n. 1). Spencer shared information about *The Reader* in his autobiography, and noted that he "regretted afterwards that I took so active a part in the business; for it ended in disappointment and loss' (H. Spencer, *An Autobiography*, 2 vols (London: Williams and Norgate, 1904), vol. 2, p. 119). As this letter suggests, Tyndall was too busy with work commitments. Spencer was still describing *The Reader*, and Tyndall's participation in it, positively in a letter to Edward Youmans on 17 December 1864 (*An Autobiography*, vol. 2, p. 118).
2. *M^r. Mill*: John Stuart Mill (1806–73), a prominent British utilitarian philosopher, Member of Parliament, and highly regarded philosopher of science (*ODNB*). He was the former editor of the liberal *Westminster Review*, and one of the proprietors of *The Reader.* Both Mill and Spencer were associated with a move towards a greater, if not exclusive, emphasis upon empiricism and materialism in relation to scientific explanation.

From William Armstrong [December 1864][1] 2208

[The British Association][2] could not, even if they proposed anything so foolish, confer upon Dr. Akin any exclusive privileges in the matter.

J. Tyndall, 'On the History of Calorescence,' *Phil. Mag.*, 29 (1865), p. 222.

1. *[December 1864]*: this letter dated by Tyndall to late December relates to Tyndall's ongoing dispute with the Hungarian physicist C. K. Akin. See letter 2162, n. 2.
2. *[The British Association]*: these words were inserted by Tyndall. The original handwritten version of this letter has not been found, and the transcription of this extract comes from Tyndall's article 'On the History of Calorescence'. The significance of this letter becomes apparent when read in conjunction with letter 2197 from Sabine and others in the series of letters relating to Tyndall's ongoing priority dispute with Akin. Akin had clearly implied that because he had received funding to perform experiments from the BAAS that he therefore had exclusive right to investigate these questions. Sabine, like Armstrong here, stated that this was by no means the case.

1865

From Thomas Archer Hirst 1 January 1865 2209

South Hindley | Wakefield | Yorkshire | Jan 1st 1865

My Dear John.

A happy and successful year to you! May you, before its close, triumph over the few who look upon your career with malevolent envy,[1] not so much by crushing and righteous protestations, as by high and true-spirited scientific achievements.

With this thought uppermost your old friend's heart warmed towards you as the new year dawned which may possibly rob him of what he feels more then ever to have been his highest earthly blessing—daily intercourse with his 'better self.'[2]

Testimonials are arriving daily[3] you shall read them shortly and judge for yourself to what use I can best put them. Rest assured I will make no further step[4] without your full approbation.

I shall reach you at latest on Wednesday next. I thought of doing so tomorrow but Harley[5] very much wishes to see me on Tuesday[6] and if I can do so without prejudice to more important matters I will. Monday morning's post will decide.

Ever yours affectionately | T. A. Hirst

RI MS JT/2/10/ 323

---

1. *with malevolent envy*: a reference to the dispute that Tyndall had had in 1864 with C. K. Akin. See letter 2162, n. 2.
2. *daily intercourse with his 'better self'*: Hirst was looking for employment, which he thought might take him away from his shared living arrangements with Tyndall. He wrote in his journal entry for 1 January 1865: 'My ordinary work and life has been entirely changed by the steps I have taken in reference to the Cork Professorship. The momentous question for me is—am I to try for it or not? The position and emolument would be acceptable enough, but the price at which I shall have to purchase both is great. It will involve my

leaving London, and changing entirely my friends and my life. To leave London is the great obstacle. For me it is the centre of all scientific interest, and there dwells, and will always dwell, my constant and true hearted friend—John Tyndall. He too is at a loss how to advise me. I have had two letters from him unfavourable to my leaving London, but leaving the final decision until we meet' (Brock and Macleod, *Hirst Journals,* p. 1712). One of the letters Hirst referred to is 2206, and the other is missing.

3. *Testimonials are arriving daily*: presumably for a position in Ireland; see letter 2206. In his journal in December 1864, Hirst wrote 'I received excellent letters and testimonials from Key, Cayley and Salmon' (p. 1711) and on 8 January 1865 he wrote 'I found splendid testimonials awaiting my return to London from Smith (of Oxford) Sylvester, Borchardt and Cremona' (p. 1713).
4. *I will make no further step*: in his journal entry for 8 January 1865, Hirst noted that as a result of meeting with Tyndall, he was persuaded not to apply after all: 'Tyndall was at Lamas and goes there again to day. He has decided for me with respect to the Cork Professorship and my mental disquietude is at an end. I shall take no further steps in the matter and remain in London' (p. 1713).
5. *Harley*: Robert Harley (1828–1910), a British mathematician and Congregational minister. He served in the ministry at at Brighouse, Yorkshire and was the chair of mathematics and logic at Airedale College from 1864–8. A focus of his mathematical research was the theory of equations (*ODNB*).
6. *to see me on Tuesday*: in his journal entry for 8 January 1865, Hirst wrote: 'I left Hiendley on Tuesday [10 January] with Rachel and Isabella and went to Brighouse where I dined with uncle and aunt Allatt and Mr Harley' (p. 1712).

From Edward Sabine 1 January 1865 2210

The Royal Society. | Burlington House | W. | Jan[uar]y. 1. 1865

Dear Tyndall:

Be but reasonably prudent, and I trust you have as many years of usefulness and happiness before you, as your friends, amongst whom I am proud to number myself could wish you. Thanks on my wife's[1] part, as well as on my own, for your kind note.[2]

I have received a paper[3] from H. de Schlagintweit[4] on a subject which connects itself very closely with your pursuits. It is on Insolation in India and Thibet; measuring Insolation by the difference of the temperature shown by two precisely similar therm[omet]ers. one in the Sun & the other in the shade; interfering causes being as much as possible avoided in both cases; and the temperature observed as nearly as may be at the times of the maxima of each day. The <u>general</u> result is a connection of a very decided character manifested between the amount of Insolation and that of the relative humidity. The insolation is greatest where the amount of transparent aqueous vapour is greatest. The facts which are adduced are derived from a very large induction. The paper is so illegible in its latter part, & his discussion of the causes of this connection so immethodically stated, that I expect it will be necessary to request him to rewrite some parts of it:—But as far as I can at present apprehend his meaning, he ascribes the very high figure which the insolator attains when much aqueous vapour is present to the diminution of loss by radiation from the sun thermometer. He regards this as in full accordance with your recent researches

Have you experiments bearing on the question of the interception or non interception by transparent aqueous vapour of heat proceeding directly from the sun itself?

When I am better informed than at present of what H. de Schlaginweits mode of explanation really is, I shall be glad to resume the subject with you; but until his paper is made intelligible, in its reasonings as well as in its facts, one can scarcely reason upon them to any purpose.

Sincerely yours | Edward Sabine.

RI MS JT/2/10/323
RI MS JT/2/13c/1276

---

1. *my wife's*: see letter 2157, n. 9.
2. *your kind note*: letter missing.
3. *a paper*: H. Schlagintweit, 'Beobaehtungen über den Einfluss der Feuchtigkeit auf die Insolation, in Indien und Hochasien', *Sitzungsberichte der Königl. Bayer. Akademie der Wissenschaften zu München*, 2 (1864), pp. 216–46. See also H. Schlagintweit, 'Numerical Elements of Indian Meteorology.—Series II. Insolation, and its Connexion with Atmospheric Moisture', *Roy. Soc. Proc.*, 14 (1865), pp. 11–2. In his work *Meteorology of India: an analysis of the physical conditions of India, the Himálaya, western Tibet, and Turkistan* (Leipzig and London: F. A. Brockhaus and Trübner & Co., 1866), Schlagintweit mentioned Tyndall's work as Sabine describes it in this letter (pp. 51–2).
4. *H. de Schlagintweit*: Hermann Schlagintweit (1829–82), an explorer of Central Asia, brother and fellow explorer of Adolf Schlagintweit (1829–1857). Together they produced a scientific study on the Alps between 1846–8. Later, on the recommendation of Alexander von Humboldt, they conducted studies of the Earth's magnetic fields in the Himalayas.

From Christophorus Henricus Diedericus Buys Ballot[1] 6 January 1865 2211

ON THE RADIANT HEAT OF THE MOON. | *To Professor Tyndall, F.R.S., &c.* | Utrecht, January 6, 1865.

SIR,

Allow me to trouble you for your duly and highly esteemed opinion on a subject to which I have given considerable time and care, as you may know from my *Changements périodiques de température dépendants du Soleil et de la Lune,*[2] and from the *Annalen*[3] of Poggendorff. I remember your having written in the Philosophical Magazine*,[4] first, that you could not, as Melloni did, find a heating effect from the full moon's rays by means of your thermoscope, but that, at the same time, you were of opinion that the smoke of London was an obstacle; and secondly, that you supposed that the moon's rays cleared the sky, and that from that cause the earth lost more heat than it gained by the direct action of the moon. I want to know whether this was a *lapsus calami,* or your real opinion.

It is very improbable to me that a source of heat should be in any way a source of cold, even if I could admit Sir John Herschel's opinion to be a true one—that the moon *does* clear the sky. I could not find, from many years' observations, that the sky is more serene when the moon is above than when she is beneath the horizon—in other words, that it is clearer about the time of full moon than it is about the time of new moon.

As for the influence on the temperature, I have added together all the observations taken on the first day of the moon, from 1729 till now, as well as all the observations taken on the second, third, fourth, and so on.

The result is not the same from 1729 to 1789 as it is from 1789 to now. In the first period I found a greater sum of temperatures for the days between the 12th and 26th than for the 27th, 28th, 29th, 30th, and the eleven first days. In the second period I find just the same as Mr. Park Harrison,[5] viz. a more elevated temperature about new moon; this is also the opinion of the observers who accompanied the late polar expeditions.

The whole series gives, however, a sum in favour of the full moon, just as it ought to be.

You will agree with me that *if the new moon really gives a more elevated temperature at the surface of the earth than the full moon, the question is still more interesting,* because the opposite *ought* to be the case, since the whole earth necessarily contains more heat about the time of full moon than about the time of new moon. I have thought, at the same time, whether the moon, when it has north declination, imparts a more elevated temperature to our thermometer than when it has south declination; and I found it to be so; but the difference is equally variable and small.

Now we all admit that other perturbations, a thousand times as great, may totally eclipse for a considerable time the small influence of the moon; but nevertheless it seems to me, when we have 135 years of continued observation, that those perturbations ought to have annihilated themselves.

Why is the mean result so small and uncertain, while we are so certain of the heating action of the moon? The explanation that has occurred to me is this:—

The moon acts partly directly, and by that part the thermometers are most elevated the second or third day after full moon (the day of the greatest north declination for our latitude); but it also acts in some measure indirectly, by elevating the temperature of the aqueous vapour in the atmosphere. By the latter action the clouds and the air have a more elevated temperature, and contain more heat, about and after full moon; but some days must elapse before those particles of air and cloud reach the surface of the earth. When fourteen days elapse before they reach the surface, they have an action wholly opposed to the direct action, and thus the two parts of the action of the moon are necessarily neutralizing each other, more or less.

Hence it is that experiments at great elevations, such as those of Prof. Piazzi Smyth[6] at the Peak of Teneriffe,[7] are of the greatest value. It is one instance among many where they are indispensable.

My letter is already too long to give you my numbers. I will publish them, but would like very much to know your opinion before doing so. Probably you have not time and leisure just now to give your attention to this question. Perhaps you may think fit, however, to publish this letter of mine, with verbal correction, in the Philosophical Magazine, together with a few remarks of your own.

Excuse me for thus applying to you; it is in order to fix your attention on the subject and to be instructed by your opinion.

I am, Most respectfully yours, | BUYS BALLOT[8]

* Vol. xxii. p. 470.

C. H. D. B. Ballot, 'On the Radiant Heat of the Moon', *Phil. Mag.*, 29 (1865), pp. 162–3

---

1. *Christophorus Henricus Diedericus Buys Ballot*: Christophorus Henricus Diedericus Buys Ballot (1817–90), a Dutch chemist and meteorologist known for Buys Ballot's Law and the Buys Ballot Table. He is most noted for his tracking of large weather system movement and solar rotation observation (*CDSB*).

2. *Changements périodiques de température dépendants du Soleil et de la Lune*: C. H. D. B. Ballot, *Changements périodiques de température dépendants du Soleil et de la Lune* (Utrecht: Kemink et fils, 1847).
3. *Annalen*: C. H. D. B. Ballot, 'Uber den Einfluss der Temperatur auf die Synaphie', *Poggend. Annal.*, 147 (1847), pp. 177–320.
4. *written in the Philosophical Magazine*: J. Tyndall, 'Observations on Lunar Radiation', *Phil. Mag.*, 22 (1861), pp. 470–2.
5. *Park Harrison*: probably James Park Harrison (1817–1901), a British church architect, archaeologist, and anthropologist ('Necrology', *The Periodical*, 1 (July 1901), p. 16). Ballot was probably here referring to: J. P. Harrison, 'Lunar Influence on Temperature as connected with Serenity of the Sky', *Phil. Mag.*, 17 (1859), pp. 153–62.
6. *Prof. Piazzi Smyth*: see letter 2002, n. 8.
7. *Peak of Teneriffe*: Tenerife is a large island in the Canary Islands chain. The tallest peak on the island is Mount Teide, which Alexander von Humboldt famously climbed and described.
8. Following this letter from Ballot, the *Phil. Mag.* included this note from Tyndall: '[In the autumn of 1861 I swept the heavens on several days and nights with my thermo-electric pile, and satisfied myself that this instrument is eminently fitted to examine "the mixed action of our atmosphere and stellar space," as regards radiation. My observations are referred to in the Philosophical Magazine, vol. xxii. p. 470. I still preserve my little observatory on the roof of the Royal Institution, in the hope of being able to carry out te idea expressed in a note at the page referred to. Until this has been done, any opinion of mine on the questions raised by my distinguished correspondent M. Buys Ballot, would be of little value; but I am glad to see the subject in such excellent hands. It also gives me pleasure to find that M. Poey has recently turned the thermo-electric pile to account in examining the atmosphere of Havana. This zealous observer thinks that his observations regarding the action of aqueous vapour are diametrically opposed to mine: I, however, entertain a strong hope that M. Poey himself will eventually clear up the apparent difference. The experiments demonstrative of the action of aqueous vapour on radiant heat of low refrangibility are, in my opinion, perfectly conclusive.—JOHN TYNDALL.]' (pp. 163–4).

From Karl Hermann Knoblauch 7 January 1865 2212

Halle, | 7 Jan 65

My dear friend,

You have made me a great and unexpected pleasure by the kindness in which you have thought of me.

I was occupied in my working-room preparing in the vacation-time a memoir on my last experiments for Poggendorff's Annals,[1] when on the

Sylvester evening[2] your sending[3] arrived. The unusual form of it made me suppose a diploma or something of that kind and my surprise was great, seeing by rolling down the paper the features of my dear old friend. All happy remembrances, especially from that time which we had spent together also at Christmas and New year in Marburg,[4] passed before my memory and <Hereafter handwritten transcript resumes> I included into my best wishes at the beginning of this year my kind friend now in London. Take with these wishes my best thanks for the joy you have prepared to me by sending your "sprechendes"[5] and "wohlgelungenes"[6] portrait.

I have had many opportunities last time to hear of you by my friends who had seen you at the Swiss-Association[7] in Zurich. First Dr. Mayer of Heilbronn sent me your kind regards. Prof. Poggendorff, all had seen you, and your toast on the "Traumeierei"[8] of sciences has run through all these circles.

It is a pity that you in your journey on the Continent never come in our neighbourhood; I would be very happy if you once would call upon me in Halle. We could not leave it the whole last summer, first on account of the birth of a little boy, called George, than for the sickness of Mrs. Knoblauch[9] and our three elder sons who got the "Scharlachfieber"[10] at the same time.

Is our friend Hirst living at present in London or in England? He was kind enough to send a paper to the "naturforschende Gesellschaft"[11] and to me. But, being not aware of his actual abode and address, we were not able to thank him. I beg you to remember me kindly to our common friend.

I must conclude my letter with a very trist notice which I just receive from Berlin, that of the decease of our dear Mistress Poggendorff.[12] She died on the 5th January by an inflammation of the lungs. We are all very sorry for this hard fate on our friend.[13]

With my best wishes for you, my dear Tyndall, I remain | ever yours | most faithfully | Herm. Knoblauch

RI MS JT/1/K/24
RI MS JT/1/TYP/7/2506–7

---

1. *Poggendorff's Annals*: possibly H. Knoblauch, 'Ueber die Diffusion der Wärmestrahlen', *Poggend. Annal.,* 125 (1865), pp. 1–37.
2. *Sylvester evening*: the German celebration of New Year's Eve, 31 December.
3. *your sending*: letter missing.
4. *Marburg*: see letter 1940, n. 2
5. *"sprechendes"*: expressive (German).
6. *"wohlgelungenes"*: well-executed (German).
7. *Swiss-Association*: see letter 2108, n. 2.
8. *"Traumeierei"*: musings (German).

9. *Mrs Knoblauch*: see letter 2044, n. 5.
10. *"Scharlachfieber"*: scarlet fever (German).
11. *"naturforschende Gesellschaft"*: Society of Natural Researchers.
12. *our dear Mistress Poggendorff*: Charlotte Eleonore Poggendorff, who died on 5 January 1865 due to inflammation of the lungs (*Johann Christian Poggendorff: Leichenrede von Emil Frommel, Hof- und Garnison-Prediger in Berlin. Nebst eigenhändigen Lebensnachrichten, Reden und Briefen* (Berlin: Dümmler, 1877), p. 12).
13. *our friend*: Johann Christian Poggendorff.

To Edward Sabine[1] 10 January 1865 2213

10th. January 1865

My dear General,

I wrote to Mr. Gassiot[2] a few days ago suggesting that the Scientific Relief Fund[3] had a very high claimant in the person of the widow[4] of the late Prof. Boole.[5] I did this knowing, that he could not have left her a proper provision, and knowing also that he had left five little children behind him. The immediate cause of my writing however was a letter recovered by a friend of mine from one of the professors in Trinity-College Dublin and in which the Scientific relief fund was referred to.

But on consideration & enquiring I fear a movement in this direction might be distasteful to Mrs. Boole. Her husband was a man of great independence of spirit, and from what I once saw of his extremely intelligent wife here in London I should judge her to be the same. Under the circumstances, and considering the great and exceptional eminence of Prof. Boole, it seems more desirable that some form of aid, more worthy of him, should be extended to his widow & children. I believe an effort will be made to obtain a pension for her, and no doubt any memorial to this effect would be backed by many of the most eminent names in this country.

Believe me | Yours ever truly | J. Tyndall

RI MS JT/2/10/324

---

1. This is a draft copy; the original handwritten version is missing.
2. *Mr. Gassiot*: Gassiot initiated the fund. See RS Minutes of Council 3 (1858–69), p. 24. See also F. G. Donnan, 'The Scientific Relief Fund and Its Committee', *Notes and Records of the Royal Society*, 7 (1950), pp. 158–71.
3. *Scientific Relief Fund*: started in 1859 by the Philosophical Club of the RS (see letter 2045, n. 3), this fund served to assist scientists by protecting their occupation or providing a spouse with assistance upon their death.

4. *widow*: Mary Boole (née Everest, 1832–1916), a self-taught mathematician and author of popular and instructional works on mathematics. She met her future husband, George Boole in 1850, and edited his *Laws of Thought*, which was published in 1854. The pair married the following year. Following her husband's death the liberal theologian Frederick Maurice offered her the post of librarian at Queen's College, London. She went on to write on evolution, spiritualism, and anarchist politics (*ODNB*).
5. *Prof. Boole*: George Boole (1815–64), an English mathematician, philosopher, and logician. He was the first professor of mathematics at Queen's College, Cork in Ireland, where he met his future wife Mary Everest. Boole was the author of *The Laws of Thought* (1854). He developed pneumonia as a result of walking three miles in heavy rain to his university and lecturing in wet clothes. His wife believed that his illness, caused by water, would be cured by water, and so subjected him to regular dousings of cold water. He died on 8 December 1864 of fever-induced pleural effusion (*ODNB*).

From James Croll 13 January 1865 2214

Andersenian University | Glasgow | Jan. 13th 1865.

Professor Tyndall

Dear Sir,

Having read most of the articles which have appeared on the subject which you have designated Negative Fluorescence, I felt somewhat surprised at Dr. Akin's remarks[1] in the Philosophical Magazine for December last. Although I consider that you were quite right in replying[2] still there does not seem to me to be anything in the subject really worth contending for.

The whole matter appears to be comprehended under the two following particulars:—

1. Heat a platinum wire for example in an oxyhydrogen flame and the wire becomes white hot.

2. The rapid vibrations of the platinum molecules are produced by the slow vibrations of the aqueous molecules constituting the flame.

As regards to No. 1. neither you nor Dr. Akin, of course, claims to be the first observer. But unquestionably you were the first to establish No. 2. For it was your famous experiments on Radiant Heat which proved that the oxyhydrogen flame consists of waves or, rather we should say, vibrations of slow period.

The point however to which most importance is attached is not these facts themselves, but a certain conclusion supposed to be derived from them, viz that when the platinum wire is heated in the oxyhydrogen flame, a change of period, or a conversion of long waves into short ones takes place.—I am sorry to say that I have been wholly unable to perceive the legitimacy of this conclusion. All that I can perceive is simply this; the vibrating molecules of the flame set the molecules of the platinum a vibrating, but owing to some physical difference in the two sets of molecules the one set vibrates more rapidly than the other.

It is certainly an abuse of language to say that because the vibrations of the platinum molecules are more rapid than those of the flame, that there is a change of period, or a breaking up of long waves into short ones.—A blacksmith will render the platinum incandescent by simply striking it with his hammer. But who would speak of a change of period in this case? And yet there is no essential difference between the two cases. In the one case the molecules of the platinum are set a vibrating by the blows imparted by the hammer, in the other case these molecules are set a vibrating by blows imparted by the molecules of the flame. But the molecules of the flame may be justly regarded as so many small hammers striking the platinum. The vibrating molecules of the platinum stand in precisely the same relation to the blows of the hammer, that they do to the blows of the molecules of the flame. And it appears to be just as proper to say that the slow blows of the hammer are converted into the rapid vibrations of the platinum molecules, as it is to say that the slow vibration of the flame molecules are converted into the rapid vibrations of the platinum molecules.

Trusting that you will pardon me for troubling you with these few random remarks

I am, | Sir, | Yours respectfully | James Croll.

RI MS JT/1/TYP/1/65–6
LT Typescript Only

---

1. *Dr. Akin's remarks*: C. K. Akin, 'On Calcescence', *Phil. Mag.*, 29 (1865), pp. 28–43. This article appeared in the January 1865 issue, not December 1864.
2. *quite right in replying*: J. Tyndall, 'On the History of Negative Fluorescence', *Phil. Mag.*, 29 (1865), pp. 44–55. This immediately followed Akin's article. Akin followed up in the March issue with 'Further Statements concerning the History of Calcescence', *Phil. Mag.*, 29 (1865), pp. 136–51. This article included a note from editor William Francis that 'the discussion of this subject must close, so far as our pages are concerned' (p. 151). On Tyndall's dispute with Akin, see letter 2162, n. 2.

From George Gabriel Stokes 13 January 1865 2215

The Royal Society | Burlington House, London. W. |
13th Jan[uar]^y 1865

My dear Tyndall,

I have put on your short paper[1] for Thursday.

It would of course be extremely interesting to see the experiment at the evening meeting,[2] but we must consult your convenience about that. Perhaps you could show it when you present your paper.

I am glad you have changed the term, which I don't think was quite correct. For if

Calorescence = - Fluorescence[3]

we ought to have

- Calorescence[4] = Fluorescence

Now I have little doubt that a miraculous concentration of blue rays would heat a solid till it became red hot. But this which would be strictly negative calorescence is not a little different from fluorescence, where blue rays of very moderate concentration cause a cold body, suppose an alcoholic solution of chlorophyll, to give out red light.

Yours very truly | G. G. Stokes

RI MS JT/1/S/239
RI MS JT/1/TYP/4/1398

1. *your short paper*: most likely a reference to J. Tyndall, 'Note on the Invisible Radiation of the Electric Light', *Roy. Soc. Proc.,* 14 (1865). pp. 33–5, which was received on 13 January 1865. It was presumably read/demonstrated at the RS on 19 January, a Thursday. See *Ascent of John Tyndall,* p. 185–6.
2. *It would of course . . . at the evening meeting*: Tyndall did demonstrate this experiment. See the Introduction to this volume; and *Ascent of John Tyndall,* pp. 185–6.
3. *- Fluorescence*: negative fluorescence.
4. *- Calorescence*: negative calorescence.

To James Croll 14 January 1865 2216

The Athenaeum[1] | 14th. Jan. 1865

My dear Sir

It is both amusing and interesting to me to trace the parallelism which has run between your thoughts & mine on the subject of "negative fluorescence." (I have changed this term to Calorescence) The very experiment to which you refer,[2] of rendering a body hot by concussion is the one which most influenced my conviction that it was possible to produce incandescence by invisible rays.

It strikes me you are rather hard on the phrase "breaking up long periods into short ones." In the case of the hammer there is a conversion of the mechanical motion into molecular motion, and in the case of the hydrogen flame there seems to me to be a conversion of the long periods into short ones. The fact at all events is that you hit the mass with waves of slow recurrence, and that you obtain, in return from the mass, waves of quick recurrence.

Excuse this hurried scrawl. Do you wish me to send your note to the Philosophical Magazine?[3] If so I would suggest one small & unimportant alteration. Instead of saying "an abuse of language" I would say an incorrect use of language.

Yours very truly | John Tyndall

Wellcome, MS. 7777/08

1. *The Athenaeum*: see letter 1907, n. 7.
2. *to which you refer*: see letter 2214 from Croll.
3. *send your note to the Philosophical Magazine?*: nothing from Croll appeared in the *Phil. Mag.* in 1865, and three short articles were published in 1866.

From Rudolf and Adelheid Clausius 15 January 1865 2217

Zürich, 15/1/65.

Lieber Tyndall,

am 2ten Januar, gerade an meinem Geburtstage, kam Dein schönes Bild hier an. Wir haben uns sehr darüber gefreut und ich habe es als ein höchst angenehmes Geburtstags-Geschenk betrachtet. Am Abend desselben Tages kam auch Dein freundlicher Brief vom 31sten December an, und vor einigen Tagen ist die Sendung mit dem schönen silbernen Becher angelangt. Du hast uns durch dieses Geschenk sehr überrascht, und es wird für unseren John, wenn er erst grösser ist, und weiss von wem es gekommen ist, immer ein theures Andenken sein. Nimm unseren herzlichen Dank für Deine Freundlichkeit.

Die Nachricht, dass die gute Frau Poggendorff, die uns im vorigen Herbst noch durch ihr munteres, lebhaftes Wesen erfreute, und für ihr Alter so merkwürdig rüstig und unternehmend war, nach einem kurzen Leiden plötzlich gestorben ist, hast Du wohl von Berlin erhalten. Es ist sehr traurig, besonders für den alten Herrn, der sich seit so langer Zeit daran gewöhnt hat, die Sorge für alle häuslichen Geschäfte ganz ihr zu überlassen, und sich selbst nur seinen wissenschaftlichen Arbeiten zu widmen. Wenn ihm nicht die liebenswürdige Tochter geblieben wäre, welche die Mutter einigermassen ersetzen kann, würde er die Einsamkeit wohl kaum ertragen können.

Mit grosser Freude habe ich gelesen dass Du die Rumford Medaille für Deine schönen Arbeiten über die strahlende Wärme erhalten hast. Es ist eine sehr wohlverdiente Auszeichnung die dazu beitragen möge, Dir die Lust zur Sache zu erhalten, trotz der Verdriesslichkeiten, wie sie Dir z.B. durch die sonderbaren Prätensionen von Akin veranlasst sind. Es ist ein merkwürdiges Verlangen, dass ein wissenschaftlicher Forscher seine eigenen selbständigen Untersuchungen unterdrücken soll, blos weil er erfahren hat, dass ein anderer sich auch mit den Sachen beschäftigt. Wenn Du von Deinen neusten Abhandlungen noch Exemplare hast so würdest Du mich sehr erfreuen, wenn Du sie mir schicken wolltest. Unter Deinen in den Phil. Transactions gedruckten Abhandlungen ist die letzte, welche ich besitze, die im Januar 1862 *[gelesene]*. Du schreibst in Deinen Briefen nichts darüber, ob Du mein Buch bekommen hast. Sollte es nicht angekommen sein, so sei doch so gut es mich wissen zu lassen, weil ich dann gleich an den Verleger, welcher die Versendung der Freiexemplare besorgt hat, schreiben müsste.

Mit besten Wünschen für das neue Jahr und herzlichen Grüssen

Dein | Clausius.

Auch ich will Ihnen meinen herzlichen Dank für Ihren freundlichen Brief und für das schöne Geschenk sagen, wodurch Sie uns so sehr überrascht haben. Johnny soll es als theures Andenken bewahren; es macht ihm schon jetzt grosse Freude, denn er *[rappelt]* mit Händen und Füssen nach dem glänzenden Becher, und hat auch neulich mit grossem Wohlbehagen *[auf Ihre]* Gesundheit daraus getrunken, was er oft wiederholen wird, sowie er erst Verstand davon hat. Nun will ich Ihnen noch die besten Glückwünsche zum neuen Jahre sagen, mögen Sie gesund und frisch bleiben, und möge Ihre wissenschaftliche Thätigkeit Ihnen immer mehr Freude und Ruhm erschaffen.

Mit herzlichen Grüssen | Adelheid Clausius.

Zurich, 15/1/65.

Dear Tyndall,

Your[1] fine picture[2] arrived on January 2nd, exactly on my birthday. We were very pleased to receive it, and I consider it a most delightful birthday gift. Your kind letter[3] from December 31st came on the evening of the same day, and a few days ago the parcel with the beautiful silver cup arrived. You surprised us very much with this gift and when John grows up and knows from whom it came, he will always hold it as a precious remembrance. Please accept our sincerest gratitude for your kindness.

You must already have received the news from Berlin that dear Mrs Poggendorff,[4] who was still delighting us with her vivacious and animated presence last autumn and who was notably active and robust for her age, died suddenly after a short ailment.[5] It is very sad, in particular for the old gentleman,[6] who had long become used to delegating all household responsibilities to her and could dedicate himself entirely to his scientific work. If it were not for his kind daughter,[7] who to a certain extent can replace the mother, he would barely be able to endure his loneliness.

I read with great pleasure that you received the Rumford Medal[8] for your beautiful work on radiant heat. This is a well-deserved award that hopefully will contribute to you continuing to enjoy the work, despite the frustrations that are caused, for instance, by Akin's strange pretensions.[9] It is a curious charge, that a researcher should suppress his own independent research solely because he learned that another researcher is working on the same matter.[10] If you have copies of your most recent papers,[11] I would be very glad if you would send them to me. The last of your papers published in Phil[osophical] Transactions that I have, is *[the one read]* in January 1862.[12] You do not write in your letters whether or not you have received my book.[13] Please let me know if it did not arrive, since I would then have to contact the publisher who was responsible for sending the free copies.

With best wishes for the New Year and kind regards,

Your | Clausius.

I, too, want to thank you from the bottom of my heart for your kind letter and the beautiful gift, which surprised us so much. Johnny[14] shall keep it as a precious remembrance; he already likes it very much: he *[wriggles]* with hands and feet for the shiny cup and recently drank *[to your]* health from it with great enjoyment. He will do so often once he understands what it is. I would also like to express my best wishes to you for the New Year, may you stay healthy and well and may you receive ever more joy and renown from your scientific work.

With kind regards | Adelheid Clausius.

RI MS JT/1/TYP/7/2273–4
LT Typescript Only

---

1. *Your*: Clausius here used the informal address form ('Du'). See letter 2047, n. 1.

2. *Your fine picture*: Tyndall had also sent a portrait of himself to Knoblauch; see letter 2212.
3. *Your kind letter*: letter 2205.
4. *Mrs Poggendorff*: Charlotte Eleonore Poggendorf.
5. *died suddenly after a short ailment*: see letter 2212, n. 12.
6. *the old gentleman*: Johnan Christian Poggendorf.
7. *his kind daughter*: Marie Poggendorf (b. 1838).
8. *the Rumford Medal*: see letter 2178, n. 1.
9. *by Akin's strange pretensions*: This is a reference to the ongoing dispute Tyndall had with the Hungarian physicist C. K. Akin. See letter 2162, n. 2.
10. *It is a curious charge . . . working on the same matter*: Akin had clearly argued that because he had funding from the BAAS to pursue research in the area of radiant heat that he therefore had exclusive right to pursue those questions. It was this claim that had evidently prompted the letters Tyndall wrote (which have not been found) that elicited the responses from Sabine and Armstrong (letters 2197 and 2208, respectively). Tyndall had clearly told Clausius of this, in a letter that is also missing.
11. *your most recent papers*: it is not clear which papers Clausius had in mind, as Tyndall had published much over the previous year. For papers that may have been included in this request, see letter 2014, n. 2; letter 2097, n. 3; letter 2098, n. 6; and letter 2154, n. 3.
12. *The last of your papers . . . January 1862*: see letter 1914, n. 1. It is notable that Clausius was very far behind in his reading of Tyndall's work, as he had published extensively on a variety of subjects since this time.
13. *my book*: see letter 2093, n. 6.
14. *Johnny*: Rudolf John ('Johnny') Clausius.

From George Gabriel Stokes 19 January 1865 2218

The Royal Society | Burlington House, London W. | 19 Jan[uar]^y 1865

Dear Tyndall,

I read your letter[1] dated July 1, 1864 to the Gov[ermen]^t Grant Committee[2] on the first possible opportunity, namely yesterday.

In it you ask permission to apply the balance (£25–19–4 ½) remaining in your hands from the Grants of 1861 and 1863[3] to certain measurements which you proposed to take in the Alps in case you thought the money could thus be turned to good account.

The Committee were willing that it should be so expended.

Yours very truly | G. G. Stokes.

RI MS JT/1/S/240
RI MS JT/1/TYP/4/1399

1. *your letter*: letter missing.
2. *Gov[ernmen]t Grant Committee*: the Government Grant Committee of the RS was established in 1849 for the purpose of granting funds to researchers. On the Government Grants, see M. Boas Hall, *All Scientists Now: The Royal Society in the Nineteenth Century* (New York: Cambridge University Press, 2002 [1984]), chapters 5 and 6; and R. M. MacLeod, 'The Royal Society and the Government Grant: Notes on the Administration of Scientific Research, 1849–1914', *The Historical Journal*, 14 (1971), pp. 323–58.
3. *the Grants of 1861 and 1863*: in 1861, Tyndall received £200 for his experiments on radiant heat and a further grant in 1863.

From Jean François Léonard Alexandre de Colnet-d'Huart[1] 29 January 1865 2219

Luxembourg le 29 janvier 1865

Monsieur,

Si je n'ai répondu plus tôt à votre aimable lettre, c'est qu'une longue maladie m'a mis pendant plusieurs mois dans un état désespéré. Maintenant que mes forces me sont revenues avec la santé, je sens plus que jamais le besoin de travailler.

Les deux premiers chapitres de la seconde partie de ma nouvelle théorie de la chaleur et de l'électricité ont été imprimés par les soins d'un de mes amis. J'ôse vous les adresser. Dans le chap.5 je crée la théorie mathématique de la diathermanie. Personne au monde n'est plus compétent que vous pour me dire ce que peut valoir mon travail. Je regrète que mon état ne m'ait pas permis de revoir les § 52, 53, 54. Plusieurs développement en série semblent superflus, mais je compte les utiliser plus tard lorsque j'examinerai l'influence du magnétisme sur la lumière et la chaleur rayonnante.

Dans le chapitre 5 je parviens à déterminer la position de l'axe de rotation (produite par la chaleur) d'une molécule m qui vibre, quelle que soit la trajectoire décrite par m.

La simplicité de ce résultat me laisse l'espoir que l'analyse aura prise sur un grand nombre de phénomènes qu'on regarde comme très complexes.

Dès que le chapitre 6me sera imprimé je prendrai la liberté de vous l'adresser.

Je termine, Monsieur, en vous avouant que dans mon cours je fais de larges emprunts à vos admirables leçons sur la chaleur. J'explique avec soin les procédés que vous avez employés dans la détermination de l'absorption de la chaleur par le gaz et les vapeurs;mais vous l'avouerrais-je, je regrette de ne posséder une partie de la clarté et de la précision que vous mettez dans vos leçons.

Je m'empresse de vous exprimer ma profonde reconnaissance pour l'offre que vous m'avez faite de m'adresser par la voie de la libraire un de vos ouvrages. Ci-joint les voies par lesquelles votre ouvrage pourrait me parvenir.

Je vous prie, Monsieur, d'agréer l'expression de la considération la plus distinguée de votre très humble serviteur. | de Colnet-d'Huart

Luxembourg, 29th January 1865

Sir,

I have not replied earlier to your kind letter,[2] because a long illness put me in a desperate state for several months. Now that my strength and my health have returned, I feel more than ever the need to work.

The first two chapters of the second part of my new theory of heat and electricity[3] have been printed by one of my friends. I have the audacity to send them to you. In chapter 5 I create the mathematical theory of diathermancy. Nobody else in the world is more competent than you to tell me what my work is worth. I regret that my health did not allow me to revise the § 52, 53, 54. Several developments in a row seem superficial, but I intend to use them later when I examine the influence of magnetism on light and radiating heat.

In chapter 5, I succeed in determining the position of the axis of rotation (produced by heat) of a molecule m that vibrates, whatever the trajectory of m.

The simplicity of this result gives me hope that analysis will be done on a great number of phenomena currently seen as complex.

As soon as chapter 6 is printed I will take the liberty of sending it to you.

I conclude, Sir, by admitting that I borrow a great deal from your wonderful lessons on heat[4] in my lectures. I carefully explain the methods you used in the determination of the absorption of heat by gas and vapours; but I assure you, I can only wish I had a fraction of the clarity and precision you use in your lessons.

I must express my sincere gratitude for the offer you made me to send me one of your works via the bookseller. I have attached instructions for the means by which your work could reach me.

Please, Sir, your humble servant's highest regards | de Colnet-d'Huart

RI MS JT/1/D/137

---

1. *Jean François Léonard Alexandre de Colnet-d'Huart*: see letter 2138, n. 1.
2. *your kind letter*: letter missing.
3. *my new theory of heat and electricity*: see letter 2128, n. 8.
4. *your wonderful lessons on heat*: see letter 1978, n. 3.

From Johann. C. Bohn[1] 28 January 1865 2220

Giessen den 28n Januar 1865

Hochverehrter Herr Tyndall!

Ich habe Sie vor allen Dingen um Verzeihung zu bitten, dass ich Ihren freundlichen Brief vom 1n October 1864 noch nicht beantwortete und Ihnen noch nicht für die Güte gedankt habe, mit der Sie die im August von mir geschriebene Note über die Geschichte der Erhaltung der lebendigen Kraft übersetzt und zur Aufnahme in das Phil. Mag. gebracht haben. Nehmen Sie nachträglich dafür meinen verbindlichsten Dank und lassen Sie allerhand häusliches Ungemach, das mich im Oktober traf und Krankheit, die sich daran schloss zu meiner Entschuldigung dienen.

Dr. Akin hat im Phil. Mag. einen Beitrag zur Geschichte der Lehre von der Umsetzung der verschiedenen Arten von Arbeit in einander beigebracht und gelegentlich auch meine Note erwähnt. Ich glaube einiges weitere Material zu dieser Geschichte liefern zu können und wäre sehr geehrt und erfreut, wenn die anliegenden Bemerkungen durch Ihre gütige Vermittlung Aufnahme im Philos. Magazine finden könnten.

Um nicht Ihre Güte zu missbrauchen habe ich, mit Beihülfe eines sprachenkundigeren Freunds, die Note englisch abgefasst, muss Sie aber freilich doch noch mit der Bitte belästigen dieselbe einer Durchsicht zu unterwerfen und die gar zu argen sprachlichen Sünden auszumerzen. Endlich wäre es mir sehr angenehm, wenn ich einige Abzüge der Note erhalten könnte; wenn es in England nicht gebräuchlich ist den Autoren solche zur Verfügung zu stellen, so bestehe ich natürlich nicht auf diesem Wunsche.—Mögten Sie mir, verehrter Herr Tyndall doch recht bald Gelegenheit geben mich Ihnen gefällig und nützlich zu erweisen. Rechnen Sie auf meine grösste Bereitwilligkeit.

Ihnen im Voraus für Ihre Güte dankend grüsse ich Sie herzlichst mit dem Ausdrucke der ausgezeichneten Verehrung als

Ihr ergebenster | Pr Dr Bohn

Giessen, 28th January 1865

Most respected Mr Tyndall!

I must ask you above all to forgive me for not yet replying to your friendly letter of 1st October 1864,[2] and for not having thanked you for the kindness with which you translated the note I wrote[3] in August on the conservation of living force and had it published in the Phil. Mag. Please receive belatedly my most sincere thanks for this and accept as my excuse all kinds of domestic disturbances that struck me in October and an illness that followed thereafter.

Dr Akin published in the Phil. Mag.[4] a contribution on the history of the theory of transformations between different types of work and mentions on occasion my note as well. I believe I can provide some additional material to this history and would be pleased and honoured if through your kind communication the enclosed comments could be published in the Philos. Magazine.[5]

To avoid abusing your generosity I have written the note in English with the help of a more linguistically competent friend, but I must still importune you with the request to review the note and eradicate those linguistic sins that are too grave. Finally, I would be glad if I could obtain a few prints of the note; of course, if it is not customary in England to make these available to authors, I will naturally not insist on this wish.—May you, Mr Tyndall, grant me quite soon an opportunity to be obliging and useful to you. You can count on my greatest eagerness.

Thanking you in advance for your kindness, I send warm greetings with expressions of great admiration,

Your humble servant | Prof Dr Bohn

RI MS JT/1/B/103

---

1. *Johann C. Bohn*: see letter 2129, n. 1.
2. *your friendly letter of 1st October 1864*: letter missing.
3. *the note I wrote*: letter 2129.
4. *Dr Akin published in the Phil. Mag*: C. K. Akin, 'On the History of Force', *Phil. Mag.*, 28 (1864), pp. 470–7.
5. *published in the Philos. Magazine*: they were published as J. C. Bohn, 'On the History of the Conservation of Energy and of its application to Physics', *Phil. Mag.*, 29 (1865), pp. 215–8.

BIOGRAPHICAL REGISTER

This register contains the names and biographical details of people who are mentioned three or more times in the letters included in this volume. Biographical information about people mentioned once or twice is contained in the notes appended to the appropriate letter(s).

Conventions

In transcribing the letters the editors have generally retained the spellings of family names given in Louisa Tyndall's transcripts. However, in a number of cases a variety of spellings were used by correspondents; for example, the family name of Thomas Bunbury, MP, was often spelled "Bumbury." Where there is a preferred spelling, as used in official documents or biographical dictionaries, the alternative is placed in parentheses; for example, "Bunbury (Bumbury), Thomas." Where there is no preferred spelling, the alternatives are given in alphabetical order, separated by a forward slash; thus, "Davey/Davy, James." Where known the dates of birth and death are given in parentheses. Nicknames and pseudonyms were widely used by Tyndall and his friends. Where such nicknames occur in the letters, the person they refer to is given in an accompanying note.

Adams, John Couch (1819–92), was a British astronomer and mathematician. He made observations of Halley's comet in 1835 and although he was only belatedly recognized, was the codiscoverer of Neptune. Born in Cornwall, he was largely self-taught in mathematics until an inheritance allowed him to study at Cambridge. He later returned there as professor of mathematics in 1859 and became director of the Cambridge Observatory from 1861.

Airy, George Biddell (1801–92), was a British astronomer. He was appointed Lucasian Professor of Mathematics in 1826, and Plumian Professor of Astronomy at Cambridge University in 1828. In 1835 he was appointed Astronomer Royal, an important government position. From the Royal Observatory in Greenwich, he advised the British government and pursued research on a number of topics, including optical diffraction, the earth's mean density, and the inequalities in the motions of Earth and Venus. In 1851 he established the new prime meridian at Greenwich, which was internationally recognized in 1884.

Akin, C. K. (1830–93), was a Hungarian physicist who came into conflict with Tyndall in 1864 over infrared experiments. He and Tyndall initially enjoyed cordial relations, but they later soured as Akin first privately and then publicly accused Tyndall of having plagiarized his research. Akin, who was less well connected in the scientific community at this time, lost out to Tyndall, who was able to marshal his large professional network in his support.

Armstrong, William George (1810–1900), was a British armaments manufacturer and industrialist who also worked on hydraulics and electrostatics. Armstrong performed extensive service for the BAAS as well as a number of engineering institutions. He was knighted in 1859 for giving the patent for his design for the Armstrong gun to the Crown. In 1887 he was elevated to the peerage as Baron Armstrong of Cragside.

Ball, John (1818–89), was an Irish politician and glaciologist. He was admitted to Cambridge University in 1835, but, as a Catholic, and unwilling to subscribe to the Protestant Articles, did not have his degree conferred. During the 1840s he began to study the glaciers of the Alps, returning to Ireland to work for the Poor Law Commission during the Irish famine (1845–52). In Ireland he noted that the geological formations of the Dingle peninsula resembled those of the Alps and proposed that the landscape must have been formed by ancient glaciers. He became Member of Parliament for Carlow County in 1852 and in 1855 was named assistant undersecretary of state in the colonial department by Lord Palmerston. He was the first president of the Alpine Club (1858–60) and edited the club's first publication, *Peaks, Passes, and Glaciers* (London: Longman, Green, Longman, and Roberts, 1859).

Baring, Louisa Caroline (1827–1903), Lady Ashburton, was a friend and associate of many artists and writers who either were or later became famous. In 1858 she married the widower William Baring, 2nd Baron Ashburton, and their family home quickly thereafter became notable on the society calendar, and Baring became a prominent and well-connected society hostess. She was an art collector and philanthropist, and the soirées she hosted attracted many prominent Victorian society figures, including the Carlyles, John Ruskin, George Campbell (Duke of Argyll), the Walter Trevelyans, Edward Lear, William Wetmore Story, Robert Browning, Samuel Wilberforce (bishop of Oxford), Richard Monkton Milnes (Lord Haughton), John Everett Millais, Charles Kingsley, and, of course, John Tyndall. Her parties, some of which lasted several days, served to bring a number of less-connected men of science into society.

Baring, William Bingham (1799–1864), Lord Ashburton, was a moderately successful politician, serving a number of electorates before ascending to the peerage in 1848. He was born in the United States, in Philadelphia, and educated at Oriel College, Oxford. Ashburton was elected FRS in 1854 and served as president of the Royal Geographical Society from 1860 to 1862. His second wife, Louisa Baring (see above), became one of the most successful society hostesses following their marriage in 1858, making their home the venue of many of the most significant and sought-after soirées during the society season.

Barlow, John (1798–1869), was an Anglican priest and the secretary of the RI from 1842 to 1860. A graduate of Trinity College, Barlow married the daughter of a Member of Parliament who made his fortune in India. He lived near the RI and entertained many of its wealthy and aristocratic supporters as well as fellow scientific practitioners, including Faraday and Tyndall, who sometimes dined with the family.

Barnard, Jane (1832–1911), was a niece of Michael and Sarah Faraday who lived at the RI and assisted the Faradays. She was the daughter of Sarah's brother, John Barnard, and Michael's younger sister Margaret. Like her aunt and uncle, she was a member of the Sandemanian Church.

Barrett, William Fletcher (1844–1925), was a physicist and psychical researcher, who studied at the Royal College of Chemistry, London. In 1863 he became Tyndall's assistant at the RI, where he met and came under the influence of Michael Faraday. He was then successively science master at the International College, London (1867–69), lecturer on physics at the Royal School of Naval Architecture (1869–73), and professor of physics at the Royal College of Science, Dublin (1873–1910). He was regarded as an excellent lecturer at both a professional and a popular level.

Becker, Ernst (1826–88), was a German chemist and founder of the Photographic Society of England. He was also a translator for *Poggend. Annal.* and served as secretary and librarian to Prince Albert.

Bennen, Johann Joseph (1824–64), known by English alpinists simply as "Bennen," was a Swiss mountain guide. Bennen made the first ascent of the Aletschorn in 1859 and participated in many of Tyndall's ascents. In August 1861 he was in the party with Tyndall and Ulrich Wenger that was the first to summit the Weisshorn at 4,505 meters, and in July 1862 Bennen, Tyndall, and guide Jean-Antoine Carrel made an attempt on the Lion Ridge of the Matterhorn. During a winter attempt at Haut de Cry in the Bernese Alps, with his clients Philipp Gosset and Louis Boissonnet, and three local guides, Nance, Rebot, and Bevard, he and Bossonnet were carried away by an avalanche on 28 February 1864 (C. D. Cunningham and W. Abney, *The Pioneers of the Alps*, 2nd ed. [London: Low, Marston, Searle and Rivington, 1888], 148–53; and Charles Gos, *Alpine Tragedy* [London: George Allen and Unwin], 18–23).

Brewster, David (1781–1868), was a Scottish editor, university administrator, and writer who worked on optics. Brewster edited or coedited a number of short-lived journals before taking on the *Edinburgh Journal of Science*; this was purchased by the London-based owner of the *Phil. Mag.*, Richard Taylor. Brewster was taken on as one of the editors, a position he held until his death. He wrote for all the major review magazines, and, through his friendship with the Evangelical minister Thomas Chalmers, was closely associated with the *North British Magazine*. A founding member of the BAAS, Brewster was concerned about the development of British science on a wider European stage. In the 1850s he became involved in a priority dispute with Charles Wheatstone over the invention of the stereoscope, which was conducted largely through the pages of the *Times*. Tyndall encountered Brewster at meetings of the BAAS and they both served as jurors at the 1855 Exposition Universelle in Paris.

Bunsen, Robert (1811–99), was a German chemist. Bunsen was born in the German university town of Göttingen, where he later studied chemistry, physics, and mathematics, as well as geology and botany. He lectured from 1833 to 1836, after which he held positions at the Polytechnikum at Kassel, the University of Marburg (1839–51), the University of Breslau, and, from 1852, the University of Heidelberg. Among his many innovations in chemistry was his discovery of cacodyl, a compound containing arsenic. He made extensive use of both electrolysis and spectrum analysis to analyze chemical substances and developed the cell (battery) that bears his name. The Bunsen burner is also named after him. Edward Frankland and Tyndall traveled to Marburg to work under Bunsen in 1848.

Busk, George (1807–86), was a surgeon and naturalist, with interests in paleontology and microscopy. He founded the Microscopical Society in 1843, and served as treasurer of the RI from 1873 to 1886. From 1857 to 1868 he was secretary of the zoological section of the Linnean Society. From 1864 he was also a member, with Tyndall, Thomas Hirst, Joseph Dalton Hooker, Thomas

Henry Huxley, and four others, of the X Club, a monthly dining club through which they promoted science and advanced scientific naturalism in British society. He was on close and friendly terms with Tyndall, and in this volume Busk suggests climbing routes in Wales for Tyndall's planned excursion there. Busk also aided Tyndall in his dispute with C. K. Akin, with whom he was in an intense priority dispute throughout the period covered by much of this volume.

Carlyle, Thomas (1795–1881), was a Scottish moralist, historian, and essayist who was one of Tyndall's major literary influences. Carlyle was a well-known figure in Victorian Britain and a leading intellectual. Tyndall greatly admired him and his work, and seems to have been very proud to move in similar social circles.

Chasles, Michel Floréal (1793–1880), was a French mathematician, noted for his work on the history of geometry. His main experimental work was in kinematics and dynamics. Chasles was a professor at the École Polytechnique, and was awarded the RS's Copley Medal in 1865 "for his historical and original researches in pure geometry."

Clausius, Adelheid Rimpham (d. 1875), was the wife of the physicist and mathematician Rudolf Clausius, and mother of Rudolf John ("Johnny") Clausius. Tyndall was very close with both of them, and as letters in this volume show, was deeply flattered when they named their son Rudolf John after him and invited him to be godfather, which he accepted.

Clausius, Rudolf (Rudolph) (1822–88), was a renowned German physicist and mathematician and one of the founders of modern thermodynamics. His famous paper on the theory of heat, "Ueber die bewegende Kraft der Warme," was published in the *Poggend. Annal.* in 1850. He taught physics at the Royal Artillery and Engineering School in Berlin in 1850 before moving to a professorship in mathematical physics at the Polytechnikum in Zurich in 1855. Tyndall translated many of his major works into English for publication in the *Phil. Mag.* and *Scientific Memoirs* in the early 1850s. Although he would later move on to the University of Würzburg in 1867, and the University of Bonn in 1869, his time at Zurich proved to be the most academically fertile period in his life. It was here that he developed thermodynamics further, and contributed to the kinetic theory of gases. Clausius was a sounding board for many of Tyndall's ideas, read through the manuscript for Tyndall's *Heat as a Mode of Motion*, and supported Tyndall in his dispute with Thomson and Tait about the Mayer-Joule priority dispute of 1862–63. Clausius was very close to Tyndall, as was his wife, and they named their son Rudolf John after Tyndall and invited him to be godfather, which he accepted. Tyndall often sent gifts to his godson.

Clausius, Rudolf John "Johnny" (1864–1915?), was the son of Rudolf and Adelheid Clausius. He was named after Tyndall, who was also his godfather. John Clausius later completed a doctorate of law in 1890 at Friedrich Wilhelms University Bonn (now University of Bonn), and became a lawyer in Cologne.

Croll, James (1821–90), was a Scottish geologist and climatologist. In 1859 he took up a position at the Andersonian Museum in Glasgow. After several publications in the *Phil. Mag.* and making connections with prominent men of science including Archibald Geikie, William Thomson, and Tyndall, he rose to a position of influence as secretary of the geological survey of Scotland. He made a number of compelling insights into the accumulation of subtle variations affecting climate over time.

Darwin, Charles Robert (1809–82), was an English naturalist, geologist, and author of *On the Origin of Species* (1859). From the 1860s he became almost a figurehead for a small group of men of science who sought to promote science and advance scientific naturalism in British society (although Darwin was not a member). The X Club, as this dining club was called, included George Busk, Edward Frankland, Thomas Hirst, Joseph Dalton Hooker, Thomas Henry Huxley, John Lubbock, Herbert Spencer, William Spottiswoode, and Tyndall.

Davy, Humphry (1778–1829), was born in Penzance, Cornwall, where he was apprenticed to a local apothecary. Largely self-educated, Davy was interested in chemistry, physics, anatomy, and poetry. He later worked at the Pneumatic Institution in Bristol under the physician and lecturer Thomas Beddoes. Davy's first individual publication on nitrous oxide was published in 1800, which led to him being offered a position as assistant lecturer in chemistry and director of the chemical laboratory at the RI in 1801. It was here that he achieved significant fame both as a lecturer and for his chemical research, during which he isolated such elements as sodium, potassium, and chlorine. Davy was famously involved in a priority dispute over the invention of the miners' safety lamp. He served as president of the RS from 1820 to 1827.

Dawes, Mary Helen (née Gordon, 1800–92), was the wife of Richard Dawes, married in 1836 (Westminster Marriages Record; the *ODNB* entry for Richard Dawes incorrectly states 1838). She was the stepdaughter of surgeon George James Guthrie. Their correspondence reveals that both Helen Dawes and her husband enjoyed Tyndall's company, although they lived in Herefordshire, quite some distance from London.

Dawes, Richard (1793–1867), was an English educationist and a graduate of Trinity College, Cambridge. From 1837 he lived in the rural parish of King's Somborne in Hampshire, where he established the King's Somborne school, which educated children from different social classes, promoted the use of simple scientific experiments in a rudimentary laboratory, and used secular reading material for language development. Dawes taught frequently at the school, as did Edward Frankland. From 1850 Dawes served as dean of Hereford, and oversaw the restoration of the cathedral. Both Richard Dawes and his wife, Helen Dawes (née Gordon), were close to Tyndall and both looked forward to and encouraged his visits to them in Herefordshire.

Debus, Heinrich (1824–1915), was a chemical assistant in Bunsen's laboratory in Marburg until early 1851 and a friend to Tyndall, Edward Frankland, and Thomas Hirst at Marburg. Debus left Marburg when Hermann Kolbe replaced Robert Bunsen as professor of chemistry, and took up a position teaching chemistry at Queenwood College, where he succeeded Frankland. He was an excellent teacher. After leaving Queenwood in 1867, he spent three years at Clifton College, Bristol, then returned to London as lecturer in chemistry at Guy's Hospital Medical School (1870–88), adding to this the position of professor of chemistry at the Royal Naval College, Greenwich (1873–88), under Hirst, who was founding director of studies at the college. Debus and Tyndall enjoyed an enduring friendship.

Dove, Heinrich Wilhelm (1803–79), was a Prussian physicist and meteorologist. He was professor at the University of Berlin and director of the meteorological institute. He received the RS's Copley Medal in 1853 "for his work on the distribution of heat over the surface of the Earth," and is considered one of the founders of meteorology.

du Bois-Reymond, Emil Heinrich (1818–96), was a German physician and physiologist who from 1840 had worked under Johannes Müller with Hermann Helmholtz. It was under Müller that he had first developed his lifelong interest in animal electricity. Upon Müller's death in 1858, du Bois-Reymond took up the position of professor of physiology at Berlin, although it was only in 1877 that the Prussian government funded the provision of a modern physiological laboratory there. He wrote extensively on animal electricity, publishing the first volume of his *Investigations of Animal Electricity* in 1848, and the last volume in 1884.

Faraday, Michael (1791–1867), was an eminent natural philosopher and the Fullerian professor of chemistry at the RI and superintendent of the House from 1852. His extensive researches focused on electricity and encompassed electrochemistry, electromagnetism, diamagnetism, and magneto-optics. He was Tyndall's main patron and helped him obtain the position of professor of natural philosophy at the RI in 1853. Despite significant differences in their views on religion and on the nature of magnetism, they worked closely together. Tyndall greatly respected Faraday and on one occasion commented to Hirst that the relationship was akin to that of father and son. Tyndall's *Faraday as a Discoverer* (London: Longmans, Green, 1868) was his tribute to his mentor following Faraday's death in 1867.

Faraday, Sarah (née Barnard, 1800–79), was the wife of Michael Faraday, and lived in their rooms at the RI. Tyndall frequently met her and commented in his journal, "She is very motherly to me, and Faraday says that were it not for her own feebleness she must become quite a mother to me, and take care of me" (2 March 1855, RI MS JT/2/13c/730).

Forbes, James David (1809–68), was a Scottish physicist and glaciologist. He was a strong advocate of the BAAS and of university reform to include rigorous examinations in mathematics and the natural sciences. Forbes held the chair of natural philosophy at the University of Edinburgh from 1833 to 1860. His early research in the 1830s focused on heat; he discovered the polarization of radiant heat and the double refraction of thermal radiation. Tyndall did not agree with all of Forbes's work in this field and in 1853 had written a paper that undermined Forbes's ideas on the vibration of heated metals. Like Tyndall, though, Forbes was an avid walker and climber, and had climbed the Jungfrau in 1841, publishing *Travels through the Alps of Savoy* in 1843. It was Forbes's later work on glaciers that brought him into more general controversy, however, as well as into direct opposition to Tyndall once more. Forbes's position on glacial movement and its physics was at odds with Tyndall's views, and in 1856 Tyndall had published rejecting Forbes's view that glacial movement was a result of what he called their "viscosity" and "plasticity." Tyndall argued instead that glacial motion could best be explained by fracture and regelation of the ice. Tyndall's dispute with Forbes was not limited to this aspect of Forbes's work, and Tyndall repeatedly and publicly accused Forbes of making unacknowledged use of Louis Rendu's earlier work. From 1860 Forbes moved to Saint Andrews University to become principal of the college of Saint Salvator and Saint Leonard (the United College).

Francis, William (1817–1904), was the illegitimate son of the printer and publisher Richard Taylor, with whom he entered the publishing partnership Taylor & Francis in 1852. Francis had learned the printing trade in Germany in the 1830s before training as a chemist with Justus von Liebig at the University of Giessen, where he took his doctorate in 1842. He made many translations for his father's *Scientific Memoirs* and the *Phil. Mag.* and became the managing editor of the latter in 1851. He first met Tyndall in October 1850 and became a close friend, facilitating his entry into the scientific community in the 1850s. The two remained close, Tyndall working as editor for the *Phil. Mag.* for several years. Tyndall oversaw the translation and publication of the work of a number of European scholars, notably in regard to this volume, including the physicist C. K. Akin, with whom he later engaged in a bitter priority dispute.

Frankland, Edward (1825–99), was initially an assistant to the chemist Lyon Playfair at the Putney College of Engineering, where he met and befriended Hermann Kolbe (1818–84), who had studied under Bunsen at Marburg. In 1847 Frankland and Kolbe traveled to Marburg, where Frankland met Sophie Fick (1821–74), whom he later married. Returning to England, Frankland obtained a teaching position at Queenwood College, where he met Tyndall and with whom he taught one of the first lab-based science courses in England. Frankland lectured not only at Queenwood but also in the nearby village, where the vicar was Richard Dawes, a pioneer in popular science education who was also a close friend and lifelong correspondent of Tyndall's. In 1848 Frankland and Tyndall spent a month-long holiday in France, during which Frankland was briefly caught up in the July uprising in Paris. In October the same year, the two men traveled to Marburg to study. Frankland later moved on to Justus von Liebig's laboratory at the University of Giessen in 1849. Returning to England, in 1850 Frankland briefly held the position of professor of practical chemistry at the Putney College of Engineering, before being appointed to the chair of chemistry at Owens College, Manchester in 1851. With Tyndall's support, Frankland was elected FRS in 1853 and was awarded the RS's Royal Medal in 1857 "for the isolation of the organic radicals of the alcohols, and for his researches on the metallic derivatives of alcohol." Following Faraday's retirement from the RI in 1861, Frankland held a temporary position there from May 1863 to 1868. From 1865 he also took on a temporary position at the Royal College of Chemistry, which became permanent in 1868. From 1865, and in consequence of this post, Frankland also became examiner of papers for the Department of Science and Art, and, along with Tyndall, Thomas Hirst, Joseph Dalton Hooker, Thomas Henry

Huxley, and four others, became a member of the X Club, a monthly dining club through which they promoted science and advanced scientific naturalism in British society.

Fresnel, Augustin-Jean (1788–1827), was a French engineer and physicist who contributed significantly to the establishment of the theory of wave optics. Fresnel studied the behavior of light both theoretically and experimentally. He is perhaps best known as the inventor of the Fresnel lens, first adopted in lighthouses while he was a French commissioner of lighthouses.

Gassiot, John Peter (1797–1877), was an electrician and wine merchant by trade. He was one of the founding members of the London Electrical Society in 1837, and served as the society's treasurer until 1841. Gassiot pursued research on electricity in the laboratory he outfitted at his home in Clapham Common, near London. During the 1860s James Clerk Maxwell worked there while establishing the unit for electrical resistance. Gassiot published papers on the functioning of the voltaic pile (battery) and showed that a spark could be produced before the circuit of a battery was closed.

Gill, Joseph (1810–76), was a civil engineer from Marsala, Sicily, who worked on steam engines and thermodynamics. Tyndall had overseen publication of Gill's letter "On the Dynamical Theory of Heat" in the *Phil. Mag.* in 1863. Gill again wrote to Tyndall on the subject in 1864, and this was also published (see letters 2062 and 2081).

Ginty, Margaret "Peb" (née Roberts, b. *c.* 1826), was the wife of Tyndall's close friend William Gilbert Ginty, married on 18 May 1846, in Killucan, Westmeath, Ireland. The Gintys later relocated to Manchester, where they appear on the 1851 English census along with three children: Kathleen, William Roberts, and Fanny; and later on, to Rio de Janeiro.

Ginty, William Gilbert (*c.* 1820–66), was born at Carrick-on-Shannon, County Leitrim, Ireland. Ginty was one of Tyndall's close friends; the two had worked and lived together in the 1840s working on the Ordnance Survey in Lancashire, Cumberland, and Westmorland, particularly in Kirby Lonsdale. Ginty later became a prominent Manchester businessman, served as honorary secretary of the Manchester Athenaeum, and was involved in various civil engineering projects, including the construction of the Liverpool-Bolton railway before, in 1854, moving to Rio de Janeiro. For his contributions to civic projects in Rio, he was awarded a knighthood in the Order of the Rose of the Brazilian Empire. He and Tyndall enjoyed a lifelong correspondence.

Graham, Thomas (1805–69), was a Scottish chemist and physicist. Graham began his studies at the University of Glasgow in 1819, and was inspired by the lectures of Thomas Thomson to focus on chemistry. He graduated with an MA from Glasgow in 1826, and became professor of chemistry at Anderson's College in 1830. Graham was particularly interested in gaseous diffusion, and it was at Anderson's that he first formulated what is now known as "Graham's Law," and elucidated what made the three phosphoric acids distinct from one another. After becoming the professor of chemistry at University College, London, in 1837, he published the textbook *Elements of Chemistry* (London: Hippolyte Bailliere, 1842), which was used both in England and on the European mainland. From 1854 onward, he succeeded John Herschel as master of the mint and was the last person to hold this position.

Grove, William Robert (1811–96), was born in Swansea, Wales. Grove studied classics at Brasenose College, Oxford, and law at Lincoln's Inn. He was called to the bar in 1835, the same year that he joined the RI and became a founding member of the Swansea Literary and Philosophical Society, which later became the Royal Institution of South Wales. In 1840 Grove was elected FRS. Despite these successes, Grove continued to struggle to find experimental space in London and complained publicly about the difficulty in getting work published in the *Phil. Trans.*, and what he saw as the closed shop that allowed only those with the right connections. He argued that British science needed thorough reform. In 1844 he became vice-president of the RI, the same year in which he was elected to the council of the RS and his influential *On the Correlation of Physical Forces* was published, a work that foreshadowed the theory of the conservation of energy.

Haast, Johann Franz Julius von (1822–87), was a German explorer and geologist. He worked for a time in England, but emigrated to New Zealand in 1858. He founded the Canterbury Museum at Christchurch, and from 1863 made a name for himself as a geologist.

Hall, John (dates unknown), was an engineering surveyor who lived in Mrs. Wright's boarding house concurrently with Tyndall in the mid-1840s. He appears to have engaged in religious conversations with Tyndall and Hirst.

Harris, William Snow (1791–1867), was an Edinburgh-trained physician who spent his life in Plymouth. After his marriage Harris had enough money to give up all but occasional private practice, and dedicated himself instead to his interest in electricity and magnetism. He invented an improved compass and the disc electrometer, but his work on lightning conductors for ships proved the most significant. Harris was well known and respected both in science and society (Darwin dined with him about a month prior to HMS *Beagle*'s departure from Plymouth). His work, which focused on static electricity, became dated in his lifetime, however, although his *Treatise of Frictional Electricity* was published posthumously.

Helmholtz, Hermann Ludwig Ferdinand von (1821–94), was a German physicist and physician. Educated in military medicine at the Friedrich-Wilhelm University in Berlin, he became the first to hold a regular post as teacher of anatomy at the Berlin Academy of Arts in 1848. In 1847 he published *Über die Erhaltung der Kraft*, in which he discussed the mathematical principles of the conservation of energy. He became professor of physiology at the University of Königsberg, and became professor for anatomy and physiology at the University of Bonn in 1855. It was at Bonn that he turned his attention to physics, a process that continued during his time at Heidelberg from 1858 and also in Berlin, where he took up the position previously occupied by Gustav Magnus. Called "the intellectual giant" by James Clerk Maxwell, Helmholtz was involved in the fields of hydrodynamics and electrodynamics, but by the early 1860s he focused on acoustics. His major work on sound, the subject of several letters in this volume, was published in 1863.

Herschel, Alexander Stewart (1836–1907), was the son of John Herschel. An astronomer and professor of physics at Glasgow, he is known primarily for his work on meteor spectroscopy and for his identification of comets. As letters in this volume show, Tyndall aided in arranging for him to lecture at the RI.

Herschel, John Frederick William (1791–1871), was a mathematician and astronomer. He was the son of the famous astronomer Frederick William Herschel and himself became one of the most eminent scientists in Britain in the early Victorian era. His *Preliminary Discourse on the Study of Natural Philosophy* (1830), in which he laid out his ideas about the practice of sound inductive science, was widely read and greatly admired. He became Sir John in 1838, when, on his return from mapping the southern skies from Cape Town, he was made a baronet. In 1849 he published his *Outlines of Astronomy*, a popular presentation of astronomy that went through eleven editions by 1871. Tyndall consulted with Herschel on a number of issues related to his research.

Hirst, Thomas Archer (1830–92), was a Yorkshire mathematician and educator. His friendship with Tyndall had begun in Halifax in 1845 when they were both working for the surveyor Richard Carter. Encouraged by Tyndall, he studied at the universities of Marburg, Göttingen, and Berlin between

1849–52. In 1853, he returned to England with a doctorate to replace Tyndall as a mathematics teacher at Queenwood College. Like Tyndall, Hirst did not get on with the headmaster, George Edmondson. He resigned in June 1856 to look after his sick wife and stayed on the Continent until 1860. He was Tyndall's closest and lifelong friend, as reflected in their extensive correspondence. In this volume, Hirst and Tyndall discussed Hirst's efforts to secure a new position.

Holzmann, Maurice (1835–1909), was a German-born member of the Royal household for almost a half century, serving Edward VII as Prince of Wales and King. Holzmann was also an avid mountaineer, and joined the Alpine Club in 1874 (*Dod's Peerage, Baronetage and Knightage of Great Britain and Ireland for 1904* [London: Sampson Low, Marston, 1904], 492; "Sir Maurice Holzmann, K.C.B., K.C.V.O., I.S.O.," *Alpine Journal* 25 [1911]: 349–50).

Hooker, Joseph Dalton (1817–1911), was a British botanist who specialized in taxonomy and geographical distribution of plants, ultimately the most famous British botanist of the Victorian era. He was the younger son of William Jackson Hooker (1785–1865), the director of the Royal Botanic Gardens, Kew, and joined his father as deputy director in 1855. Hooker was an intimate friend of Charles Darwin and Thomas Henry Huxley. He was first introduced to Tyndall by Huxley in November 1855 and the three became close friends. Like Tyndall, Hooker was a keen and talented mountaineer, but unlike Huxley and Tyndall, Hooker came from a socially well-placed family, and this gave him many connections both in science and society that came less easily to his friends. With Charles Lyell, Hooker was instrumental in bringing Darwin's theory of natural selection to public attention. Hooker played a significant role in making the case that men of science should receive government support for their work. In 1864, with Thomas Henry Huxley, Tyndall, Thomas Hirst, and five others, he established the X Club, a monthly dining club through which they promoted science and advanced scientific naturalism in British society.

Hopkins, William (1793–1866), was an English mathematician and geologist. He entered Cambridge University as an undergraduate at Peterhouse in 1822 after the death of his first wife. He graduated the Mathematical Tripos as seventh wrangler in 1827 and received his MA in 1830. Settling in Cambridge, Hopkins became a private mathematics tutor training undergraduates for the Tripos exams. He was one of the most successful tutors of his time and became known as the "senior wrangler maker." In his research he was one of the most significant proponents of the solidity of the earth and, as seen in

Tyndall Correspondence, vol. 7, supported Tyndall in his dispute with James David Forbes over their conflicting theories of glacial motion.

Humboldt, Alexander von (1769–1859), was a world-renowned Prussian naturalist and explorer, and by the 1850s one of the most notable scientific figures in Europe. From 1799 to 1804 Humboldt journeyed through South America with the French botanist Aimé Bonpland (1773–1858), and published thirty volumes upon his return. His meticulous attention to climate, temperature, and the ecology of his discoveries was made possible by the many scientific instruments that he took with him, and he defined a new era of accuracy in science. His wide-ranging *Kosmos* appeared in five volumes between 1845 and 1862. Financial constraints stemming from the publication of his works forced Humboldt to spend the last thirty years of his life in Berlin working for the Prussian government.

Huxley, Henrietta "Nettie" Anne (née Heathorn, 1825–1914), was the wife of Thomas Henry Huxley. She was born in the West Indies to Henry Heathorn and Sarah Henrietta Richardson (née Harris), and schooled in Neuwied, Germany, for two years in her youth. In 1843 she, her mother, and half sister moved to Australia to join her father. She was keeping house for her sister Oriana and brother-in-law William Fanning in Sydney, Australia, when she met Huxley in 1847, who was then a surgeon and zoologist on HMS *Rattlesnake*. After an eight-year engagement she arrived in London in 1855, and they were married. She assisted her husband in his work by translating German and drawing diagrams for his lectures. She was also a poet, and published a volume of her collected poems at the age of eighty-six, *Poems of Henrietta A. Huxley with Three of Thomas Henry Huxley* (London: Duckworth, 1913). As with many of the wives of this small group of scientific friends, Henrietta was very much a part of their community; she corresponded with the Darwin's as well as with Tyndall. The wives of the X Club members were referred to as the Y-ves ("Death of Mrs. Huxley," *Times*, 6 April 1915, 12).

Huxley, Thomas Henry (1825–95), was a British zoologist, anatomist, educator, and science popularizer. Following medical training at Charing Cross Hospital in London, Huxley acted as a ship's surgeon (1846–50) on HMS *Rattlesnake*. After eking out a living by writing for Chapman's *Westminster Review*, in 1854 he was appointed lecturer on natural history at the Royal School of Mines, and paleontologist to the Geological Survey in 1855. In the same year he joined Tyndall (whom he first met in 1851 when traveling to the Ipswich meeting of the BAAS), at the RI as Fullerian Professor of Natural History. Huxley was also a lecturer at Saint Thomas's Hospital. While in Australia he met Henrietta ("Nettie") Anne Heathorn, whom he married in July

1855 after an eight-year engagement. Along with Tyndall and others, Huxley was interested in establishing a scientific journal—the *Reader*—as shown in letters in this volume, and was a founding member of the X Club, a monthly dining club through which with he, Tyndall, Thomas Hirst, Joseph Dalton Hooker, and five others promoted science and advanced scientific naturalism in British society.

Jewsbury, Geraldine Endsor (1812–80), was a novelist, reviewer, and writer. She became acquainted with Thomas and Jane Carlyle after having written to Thomas for advice on developing a literary career, and subsequently began an enduring and intimate relationship with Jane. Her novels reflect much of her feelings about this relationship and her advocacy for women's active role in society. She corresponded with Tyndall across a number of years.

Jones, Henry Bence (1813–73), was a British physician and medical chemist. He was educated at Harrow and Trinity College, Cambridge, before studying medicine at Saint George's Hospital, London. Attracted by the potential applications of chemistry to clinical medicine, he studied organic chemistry with Justus von Liebig at Giessen (1841–43), before assuming a career as a successful private physician and physician at Saint George's Hospital. Though not elected secretary of the RI until 1860, as a manager from 1853 and close friend of Faraday's, he supported Tyndall's appointment there in 1853. He frequently entertained Tyndall at both his London home and Folkestone, where he kept a summer house.

Joule, James Prescott (1818–89), was a physicist and mathematician. The son of a brewer, he studied for a brief time under John Dalton before focusing his own work on electromagnetism. In 1838 Joule published his first paper on electromagnetic engines. This research led to his formulation of the mechanical equivalent of heat, which he presented to the BAAS in 1843, and for which the RS awarded him a Royal Medal in 1852. However, as we see at the start of this volume (and in letters across *Tyndall Correspondence*, vol. 7), Tyndall had come to believe that the German physicist Julius Mayer deserved credit for the discovery rather than Joule alone. Tyndall appealed to his friends and correspondents in Germany to send him copies of Mayer's works, which convinced him of the truth of his position and he then set about publishing and promoting Mayer's work in England, a move that upset Joule and his colleagues.

Kirchhoff, Gustav Robert (1824–87), was a German physicist. In Königsberg, Kirchhoff studied physics under Franz Neumann and mathematics under Friedrich Julius Richelot. His formulation of the laws of electrical circuits was part of his doctoral dissertation, which he finished at Königsberg in 1847. During an associate professorship at the University of Breslau in 1850, he became acquainted with Robert Bunsen, who made sure that Kirchhoff followed his move to Heidelberg in 1854. Together with Bunsen, he worked on thermal radiation and spectrum analysis. Regarding his law of thermal radiation, he would fight a long struggle for recognition of priority with the Scottish physicist Balfour Stewart (1828–87). Kirchhoff and Tyndall were close friends and colleagues. Tyndall had recommended him for the RS's Rumford Medal, which he received in 1862; Kirchhoff was unable to attend the ceremony and Tyndall accepted it on his behalf.

Knoblauch, Karl Hermann (1820–95), was a German physicist most notable for his studies of radiant heat. He was one of the six founding members of the Deutsche Physikalische Gesellschaft at Berlin on 14 January 1845. As a researcher and teacher at the University of Marburg (1849–53), he produced experimental demonstrations about the nature of diamagnetism. Tyndall was Knoblauch's student and collaborator on the diamagnetism work during his time at Marburg. They kept up a correspondence for over twenty-five years.

Lamé, Gabriel (1795–1870), was a French mathematician who contributed to the theory of partial differential equations by the use of curvilinear coordinates, and the mathematical theory of elasticity.

Laplace, Pierre-Simon, Marquis de (1749–1827), was an influential French scholar whose work was important to the development of mathematics, statistics, physics, and astronomy.

Leslie, Sir John (1766–1832), was a Scottish mathematician and physicist who performed significant research into heat.

Linth, Arnold Escher (von der) (1807–72), was a Swiss geologist and son of Hans Conrad Escher von der Linth (1767–1823). A student of Horace-Bénédict Saussure in Geneva, and Alexander von Humboldt in Berlin, he became professor of geology at the École Polytechnique in Zurich in 1856. In 1842 he was a member of the first party to ascend the Lauteraarhorn. He produced, in collaboration with Bernhard Studer, a detailed geological map of Switzerland in 1853. For his father's work diverting the Linth River the family was given the surname "von der Linth."

Lubbock, Ellen Frances (née Horden, 1835–79), the daughter of a Lancashire Anglican minister, was the wife of John Lubbock, married in 1856. They had three sons and three daughters. Ellen Lubbock and Tyndall enjoyed a long and intimate correspondence, and she, perhaps even more so than Henrietta Huxley, was an enthusiastic supporter and facilitator of the small group of men who, from 1864, made up the X Club, a monthly dining club that included Tyndall, Thomas Hirst, Joseph Dalton Hooker, Thomas Henry Huxley, and five others, that worked to promote science and advance scientific naturalism in British society.

Lubbock, John (1834–1913), was a banker, anthropologist, natural historian, and politician. In his youth he befriended Charles Darwin and John Tyndall. During the 1860s Lubbock published a number of influential archeological texts, one of which was *Pre-historic Times: As Illustrated by Ancient Remains.* His research also included biology, which helped him into the X Club, of which he was the youngest member (this monthly dining club, which also included Tyndall, Thomas Hirst, Joseph Dalton Hooker, Thomas Henry Huxley, and four others, worked to promote science and advance scientific naturalism in British society). The Lubbocks were among Tyndall's closest friends; Lubbock joined Tyndall and Huxley on their climbing adventures in the Alps on a number of occasions, and he corresponded with them frequently.

Lyell, Charles (1797–1875), was a highly influential geologist. Trained as a lawyer, Lyell become a professor of geology at King's College. He was most noted for his controversial three-volume work *Principles of Geology* (1830–33), in which he argued that the only legitimate type of causal explanation for the past had to be of the kind seen in operation in the present; natural laws were nothing if not consistent. In consequence of this philosophical commitment, Lyell was forced to argue that the earth was in fact many millions of years older than even the most generous of his contemporaries had hitherto been willing to admit; the "catastrophism" of Georges Cuvier was then the dominant philosophy among British geologists. Although controversial, Lyell was highly regarded, both as a philosopher of science and as a geologist. He had mentored Charles Darwin from his return to England from his voyage on HMS *Beagle* in 1836, and in 1858, along with Joseph Dalton Hooker, had pressed Darwin to publish an abstract of his big species book, which became *On the Origin of Species* (1859). By the 1860s Lyell was one of the elder statesmen of science.

Magnus, Heinrich Gustav (1802–70), was a German chemist and physicist, who received his degree from the University of Berlin in 1827 under Mitscherlich. Magnus spent brief periods in Stockholm and Paris, but returned to the University of Berlin (*Privatdozent* in 1831, extraordinary professor in 1833, and ordinary professor in 1845). He was an extremely popular teacher and established a world-renowned laboratory in Berlin. Tyndall worked in Magnus's laboratory for several months in 1851.

Maxwell, James Clerk (1831–79), was a Scottish physicist and mathematician. From the late 1850s onward he developed Michael Faraday's "field theory," which had been put into mathematical expression by William Thomson some years previously. His formulation of what are now known as Maxwell's equations set the foundation for the classical theory of electromagnetism. Maxwell's work culminated in his 1873 publication *A Treatise on Electricity and Magnetism.*

Mayer, Julius Robert von (1814–78), was a German physicist, chemist, and physician. In 1842 he published a short text that contained one of the first statements on the conservation of energy and the mechanical equivalent of heat. Mayer was a scientific outsider; his arguments in this paper were based on philosophical principles and thought experiments, and his papers on heat and the conservation of force received little recognition in the 1840s. In the late 1850s and early 1860s, Tyndall sought to revive Mayer's work and promote him as the co-discoverer of energy conservation along with James Joule. The ensuing priority dispute between Mayer and Joule, which was largely fought out between their proxies—Tyndall on the one hand and William Thomson and Peter Guthrie Tait on the other—was carried on in increasingly acrimonious terms both in private correspondence and then in the public arena.

Melloni, Macedonio (1798–1854), was an Italian physicist, notable for demonstrating that radiant heat has similar physical properties to those of light, for which he received the RS's Rumford Medal in 1834.

Mitchell, Elizabeth (née Tyndale, 1812–66), was a sister of Hector Tyndale who married Edward Mitchell ("Died," *Philadelphia Inquirer*, 3 November 1866, 5).

Moigno, François-Napoléon-Marie (1804–84), was a French Catholic priest, physicist, writer, and editor. Educated at a Jesuit seminary, he studied science alongside theology and was given the Mathematical Chair at Jesuit house in Paris in 1836. After leaving the order in 1843 he became the science editor of *L'Epoque*, and later became the editor of the scientific journal *Cosmos* (1852–62). He founded the journal *Les Mondes* in 1863 and undertook the translation and publication of many of Tyndall's works in French, including *Radiation* (1867), *Le son* (1869), *La Chaleur* (1864), and *Faraday Inventeur* (1868) ("L'Abbé Moigno," *Nature* 30 [1884]: 291–92).

Moore, Harriet Jane (1801–84), was a British watercolor artist who painted scenes of Faraday's apartment, study, and laboratory at the RI during the 1850s. Moore was the eldest child of surgeon James Carrick-Moore and Harriet Moore (née Henderson). She was elected a member of the RI in 1852. See Frank A. J. L. James, "Harriet Jane Moore, Michael Faraday, and Moore's Mid-Nineteenth-Century Watercolours of the Interior of the Royal Institution," in *Fields of Influence: Conjunctions of Artists and Scientists, 1815–1860*, ed. James Hamilton (Edgbaston: University of Birmingham Press, 2001), 111–28.

Moore, John Carrick (1805–98), was a field geologist and the eldest child of surgeon James Carrick-Moore and Harriet Moore (née Henderson). He entered Queen's College, Cambridge, in 1823, receiving his BA in 1827 and MA in 1830. He was called to the bar in 1831, but became interested in geology and devoted much of his life to research. He was an active member of the Geological Society, becoming a fellow in 1838 and serving as secretary (1846–52). He was also a member of the RI from 1836, becoming a visitor in 1853 and a manager in 1857. Later in life he would write *Recollections of an Octogenarian* (London: Chiswick Press, 1888).

Moore, Julia (1804–1904), the youngest of the three daughters of surgeon James Carrick-Moore and Harriet Moore (née Henderson). The wealthy Moores had close connections with the RI and frequently entertained Tyndall at their Wimbledon home.

Murchison, Charlotte Hugonin (1788–1869), was a Scottish geologist who married the geologist Roderick Impey Murchison. She had a strong influence in the career of her husband; indeed, it has been suggested that it was through her influence that he took up geology. Murchison was noted for her own skills both in fossil collecting and in sketching landscapes. In addition to her work, she moved in some of the most learned circles of Victorian Britain, and gained the friendship of people like the geologist Charles Lyell and the Scottish science writer Mary Somerville (1780–1872).

Murchison, Roderick Impey (1792–1871), was a geologist and geographer. He was elected fellow of the Geological Society in 1825, and was twice its president. In 1826 he was elected FRS. He was also one of the founders of the BAAS, and was president of the geographical and geological sections ten times in total. One of his important contributions to geology was his book *The Silurian System* (1839), which established the oldest fossil-bearing classification then known. In January 1859 Murchison proposed Tyndall for membership of the Athenaeum Club.

Murray, John (1808–92), was an English publisher and the eldest child of Anne Elliot and publisher John Murray. This Murray was the third generation to run the publishing house. He was educated at Charterhouse School (1819–26) and for a year (1827) at Edinburgh University. He began his career in the family firm at 50 Albemarle Street, London, in 1828 and became head of the firm after his father's death in 1843. Murray published the *Quarterly Review*, a popular series of *Handbooks for Travellers*, and numerous scientific publications. On 24 November 1859, Murray published both Darwin's *On the Origin of Species* and Samuel Smiles's *Self-Help*. He became a member of the Alpine Club in 1858 and in 1860 published Tyndall's *Glaciers of the Alps*. Murray's house is at the opposite end of the street and across the road from the RI.

North, Frederick (1800–69), was a British Liberal politician who served as a Member of Parliament for Hastings. The highest point in Hastings is called North's Seat in his honor. On occasion he climbed with Tyndall, and along with Tyndall and the rest of their party, was caught in an avalanche, although all escaped serious injury.

Owen, Richard (1804–92), was a comparative anatomist and paleontologist. Despite being a controversial figure, Owen was an outstanding naturalist with a remarkable gift for interpreting fossils; indeed, he was often referred to as the "the English Cuvier." It seems clear that he was an awkward man socially, and although initially encouraging of younger scholars was also quickly jealous of their successes. From the 1860s Owen fell afoul of Charles Darwin, Thomas Henry Huxley, and other soon-to-be members of the X Club following his critical review of Darwin's *On the Origin of Species* (1859). Owen and Huxley were bitter rivals, Huxley leading the Darwinians in opposition to Owen's efforts to establish a national museum; Owen eventually prevailed, and the Natural History Museum was founded in South Kensington in 1881.

Poggendorff, Charlotte Eleonore (née Kneser, 1805–65), was the wife of Johann Christian Poggendorff. Emil du Bois-Reymond referred to her as "Mother of the Annals" (*Poggend. Annal.*), which suggests that she played a significant role assisting her husband in this endeavor as an editor for the journal. She died on 5 January 1865 due to inflammation of the lungs, as noted in letters in this volume from Hermann Knoblauch and Rudolf Clausius (E. Frommel, *Johann Christian Poggendorff: Leichenrede von Emil Frommel, Hof- und Garnison-Prediger. Nebst eigenhändigen Lebensnachrichten, Reden und Briefen* [Berlin: Dümmler, 1877], 12; G. Finkelstein, *Emil du Bois-Reymond: Neuroscience, Self, and Society in Nineteenth-Century Germany* [Cambridge: MIT Press], 147).

Poggendorff, Johann Christian (1796–1877), was a German physicist at the University of Berlin. His major contribution to science was as editor for fifty-two years (1824–76) of *Annalen der Physik und Chemie*, which became known as *Poggendorff's Annalen* during his tenure as editor. His wife, Charlotte Eleonore Poggendorff, aided in his editorial work. He used his broad knowledge of science and scientific men in producing a *Biographisch-literarisches Handworterbuch zur Geschichte der exacten Wissenschaften,* the first two volumes of which were published in 1863. At Tyndall's invitation, in 1862 Poggendorff and his daughter visited London and stayed at Waverley Place in Tyndall's absence. Poggendorff's wife died at the beginning of 1865.

Pollock, Frederick (1845–1937), third baronet, was the eldest son of William Frederick and Juliet Pollock. He attended Eton, where he was King's Scholar, attended Trinity College, Cambridge, where he became a Cambridge Apostle, and was elected fellow of Trinity in 1868. Frederick became Corpus Professor of Jurisprudence at Oxford (1883–1903), and was the editor of the *Law Reports* and the author of *The History of English Law.* He was known among family and friends as "Fred."

Pollock, Jonathan Frederick (1783–1870), first baronet, was a British judge, politician, and father of William Frederick Pollock (1815–88). He was educated at Cambridge University, graduating in 1806 as senior wrangler. He was Member of Parliament for Huntington (1831–44), and served twice as attorney general (1834–35, 1841–44). He was Lord Chief Baron of the Exchequer from 1844 to 1866, and received a baronetcy on his retirement. Pollock was referred to within the family as "the Chief Baron."

Pollock, Juliet (née Creed, 1819–99), was the daughter of Catherine Herries and the Reverend Henry Creed, vicar of Corse, Gloucestershire. She married William Frederick Pollock in 1844, with whom she had three sons: Frederick Pollock, Walter Herries Pollock, and Maurice Emilius Pollock. She and her husband were loyal supporters of the RI, attended Tyndall's lectures, frequently entertained Tyndall at dinner, and were among his closest friends during the period covered by this volume. Juliet Pollock was the author of several books, including *Julian and His Playfellows* (London: Grant and Griffith, 1852), *New Friends* (London: John W. Parker and Son, 1858), and *Macready as I Knew Him* (London: Remington, 1885). It was through Pollock that Tyndall met Emily Tennyson (née Sellwood, 1813–96), wife of the poet Alfred Tennyson.

Pollock, Maurice Emilius (1857–1932), third son of William Frederick and Juliet Pollock. He became a sculptor and married Lydia Helen Roberts

(1853–84) in 1880, and Mabel Mary McPherson (1860–1937) in 1889. He was known within the family circle affectionately as "Moss."

Pollock, Walter Herries (1850–1926), second son of William Frederick and Juliet Pollock. He was educated at Eton College (1862–65) and was admitted to Trinity College, Cambridge, where he obtained a BA (1871) and MA (1875). He was admitted to the inner temple in 1870 and called to the bar in 1873. Pollock became a writer, poet, and journalist, and was editor of the *Saturday Review* (1884–94). He married Emma Jane Pipon in 1876.

Pollock, William Frederick (1815–88), second baronet, was a lawyer, author, and the eldest son of Jonathan Frederick Pollock and Frances Rivers (d. 1827). He married Juliet Creed in 1844, and together they had three sons: Frederick Pollock, Walter Herries Pollock, and Maurice Emilius Pollock. Pollock was admitted to Trinity College, Cambridge, in 1832, obtaining his BA (1836) and MA (1840). He was admitted to the inner temple in 1833, and called to the bar in 1838. He became a master of the court of exchequer in 1846 and was appointed queen's remembrancer in 1874. Pollock's two-volume *Personal Remembrances* (London: Macmillan, 1887) illustrates how the family circulated among London's elite and entertained at their Montagu Square home. Both William and Juliet were loyal supporters of the RI, and bought into the *Reader* with Tyndall and others at the end of 1864. Tyndall was frequently among their household guests, both in London and on the Isle of Wight.

Ramsay, Andrew Crombie (1814–91), was a Scottish geologist born in Glasgow. His study of geology was encouraged by John Nichol, and through this support Ramsay's abilities were recognized at the 1839 BAAS meeting in Glasgow. After working as a local director for the Geological Survey of Great Britain he became professor of geology at University College London (1848). He held the chair of geology at the School of Mines from 1851 to 1872 (the Royal School of Mines from 1863), and in 1872 succeeded Roderick Murchison as director of the geological survey. He married Louisa Williams in 1852 and on their honeymoon in Europe became interested in glaciers. Quickly becoming a leading expert in the field, Ramsay argued that glaciers formed some lake basins. He accompanied Tyndall to the Alps in 1858.

Rankine, William John Macquorn (1820–72), was a Scottish mechanical engineer. He did extensive work on steam engines, and was a founding contributor, with Rudolf Clausius and William Thomson, to the science of thermodynamics, particularly focusing on the first law of thermodynamics, also known as the law of energy conservation.

Regnault, Henri Victor (1810–78), was a French chemist, physicist, and photographer. A pupil of Justus von Liebig's at the University of Giessen, he was professor of physics at the Collège de France (1841–54), and director of the famous porcelain factory at Sèvres (1854–70). He suffered a serious accident in 1856, to the great concern of Faraday.

Rendu, Louis (1789–1859), was a French glaciologist and Roman Catholic bishop of Annecy. In 1840 he published *Théorie des glaciers de la Savoie* (Chambéry: Chez Puthod), in which he explained glacial motion in terms of flowing ice. The Scottish glaciologist and physicist James David Forbes was later accused of borrowing work from Rendu in his suggestion that glaciers behaved like viscous bodies. Tyndall was one of Forbes's most outspoken opponents, arguing that ice, being a hard, crystal-like solid, could not have viscous properties.

Reusch, Friedrich Eduard (1812–91), was professor of physics at the University of Tübingen.

Roche, Annette (dates unknown), wife of James Roche (1770–1853), was an active supporter of literary, philanthropic, and mercantile movements in Cork. She was a good friend of Tyndall's and cared for his mother and sister, and as shown in letters in this volume, arranged for the erection of a tombstone over the grave of his father in Ireland.

Sabine, Edward (1788–1883), was an Anglo-Irish astronomer, geophysicist, military officer, and science administrator who led the British effort to establish magnetic observatories all over the globe. He was graduate of the Royal Military Academy and was the astronomer for the Ross (1818) and Parry (1819–20) expeditions to the Arctic in search of the Northwest Passage, and received the RS's Copley Medal in 1821 for his magnetic observations in the Arctic. Sabine became an outstanding lobbyist for science and an effective administrator, and from 1851 was a valuable patron to Tyndall. In the early 1850s he was treasurer of the RS, general secretary of the BAAS, and held the rank of lieutenant-colonel in the army. As president-elect of the BAAS in 1852, Sabine had offered Tyndall one of the secretary positions at the Belfast meeting, and later proposed Tyndall for election to the RS, using his influence to help garner support and gather signatures. He served as president of the RS from 1861 to 1871. Sabine published a series of articles titled "Contributions to Terrestrial Magnetism" in the *Phil. Trans.* across many decades (1836–78).

Seguin, Marc (1786–1875), was a French engineer and physicist. Seguin developed an interest in the science of heat from his granduncle Joseph Montgolfier, the famous balloonist. As an engineer Seguin oversaw the construction of France's first suspension bridge, developed France's rail network, and made improvements to Stevenson's steam engine. Although he was a significant figure in French science, it was not until 1845 that he was elected a correspondent of the Académie des Sciences. He was well known to John Herschel, Michael Faraday, and Humphry Davy. His most important work was his 1839 *De l'influence des chemins de fer*, a handbook for railroad construction and design. It was in this work that he rejected the caloric theory of heat, then popular in France, and suggested instead that a certain quantity of heat must be lost in the production of force and mechanical power. This was the basis for his later claim to priority over Joule and Mayer for the discovery of the mechanical equivalent of heat. Although his priority claim was ultimately unsuccessful, in 1853 he established the journal *Cosmos* under the editorship of Francois Moigno in part to advance his own views. *Cosmos* became a significant journal in French science.

Spencer, Herbert (1820–1903), was a radical philosopher, social theorist, journalist, and sociologist. Spencer believed there to be underlying principles that connected the social, political, and moral aspects of the universe, and it was to the elucidation of this belief that he dedicated much of the rest of his life in the publication of his multivolume synthetic philosophy. He met and befriended Tyndall in the 1850s through their mutual connection, John Chapman, then owner of the radical *Westminster Review*. Although Spencer was increasingly hampered by poor health and social anxiety, he still enjoyed walking tours with both George Henry Lewes and Tyndall. In 1865 he joined with Tyndall, Thomas Hirst, Joseph Dalton Hooker, Thomas Henry Huxley, and four others to form the X Club, a monthly dining club through which they promoted science and advanced scientific naturalism in British society.

Spottiswoode, William H. (1825–83), was a mathematician, physicist, and linguist, and partner of Eyre and Spottiswoode, the queen's printers. In 1847 he published his first work in mathematics, and later published on determinants, invariant theory, geometry, and calculus. In the 1870s Spottiswoode would turn to the experimental study of polarized light. In 1861 he married Eliza Taylor Arbuthnot, eldest daughter of William Urguhart Arbuthnot, member of the council for India, whose parties were considered the height of the London season, attracting men of science, politics, and society. In 1865 he, along with Tyndall, Thomas Hirst, Joseph Dalton Hooker, Thomas Henry Huxley, and four others formed the X Club, a monthly dining club through which they promoted science and advanced scientific naturalism in British society. Spottiswoode was very active in promoting science and served in an administrative role in a number of the major scientific societies. He was treasurer of the BAAS (1861–74); secretary of the Royal Geographical Society; president of the mathematical section of the RS (1865); and later, in 1878, president of the BAAS, and then of the RS.

Steuart/Stewart, Elizabeth Dawson (née Duckett, 1802–93), was born into a wealthy family with an estate of twelve thousand acres and a gothic mansion (Duckett's Grove). She married William Richard Steuart in 1820, and they lived at Steuart's Lodge, Leighlinbridge, in County Carlow, Ireland. She took a keen interest in Tyndall's career and was probably his most consistent and long-lived patron. She continued to support Tyndall after her husband died in 1852.

Stokes, George Gabriel (1819–1903), was an English physicist and mathematician who held many distinguished positions throughout his life, including the Lucasian Chair of Mathematics at Cambridge in 1849, and secretary (1854–83) and president (1885–90) of the RS. Stokes's earliest research focused on hydrodynamics, including a significant paper in 1850 in which he applied his work on the friction of fluids to the behavior of pendulums. In the early 1850s he turned to questions of optics, and in 1852 he published on fluorescence, a term he coined. Perhaps his major contribution to contemporary science was his oversight, for decades, of the refereeing of articles for the *Phil. Trans.* Tyndall consulted Stokes on a number of issues related to his research.

Sylvester, James Joseph (1814–97), was an English mathematician and professor. He studied mathematics at Cambridge, but was denied a degree because he was Jewish (Oxford and Cambridge required graduates adhere to the thirty-nine articles of the Anglican faith). He had an academic career that took him to Ireland, the United States, and, from 1844, back to England. He made a number of significant contributions in mathematics, including to matrix theory, invariant theory, number theory, partition theory, and combinatorics.

Tait, Peter Guthrie (1831–1901), was a Scottish mathematician and physicist, best known for his textbook *Treatise on Natural Philosophy*, which he cowrote with William Thomson. Throughout the period covered by this volume Tait and Thomson were involved in the priority dispute between Joule and Mayer regarding the discovery of the mechanical equivalent of heat. Tyndall advanced Mayer as co-discoverer, while Tait and Thomson argued that Joule had sole claim to that accolade. Tait later coauthored *The Unseen Universe* (1875) with Balfour Stewart, a response to Tyndall's 1874 "Belfast Address," in which they argued that immortality and an afterlife were logically consistent with the laws of thermodynamics.

Tennyson, Alfred (1809–92), was a Lincolnshire poet educated at Trinity College, Cambridge, where he was a member of the Apostles. Tennyson was appointed poet laureate in succession to Wordsworth in 1850, the same year that he published *In Memoriam.* In 1853 he moved to Farringford on the Isle of Wight, from where he published *Maud* in 1855. Like most nineteenth-century men of science, Tyndall greatly admired Tennyson's poetry, and was clearly happy to mix in the same social circles and visit the Tennyson family on short trips to the Isle of Wight.

Thompson, Benjamin (1753–1814), Count Rumford, was an American-born British physicist and inventor whose challenges to established physical theory were part of the nineteenth-century development of thermodynamics. He was one of the founders of the RI, and the RS's Rumford Medal was named after him.

Thomson, William (1824–1907), Lord Kelvin (from 1892), was a distinguished Irish-Scottish mathematician and physicist. He was educated at the University of Glasgow and Peterhouse College, Cambridge. Thomson was professor of natural philosophy in Glasgow from 1846, paying frequent visits to London. He first met Tyndall at the Edinburgh meeting of the BAAS in 1850 and from the start the two physicists had a strained relationship. Thomson was among the most significant contributors to thermodynamics, and came to an expression of the second law of thermodynamics around the same time as Clausius. He supported Joule in the 1840s as Joule was presenting his first papers on the conservation of force and the mechanical equivalent of heat, and, along with Peter Guthrie Tait, championed Joule in the priority dispute with Mayer.

Tyndale, George Hector (1821–80), was Tyndall's cousin and an American businessman. He was the son of Tyndall's father's brother, Robinson Tyndale, who had emigrated to Philadelphia from Leighlinbridge around 1804 and founded a glass and ceramics import business. Hector inherited the business and first met his physicist cousin in 1854 on one of his frequent business trips to Europe. They immediately struck up an intellectual rapport and began an occasional correspondence. He became involved in Republican politics in 1856, but held no sympathy for the abolitionist cause espoused by John Brown. He served as a Union general during the American Civil War

(1861–65), which was the topic of much of their correspondence during this volume.

Tyndale, Julia (née Nowlen, 1823–97), was the wife of George Hector Tyndale. They married in 1842 ("Deaths," *North American,* 13 August 1897, 6).

Tyndale, Robert (1837–1917), was a "Pensioner of Indian Public Works Department late of Oaklands New Ross County Wexford" (Calendars of Wills and Administrations, 1917, p. 758).

Tyndale, William (1494–1536), "Old Martyr Tyndall," was a religious reformer during the Protestant Reformation, and translator of the Bible. A Protestant martyr, he was executed by strangulation and then burned at the stake. It was a family tale that Tyndall was descended from him, and he wrote in his journal in 1860 that he "was struck by the remarkable reproduction of his temper amongst us;—give us the same circumstances and I am sure either Hector or myself would act as he did tomorrow" (RI MS JT/2/13c/1187). In 1869 Tyndall visited the supposed birthplace of William Tyndale, and the monument to him erected at Nibley Knoll in Gloucestershire.

Tyndall, Emma (1817–1904), was Tyndall's sister, whose education he encouraged. She does not appear to have married and later in life suffered from a mental illness, which was manifested as an extreme form of religiosity.

Tyndall, John (*c.* 1815–75), was a relative of Tyndall's in Gorey, County Wexford, Ireland. A doctor and member of the Royal Historical and Archaeological Society of Ireland, he died in 1875 of acute bronchitis ("Obituaries," *Lancet*, 1 May 1875, 633).

Tyndall, Sarah (née Macassey/Macasey/McAssey, 1793–1867), was Tyndall's mother, and was widowed in March 1847. She remained in the family cottage at Leighlinbridge, County Carlow, Ireland, supported financially by Tyndall. Her declining health was a major concern for Tyndall throughout the period covered by this volume.

Verdet, Marcel Émile (1824–66), was a French physicist. He worked in magnetism and optics, editing the works of Augustin-Jean Fresnel. Verdet did much to champion the early theory of the conservation of energy in France through his editorial supervision of the *Annales de chimie et de physique.*

Wheatstone, Charles (1802–75), was an English experimental physicist and prolific inventor. Wheatstone is particularly well-known for his collaboration with Josiah Cooke on the development of the telegraph and his contribution to developing the five-needle telegraph. In 1834 he accepted a professorship of experimental physics at King's College, London. In the same year he used a revolving mirror to estimate the speed of electricity moving through a wire, a similar experimental design that he later used to estimate the speed of light. Tyndall was especially interested in Wheatstone's invention of electromagnetic clocks, which could be used to measure small intervals of time.

Wynne, Frank George (dates unknown), was the third son of George Wynne. He would become a civil engineer.

Wynne, George (1804–90), was an Irish surveyor, engineer, and military officer. Born in Dublin, he enlisted in the Royal Engineers (*c.* 1825) and was commissioned as lieutenant in 1826. Wynne met Tyndall while employed in the Irish Ordnance Survey (1835–40). Wynne later served with distinction in Greece, China, and the West Indies, rising to general by the time he retired in 1877 ("The Late General Wynne," *Times*, 12 July 1890, 11). Wynne had offered to fund Tyndall's studies in Germany, although Tyndall declined (see letter 0423 in R. Barton, J. Rankin, and M. S. Reidy, eds., *The Correspondence of John Tyndall, Volume 3: The Correspondence, January 1850–December 1852* [Pittsburgh: University of Pittsburgh Press, 2017]).

Youmans, Edward Livingston (1821–87), was an American scientific writer, editor, and popularizer of science. Youmans was a conduit for British popular science in the United States through the publisher Appleton and was crucial for arranging for the US publication of the works by Herbert Spencer, Tyndall, and others, for a general audience. He also lectured and published on popular science himself, and founded *Popular Science Monthly* in 1872.

INDEX

The index refers to the letters using their serial numbers. The letter 'n' denotes a footnote reference and appears if an item is not otherwise mentioned in the letter to which the footnote pertains. A number in **bold** signifies a letter to or from a correspondent.

Abbott, Frederick K., 1994
Académie des Sciences, 2121
Adair, Henrietta Mary, **2009**
Adams, Eliza (née Bruce), 2005
Adams, John Couch, **1930**, **1933**, **1939**, **1950**, **2005**
Addison, Joseph Edward, **1994**
Airy, George Biddell, 1943, **2029**, **2069**
Airy, Richarda (née Smith), 2112
Akin, C. K., **1925**, **2075**, **2076**, 2077, **2079**, 2093, 2114, 2162, **2165**, **2167**, **2170**, 2176, 2190, 2197, 2198, 2201, 2204, 2208, 2214, 2217, 2220
Albert, Arthur William Patrick, 2052, 2054
Albert, Leopold George Duncan, 2056
Alford, Henry, 2126
Alpine Club, 1964
Alpine Journal, 2113
Alps, the, 1964, 1971, 2021, 2025, 2045, 2051, 2087, 2106, 2122, 2124, 2127, 2132, 2142, 2145, 2156, 2157, 2161, 2218
American Civil War, 1912, 1946, 1951, 1952, 1988, 2019, 2026, 2095
American Journal of Science, 2067
Anderson, Charles, 1912
Andrews, Emily Augusta, 1936
Annalen der Chemie und Pharmacie, 2002
Annalen der Physik und Chemie, 1927, 2211, 2212
Appleton and Company, 2001
aqueous vapour, 1942, 1947, 1958, 1979, 2012, 2029
Archives des Sciences Physiques et Naturelles, 1927, 1964
Armstrong, William, 2031, 2032, **2208**
Askham, William, 2085
Athenaeum, 2004, 2140
Athenaeum Club, 1907, 1922, 2202
Atlantic Monthly, 2090

Bacon, Francis, 1920, 2062
Bailey, Joseph, 1946
Bakerian Lecture, 2067
Ball, John, 2110, 2121, 2122, 2139, 2152, 2174, 2175
Ballot, Christophorus Henricus Diedericus Buys, **2211**
Balmat, Auguste, 1934
Banks, Nathaniel, 1946
Barlow, Cecilia Anne (née Law), 2178
Barlow, John, 1915, 1917n, 2126, **2178**
Barnard, Jane, 1915, 2124, 2125
Barnett, George, 1913
Barnett, Henry, **1913**
Barrett, William Fletcher, 2043, 2125, 2136, 2189
Barrett, William Garland, **2189**
Basire, James, 1992
Baudry, Jules, 2143
Baxendell, Joseph, 1985
Beaumont, Jean-Baptiste Armand Louis Léonce Élie de, 1999
Becker, Bernard Henry, **1948**
Becker, Ernst, 1910, 1911, 2073, 2123
Becker, Fraulein, 2060

Beek, Albert van, 2159
Bennen, Johann Joseph, 1915, 1918, 1919, 2023, 2113
Bennet, Charles, 2031, 2032
Bennet, Lewis, 1939
Bérard, Jacques Etienne, 1966, 2093
Berger, Capel Henry, **2104**
Bernoulli, Johann, 2129
Bohn, Johann, **2129**, **2220**
Boole, George, 2213
Boole, Mary (née Everest), 2213
Boys Own Magazine, 1947
Bradley, George Granville, 2122
Brewster, David, 2002
Bridgewater Treatises, 1952
British Association for the Advancement of Science, Cambridge meeting of (1862), 1904, 1905, 1932, 1933, 1937, 1972; Newcastle Upon Tyne meeting of (1863), 2024, 2032, 2035, 2076; Bath meeting of (1864), 2106, 2138, 2143, 2144, 2163
Brown, John, 1988
Bunsen, Robert, 1943, **2033**, 2040
Burke, Edmund, 2126
Busk, George, 1945, 2110, **2173**, **2194**

caloresscence, 2075, 2076, 2165, 2167, 2170, 2177, 2197, 2208, 2215, 2216
Cambridge Philosophical Society, 2042
Cameron, Julia Margaret, 2135, 2196
Campbell, Elizabeth Georgiana (née Leveson-Gower), 1947
Campbell, George Douglas, 1944, 1947
Carlyle, Thomas, 2122, 2133, 2134, 2196
Carpenter, William Benjamin, 2068, 2110
Cauche and Bardou, 2179
Cayley, Arthur, 2138
Chancourtois, Alexandre-Eugène Béguyer de, **2039**, **2046**
Charpentier, Gervais-Hélène, 2156
Chasles, Michel Floréal, 1994, 2138
Chemical News, 1990, 2029
Christmas Lectures, 2055
Clausius, Adelheid Rimpham, 1906, 1907, 1962, 1963, 1965, 1972, 1979, 2006, 2035, 2093, 2147, 2153, 2191, **2192**, **2205**, **2217**
Clausius, Helene, 1972
Clausius, Rudolf (Rudolph), **1904**, **1905**, **1906**, **1907**, 1920, **1962**, **1963**, **1965**, **1966**, **1967**, **1969**, **1972**, **1979**, **1998**, **2006**, 2013, 2028, **2035**, **2048**, **2049**, 2062, **2093**, 2096, **2108**, 2121n, **2147**, **2153**, **2191**, **2217**
Clausius, Rudolf John "Johnny", 2147, 2153, 2192, 2205, 2217
Colding, Ludwig August, 2002, 2129
Colenso, John William, 2138
Collet, Sophia Dobson, 2087
Colnet-d'Huart, Jean François Léonard Alexandre de, **2219**
Comptes Rendus de l'Académie des Sciences, 1994, 2093
conservation of energy, 2129
Conwill, John, 2137
Cooke, Josiah Parsons, **1955**
Copley Medal, 1949, 2169
Cosmos, 1964, 1984
Council of Military Education, 1993, 1994
Couttet, Francois, 1913
Coxe, James, 1944
Coxe, Mary Anne, **1944**, **1947**
Croll, James, **1970**, **2072**, 2081, **2214**, **2216**
Cromwell, Oliver, 1956
Crookes, William, 2030
Cumming, Alexander (Alick/Alix), 1944

Dana, James Dwight, 2067
Darwin, Charles Robert, 2169
Davy, Humphry, 1920, 2002
Davy, John, 2027
Dawes, Mary Helen (née Gordon), 1902, 1922, **1957**, **1959**, **1961**, 2106, 2163
Dawes, Richard, **1902**, **1922**, 1957, 1959, 1961, **2106**, 2107, **2163**
Debus, Heinrich, 1912, **1926**, 1946, **1952**, 1988, 2010, 2026, 2078, 2091, 2096, **2137**, **2141**, **2144**, **2172**, **2182**, **2183**
Dedekind, Julius Wilhelm Richard, 2028
Deerfoot Red-Jacket, 1939
de la Rive, Auguste Arthur, 1927
Delaroche, François-Étienne, 1966
Descartes, René, 2129
Desor, Pierre Jean Édouard, 2028
Despretz, César-Mansuète, 1920, 1967
Deville, Henri Étienne Sainte-Claire, 2040

Dewitt, Dr., 2030
d'Huart, Jean François Léonard Alexandre de Colnet, **2128**
Diegan, Mr., 2142, 2171
Dirichlet, Johann Peter Gustav Lejeune, 1994
Disraeli, Benjamin, 2200
Dobson, Arthur Dudley, 2087
Donnelly, John Fretcheville Dykes, 2091
Dove, Heinrich Wilhelm, 1905, 1948
Draper, John William, **2020**
du Bois-Reymond, Emil Heinrich, 2064n
Duboscq, Louis Jules, 2030
Dulong, Pierre Louis, 1969, 2049
Dumas, Jean-Baptiste, 2083
Dupuis, Mr., 1934
Dutch Academy of Sciences, 1986
Dyer, Joseph Chessborough, 1914, 1985

Edinburgh New Philosophical Journal, 1965
Edinburgh Review, 2066
Edmondson, George, 2051
Eclectic Review, 2189
electrical force, 2185
electricity, 1984
Elliot Brothers, 2073, 2123
Ellis, Alexander John, **1991**, 2138
Emerson, Ralph Waldo, 2026
Emmanuel, Francis Albert Augustus Charles, 2056
Engineer, 2002
Ericsson, John, 2062
Essays and Reviews, 2122
examinations, 1926n, 2078, 2118, 2141, 2144

Falconer, Hugh, 2120
Faraday, Michael, 1905, 1907, 1911, **1915**, 1917, 1954, **1977**, 1999, **2036**, **2043**, 2044, 2047, 2054, 2110, **2124**, **2125**
Faraday, Sarah (née Barnard), 2036, 2124, 2125
Favre, Jean Alphonse, **1964**
Fitzroy, Robert, **2166**
fluorescence, 2075
Forbes, James David, 2006
force, 2002
Forster, William Edward, 2098
Foucou, Félix, **2143**, **2156**
Fourier, Joseph, 2128
Francis, William, 2030, 2072, **2082**, **2085**, 2086, 2147, **2198**, **2201**, **2204**, 2206
Fraunhofer, Joseph von, 2179
Frankland, Edward, 1954, 1994, 2015, **2017**, 2042, 2091
Frankland, Lucy, 2135
Fresnel, Augustin-Jean, 2128, 2129

Galilei, Galileo, 2002
Gamaliel the Elder, 2104
Gardiner, Luke, 1956
Gassiot, John Peter, 1990, 2213
Gay-Lussac, Joseph Louis, 2062, 2159
Geographical Society of London, 1928
Gesellschaft Naturforschender Freunde zu Berlin, 1911
Gill, Joseph, **2062**, 2072, **2081**
Ginty, Margaret "Peb" (née Roberts), 2051, 2148
Ginty, William Gilbert, **2051**, 2148, 2158
glaciers, 1908, 1927, 1928, 1929, 1938, 1973, 1979, 1983, 2002, 2023, 2025, 2028, 2046, 2087, 2100, 2122, 2124, 2130, 2164
Glaciers of the Alps, 1924, 1927, 2053, 2054, 2071, 2103, 2127, 2132, 2143, 2156
Glaisher, James, 1933
Good Words, 2002, 2006, 2029
Gould, John, 2031
Government Grant Committee, 2218
Graham, Thomas, 2123
Grant, Francis, 2054
Greathed, William Wilberforce Harris, 1913
Greenfield, B. W., 2095
Griffith, George, 2079, 2165, 2197
Grove, William Robert, 1949, 1990, 2111
Guthrie, Anna Leonora, 1961
Gyde, Charles, 2085, **2086**

Haast, Johann Franz Julius von, **1928**, **2021**, **2045**, **2087**, **2120**
Hachette, Louis Christophe François, 2156
Hall, James, **1968**
Hamilton, Miss, 2009
Hampshire Advertises, 1968
Harley, Robert, 2209
Harrison, James Park, 2211

Harris, William Snow, 1961, **1984**, **1986**, **2066**, **2185**
Haughton, Samuel, **2007**
Head, Edmund, 2164
heat, 1920, 1927, 1958, 1985, 1995, 1997, 2002, 2012, 2014, 2017, 2044, 2049, 2062, 2066, 2067, 2081, 2093, 2097, 2114, 2115, 2117, 2128, 2131, 2182, 2183, 2195, 2211, 2214, 2219
Heat as a Mode of Motion, 1904, 1954, 1958, 1962, 1963, 1965, 1966, 1967, 1969, 1972, 1973, 1976, 1977, 1979, 1981, 1990, 1992, 1996, 1998, 1999, 2000, 2001, 2004, 2007, 2009, 2033, 2039, 2041, 2045, 2062, 2068, 2073, 2083, 2120, 2121, 2123, 2134, 2139, 2140, 2159, 2160
Heathfield, Mr., 1947
Hector, James, 2045
Heer, Oswald, 2108
Helmholtz, Anne (née Mohl), 2060
Helmholtz, Hermann Ludwig Ferdinand von, 1991, 2002, 2040, **2060**, **2064**, 2129
Hentsch, Lütscher & Co., 2040
Herries, Isabella, 2011
Herschel, Alexander Stewart, 1958, 1960, **1980**, 1981, **1982**, 1983
Herschel, John Frederick William, 1943, **1958**, **1960**, 1980, **1981**, 1982, **1983**, 2002, **2111**, 2134, 2135, 2139, **2140**, **2176**, **2177**, **2179**, **2180**, **2184**, **2186**, 2211
Hessel, Johann Friedrich Christian, **1940**
Hessel, Phil. C., 1940
Hill, Henry, **2109**, 2110
Hirn, Gustave-Adolphe, 2062, 2081
Hirst, Thomas Archer, 1912, **1916**, **1921**, 1931, 1952, 1988, **1993**, 1994, 1998, 2010, **2025**, **2028**, 2035, 2039, 2044, 2049, **2080**, **2083**, **2084**, **2088**, **2090**, **2091**, 2093, 2095, **2096**, **2098**, 2108, 2109, **2110**, 2130, **2138**, 2147, 2156, 2169, **2199**, **2206**, **2209**, 2212
Hofmann, August Wilhelm von, 2035, 2137
Hogan, Jane (née Roberts), **2148**, **2158**
Holland Fund, 2015
Holmes, Mr., 1928
Holzmann, Maurice, **2052**, **2055**, **2056**
Hooker, Charles Paget, **2034**
Hooker, Joseph Dalton, **1941**, **1942**, 2034, 2045, 2087, 2120
Hopkins, William, 1908, 2073, **2078**, 2100, 2152
Hüber, William, **2127**, **2132**
Humboldt, Alexander von, 1911, 2111
Hunt, Edward Bissell, 2067
Hunt, Thomas Sterry, **2067**
Hutchinson, Thomas Neville, 2145
Hutton, James, 2152
Huxley, Henrietta "Nettie" Anne (née Heathorn), 1974
Huxley, Thomas Henry, 1907, 1909, 1915, 1916, 1917, **1918**, 1952, 1954, 1974, 2001, 2027, 2031, 2110, 2114, 2184, **2200**, 2202

ice, 2071, 2073, 2103
Illustrated London News, 2002
International Exhibition of London (1862), 1904

Jenkins, Thomas E., **2030**
Jewsbury, Geraldine Endsor, **1974**, **1987**, **2003**
Jones, Henry Bence, 1915, 1917n, 1958, **2015**, **2023**, 2031, **2047**, 2084, **2181**, **2187**
Joule, James Prescott, **1914**, **1920**, 1921, 1962, 1965, 1966, 1969, 1984, **1985**. 1990, 2002, 2007,2013, 2016, 2022, 2024, 2035, 2062, 2068, 2081, 2093, 2137
Journal für die reine und angewandte Mathematik, 1994
Journal of Tortolini: *Annali di scienze matematiche e fische,* 1994
Jukes, Joseph, 2152

Kenngott, Gustav Adolph, 2147
Kenngott, Madame, 2147, 2153, 2191
Kirchhoff, Gustav Robert, **1943**, 1949, 2033, 2040, 2064, **2065**, **2070**, 2114
Klaproth, Martin Heinrich, 1911
Knoblauch, Bertha Marie Emilie (née Winter), 2044, 2212
Knoblauch, Karl Hermann, 2041, **2044**, **2212**
Koenig, Rudolph, 2030
Kolb, Karl, 2127
Kolk, Hendrik Willem Schroeder van der, **2159**

Lamé, Gabriel, 2128
Laplace, Pierre-Simon, 1995n, 2048, 2057, 2086
Larcom, Thomas Aiskew, 2174
Lardner, Dionysius, 2062
Laughton, Johanna, 2126
Laveleye, Émile Louis Victor de, **2161**
Lawson, Dr., **1992**
Leader, The, 2090, 2116
Lee, Robert E., 2026
Lee-Warner, Henry, 2145
Leibnitz, Gottfried Wilhelm, 2129
Leslie, John, 2012
Les Mondes, 1978, 1990, 1999, 2068
Lewes, George Henry, 2090
Lewis, Richard Thomas, 1935
Liebig, Justus von, 2002
Lincoln, Abraham, 1988
Linth, Arnold Escher (von der), 1906, 1907, 1972, 2028
Liouville, Joseph, 1994
Livingstone, David, 2144
Locke, John, 1920
Logan, William Edmond, 2164
Longmans, 1902, 1992, 1996, 2001, 2083
Lowell Institute, 2160, 2188
Lowell, John Amory, **2160**, **2188**
Lubbock, Amy Harriet, 1945, 1976
Lubbock, Constance Mary, 1945
Lubbock, Ellen Frances (née Horden), 1909, 1954, **1975**, **1976**, 2083, 2090, 2098, **2130**, 2138
Lubbock, Harriet Caroline (née Hotham), 1975
Lubbock, John, **1909**, 1913, 1915, **1945**, 1954, 1975, 1976, 2090, 2098, 2125, 2130
Lubbock, John Birkbeck, 1945, 1975
Lyell, Charles, **1929**, **1938**, 1976, 1981, 2160
Lyster, James, 2090

Macleod, Norman, **2118**
Macmillan's Magazine, 2002
Macready, William Charles, 2133
magnetic experiment, 2111
Magnus, Heinrich Gustav, **1911**, 1927, 1931, 1943, **2000**, 2029, 2040
Mallet, Robert, 2152
Mariotte, Edme, 1920, 2159
Marshall, John, 1936, 2042
Martins, Charles Frédéric, 2028
Maskelyne, Nevil Story, 2073
Mathews, William, 1964
Matthiessen, Augustus, 2050, 2059
Mattus, Mr., 2199
Maunoir, Charles, 2127
Maxwell, James Clerk, **2089**
Maxwell, Theodore, **2131**
Mayer, Julius Robert von, 1904, 1905, 1906, 1920, 1921, 1962, 1965, 1966, 1969, 1979, 1998,1999, 2002, 2006, 2007, **2013**, **2016**, **2022**, 2023, **2024**, 2035, **2041**, 2073, 2093, 2096, 2098, 2108, 2111, 2129, 2212
May, Thomas Erskine, 2126
McLeod, Herbert, 2137
Mechanic's Magazine, 1990
Melloni, Macedonio, 1985, 2211
Memoires de L'Académie de Sciences de l'Institut de France, 1964
Merivale, Charles, 1936
meteorology, 2029, 2166n
Michel, Christian, 2145
Millais, John Everett, 1971
Miller, William Allen, 1978, 1990
Miller, William Hallowes, 1943, 2000
Mill, John Stuart, 2207
miracles, 1922
Mitchell, Elizabeth, 1946, 1952, 1956, 2010, 2095
Mitchell, John Grant, 1946
Mitchel, Ormsby Macknight, 1946
Moigno, François-Napoléon-Marie, **1978**, 1984, **1990**, **1996**, **1999**, **2068**, 2090, 2127, 2128
Moll, Gerard, 2159
Monthly Notices of the Royal Astronomical Society, 1999
moon, 1935, 2211
Moore, Harriet Jane, 2074, 2125, 2126, 2134
Moore, Harriet (née Henderson), **2074**
Moore, John Carrick, 2146, 2152
Moore, Julia, 2074, 2125, 2126, 2134
Morlot, Charles Adolph, 1909
Mortillet, Louis Laurent Gabriel de, 1929, 1938
Mountaineering in 1861, 1902, 1971, 2132
Mountjoy, Errell, 1956

Mousson, Johann Rudolf Albert, 1906, 1907
Müller, Friedrich Max, 2011
Murchison, Charlotte Hugonin, **1971**
Murchison, Roderick Impey, 1928, 1971, **2100**, 2120, 2145, 2146, 2156, 2164n
Murray, John, **1924**, **2053**, 2143
muscular action, 2007
Mussy, Noël-Francois-Odon Guéneau de, 2196

Napier, William Francis Patrick, 2090
Natterer, Johann August, 1967
Naturforschende Gesellschaft, 1906, 1907, 2028, 2108, 2130, 2212
Neumann, Franz Ernst, 1969
North British Review, 2093
North, Frederick, 2130, 2145, **2157**
North, Janet Catherine, 2157
North, Marianne, 2157

optics, 2068
Orleans, Prince Philippe of, 2095
Ørsted, Hans Christian, 1920
Osenbrüggen, Eduard, 2192
Owen, Richard, **2150**
Owen, Robert, 1952, 2031

Patmore, Coventry, 1936
Pembroke College, 1930
Percy, Algernon, 2031, 2032, 2054
Percy, Eleanor, 2031, 2053, **2054**
Percy, John, 2143
Petit, Alexis Thérèse, 1969
Phillips, Dr., 2101, 2121
Phillips, Wendell, 1988
Philomatic Society of Paris, 2042
Philosophical Club, 2045, 2089, 2173, 2194
Philosophical Magazine, 1905, 1914, 1920, 1921, 1925, 1927, 1935, 1965, 1984, 1994, 1998,1999, 2002, 2008, 2012, 2013, 2016, 2029, 2035, 2040, 2048, 2049, **2057**, 2059, 2066, 2067, 2073, 2081, 2123, 2129, 2137, 2143, 2147, 2152, 2154, 2156, 2159, 2161, 2190, 2211, 2214, 2216, 2220
Philosophical Transactions, 1908, 1927, 1986, 1994, 2017, 2029, **2050**, 2112, 2159, 2185, 2195, 2217
Phipson, Thomas Lamb, 1984, 1986
Plateau, Félix Auguste Joseph Plateau, 2038
Plateau, Joseph, **2037**, **2038**
Playfair, Jean Ann (née Millington), 2103
Playfair, John, 2152
Playfair, Lyon, **2103**
Plücker, Julius, 2128
Poggendorff, Charlotte Eleonore (née Kneser), 1931, 2191, 2212, 2217
Poggendorff, Johann Christian, 1905, 1906, **1910**, 1911, 1921, 1927, **1931**, 2086, 2212, 2217
Poggendorff, Marie, 1921, 1931
Poisson, Siméon Denis, 2117
Pollock, Frederick, 1919, 1932, 1936, 1954, 1973, 2063, 2134, 2151, 2202
Pollock, Jonathan Frederick, 2126
Pollock, Juliet (née Creed), **1917**, **1919**, **1923**, **1932**, **1936**, **1937**, **1954**, 1973, **1989**, 2008, **2010**, **2027**, **2031**, 2063, **2105**, **2107**, **2122**, **2126**, **2133**, **2134**, **2135**, **2139**, 2140, **2149**, 2154, **2196**, 2202, **2203**
Pollock, Maurice Emilius, 1919, 1932, 1936, 1954, 2031, **2058**, 2063, 2122, 2126, 2134, 2202
Pollock, Walter Herries, 1919, 1923, 1932, 1936, 1937, 1954, 2031, 2063, 2122, 2126, 2133, 2134, 2151, 2196, 2202
Pollock, William Frederick, 1915, 1917n, 1919, 1932, 1954, **1973**, 1976, **2008**, 2031, 2063, 2122, 2126, 2134, 2139, **2151**, **2154**, **2155**, **2190**, **2193**, 2196, **2200**, **2202**
Polytechnic Institute of Zurich, 1906
Pope, John, 1946
Prevost, Mr., 2040
Prevost, Pierre, 2114
Proceedings of the American Association for the Advancement of Science, 2067
Proceedings of the Royal Institution, 2002
Prout, William, 1952
Pridie, William Roby, 2109, 2110

Queenwood College, 2051, 2172, 2182, 2183

radiant heat, *see* heat
radiation, 1981, 1999
Ramsay, Andrew Crombie, 1929, 2021, 2100, 2110, **2146**, **2152**, **2164**

Rankine, William John Macquorn, 1965, 2062, 2081, 2093
Reader, The, 2075, **2077**, 2156, 2181, 2184, 2186, 2187n, 2189, 2193, 2196, 2200, 2202
refrangibility, 2114, 2136, 2177, 2182, 2183, 2195, 2211n
Regnault, Henri Victor, 1990, 2062, 2081, 2093, 2159
religion, 1902, 2137, 2140, 2200n
Rendu, Louis, 2002
Rennie, George, 2066
Reusch, Friedrich Eduard, **2071**, **2073**, **2123**
Revue des Deux Mondes, 2161
Revue Nationale, 2156
Rice, Marianne Spring, 2202
Riess, Peter Theophil-Pierre, 2000
Rink, Hinrich (Henrik) Johannes, 2046
Roberts, John, 2148
Roche, Annette, **2061**, **2094**, **2101**, **2113**, **2121**, **2142**, **2171**
Roche, François-Étienne de la, 2093
Royal Astronomical Society, 2069
Royal Family, 2052, 2112
Royal Institution, 1973, 1975, 1980, 1983, 1990, 2001, 2013, 2014, 2015, 2037, 2038, 2052, 2054, 2055, 2056, 2074, 2105, 2125, 2165
Royal Military Academy at Woolwich, 1994
Royal Observatory, Greenwich, 2112
Royal Prussian Academy of Sciences, 2000
Royal School of Mines, 2042, 2064
Royal Society, 1943, 1949, 1984, 1986, 1992, 1994, 2000, 2014, 2037, 2038, 2044, 2045, 2079, 2089, 2147, 2162, 2169, 2171, 2172, 2174, 2175, 2178, 2218
Ruhmkorff coil, 1980, 1982
Rumford Medal, 1943, 1949, 2169, 2172, 2178, 2217

Sabine, Edward, 1905, **2174**, **2175**, **2197**, **2210**, **2213**
Sabine, Elizabeth Juliana Leeves, 2157
Saturday Review, 2076, 2079
Sauerwald, Ferdinand, 2064
Savory, William Scovell, 2063
Schlagintweit, Hermann, 2210
Schönbein, Christian Friedrich, 2041
science and religion, 2104, 2137, 2140, 2200n
Scientific American, 2013
Scientific Relief Fund, 2213
Sclater, Philip Lutley, **2023**
Séguin, Marc, 1990, 1996, 2002, 2006, 2035, 2062, 2093, 2096, 2108, 2137
Sharpey, William, **1949**
Siemens, Charles William, 2062
Simon, Étienne Jules Adolphe Desmier de Saint, 1964
Simons, Gerard, 2159
sky, colour of, 1960
Smith, John Stores, 1974
Smyth, Charles Piazzi, 2002, 2211
Society of German Natural Scientists and Physicians, 2191
Somerville, Mary, **1995**, **1997**
Soret, Jacques-Louis, **1927**, **2040**
sound, 1948, 2001, 2030, 2057, 2176
spectral analysis, 1978, 2176, 2179, 2197
spectroscope, 2064, 2065, 2070
Spedding, John James, 1936
Spedding, Sarah, 1936, 2031, 2063
Spencer, Herbert, 2001, **2207**
Spottiswoode, William H., 2090, 2110, 2138
Stanley, Arthur Penrhyn, 2202
Steiner, Jakob, 1994
Steinheil, Carl August von, 2065, 2070
Stephenson, George, 2111
Steuart/Stewart, Elizabeth Dawson, **1903**, **2032**, **2116**
Stewart, Balfour, 2114
Stirling, Robert, 2062
St. James's Magazine, 2189
Stokes, George Gabriel, **1908**, 1925, 1933, 1949, **2010**, **2014**, **2050**, **2059**, 2076, 2077, 2079, **2097**, **2099**, **2112**, **2114**, **2115**, **2117**, **2119**, **2136**, **2162**, **2168**, 2173, 2176, 2182, **2185**, **2195**, **2215**, **2218**
Strutt, Edward, 2060
Strutt, Mary, 2060
Stuart, L. C., 2159
Studer, Bernhard, 2028
Swiss Education Council, 1972
Sylvester, James Joseph, 2098, 2110, **2169**
Symonds, John, 2157

Tairraz, Joseph, **1934**
Tait, Peter Guthrie, 2002, 2006, 2013, 2035, 2093
Taylor and Francis, 2086
Taylor, David, 2030
Taylor, Henry, 2149
Taylor, Tom, 2200
Tennyson, Alfred, 1918n, 1946, 1954, 2122, 2133, 2135, 2171
theology, 2139
Thompson, Benjamin, 1920, 2172
Thomson, William, 1979, 1984, 1986, 1998, **2002**, 2006, 2007, 2008, 2013, 2024, 2033, 2035, 2066, 2068, 2090n, 2093, 2108, 2129
Thury, Louis-Étienne François Héricart-Ferrand vicomte de, 1927
Times, The, **2145**, 2146, 2156, 2157
Tindall, William, 1956
Transactions of the Royal Society of Edinburgh, 2002
Tremblay, Mr., 1990
Trench, Richard Chenevix, 2027
Tunstall, Mrs., 2028, 2083
Turner, Charles Tennyson, 2126, 2135, 2139, 2154n
Tyndale, George Hector, **1912**, **1946**, 1951, 1952, **1953**, **1956**, **1988**, 1997, **2010**, **2019**, **2026**, **2095**
Tyndale, Harold, 1946
Tyndale, John Warre, 2095, 2110
Tyndale, Julia (née Nowlen), 1946, 1952, 1956, 1988, 2010, 2019, 2026, 2095
Tyndale, Robert, 2032, 2096
Tyndale, Robinson, 2026
Tyndale, Sharon, 1946
Tyndale, William, 1923n, 1952n, 2019, 2095
Tyndall, Emma 2094, 2116, 2121
Tyndall, John (of Gorey), 1951, 1956, 2019, 2095, 2096
Tyndall, Joseph, 2019
Tyndall, Margaret, 2110
Tyndall, Sarah (née Macassey/Macasey/McAssey), **1951**, 2088, 2094, 2096, 2142, 2171

University College, 2080, 2083

Verdet, Marcel Émile, 2002, 2035, 2081
Vere, Aubrey Thomas de, 1936
Victoria, Queen, 2052, 2055, 2056
Vogt, Karl [Carl] Christoph, 2050, 2059

Wall, Miss, 2094, 2101
Walters, Anton, 1915, 1918
Wartmann, Elie, 2028
Weber, Wilhelm Eduard, 2191
Wheatstone, Charles, 1905, 1990
Whewell, William, 1984
White, Walter, **2042**, 2117
Whymper, Edward, 1918
Wilson, Effingham, 2111
Wislicenus, Johannes, 2191
Wotton, Henry, 2126
Wright, Fanny (née Bellamy), 2051
Wright, Richard Pears, 2051, 2137
Wyndham-Quin, Augusta Charlotte (née Goold), 2018
Wyndham-Quin, Caroline, 2018
Wyndham-Quin, Edwin, **2018**, 2174
Wynne, Anne, 2205
Wynne, Frank George, 2102, 2147, 2191, 2192, 2205
Wynne, George, 2051, **2102**, 2205

Yorke, Philip James, 1912
Youmans, Catherine Newton Lee, 2001
Youmans, Edward Livingston, **2001**
Young, Thomas, 2128

Zeno, T., **1935**